ARSENIC TOXICITY

Prevention and Treatment

ARSENIC TOXICITY

Prevention and Treatment

Edited by

Narayan Chakrabarty

Scientific Consultant & Director
West Bengal Biotechnology Development Corporation Ltd
Government of West Bengal, India

Visiting Professor
Post-Graduate Department of Zoology
Maulana Azad College
West Bengal, India

CRC Press
Taylor & Francis Group
Boca Raton London New York

CRC Press is an imprint of the
Taylor & Francis Group, an **informa** business

CRC Press
Taylor & Francis Group
6000 Broken Sound Parkway NW, Suite 300
Boca Raton, FL 33487-2742

First issued in hardback 2019
First issued in paperback 2021

© 2016 by Taylor & Francis Group, LLC
CRC Press is an imprint of Taylor & Francis Group, an Informa business

No claim to original U.S. Government works

ISBN 13: 978-1-03-209840-1 (pbk)
ISBN 13: 978-1-4822-4196-9 (hbk)

Contents

SECTION I Arsenic Toxicity, Propagation, and Proliferation

SECTION II Remediation

SECTION III   Treatment

SECTION IV   Remediation of Arsenic by
Nutraceuticals and Functional Food

Preface

Arsenic is the most talked about metalloid in the modern world. It is a well-established fact that the toxicity of this element covers two-thirds of the world. Initially, its toxicity was found in Third World countries, particularly Southeast Asian countries. In the early part of the twentieth century, people were not aware of arsenic's toxicity. It was used in drugs such as Fowler's solution. Later, its toxic manifestation on the skin was noticed.

Arsenic was used as a chemical for slow poisoning, leading to murder in the Middle Ages. Some notable cases are the deaths of King Faisal bin Al Husain in 1938 [1], George III of England [2], Napoleon Bonaparte [3], and Emperor Guangxu of China [4].

Arsenic affects the liver, kidney, and lungs [5]. It also leads to cardiovascular diseases [6], cancer [7], and diabetes [8]. Longtime exposure to arsenic may lead to night blindness [9]. It interferes with the metabolic activity by inhibiting the function of pyruvate dehydrogenase and thereby inhibits cell functions. Arsenic also acts as a catalyst in converting normal stem cells into carcinogenic stem cells [10].

In this book, we have attempted to present a vivid picture of the source of arsenic. The main source is groundwater, followed by exposure through food, the most affected of which is rice. There are some occupational exposures to arsenic as well. Inorganic arsenic is used in glass production, nonferrous alloy manufacturing, wood preservation, and semiconductor manufacturing units.

The book then presents methods for arsenic removal, mainly from soil and water. Some modern, eco-friendly bioremediation techniques are discussed in detail. Classical as well as modern treatment methods are discussed. Special emphasis is given to using nutraceuticals and functional foods for the remediation of arsenic toxicity.

I dedicate this work to my father, the fountain of inspiration in my life, the late Bhabani Prasad Chakrabarty. I express my sincerest gratitude and thanks to my beloved friend, Dr. Debasis Bagchi. Without his inspiration and guidance, I would not have completed this book. Special thanks to my daughter, Diotima Chakrabarty, a software engineer by profession, who has devoted her invaluable time to coordinate the chapters and format the book. My sincerest thanks to all my eminent contributors who have enriched the book.

REFERENCES

1. Dr. Mohammad Al Janabi, 78 years after the murder of King Faisal the first, Iraq Law Network. http://www.qanon302.net/news/news.php?action=view&id=7230. Accessed on December 24, 2012.
2. King George III: Mad or misunderstood? *BBC News*, July 13, 2004. http://news.bbc.co.uk/l/hi/health/3889903.stm. Retrieved April 25, 2010.
3. Whorton JG. *The Arsenic Century*. Oxford University Press, New York, 2011, 226–228.

4. Forensic scientists: China's reformist second-to-last emperor was murdered. *Xinhua*, November 3, 2008. http://news.xinhuanet.com/english/2008-11/03/content_10301467.htm.

5. Test ID: ASU. Arsenic, 24 hour, urine, clinical information. *Mayo Medical Laboratories Catalog*, Mayo Clinic. http://www.mayomedicallaboratories.com/test-catalogue/Clinical+and+Interpretive/8644. Retrieved September 25, 2012.

6. Tseng CH, Chong CK, Tseng CP et al. Long term arsenic exposure and ischemic heart disease in arseniasis-hyperendemic villages in Taiwan. *Toxicol. Lett.* **137**(1–2): 15–21, January 2003. doi: 10.1016/S0378-4274(02)00377-6 (http://dx.doi.org/10.1016%2FS0378-4274%2802%2900377-6). PMID 12505429 (/www.ncbi.nlm.nih.gov/pubmed/12505429).

7. Smith AH, Hopenhayn-Rich C, Bates MN, et al. Cancer risks from arsenic in drinking water. *Environ. Health Perspect.* **97**: 259–267, July 1992. doi: 10.2307/3431362 (http://dx.doi.org/10.2307%2F3431362). PMC 1519547 (//www.ncbi.nlm.nih.gov/pmc/articles/PMC1519547). PMID 1396465 (//www.ncbi.nlm.nih.gov/pubmed/1396465).

8. Kile ML, Christiani DC. Environmental arsenic exposure and diabetes. *JAMA* **300**(7): 845–846, August 2008. doi: 10.1001/jama.300.7.845 (http://dx.doi.org/10.1001%2Fjama.300.7.845). PMID 18714068 (//www.ncbi.nlm.nih.gov/pubmed/18714068).

9. Hsueh YM, Wu WL, Huang YL, Chiou HY, Tseng CH, Chen CJ. Low serum carotene level and increased risk of ischemic heart disease related to long-term arsenic exposure. *Atherosclerosis* **141**(2): 249–257, December 1998. doi: 10.1016/S0021-9150(98)00178-6 (http://dx.doi.org/10.1016%2FS0021-9150%2898%2900178-6). PMID 9862173 (//www.ncbi.nlm.nih.gov/pubmed/9862173).

10. Tokar EJ, Qu W, Waalkes MP. Arsenic, stem cells, and the development basis of adult cancer. *Toxicol. Sci.* **120**(S1), S192–S213, 2011.

Editor

Dr. Narayan Chakrabarty, PhD, FICS, is a visiting professor in the Post-Graduate Department of Zoology, Maulana Azad College, Calcutta University, Calcutta, India. He is also a scientific consultant and director, Department of Biotechnology, West Bengal Biotech Development Corporation, Government of West Bengal, India. Dr. Chakrabarty is a member of the Departmental Technical Committee for Rural Biotechnology of the Government of West Bengal. He authored a book on organic reaction mechanisms. He has vast experience in conducting programs for estimating and remediating arsenic toxicity in rural Bengal. He was also a resource person in various such programs under the sponsorship of UNICEF and the Government of India.

Contributors

Tetsuro Agusa
Center for Marine Environmental
  Studies
Ehime University
Matsuyama, Japan

Ok-Nam Bae
College of Pharmacy
Hanyang University
Ansan, South Korea

Debasis Bagchi
Research & Development Department
Cepham Research Center
Piscataway, New Jersey

and

Department of Pharmacological and
  Pharmaceutical Sciences
College of Pharmacy
University of Houston
Houston, Texas

Manashi Bagchi
Research & Development Department
Cepham Research Center
Piscataway, New Jersey

Dayene C. Carvalho
Institute of Chemistry
Federal University of Uberlândia
Uberlândia, Brazil

Narayan Chakrabarty
West Bengal Biotech Development
  Corporation Limited
and
Post-Graduate Department of Zoology
Maulana Azad College
Government of West Bengal
Kolkata, India

Rajdeep Chowdhury
Department of Biological Sciences
Birla Institute of Technology and
  Science
Pilani, India

Luciana M. Coelho
Department of Chemistry
Federal University of Goiás
Catalão, Brazil

Luciene M. Coelho
Department of Chemistry
Federal University of Goiás
Catalão, Brazil

Nívia M.M. Coelho
Institute of Chemistry
Federal University of Uberlândia
Uberlândia, Brazil

Jin-Ho Chung
College of Pharmacy
Seoul National University
Seoul, South Korea

Anupam Das
Department of Dermatology
Kolkata Medical College
Kolkata, India

Nilay Kanti Das
Department of Dermatology
Kolkata Medical College
Kolkata, India

Nidhi Dwivedi
Division of Regulatory Toxicology
Defence Research and Development
  Establishment
Gwalior, India

Elvira Esteban
Faculty of Sciences
Department of Agricultural Chemistry
    and Food Science
Autonomous University of Madrid
Madrid, Spain

S.J.S. Flora
Division of Regulatory Toxicology
Defense Research and Development
    Establishment
Gwalior, India

Nguyen Van Hue
Department of Tropical Plant and Soil
    Sciences
College of Tropical Agriculture and
    Human Resources
University of Hawaii
Manoa, Hawaii

Michael F. Hughes
National Health and Environmental
    Effects Research Laboratory
Office of Research and Development
U.S. Environmental Protection Agency
Research Triangle Park, North Carolina

Ioannis A. Katsoyiannis
Laboratory of General and Inorganic
    Chemical Technology
Department of Chemistry
Aristotle University
Thessaloniki, Greece

Piyush Kumar
Department of Dermatology
Katihar Medical College and Hospital
Bihar, India

Yuhu Li
School of Metallurgy and Environment
Central South University
Changsha, Hunan, People's Republic of
    China

Kyung-Min Lim
College of Pharmacy
Ewha Womans University
Seoul, South Korea

Zhihong Liu
School of Metallurgy and Environment
Central South University
Changsha, Hunan, People's Republic of
    China

Rebeca Manzano
Division of Environmental and Food
    Sciences and Technologies
Department of Agriculture
University of Sassari
Sassari, Italy

Manassis Mitrakas
Laboratory of Analytical Chemistry
Department of Chemical Engineering
Aristotle University
Thessaloniki, Greece

Eduardo Moreno-Jiménez
Faculty of Sciences
Department of Agricultural Chemistry
    and Food Science
Autonomous University of Madrid
Madrid, Spain

Sudeshna Mukherjee
Department of Biological Sciences
Birla Institute of Technology and
    Science
Pilani, India

Thais S. Neri
Institute of Chemistry
Federal University of Uberlândia
Uberlândia, Brazil

Vidhu Pachauri
Division of Regulatory Toxicology
Defence Research and Development
    Establishment
Gwalior, India

Dionisios Panagiotaras
Department of Mechanical Engineering
Technological-Educational Institute of
    Western Greece
Patras, Greece

Dimitrios Papoulis
Department of Geology
University of Patras
Patras, Greece

Rudrajit Paul
Department of General Medicine
Kolkata Medical College
Kolkata, India

Helen C. de Rezende
Institute of Chemistry
Federal University of Uberlândia
Uberlândia, Brazil

Domingo A. Román-Silva
Faculty of Basic Sciences
Bio-Inorganic and Environmental
    Analytical Chemistry Laboratory
Department of Chemistry
University of Antofagasta
Antofagasta, Chile

Amrita Sil
Department of Pharmacology
Institute of Post-Graduate Medical
    Education and Research
Kolkata, India

Elias Stathatos
Department of Electrical Engineering
Technological-Educational Institute of
    Western Greece
Patras, Greece

Anand Swaroop
Corporate Administration
Cepham Research Center
Piscataway, New Jersey

Sanjay Kumar Verma
Department of Biological Sciences
Birla Institute of Technology and
    Science
Pilani, India

Anastasios I. Zouboulis
Laboratory of General and Inorganic
    Chemical Technology
Department of Chemistry
Aristotle University
Thessaloniki, Greece

Section I

Arsenic Toxicity, Propagation, and Proliferation

1 Introduction to Arsenic Toxicity

Narayan Chakrabarty

CONTENTS

1.1 INTRODUCTION

Arsenic is one of the most hazardous chemical pollutants in human life in this century. The famous Alchemist Albertus Magnus discovered the element arsenic. From the position of arsenic in the periodic table its electronic configuration is $[Ar]3d^{10}4s^24p^3$. It is a metalloid, that is, it has the characteristics of both metal and nonmetal. Atomic number of arsenic is 33 and atomic weight is 74.91. Its ionization energy is very high. It does not exist in cationic state; hence, it mainly exists in covalent compounds and in the anionic part of a salt. Its oxidation states are +3 and +5. The dreaded pollutant arsenic is naturally present in the environment but our activities control and proliferate its distribution throughout our globe.

1.2 ORIGIN OF THE PROBLEM

A recent study found that over 200 million people in more than 70 countries are probably affected by arsenic poisoning in drinking water. Arsenic contamination in groundwater is found in many countries in the world including the United States [1]. Worst affected places are East Asian countries [2].

Salient Features of Arsenic

Symbol	As
Atomic number	33
Atomic mass	74.91
Position in the Periodic Table	Group 15 and fourth period
Electronic configuration	$1s^2 2s^2 2p^6 3s^2 3p^6 3d^{10} 4s^2 4p^3$
Ionization potential	First: 947 kJ mol^{-1}
	Second: 1950 kJ mol^{-1}
	Third: 2730 kJ mol^{-1}
Common oxidation states	+3 and +5

Naturally Occurring Compounds of Arsenic and Its Radicals

Arsenic chloride	$AsCl_3$ and $AsCl_5$
Arsenic nitrate	$As(NO_3)_3$ and $As(NO_3)_5$
Arsenic sulfate	$As_2(SO_4)_3$
Arsenate	AsO_4^{3-}
Arsenite	AsO_3^{3-}
Arsenic oxide	As_2O_3 and As_2O_5
Arsenic sulfide	As_2S_3
Arsine	AsH_3

The presence of arsenic is mainly due to biogeochemical distribution from its sources inside the soil. At some places it is considered to come from volcanic eruption and at some other places from pyrite oxidation. Microorganisms have crucial role in these transformations. Biotransformation in marine ecosystem has a direct hand in the presence of arsenic in sea food.

Arsenic present in the soil goes into groundwater and consequently penetrates into the drinking and all other usable water. It is also found in both sea water and river water. Due to its presence in groundwater, percolated through different water aquifers, arsenic ultimately goes into agricultural products. Contaminations are found even in poultry and livestock. Of all the different oxidation states of arsenic, the common states are +3 and +5. Of these, +5 state is less toxic than +3 state. The diversity of arsenic as a contaminant is its metalloid character. Because of this character, it can form salts with chloride and nitrate in favorable atmosphere. Again, reacting with soil alkali, it forms arsenite:

$$2As + 2OH^- + 4H_2O \rightarrow 2AsO_3^{3-} + 5H_2$$

By expanding its valence shell, arsenic can form addition compound with amines, ethers, and halides. Exploiting its character, arsenic forms compounds with organic grouping. Therefore, arsenic is pollutant in both inorganic and organic form.

Inorganic arsenic forms toxic compounds. It is found that the gaseous form known as arsine is highly toxic. Inorganic arsenic is the natural arsenic compounds in forms of oxide, chloride, sulfide, and sulfate. Inorganic arsenic is found in the soil, ores of copper and lead, and, of course, in groundwater. Arsenic is also present in water in a low concentration. But it becomes more concentrated when used in alloys, organic pesticides, and in wood preservative such as chromated copper lumber.

Organic arsenic is arsenic compounds made using organic species and arsenic.

Adsorption reactions between arsenic and mineral surfaces act as a controlling factor in the removal or dissolution of arsenic in groundwater. Fe_2O_3 and MnO_2 are dormant sources for this [3].

Oil spill in Gulf might have developed arsenic build up. Usually, oceans can filter out arsenic on the sea floor as sediments. Oil can act as an inhibitor for this sedimentation. Therefore, the more the oil allowed into the water worldwide the less the ability of the ocean to act as a natural filter for arsenic [4].

A more recent paper that described the contamination in Perth, Australia, showed that pyrite oxidation was the source of arsenic.

1.3   CONCEPT OF REMEDIATION

Inorganic arsenic is much more toxic than organic arsenic. Arsenic toxicity is found in the form of arsenic salt, arsenite, and organic arsenic. One redeeming feature of arsenic toxicity is that arsenic in +3 state is readily oxidized to less toxic arsenic +5 state. Arsenic +3 state does not exist in aqueous solution; they are readily hydrolyzed to AsO_3^{3-} ion. Arsenic +3 state is thermodynamically unstable in aerobic environment and also oxidized to arsenic +5 state. Here, the first stage of remediation chemistry is explored.

In the presence of iron oxide and manganese dioxide this oxidation is faster. Of late, it has been noticed that some microorganisms catalyze this activation process. The initial filtrations were primarily based on filter bed comprising iron oxide and iron. But in Bangladesh and in the eastern part of India it was noticed that the equilibrium between arsenic and iron scavenger moves backward with time. So this method lost its usefulness after a period of time.

Then, different filtrations using numerous techniques were applied. These techniques were further refined and modified after nanotechnology came into effect. A better and more ecofriendly technique of phytoremediation came at a later stage.

Remediation can be done in two different ways:

1. By removing arsenic from drinking water, agricultural products, and other foods including sea foods, chicken, etc.
2. By changing the food habits with the use of nutraceuticals and functional foods through nutritional intervention.

All the removal technologies rely on some basic processes:

1. *Oxidation reduction reactions*: In this process, arsenic is not removed from the solution but may be immobilized to prevent its further adverse effect.
2. *Adsorption and ion exchange*: Some arsenic adsorbents like alumina, iron, and FeO that have strong affinity for arsenic may be used as adsorbent.

Ion exchange and ultracentrifuge are also basically adsorption process. Disposal of the solid waste containing arsenic is a problem.

3. *Precipitation*: This can be done by forming insoluble arsenate from soluble arsenic. Sometimes, coagulants are added for coprecipitation of arsenic. The colloidal arsenic obtains flocculates by adding strong electrolytes like calcium chloride. Here also the disposal problem persists.

4. *Solid/liquid separation*: In this process, sedimentation by gravity settling is an effective method. A crude method is sand filtration and sophisticated method is membrane filtration.

5. *Biological removal*: Suitable bacteria can be deployed for catalyzing the processes by which soluble arsenic from aqueous or alcoholic system is converted to insoluble form.

6. Nanotechnology [5] for the removal of arsenic.

1.4   GEOCHEMISTRY

Geochemical characteristics of groundwater show that arsenic is present mostly in groundwater of medium depth. Both deep tube wells and shallow waters of up to 20–30 ft scarcely contain arsenic. It is a common characteristic of arsenic that it is always associated with iron and manganese. Arsenicals in many cases are associated with soil phosphates. High levels of geogenic arsenic and manganese in drinking water in India may be attributed to the bacterial oxidation of pyrites, but the same oxidation of pyrite with different modes and bacterial activity leads to bringing arsenic to the surface in the United States where arsenic has been found around Alaska, Crater Lake, Mono Lake, and Searle's Lake and the volcanic lakes of Nicaragua and Costa Rica. High arsenic with low manganese is common in waters from Holocene gray sediments aquifer in India while low arsenic and high manganese was found in Pleistocene reddish-brown aquifer in West Bengal, India [6].

Multivariate geostatistical analysis suggests that the dissolution of FeO and MnO was the dominant process of the release of arsenic in groundwater in Indian soil. Geochemical model also suggests that the concentration of arsenic, manganese, iron, and phosphates in groundwater are affected by secondary mineral phases [7].

Investigations by scanning electron microscopy coupled with energy dispersive X-ray of aquifer materials showed that arsenic is found on the mineral grains in the clay of silt quality as well as in the sand layers. Excessive extraction of groundwater for drinking and irrigation purposes is a cause for the fluctuation of the water table in West Bengal, India. Aeration in the groundwater, which may be caused by the fluctuation of the water table, may be responsible for the formation of carbonic acid. This carbonic acid in turn is responsible for the weathering of silicate minerals, and the clay thus formed release arsenic into the groundwater of the area [8].

1.5   HEALTH EFFECTS

In a recent paper it has been demonstrated that arsenic coming from groundwater and irrigation water in the state of Bihar, India, is responsible for skin pigmentation and a higher risk of cancer [9]. The respiratory effects of chronic low level arsenic exposure from groundwater in India has been analyzed, and the result is that 20.6% of victims have malfunctioning of lung, both restrictive and combined types, whereas, 13.6% of nonexposed (controlled) have this defect. So it can be said that very low levels of exposure to arsenic has deleterious effect on respiratory system [10]. Another study among victims of arsenic toxicity in north China shows that in victims with skin lesions, higher levels of inorganic arsenic in urine samples and organic arsenic in the form of monomethylarsonic acid (MMA) have been found. The arsenic victims without skin lesions have higher MMA but lower inorganic arsenic in their urine samples. So it can be suggested that the victims with skin lesions have a lower arsenic methylation capacity than victims without skin lesions [11].

In India, a study showed that higher intake of water in summer leads to higher body arsenic than in winter when water intake is low. An investigation in the state of Uttar Pradesh, India, showed that an older source of drinking water has higher chance of contamination and it may not be depth based. But in shallow water source the arsenic exposure is absent or very less [12].

It is high time to reconsider arsenic regulations because currently arsenic regulations are focused mainly on drinking water but it does not positively consider the arsenic uptake through food and drinks. The exposure to arsenic from rice consumption is equal or more than that from drinking water if a person takes two meals constituting mainly rice in a day. So food safety regulations should be reconsidered for arsenic contamination through foods and drinks [13].

Optimum limit of arsenic in drinking water and soil threshold values for arsenic are summarized here:

In water: 10 ppb (WHO), 50 ppb (EU), 10–60 ppb (the Netherlands), 10 ppb (German drinking water), and 10–30 ppb (German surface water).
Soil: 500 mg/kg (Californian manual), 29–55 mg/kg (the Netherlands), and 20 mg/kg (Germany) [14].

1.6 ARSENIC TOXICITY IN FOOD AND FOOD PRODUCTS

Though arsenic contamination comes primarily from groundwater, it has been observed that the water-logged situation typically favorable for growing rice and some other cereals mobilizes arsenic in them. Irrigation water containing arsenic may contribute further to the contamination. Rice contains mainly inorganic arsenic in the form of arsenide and arsenate along with organic arsenic such as dimethylarsinic acid (DMA) and MMA [15].

Arsenic-contaminated food is a major problem throughout the world. Arsenic is not only found in rice and wheat but also its alarming presence is noticed in many vegetables, apple juice, and chickens as well. For the faster rate of growth of chickens, roxarsone is commonly injected into them. This is a secondary contamination and can be avoided easily. But the paddy field is contaminated with soil arsenic and also soluble arsenic from water—both groundwater and irrigation water. The vegetables also get primary contamination as in rice and it has been found in alarming proportions in apple juice. The contaminants are arsanilic acid, nitarsone, cabarsone, and inorganic arsenic. In many pesticides arsenic is present and that is also a direct source of the contaminant in foods. Phosphorus and arsenic being the members of same group in the Periodic Table and having almost the same atomic volume, they are present together in nature at many places. Soil phosphates have arsenate as a common contaminant. Mining-related arsenic is a common problem in South America.

In the study, arsenic content has also been observed in fish and between the two types of fishes, that is, sweet water fish and saline water fish, sweet water fish has lesser arsenic content [16].

It is generally believed that arsenic contaminations in food are mainly in the organic form that has much less toxicity than inorganic arsenic. Inorganic arsenic is an environmental contaminant in drinking water. This is true in the case of sea food where the main arsenic contaminants are arsenosugars, but there are some foods where inorganic arsenic is also present as contaminant. This is particularly true in the case of rice. MMA and DMA were supposed previously to have insignificant toxicity and were not considered potential health hazards. But the mechanism of arsenic toxicity shows that their transformation in vivo leads to the formation of highly toxic arsenic; also, they can catalyze and sometimes participate actively in hindering enzymatic activity in the physiological system.

1.7 REMEDIATION BY FILTRATIONS

The SONO filtration technique, first introduced in Bangladesh got admiration from UNICEF and this filter though basically contains iron oxide and iron fillings, some silica and charcoal absorbents are additionally used. The main difficulty of SONO filtration is that it is just a household process of removing arsenic from water and cannot be used on a large scale.

The membrane filtration using cationic surfactant *cetylpyridinium chloride* (CPC) and *polyethersulfone* (PES) membrane are tried successfully by commercial water filter manufacturers with more and more sophisticated membranes [17]. Their commercial use in mass scale as well as in rural area is not feasible because of the high price and handling problems.

Remediation of arsenic toxicity encounters big problems due to institutional weakness and lack of accountability and political mileage given more importance than scientific application. These are the major hindrances for the remediation of arsenic toxicity in Southeast Asia.

It has been suggested that amorphous iron oxide adsorbent has very high adsorbing capacity of arsenic, which has been tried in Cambodia [18]. A thiol-modified sand cartridge has been used for separation of As(V) and As(III) [19].

A hybrid anion exchanger resin has been developed [20] to remove arsenic, but the disposal problem of the resin has not been taken up in the paper.

As it is evidenced that phosphate ion competes with arsenate and arsenide ions for the adsorption on FeOOH (iron oxide), this characteristic can be used for the mobilization of arsenic in groundwater in Bengal Basin in India. The competing ability is in the order of

$$HCO^- = H_4SiO_4 > Fe(II) \gg PO_4^{3-}$$

The adsorption is pH dependent as in acidic pH the concentration of As(III) increases but in basic pH the concentration of As(V) increases. From the studies, it can be said that the reductive dissolution of FeOOH cannot explain the high concentration of arsenic in the groundwater in Bengal Basin. Hence, along with the dissolution of FeOOH, competitive adsorption reactions with the aquifer sediments are responsible for the enrichment of arsenic in groundwater in Bengal Basin [21].

1.8 MECHANISM OF ARSENIC TOXICITY AND ITS EFFECT

Recent advances in arsenic toxicity treatment are multifarious. About 80%–90% soluble arsenic compounds are absorbed from the gastrointestinal tract. After absorption, arsenic attacks the liver, heart, lungs, and kidney and also presents itself in muscles and neural tissues. The main danger of arsenic toxicity is that pentavalent arsenic when absorbed is biologically transformed to toxic trivalent arsenic. Trivalent arsenic thus transformed undergoes methylation. MMA and DMA are less toxic and are excreted from the body more easily than inorganic arsenic.

Arsenic (III) binds with thiol group containing enzyme. Reduction of arsenic +5 to arsenic +3 is done by enzymes and also nonenzymatic reduction where glutathione has an important role and in the process arsenic thiols are formed in the physiological system. The interaction of arsenic with glutathione and other enzymes may change their biological functions [22].

MMA and DMA (both in pentavalent state of arsenic) are transformed to trivalent arsenic when they react with thiol groups that are in a highly toxic state. Arsenic +3 state also inhibits pyruvate dehydrogenase, which is needed for enzymatic activity. Piruvate dehydrogenase oxidizes pyruvate to acetyl coenzyme A (acetyl CoA), which is an intermediate of the citric acid cycle. It affects the electron transform system for ATP production. So it decreases the production of ATP, which affects the energy metabolism seriously.

Oxidative stress theory of arsenic toxicity states that dimethyl arsine (trivalent arsenic) produced by the reduction of arsenic +5 to arsenic +3 reacts with molecular oxygen forming DMA radicals and superoxide anions. Exposure to superoxide anions may cause DMA damage. The ability of arsenic toxicity to develop cancer in the lungs, skin, and bladder may also be attributed to oxidative stress theory. Human lungs are responsive to arsenic carcinogenesis because dimethyl arsine, a toxic gas, is excreted via lungs. This may be due to high partial pressure of oxygen. In human bladder, high concentration of DMA and MMA may lead to arsenic-induced carcinogenesis in the lumen of the bladder [23].

Suspected arsenic toxicity in physiological system is found in blood cell counts and serum electrolytes studies like sodium and potassium. Clinical diagnosis of arsenic toxicity is first manifested in the change of pigmentation of the skin. Among various skin lesions, most predominant are keratosis and both hyper- and hypopigmentation of hands and feet. The clinical diagnosis is done by taking samples of blood, urine, hair, and nail. These diagnoses are sensitive to the time of exposure as well as the time of measurement. Blood samples can hold arsenic concentration within a reliable range for hours. Urine samples needs more concentration and are also not reliable after even a day. Only hair samples have higher retention time for arsenic (few days); moreover, it is more easily available in sufficient quantities. So, hair sample is the best for measuring arsenic toxicity in human body.

Arsenic exposure influences the activity of different enzymes of haem biosynthesis. So it affects hæm synthesis pathway.

Arsenic directly or indirectly propagates (1) cardiovascular diseases, (2) respiratory diseases, (3) hepatic diseases, (4) bone marrow problem, and (5) renal problem; it also affects nervous system and propagate malignancy.

It affects central nervous system particularly in children with chronic exposure to arsenic. Arsenic may cross blood–brain barrier and alteration in biogenic amines of rat brain has been reported.

Accumulated arsenite in liver inhibits oxidation of pyruvate and ketoglutarate. Trivalent arsenic reacting with vicinal thiols stops the supply of thiols necessary for the oxidation of pyruvates and ketoglutarate.

Skin cancer has been confirmed with chronic inorganic arsenic exposure. One of the most severe adverse manifestations of arsenic toxicity appears to be cancer. Arsenic acts as a catalyst to develop stem cell cancer. As it attacks the stem cell, there may be evidences that arsenic exposure in tender age may be the cause of cancer in adulthood. The contact of arsenic toxicity during utero- or postnatal life can emerge as disease at a much later life. So arsenic has been proved as a multiside carcinogen [24–28].

Arsenic(III) produces reactive oxygen species in the cell. Iron in the physiological system reacts with it to produce hydroxyl radical that interferes with the potassium channels and ultimately disrupts the cellular electrolytic functions. It leads to high blood pressure, cardiovascular problems, and affects the central nervous system.

As arsenic toxicity triggers Vitamin A deficiency, it may lead to night blindness.

1.9   TREATMENT

Chelation therapy was the first application in the treatment of arsenic poisoning that has got some credence. The process involves the formation of a chelate complex with arsenic and the formed chelate is then more easily eliminated from the body. 2,3-Dimarcaprol (BAL) was used traditionally since 1949. However, it rapidly mobilizes body arsenic and increases the brain arsenic significantly. It leads to serious side effects like profuse sweating, vomiting, headache, salivation, chest pain, etc. Then, BAL was replaced by a chemical derivative DMSA (meso-2,3-dimercaptosuccinic acid). Its advantage is that it is an orally active chelating agent and less toxic than BAL. It has been proved that DMPS (sodium 2,3-dimercapto-1-propanesulfonic acid) is another chelater that is considered a better one because in physiological system DMSA is biotransformed almost completely to DMSA:CySH (1:2) mixed disulphide. But DMPS:CySH mixed disulphide is found only in minute amounts. Another difference is that DMPS is distributed in both extracellular and intracellular manner but DMSA is distributed only in extracellular manner.

D-Penicillamine in combination with dimarcaprol was also used for the removal of arsenic but at the same time with serious side effects causing leukopenia, anemia, and agranulocytosis and damaging renal, gastrointestinal, and pulmonary system; hence, it has been discarded.

The most widely used remediation is by filtration. These are done by different techniques suitable for different environment. Among these are filtrations like the old SONO filters used in households of Bangladesh, iron filters such as zerovalent iron, membrane filtrations, and ultrafiltrations using nanoparticle. In some techniques, nanotechnology polymer beads of nanodimension impregnated iron oxides are used. It has been reported that this technology can absorb 10–12 mg of arsenic on 1 g of polymer bead.

It is very important to note that arsenic cannot be removed from water by boiling.

1.10   REMEDIATION BY NANOTECHNOLOGY

Nanotechnology is now employed for remediation of arsenic toxicity. The nanosized polymer beads fused with iron oxide acts as filter bed for the removal of arsenic from groundwater. More recently, nanosized zero-valent iron (nZVI) and carbon nanotubes (CNTs) containing nano carbon beds are used for remediation, which is often called nanoremediation of arsenic toxicity. The revolutionary nanotechnologies in biomedical industries have forced open a new vista in remediation, prevention, and therapy to the toxicologists. It has been proved an effective alternative to the conventional site-remediation technology. The large specific surface area (SSA) of nanoparticle iron can effectively remove arsenic from groundwater. Nanoparticles are 5–10 times more effective in removing arsenic than microsized iron. One thing to be remembered is that the use of nZVI in filtering arsenic might have a potential threat on us. This is due to the insufficient study about the effect of the accumulation of nanoparticles including arsenic-bonded iron on our health. It is presumed that Fe^{2+} readily oxidizes in air to form Fe^{3+}, which precipitates out as iron oxide and oxyhydroxides. To get the maximization of the surface mobility of nZVI, the possibility of the nanomaterials migrating beyond the contaminated area may be increased. It may have a negative impact on the remediation process. When nanoparticles are inhaled, it may affect the lungs. CNTs having carbon atom layers in hexagonal arrays of carbon atoms (graphite arrangement) are effective in arsenic removal as well as it is utilized to remove heavy metals like lead, zinc, chromium, etc., as well as organic and biological impurities. The CNTs are available as single-walled as well as multiwalled tubes. The nanoabsorbent carbon effectively absorbs arsenic from water.

In nanotechnology, the large surface area of the nanoadsorbent (zerovalent iron and nanotubes of carbon) is more effective in absorbing the arsenic contaminant in water. The insight remediation of groundwater is costly and cumbersome in nanotechnology. This is because it needs a permeable reactive barrier in the treatment zone where the contaminants become immobilized and only groundwater passes through the barrier. In many cases it needs pressure building operations from outside for an appreciable flow of water. Moreover, the serious limitation in using nanotechnology is that the permeable reactive barrier is effective only in contaminant plumes that pass through it. The nanotechnology provides an effective and fast site remediation. But the nanoparticles that absorb the pollutant may not be easily disposable. Before its wide use in water purification, the disposal problem of the nanomaterials used for the nanoremediation must be properly addressed and further research on their character and extent of the mobility of arsenic from them is needed. As the filtration by absorption or adsorption depends on the pH of water, all types of water cannot be treated to the same extent by the same filtration technique. For example, the absorption of arsenic in nanosized zerovalent iron and iron oxides decreases with increasing pH of water.

Commercial nanofiltration membranes operate more than satisfactorily for the removal of arsenic and pesticides [29].

1.11   PHYTOREMEDIATION

A pH-independent method of arsenic removal was a long-cherished desire of researchers that was made possible in the form of phytoremediation. Among the different processes there is phytoextraction where plants uptake arsenic from soil and water and store it in its harvestable biomass. It is usually observed in plants resistant to arsenic pollution. Phytostabilization is another process where plants stop or reduce the phytoavailability and mobility of the arsenic in the environment. This process does not remove arsenic from the site but reduces its mobility and excludes arsenic from plant uptake. Phytotransformation process is the use of plants ability to modify, immobilize, or inactivate the arsenic through its metabolism. Phytovolatilization is the ability of some hyperaccumulating plants to absorb the arsenic from soil and water and to translocate it to the aerial parts of the plant to volatilize the pollutant arsenic in the air. Rhizofiltration is done by aquatic plants. In this process the hyperaccumulating aquatic plants sorb the pollutant arsenic from aquatic environment. The aquatic plants for remediation of arsenic toxicity are many. Some examples are *Azolla caroliniana*, *Myriophyllum*, and Brazilian water weed such as *Veronica aquatica* and water pepper.

Phytoremediation [30] are helpful tools for the removal of arsenic. In natural water arsenic compounds are present in microorganisms such as phytoplankton and the arsenic is converted to organic arsenicals. The organic arsenicals can also be taken up by bacteria.

There are some floating hyperaccumulating plants that are capable of absorbing arsenic through their roots and submerged plants that can accumulate arsenic by their whole body. Phytoremediation of arsenic has five processes:

1. *Phytoextraction*: It removes arsenic from the contaminated site. These are actually hyperaccumulating plants.
2. *Phytostabilization*: It does not remove pollutants from contaminated sites but reduces the mobility of the contaminant, in this case arsenic, and consequently stops the plant uptake it.
3. *Phytovolatalization*: Some hyperaccumulating plants can uptake arsenic and translocate it to their aerial portions and consequently volatilize the arsenic in air as arsine. So, this should not be tried from an environmental point of view.
4. *Phytotransformation*: This is partly phytodegradation and partly phytostabilization through plant metabolism.
5. *Rhizofiltration*: Hyperaccumulating plants in aquatic environment absorbs arsenic.

Chinese brake fern (*Pteris vittata*) [31,32] has high potential to accumulate arsenic by its root and then translocate it to shoot. Water hyacinth a floating aquatic plant found in South America can accumulate arsenic. White lupin (*Lupinus albus*) [33] can absorb arsenic by rhizofiltration. It is a nitrogen fixing legume and can grow in acidic soils with low nutrient availability.

Moringa oleifera seeds are well known arsenic absorbent. It has a potential to use extensively in phytoremediation of arsenic [34]. It is a well-known plant grown in alluvial soils.

1.12 REMEDIATION USING NUTRACEUTICALS AND FUNCTIONAL FOODS

From the mechanism of arsenic toxicity, it is clearly evidenced that the free radicals and the active oxygen including peroxy radicals are the culprit in propagating diseases. These are produced in the reaction between body proteins and arsenic and enzymes.

It is here where the remediation of arsenic toxicity can be affected by using antioxidants that has the potential to scavenge out free radicals as well as the active oxygen radicals. The typical antioxidants that are naturally occurring and have the potential to mitigate the arsenic toxicity are ascorbic acid, carotene, cinnamic acid, flavonoids, peptides, quercetin, tocopherol, and phenolic compounds. Quercitine can scavenge superoxide radicals and also protect from lipid peroxidation. Omega-3 and omega-6 fatty acids, folic acids, biotin, and PUFA have proven their ability to lower arsenic concentration in human body. Polyphenolic antioxidants like flavanan-3-ols, anthocyanidins, and proanthocyanidins are also effective. Tannins and catechins are novel antioxidants and have proven effectiveness against arsenic-induced cellular injury and carcinogenesis.

Functional foods that can scavenge out free radicals have the potential to lower down arsenic level in human body. Sunflower oil and nuts oil having rich quantity of PUFA fortified with omega-3 and omega-6 fatty acids are good in combating chronic arsenic toxicity. Jaggery and sugarcane juice fortified with vitamin C is another functional food against arsenic toxicity. Edible oils that are rich with polyphenolic antioxidants are also helpful. Cranberry and grapes fortified with vitamin A and vitamin E are good for alleviating arsenic toxicity. It has been proved that addition of vitamin E partly prevents the arsenic-induced health diseases. Apple juice and orange juice fortified with folic acid (Vit. B9) can be used to effectively lower the arsenic level in vivo. This is particularly useful for skin lesions. Root extracts and crushed root of *M. oleifera* is now used for treatment of arsenic toxicity.

In the management of arsenic toxicity, a helpful method is intake of spirulina extract (250 mg) + zinc (2 mg) twice daily for 16 weeks.

Antioxidant characteristic: C-phycocyanin (C-PC) is one of the major biliproteins of spirulina with antioxidant and scavenging characteristic. C-PC also induces apoptosis in lipopolysaccharide–stimulated RAW26.47 macrophages. Additionally, it has anticancer and anti-inflammatory properties [35].

Daily dose of 1 g/day of chlorogenic acid (CGA) is good for remediation of arsenic toxicity because of its anticarcinogenic hypoglycemic and antioxidant property. It also increases blood level homocysteine. It is found in coffee beans, tomatoes, blueberry, peanuts, peers, apples, etc. Crushed garlic (*Allium sativum*) or onion (*Allium sepa*) are antioxidant and antihypertensive because of the presence of quercetin and polyphenols in them and are good for remediation of arsenic toxicity [36].

Charts of nutraceuticals and functional foods for the treatment of arsenic toxicity are listed in Chapter 15.

1.13   CONCLUDING REMARKS

From the provided information, we can say that the treatment of arsenic toxicity may be better effected by combination therapy. That is, taking proper nutraceutical and functional foods in accurate dose along with chelation therapy. In this book, the methods of combination therapy and their plausible mechanism of action are discussed in Sections III and IV.

REFERENCES

1. Ravenscroff, P. Predicting the global distribution of arsenic pollution in ground water. Paper presented at *Arsenic: The Geography of a Global Problem*, *Arsenic Conference*, Royal Geographic Society, London, U.K., 2007.
2. WHO. *Guidelines for Drinking Water Quality*. Geneva, Switzerland: World Health Organization, 2006.
3. Jain, A., Loeppert, R.H. Effect of competing anions on the adsorption of arsenate and arsenite by ferrihydrite. *J. Environ. Qual.*, 29(5) (2000) 1422–1430.
4. Bhattacharya, P., Welch, A.H., Ahmed, K.M., Jacks, G., Naidu, R. Arsenic in groundwater of sedimentary aquifers. *Appl. Geochem.*, 19(2) (February 2004) 163–260.
5. Rajan, C.S. Nanotechnology in ground water remediation. *Int. J. Environ. Sci. Develop.*, 2 (2011) 3.
6. Sankar, M.S., Vega, M.A., Defoe, P.P. et al. Elevated arsenic and manganese in groundwaters of Murshidabad, West Bengal, India. *Sci. Total Environ.*, 488 (2014) 574–583.
7. Halim, M.A., Majumder, R.K., Rasul, G. et al. Geochemical evaluation of arsenic and manganese in shallow groundwater and core sediment in Singair Upazila, Central Bangladesh. *Arab. J. Sci. Eng.*, 39 (2014) 5585–5601.
8. Singh, N., Singh, R.P., Mukherjee, S. et al. Hydrogeological processes controlling the release of arsenic in parts of 24 Parganas District, West Bengal. *Environ. Earth Sci.*, 72(1) (2014) 111–118.
9. Singh, S.K., Ghosh, A.K., Kumar, A. et al. Groundwater arsenic contamination and associated health risks in Bihar, India. *Int. J. Environ. Res.*, 8(1) (2014) 49–60.
10. Das, D., Bindhani, B., Mukherjee, B. et al. Chronic low-level arsenic exposure reduces lung function in male population without skin lesion. *Int. J. Public Health*, 59(4) (2014) 655–663.
11. Zhang, Q., Li, Y.F., Liu, J. et al. Differences of urinary arsenic metabolites and methylation capacity between individuals with and without skin lesions in inner Mongolia, Northern China. *Int. J. Environ. Res. Public Health*, 11(7) (2014) 7319–7332.
12. Katiyar, S., Singh, D. Prevalence of arsenic exposure in population of Balla district from drinking water and its correlation with blood arsenic level. *J. Environ. Biol.*, 35(3) (2014) 589–594.

13. Sauve, S. Time to revisit arsenic regulations: Comparing drinking water and rice. *BMC Public Health*, 14 (2014) 465.
14. Matschullat, J., Perobelli Borba, R., Deschamps, E., Figueiredo, B.R., Gabrio, T., Schwenk, M. Human and environmental contamination in the Iron Quadrangle, Brazil. *Appl. Geochem.*, 15 (2000) 181–190.
15. Biswas, A., Biswas, S., Lavu, R.V.S. et al. Arsenic-prone rice cultivars: A study in endemic region groundwater flow dynamics and arsenic source characterization in an aquifer system of West Bengal, India. *Paddy Water Environ.*, 12(3) (2014) 379–386.
16. Marcussen, H., Alam, M.A., Rahman, M.M. et al. Species-specific content of As, Pb, and other elements in pangas (*Pangasianodon hypophthalmus*) and tilapia (*Orechromis niloticus*) from aquaculture ponds in southern Bangladesh. *Aquaculture*, 426 (2014) 85–87.
17. Ergican, E., Gecol, H. The effect of co-occurring inorganic solutes on the removal of arsenic(V) from water using cationic surfactant micelles and an ultrafiltration membrane. *Desalination*, 181 (2005) 9–26.
18. Kang, Y., Takeda, R., Nada, A. et al. Removing arsenic from groundwater in Cambodia using high performance iron adsorbent. *Environ. Monitor. Assessm.*, 186(9) (2014) 5605–5616.
19. Du, J.J., Che, D.S., Zhang, J.F. et al. Rapid on-site separation of As(III) and As(V) in waters using a disposable thiol-modified sand cartridge. *Environ. Toxicol. Chem.*, 33(8) (2014) 1692–1696.
20. German, M., Seingheng, H., Sengupta, A.K. Mitigating arsenic crisis in the developing world: Role of robust, reusable and selective hybrid anion exchanger (HAIX). *Sci. Total Environ.*, 488 (2014) 551–557.
21. Biswas, A., Gustafsson, J.P., Neidhardt, H. et al. Role of competing ions in the mobilization of arsenic in groundwater of Bengal Basin: Insight from surface complexation modeling. *Water Res.*, 55 (2014) 30–39.
22. Flora, G.J.S. Recent advances in arsenic toxicity treatment. *Pharmabiz.com*, 2012.
23. Shi, H., Shi, X., Liu, K.J. Oxidative mechanism of arsenic toxicity and carcinogenesis. *Mol. Cell. Biochem.*, 255 (2004) 67–78.
24. Liaw, J., Marshall, G., Yuan, T., Ferreccio, C., Steinmaus, C., Smith, A.H. Increased childhood liver cancer mortality and arsenic in drinking water in Northern Chile. *Cancer Epidemiol. Biomark. Prev.*, 17 (2008) 1982–1987.
25. Marshall, G., Fetteccio, C., Yuan, Y., Bates, M.N., Steinmaus, C., Selvin, S., Liaw, J., Smith, A.H. Fifty year study of lung and bladder cancer mortality in Chile related to arsenic in drinking water. *J. Natl. Cancer Inst.*, 99 (2007) 920–928.
26. Smith, A.H., Marshall, G., Yuan, Y., Ferreccio, C., Liaw, J., von Ehrenstein, O., Steinmaus, C., Bates, M.N., Selvin, S. Increased mortality from lung cancer and bronchiectasis in young adults after exposure to arsenic in utero and early childhood. *Environ. Health Perspect.*, 114 (2006) 1293–1296.
27. Yuan, Y., Marshall, G., Ferreccio, C., Steinmaus, C., Liaw, J., Bates, M., Smith, A.H. Kidney cancer mortality: Fifty year latency patterns related to arsenic exposure. *Epidemiology*, 21 (2010) 103–108.
28. Yorifuji, T., Tsuda, T., Grandjean, P. Unusual cancer excess after neonatal arsenic exposure from contaminated milk powder. *J. Natl. Cancer Inst.*, 102 (2010) 360–361.
29. Kosutic, K., Furac, L., Sipos, L., Kunst, B. Removal of arsenic and pesticides from drinking water by nanofiltration membrane. *Separat. Purif. Technol.*, 42(2) (2005) 137–144.
30. Mc Cuttcheon, S.C., Schnoor, J.L. (eds.). *Phytoremediations Transformation and Control of Contaminants*. Hoboken, NJ: Wiley-Interscience, Inc., 2003.
31. Tu, S., Ma, L.Q., Fayiga, A.O., Zillioux, E.J. Phytoremediation of arsenic-contaminated groundwater by the arsenic hyperaccumulating fern *Pteris vittata* L. *Int. J. Phytoremed.*, 6(1) (2004) 35–47.

32. Ma, L.Q., Komar, K.M., Tu, C., Zhang, W.H., Cai, Y., Kennelley, E.D. A fern that hyper-accumulates arsenic: A hardly fast growing plant helps to remove arsenic from contaminated soil. *Nature*, 409 (2001) 579.

33. Martínez-Alcalá, I., Clemente, R., Bernal, M.P. Efficiency of a phytoimmobilisation strategy for heavy metal contaminated soils using white lupin. *J. Geochem. Explor.*, 123 (2012) 95–100.

34. Gupta, R., Dubey, D.K., Kannan, G.M., Flora, S.J.S. Biochemical study on the protective effects of seed powder of *Moringa oleifera* in arsenic toxicity. *Cell Biol. Interact.*, 31 (2006) 44–56.

35. Reddy, M.C., Subhasini, J., Mahipal, S.V.K. et al. C-phycocyanin, a selective cyclooxygenase-2 inhibitor, induces apoptosis in lipopolysaccharide-stimulated RAW 264.7 macrophages. *Biochem. Biophys. Res. Commun.*, 304(2) (2003) 385–392.

36. Chen, X.Y., Peng, C., Jiao, R. et al. The Chinese University of Hong Kong, China. Anti-hypertensive nutraceuticals and functional foods. *J. Agric. Food Chem.*, 57(11) (June 10, 2009) 4485–4499.

2 Arsenic Poisoning and Its Health Effects

Ok-Nam Bae, Kyung-Min Lim, and Jin-Ho Chung

CONTENTS

2.1 INTRODUCTION

Arsenic is a naturally occurring metalloid element found in foods and environmental media that include soil, air, and water. Exposure to arsenic is a major concern for public health worldwide, and accordingly, arsenic is ranked first in the current U.S. ATSDR substance priority list beating other important environmental toxicants. Here, the biotransformation of arsenic and the epidemiological/experimental evidence of arsenic toxicity currently available will be discussed. We will also briefly discuss the suggested mechanisms underlying the toxic effects of arsenic.

2.1.1 ENVIRONMENTAL EXPOSURE TO ARSENIC

The most common route of human exposure to arsenic is through the consumption of drinking water that is contaminated with inorganic arsenic [1]. Naturally occurring geogenic arsenic is found in volcanic ash deposits and rocks. It is suggested that the oxidation and dissolution of arsenic-bearing minerals might be the major natural source of arsenic release into the environment. Arsenic contamination of drinking water can occur by reduction and oxidation of arsenic-containing soils and water by microorganisms such as chemolithoautotrophic arsenite oxidizers or dissimilatory arsenate-reducing prokaryotes [2]. It has been reported that the groundwater in West Bengal in India, Inner Mongolia in China, areas of Bangladesh, Argentina, Australia, Chile, Mexico, Vietnam and Taiwan, and the Western region of the United States is contaminated with high levels of arsenic [3]. The World Health Organization is recommending that the levels of arsenic in drinking water should not exceed 10 ppb, but nearly 100 million people worldwide are estimated to be dependent on drinking water contaminated with higher level of arsenic [4], and more than 30 million people in India and Bangladesh are known to be exposed to drinking water with extremely high levels of arsenic (>50 ppb).

Although it is well established that arsenic-contaminated drinking water is the main route for arsenic exposure, exposure from arsenic-contaminated food, beverage, and air cannot be neglected [5]. It is also reported that arsenic accumulation in the environment has been increased by industrial and human activities that include metal smelting and mining and fossil fuel combustion and semiconductor and glass industries. Extending industrial applications of arsenic enhances environmental as well as occupational exposure to arsenic. Chronic exposure to high concentrations of arsenic is associated with diverse disease conditions such as skin lesions, cardiovascular diseases, blackfoot disease (BFD), and cancers. Many efforts are being directed to elucidate the source of arsenic exposure in the hope of reducing arsenic-associated diseases.

2.1.2 TYPES OF ARSENICALS

The predominant form of arsenic in drinking water is inorganic. In most cases, arsenic occurs as arsenate (pentavalent inorganic arsenic [iAsV]) or as arsenite (trivalent inorganic arsenic [iAsIII]) in anaerobic conditions [6]. Following consumption, absorbed inorganic arsenic compounds are metabolized to monomethyl- or dimethylarsenic compounds either in pentavalent or trivalent forms. The details of the biotransformation process of inorganic arsenic will be discussed in the later part.

High levels of arsenic were found in edible plants cultivated in arsenic-contaminated areas such as Chile and Brazil, through the bioaccumulation of inorganic arsenic from arsenic-contaminated soils and water [5]. The major form of arsenic in plants is inorganic, which is mainly due to low methylation capacity of the plants. While inorganic arsenic is predominant in drinking water and plants, the major arsenic forms in fish or seafood are organic arsenicals such as arsenobetaine or arsenocholine. Fish is known to have the highest rates of arsenic bioaccumulation, but the proportion of inorganic arsenic in fish and shellfish is typically less than 15% with the remainder being organic forms. However, since these organic arsenicals in food do not significantly contribute to arsenic-associated health problem, we will focus on the metabolism and the potential toxicities of inorganic arsenic.

2.2   BIOTRANSFORMATION OF INORGANIC ARSENIC

The toxicity of arsenic depends on the methylation levels (mono-, dime-, or trimethyl) and the oxidation/reduction states (trivalent or pentavalent) that are changed during the biotransformation processes in the body. It is prerequisite to understand the biotransformation process of arsenicals before discussing their health effects and the mode of action.

2.2.1   ABSORPTION AND EXCRETION OF ARSENICALS

At least 90% of ingested inorganic arsenic (iAs) is absorbed through the intestine. The metabolites of inorganic arsenic including monomethylarsonic acid (MMAV) or dimethylarsinic acid (DMAV) can also be easily absorbed through the gastrointestinal tract [6]. The absorbed arsenic is mainly metabolized in the liver, where inorganic arsenic is converted to monomethyl- or dimethylarsenicals by arsenic methyltransferase (AS3MT) [7]. Most of the arsenic and their metabolites such as iAsV, iAsIII, MMAV, and DMAV are excreted in the urine, with DMAV being the most predominant urinary metabolite (DMAV>MMAV>iAs). The concentrations of iAs were highest in the liver and kidney at 1 h after an oral administration of iAsV, and at 4 h after administration, DMAV became the predominant form in the liver [8]. While DMAV and MMAV have long been suggested as the major metabolites in urinary excretion, recent studies have shown that trivalent methylated arsenicals including monomethylarsonous acid (MMAIII) or dimethylarsinous acid (DMAIII) and thioarsenicals were also detected in the urines from arsenic-exposed humans [9]. Through specific assessment of trivalent methylated metabolites, Valenzuela et al. have shown that DMAIII represented the major metabolite (DMAIII>DMAV> iAsV>iAsIII>MMAIII>MMAV) in the urine samples from arsenic-exposed people in Mexico [9]. Exact profiling of urinary arsenic metabolites using more advanced techniques may be required to know the toxicological importance and contribution of the respective metabolites.

2.2.2 Transport of Arsenic

There have been numerous studies to elucidate how arsenic is taken up by the cells and how the metabolites are excreted from the cells. Intestinal absorption of iAsV is suggested to be mediated by phosphate transporters, while glucose transporter (GLUT), aquaporins (AQP), and organic anion transporters (OAT) may be involved in the intestinal absorption of iAsIII [10]. In hepatocytes, where the biotransformation of arsenicals actively occurs, iAsIII is taken up by the cell mediated by aquaporins 7/9 or GLUT [11]. After methylation and glutathione (GSH) conjugation, these arsenic metabolites are excreted by the members of the ATP-binding cassette transporters such as multidrug resistance–associated proteins (MRPs). MRP1 is localized to the basolateral membrane, playing key roles in the excretion of GSH-conjugated arsenicals into the blood, whereas MRP2 is localized to the apical membrane and is responsible for the efflux of arsenic metabolites into bile fluid [12]. MRP4 is recently reported to be involved in cellular efflux of methylated arsenicals in hepatic and urinary excretion. It is known that the MRP expression pattern affects arsenic efflux and accumulation, and, subsequently, modulates the cellular sensitivity to arsenic cytotoxicity [13]. Interestingly, the expression pattern of these inward or outward arsenic transporters is modulated by arsenic exposure [10].

2.2.3 Arsenic Methylation

Methylation of iAs is the main biotransformation pathway of iAs in many mammalian species including humans, although it is not necessarily a detoxification pathway. As we have discussed the urinary excretion profile of arsenic, there are four major methylarsenic metabolites, MMAIII, DMAIII, MMMV, and DMAV [11]. The classical biomethylation pathway consists of two distinct reactions: (1) the reduction of pentavalent arsenicals and (2) the oxidative methylation of trivalent species [14]. Interestingly, the methylation of arsenic does not occur in some animal species such as the guinea pig, marmoset, and squirrel monkey, whereas it is predominant in the hamster, rat, mouse, rabbit, rhesus monkey, and humans, which have a specific enzyme, AS3MT, that converts iAs to their methylated forms [11]. AS3MT is an S-adenosylmethionine (SAM)-dependent methyltransferase (originally Cyt19) [7] and requires dithiol-containing reductant such as glutathione (GSH) or dithiothreitol for its enzymatic activity. iAsIII is converted in sequence to MMAV and DMAV in both humans and in rodents by oxidative methylation, in which a coordination of the unshared electrons on arsenic with the methyl group from SAM results in the generation of neutral methylated arsenicals such as MMAV and DMAV by the oxidizing activity of the methyl group in converting trivalent arsenicals to pentavalent arsenicals.

Another theory (i.e., reductive methylation) on arsenic methylation has been recently suggested, in which thiol-containing complexes of trivalent arsenicals are used for sequential methylation reactions [15]. Here, conjugation of trivalent arsenicals with GSH occurs first, and then arsenic–GSH conjugates (thioarsenicals) serve as the substrate by AS3MT. Notably, the involvement of arsenic–sulfur conjugates as intermediates was also suggested in oxidative methylation, and then the oxidative and reductive methylation would be basically the same.

For the subsequent methylation, the reduction of pentavalent arsenicals to trivalent form is required. Glutathione S-transferase omega (GSTO) is involved in the reduction and also in subsequent conjugation of GSH to trivalent arsenicals [6]. The isotype hGSTO1 was suggested to have the enzymatic activities as MMAV reductase or DMAV reductase [16]. As discussed earlier, thioarsenicals such as AsIII(SG)3 can serve as a substrate for AS3MT. It is suggested that GSH conjugation might be possible either enzymatically by GSTO or nonenzymatically in the presence of hepatic levels of GSH [15]. The formation and the stability of As-GSH conjugates are found to be dependent on the cellular GSH level. As several types of thioarsenicals such as monomethylmonothioarsonic acid and dimethyldithioarsonic acid have been detected in the urine of arsenic-exposed humans and animals [17], the studies on these thioarsenicals are being actively conducted. Further elucidation of the metabolism and the biological activity of thioarsenicals would be important. The biotransformation of arsenic is summarized in Figure 2.1.

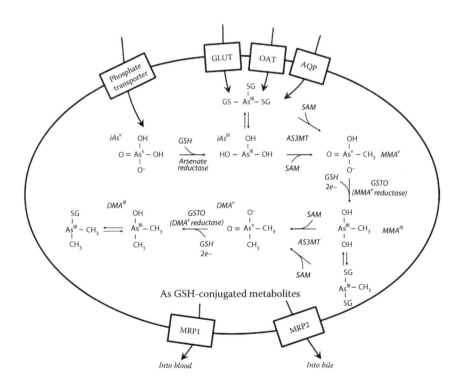

FIGURE 2.1 Arsenic uptake, biotransformation, and excretion in hepatocytes. iAsV, pentavalent inorganic arsenic (arsenate); iAsIII, trivalent inorganic arsenic (arsenite); MMAV, monomethylarsonic acid; MMAIII, monomethylarsonous acid; DMAV, dimethylarsinic acid; DMAIII, dimethylarsinous acid; AQP, aquaporins; GLUT, glucose transporter; MRP, multidrug resistance–associated proteins; OAT, organic anion transporters; AS3MT, arsenic methyltransferase; GSH, glutathione; GSTO, glutathione S-transferase omega; SAM, S-adenosylmethionine.

2.2.4 RELATIVE TOXICITY OF ARSENIC METABOLITES

It is well established that the toxicity of arsenic metabolites is different depending on their oxidoreduction and methylation states. The toxicity of the trivalent form of arsenicals (iAsIII, MMAIII, and DMAIII) has been found to be generally higher than the pentavalent arsenicals (iAsV, MMAV, and DMAV), reflecting that these trivalent arsenicals may play major roles in arsenic-associated health problems. In cells in vitro and animals in vivo, methylated trivalent arsenicals of MMAIII and DMAIII have been consistently more toxic than inorganic arsenic or other arsenic metabolites [18]. In addition to the strong toxicity on cells and animals, MMAIII and DMAIII have shown the most potent biological effects on enzyme activities and signal transduction [11]. The effective concentration of MMAIII or DMAIII for cellular dysfunction, oxidative stress, or malignant transformation is much lower than the parent inorganic form of iAsIII [19]. Interestingly, it is recently suggested that the sulfur-containing arsenic metabolites, such as DMMTAV, are also highly reactive. Although DMMTAV is pentavalent, it showed stronger toxicity against mammalian cells than the methylated pentavalent MMAV or DMAV [20], while its toxicity is weaker than MMAIII or DMAIII. In a recent study, DMMTAV demonstrated a high reactivity with thiol compounds, which is comparable to that of MMAIII [21], challenging the general idea that pentavalent arsenicals are much less toxic than trivalent arsenicals.

Notably, there have been attempts to correlate levels of arsenic metabolites in urine with the severity of arsenic-associated health problem. For example, the concentration of MMAIII in the urine of iAs-exposed population was significantly higher in the individuals with skin lesions when compared with those without skin lesions [9], suggesting that urinary levels of MMAIII may serve as an indicator to identify individuals susceptible to arsenic toxicity. Recent studies have shown the relationship of the production and urinary excretion of DMAIII with bladder carcinogenesis in rats [22]. However, there still remain several points to be established to fully understand the contribution of arsenic biotransformation to arsenic-associated chronic disorders. The precise mechanism of arsenic biotransformation, the factors influencing the ratio and the rate of production of methylated arsenic metabolites, and the biological effects of these methylated tri- or pentavalent arsenicals need to be addressed further.

2.3 ARSENIC HEALTH EFFECTS

Chronic ingestion of arsenic-contaminated drinking water is closely associated with diverse disease conditions, including skin disorder, cardiovascular disease, and high risk of cancer. Arsenicosis refers to the effects of arsenic poisoning, usually over a long period up to 20 years. We will briefly discuss the contribution of arsenic to these disease conditions, by reviewing epidemiological and experimental studies.

2.3.1 SKIN TOXICITY

Arsenic-associated cutaneous abnormality is one of the most predominant manifestations of arsenic exposure–related health effects. Diagnosis of arsenicosis is usually established by simple examination of the skin. Among the people who were exposed to arsenic-contaminated environment, about 20% of the people in Bangladesh and 10% of the people in West Bengal showed dermatological problems. Representative arsenic-associated cutaneous problems include hyperkeratosis, melanosis (pigmentary changes), BFDs (a type of peripheral vascular diseases [PVDs]), and skin cancers such as Bowen's diseases.

2.3.1.1 Premalignant Skin Lesions

Premalignant skin lesions are hallmark signs of arsenic poisoning. While other arsenic-associated diseases including internal cancers and cardiovascular diseases have shown to develop chronically with long latencies, these premalignant skin lesions occur within relatively short period of arsenic exposure. Pigmentary disorder and hyperkeratosis are known to be early stage dermatological symptoms, preceding the occurrence of skin cancer [23]. In cross-sectional analyses on arsenic-associated skin disorder, the obvious dose–response effects of arsenic on the occurrence of skin lesions were observed [24]. The positive association between arsenic exposure and skin lesions was also confirmed in a prospective case–cohort analysis [25].

Melanosis is the earliest and the most common manifestation among all dermatological alterations, which appears as hyperpigmentation throughout the body in either disseminated or localized area. Depigmentation on normal or hyperpigmented skin could be developed, producing the unique appearance of leukomelanosis. Arsenic-associated hyperkeratosis is also suggested as a sensitive early indicator of arsenic poisoning [26], as the palmar keratosis occurs early before people develop arsenic-related cancers of the bladder and lungs. Hyperkeratosis predominantly appears on the palms and soles, typically as indurated skin with papules or wartlike keratoses.

The cutaneous toxicity associated with arsenic may be mediated by increased expression of various cytokines and growth factors that regulate differentiation and proliferation of keratinocytes. Arsenic has been found to stimulate expression and secretion of tumor necrosis factor-α (TNF-α), interleukin-1 and interleukin-8, and growth factors such as tumor growth factor-β and granulocyte–monocyte colony-stimulating factor [27]. Arsenic also modulates signal transduction in keratinocytes by inhibition of AP1 and AP2, reducing expression of a pivotal keratinocyte differentiation marker, involucrin [28]. The expression levels of keratin-16 or keratin-8/-18, which are found in hyperproliferative states or in less differentiated epithelial cells, respectively, are increased by arsenic exposure [29].

2.3.1.2 Skin Cancer

Skin is suggested as the most sensitive target for arsenic-associated cancer [30]. Since carcinogenic effects of arsenic will be discussed in detail in Section 2.3.2, we will briefly introduce the unique characteristics of arsenic-associated skin cancer here. Arsenic-associated skin malignancies can be found in either hyperkeratotic areas or in nonkeratotic areas [26]. While the skin cancers are generally isolated and mostly occur on the sun-exposed parts of the skin, arsenic-associated skin cancer is usually multifocal and found throughout the whole body [31].

Bowen's disease, a skin carcinoma in situ, is a typical type of arsenic-associated skin cancer. Three types of skin cancer including Bowen's disease, basal cell carcinoma (BCC), and squamous cell carcinoma (SCC) are found to be linked with arsenic exposure. Arsenic-induced Bowen's diseases are capable of malignant transformation into invasive BCC and SCC [32]. Full layers of epidermal dysplasia, hyperkeratosis, acanthosis, individual dyskeratosis, and cellular dysplasia were observed in arsenic-associated Bowen's disease. It is also reported that the risk of melanoma is positively related to arsenic concentrations in toenail in arsenic-exposed population [33], supporting that arsenic may be associated with the development of melanoma.

2.3.1.3   Immunological Dysfunction in Skin

Arsenic exposure is associated with G2/M cell cycle arrest and DNA aneuploidy in both cultured keratinocytes and in the skin lesions of arsenic-associated Bowen's disease [32]. Up to now, several key mechanisms have been suggested for arsenic-associated health effects such as (1) generation of oxidative stress, (2) binding to protein thiol groups, (3) depletion of methyl groups affecting epigenetic regulation, (4) alteration in signal transduction, and (5) reduced DNA repair and chromosomal abnormality [6]. Although these mechanisms clearly contribute to arsenic-associated skin abnormalities, there may exist other contributing factors, considering that only 10% of the population exposed to arsenic develops cutaneous problems including melanosis, hyperkeratosis, Bowen's disease, and invasive skin cancers [34].

Among several potential factors, cellular immune dysfunction may play important roles in determining the susceptibility to arsenic-associated skin damage [32]. Arsenic is known to decrease contact hypersensitivity, and T-cell proliferation is impaired in arsenic-induced Bowen's disease. CD4+ cells are the preferential target affected by arsenic. Arsenic induces apoptosis of CD4+ cells through an autocrine TNF-α loop [35]. Expression of IL-2/IL-2R was significantly inhibited with functional impairment of Langerhans cells in arsenic-associated Bowen's diseases [36]. Decreased cell-mediated immunity after arsenic exposure might be closely linked with the pathogenesis of disseminated occurrence of arsenic-induced skin cancer.

2.3.1.4   Blackfoot Diseases

The occurrence of BFD is closely related to the level of arsenic contamination in the drinking water in endemic area of Taiwan. BFD is characterized by progressive darkening discoloration of the skin extending from the toes toward the ankles [37]. Typical symptoms include numbness or coldness of lower extremities, intermittent claudication, progressive development of gangrene, and ulceration, ultimately leading to spontaneous amputation. In epidemiological studies, the co-occurrence of BFD with arsenic-related skin lesions such as hyperpigmentation, hyperkeratosis, and skin cancer has been observed [23]. The dose–response relationship in epidemiological data also revealed that there is a significant association between the arsenic contamination level and the prevalence of BFD [37].

BFD is a unique type of PVD, and the pathological changes of BFD are basically atherosclerosis and thromboangiitis. A destruction of vascular endothelial cells (ECs) takes place at an early stage in the affected limbs [38]. The mechanisms underlying BFD still remain unclear; the toxic effects of arsenic on the vascular system, which we will discuss later, may contribute to the development of BFD. Interestingly, anti-EC IgG antibody was found in the sera of BFD patients, which may play key roles in EC destruction/recovery in BFD [39]. It has been frequently questioned why the occurrence of BFD is limited in endemic area in Taiwan, not in other arsenic-contaminated areas. Differences in ethnicities, duration, nutritional state, arsenic metabolism, and/or the existence of cocontaminants have been suggested. Individual susceptibility to produce anti-EC antibody is also considered to play important roles [38].

2.3.2 CANCER

Arsenic is regarded as a group I carcinogen causing tumors in the bladder, lungs, and skin, but its carcinogenic mechanisms are not well understood yet. Unlike other carcinogens, arsenic itself does not belong to either tumor initiator or promoter, and the mechanism underlying arsenic-induced cancers is still far from clear in spite of extensive efforts. The complexity of carcinogenic mechanism of arsenic can be clearly illuminated by the absence of relevant animal models simulating the carcinogenicity of arsenic in clear dose- and time-dependent manner. However, lots of information from in vitro, ex vivo, and in vivo studies have been accumulated that may shed some light on the mechanism of arsenic-induced cancers. A couple of recent review articles have summarized the publications on the carcinogenicity of arsenic in details that will be instrumental to understand the mechanism underlying arsenic-induced cancer [40,41]. In this section, we will discuss arsenic-induced carcinogenicity focusing on epidemiological studies, mechanistic explanation concerning arsenic-induced genetic and epigenetic alteration and cocarcinogen theory, and finally regulation issue regarding drinking water standard based upon the risk of cancer associated with arsenic exposure.

2.3.2.1 Epidemiological Studies

To date, many lines of epidemiological evidence strongly indicate that chronic exposure to drinking water contaminated with high levels of arsenic is associated with increased incidence of cancers of the bladder, lungs, and skin in human. Chen groups published a series of papers in the 1980s that the exposure to iAs through drinking water was associated with increased incidence of bladder cancer in southwestern Taiwan in the 1950s [42,43]. Chiou et al., however, failed to find a statistically significant increase in bladder cancer in a northwestern Taiwanese population who consumed drinking water containing iAs at concentrations less than 100 μg/L [44]. A cohort study in Denmark [45] and a case–control study in Argentina [46] also failed to show increases in bladder cancer among populations exposed to drinking water with arsenic concentrations less than 100 μg/L. After reviewing literature, Schoen et al. suggested that an increase in bladder cancer is observed only in humans who were exposed to drinking water contaminated with arsenic at high concentrations, usually several hundred μg/L ranges [47]. This conclusion was further supported by a cohort study in Taiwan [48] and an ecological study in Argentina [49].

Increased occurrence of lung cancer in association with the ingestion of arsenic-contaminated drinking water has been also observed in Taiwan and in South America [50]. Smith et al. found a significant increase in lung cancer–associated mortality in a region of Chile where arsenic levels in drinking water were 90–1000 µg/L [50]. However, in a case–control study in Bangladesh, nonsmokers who are exposed up to 400 µg/L levels of arsenic in drinking water did not show an increase in the incidence of lung cancer [51]. Chen et al. found no significant association between arsenic exposure up to 300 µg/L in drinking water and the incidence of lung cancer among never-smokers [52]. Similar to bladder cancer, all these epidemiological studies indicate that lung cancer was not observed in the populations exposed to less than 100 µg/L of arsenic in drinking water.

The association between iAs in drinking water and skin cancer has been observed in a number of studies conducted in Taiwan and some other countries including Argentina, Chile, and Mongolia [42]. These epidemiological studies indicate that skin cancer was observed only in the populations exposed to relatively high concentrations (higher than 300 µg/L) of arsenic in drinking water. In a cohort study with 57,053 Danish people, Baastrup et al. found no significant association between arsenic concentrations in drinking water (0.05–25.3 µg/L) and risks for cancers of the lungs and bladder [45]. Interestingly, low concentration of arsenic in drinking water paradoxically decreased the risk for skin cancers, suggesting that arsenic might have anticancer effects at low concentrations. In an ecological study in Inner Mongolia, Lamm et al. found that skin cancer cases were found only for those exposed to arsenic at concentrations greater than 150 µg/L [53]. In a case–control study in the residents of Eastern Europe (Hungary, Romania, and Slovakia), Leonardi et al. observed statistically significant increases in the risk of skin cancer (BCC) in the group exposed to drinking water contaminated with arsenic (19.54–167.29 µg/L) [54]. Considering the contribution of arsenic intake from food, they suggested that there is a significant association between BCC and arsenic in the subgroup exposed to lifetime average water arsenic concentration <40 µg/L. It should be noted that Leonardi et al. suggested that increased incidence of BCC could be observed at levels of iAs less than 50 µg/L in drinking water, which is inconsistent from other studies [54]. Therefore, concerns were raised to apply this result to calculate arsenic-induced cancer risk [41].

2.3.2.2 Genetic Toxicity

Arsenic has been suggested as a genotoxic carcinogen in numerous studies including in vitro and in vivo experiments and epidemiological studies [40]. The genotoxic effects in carcinogenesis can be explained as potential causes for direct or indirect DNA damage, inappropriate DNA repair, and cell transformation [41].

DNA-reactive carcinogens that directly interact with DNA and subsequently cause DNA damages are mostly reactive electrophiles. However, arsenic itself has an anionic structure that cannot easily react with DNA directly. Representing this, an overwhelming number of studies have purported that arsenic does not cause direct damage to DNA. However, there are studies suggesting that arsenic can cause direct DNA damage. In DNA nicking assay, trivalent methylated arsenic metabolites, MMAIII and DMAIII, directly interacted with DNA, which suggests the capability of arsenic as a DNA-reactive carcinogen [55].

Indirect DNA damage by arsenic has been reported in numerous in vitro studies as determined by micronucleus formation, sister chromatid exchange, chromosomal aberration, and DNA damage. Arsenic-induced micronucleus formation was observed in normal human cells such as peripheral blood leukocytes and epithelial cells [56], as well as cancer cells like colorectal carcinoma cells and epithelial lung adenocarcinoma cells [57]. Interestingly, when the effects of arsenic and its metabolites on micronucleus formation were compared, trivalent arsenics (iAsIII and MMAIII) showed a more potent effect, whereas pentavalent methylated arsenics (MMAV and DMAV) failed to induce micronucleus formation [58]. Arsenic can also enhance sister chromatid exchange in human peripheral blood leukocytes and chromosomal aberration such as chromatid gaps, chromatid breaks, and chromatid exchanges in human umbilical cord fibroblast and peripheral blood leukocytes. Severe genetic modification by arsenic was also revealed as DNA damage in various in vitro assays. Comet assay showed that arsenic caused severe DNA damage in human colon cancer cells, HepG2 cell, and $CaCO_2$ cell. Arsenic also induced damages in nuclear and mitochondrial DNA in human prostate epithelial cells and keratinocytes.

Various animal models were introduced to confirm the in vitro results of arsenic-induced DNA damage. Induction of micronucleus was observed in ovarian tissue of the rats exposed to arsenic in drinking water [59], and the frequency of micronucleated reticulocytes increased in Golden Syrian hamsters when exposed to arsenic-contaminated drinking water [60]. These results support the correlation between the exposure to arsenic-contaminated drinking water and DNA damages. Increase in both micronucleus formation and chromosomal aberration was observed in arsenic-exposed mice [61,62]. Interestingly, increased micronucleus formation and chromosomal aberration were shown at the human reference dose (0.3 µg/kg/day) established by the U.S. EPA. In addition, DNA damages in blood cells from mice orally administered with arsenic were demonstrated by comet assay, and oxidative stress–mediated DNA fragmentation was also observed in the bone marrow cells of Swiss albino mice treated with arsenic.

Several epidemiological studies were conducted in the region highly contaminated with arsenic such as West Bengal and Mongolia, to confirm arsenic-induced DNA damage. Increased micronucleus formation was observed in the population of these regions [63]. In addition, increased micronucleus frequency, chromosomal aberration, and sister chromatid exchange were shown in the population of West Bengal that was suspected from chronic exposure to high level of arsenic [64]. In a study employing comet assay and chromosomal aberration assay, population with arsenic-induced premalignant hyperkeratosis exhibited significantly higher chromosome aberration and DNA strand breaks supporting the correlation between arsenic-induced damage and carcinogenesis [65]. According to these results, micronucleus formation and chromosomal aberration are suggested as biomarkers for arsenic-induced DNA damage and carcinogenesis.

It is thought that oxidative stress plays a key role in arsenic-induced DNA damage. During biotransformation of arsenic, reactive oxygen species can be generated, which can react with DNA resulting in oxidative DNA damage. In human keratinocytes, arsenic induced oxidative DNA damage with increased expression of antioxidant enzymes [66]. In addition, vitamin C, an antioxidant, blocked the mitochondrial DNA damage by arsenic. Moreover, oxidative DNA damage along with enhanced expression of antioxidant enzymes was observed in skin cancer tissues. Arsenic-induced oxidative DNA damage has also been studied in superoxide dismutase and catalase knockout mice. Splenocytes from superoxide dismutase knockout mice manifested higher vulnerability toward arsenic-induced DNA damage than wild-type mice [67]. Similarly, catalase-deficient lymphocytes and hepatocytes were more susceptible to arsenic than the wild types [68]. These results mean that oxidative stress plays a pivotal role in arsenic-induced DNA damage.

Effects of arsenic on DNA repair, which is closely related to genotoxicity, have been evaluated in several in vitro and in vivo studies. In human lung cells, 8-oxoguanine DNA glycosylase 1 activity, DNA ligase IIIα protein level, and XRCC1 content that play key roles in base excision repair were affected by arsenic and its metabolites [69], indicating that inappropriate base excision repair is involved in arsenic-induced genotoxicity. Supporting this, increased DNA damage and impaired DNA repair were observed in arsenic-exposed DNA polymerase β-deficient mouse embryonic fibroblasts, which resulted in lower cell viability [70]. In a study with human keratinocytes, arsenic-inhibited poly(ADP-ribose) polymerase-1 leading to suppression of DNA repair, which could increase the chance of survival of initiated carcinogenic cells [71]. Inhibitory effects of arsenic on poly(ADP-ribose) polymerase-1 are due to arsenic binding to the zinc finger, which is critical for recognition and binding to the damaged DNA site [72]. Interaction of trivalent arsenic with sulfhydryl groups of DNA repair enzymes could also affect DNA repair [73].

However, the effects of arsenic on DNA repair have not been reproduced in in vivo studies. In F344 rats exposed to arsenic through drinking water, arsenic did not affect micronucleus frequency in bone marrow. Either DNA repair or DNA damage in urinary bladder transitional cells was not affected by arsenic [74]. Decreased gene expression was observed in C57BL/6J mice acutely exposed to low-level arsenic; however, these effects were reversed in the group subacutely exposed to high-level arsenic [75]. Even in the animal model using cytosine arabinoside, MMAIII enhanced base excision repair leading to a quick reversal of DNA damages.

Several studies demonstrated that arsenic induces cell transformation, which could lead to malignant transformation. In human urothelial cells, arsenic and MMAIII induced irreversible cell transformation with increased tumorigenicity in immunodeficient mice [76]. Inflammatory cytokines, autophagy, and matrix-associated proteins have been suggested to be involved in arsenic-induced malignant transformation of urothelial cells. Malignant transformation of human keratinocytes by arsenic resulted in increased production of cancer stem cells with elevated MMP-9 secretion and colony formation [77]. It has been suggested that suppression of let-7c following activation of Ras/NF-κB signaling is involved in this process [78]. HIF-2α-mediated inflammation was also involved in malignant transformation of human bronchial epithelial cells, which promoted angiogenesis through β-catenin and VEGF [79]. However, it is unclear whether these arsenic-induced cell transformations are relevant to human carcinogenesis as they are mostly performed in cell lines in which DNA repair is suppressed.

2.3.2.3   Epigenetic Alteration

In addition to genetic mutation, epigenetic modification is an important process for malignant transformation. Epigenetic alterations refer to the changes induced in cells that modify the expression of genes at transcriptional, translational, or posttranslational levels without changes in DNA sequence. Mechanisms of epigenetic alterations in general include DNA methylation, histone modifications, and micro RNA gene splicing, ultimately leading to the inactivation of tumor suppressor genes and activation of oncogenes. First of all, DNA methylation is an important regulator of gene transcription, and a large body of evidence has demonstrated that aberrant DNA methylation patterns have been associated with a large number of human malignancies. Aberrant DNA methylation is manifested in two distinct forms: hypermethylation and hypomethylation compared to normal tissue.

Arsenic undergoes biotransformation consuming SAM as the major methyl donor for DNA methyltransferase. Depletion of SAM by arsenic may contribute to the alteration of methylation of both global DNA and specific cancer gene promoters [40]. In vitro studies showed that treatment with inorganic arsenic resulted in hypomethylation of global DNA in various cancer cell lines [80]. Significant hypomethylation of global DNA was also observed in animal models administered with arsenic through drinking water [81]. In addition, gene-specific studies demonstrated that hypomethylation of oncogenes was observed such as H-ras and c-myc [82,83].

In addition, arsenic also induces hypermethylation that typically occurs at CpG islands in the promoter region resulting in the inactivation of genes that include tumor suppressor genes. Human lung adenocarcinoma cells treated with arsenic displayed hypermethylation in the promoter region of p53 and CpG sequences [84]. Gene-specific hypermethylation was also found in the genes such as death-associated protein kinase (DAPK), Deleted in Bladder Cancer 1, and h-MLH1 (DNA mismatch repair) following the treatment of arsenic to cells [56,85]. Chronic arsenic exposure through drinking water also induces the hypermethylation of DAPK genes in urothelial carcinoma, resulting in the decrease of DAPK protein, which plays a role as a tumor suppressor [86]. Another study in West Bengal, India, demonstrated that significant hypermethylation was found in the promoters of both DAPK and p16 genes in arsenic-induced skin lesions, which may result in the downregulation of both tumor suppressor genes [87].

While DNA methylation is a well-characterized epigenetic alteration, posttranslational modification of histone proteins and chromatin remodeling are other important determinants in the regulation of gene expression [88]. By specific histone-modifying enzymes, the N-terminal tails of histones can undergo a variety of posttranslational modification, including methylation, acetylation, ubiquitination, and phosphorylation. Several studies suggest an association between histone acetylation and arsenic exposure in cell culture system. Li et al. demonstrated that arsenite significantly stimulated histone H3 acetylation in lung fibroblast cells in vitro [89]. Jensen et al. reported that changes in histone H3 acetylation in gene promoter regions were observed during arsenic-induced malignant transformation of urothelial cells [85]. They also observed that arsenic exposure increased DNA hypermethylation in hypoacetylated promoter regions, suggesting that epigenetic alterations are associated with arsenic-induced malignant transformation. More recently, significant reduction of histone H4K16 acetylation was observed in human UROtsa cells, following exposure to inorganic arsenic and its metabolite, MMAIII [90].

Arsenic exposure in lung carcinoma cells can lead to differential effects on the methylation of H3 lysine residues depending on the site such as increased H3K9 dimethylation and H3K4 trimethylation and decreased H3K27 trimethylation [91]. Another study by Zhou et al. demonstrated that inorganic arsenic significantly increased dimethylation and trimethylation of H3K4 and decreased monomethylation of H3K4 in lung carcinoma cells [92]. Interestingly, histone H3K4 trimethylation remained elevated and apparently inherited through cell divisions, even 7 days after the removal of iAs.

A couple of studies suggested the association between histone phosphorylation and arsenic carcinogenesis. Li et al. suggest that arsenic trioxide induces histone H3 phosphoacetylation at the caspase-10 gene, which may play a role in apoptosis contributing to the anticancer effects of arsenic trioxide on acute promyelocytic leukemia [93]. One year later, they showed that arsenite stimulates histone H3 phosphoacetylation of proto-oncogenes c-fos and c-jun chromatin in lung fibroblasts [89]. All these studies clearly demonstrated that histone modification is dysregulated by arsenic exposure, but further work is required to understand the overall effects of altered histone modification on arsenic-induced toxicity or carcinogenesis.

A microRNA (miRNA) is a small noncoding RNA molecule found in animals, which functions in transcriptional and posttranscriptional regulation of gene expression in a sequence-specific manner. The first human disease known to be associated with miRNA dysregulation was chronic lymphocytic leukemia, and thereafter, many miRNAs have been found to have links with some types of cancer [94].

Marsit et al. reported that treatment with arsenic to human lymphoblast TK6 cells grown in folate-deficient media resulted in decreased miR-210 expression but increases in the expression of miR-22, miR-34a, miR-221, and miR-222s [95]. The altered miRNA expressions were returned back to baseline upon removal of the stress conditions, suggesting that chronic exposure to arsenic-contaminated drinking water may be required to permanently alter the expression of miRNAs. Similar results were observed in recent studies showing that arsenic treatment to various cultured cells resulted in altered expression of miRNAs that were upregulated or downregulated.

So far, in vitro studies suggest that epigenetic mechanisms on altered miRNA expression may be associated with arsenic-induced carcinogenesis. Some questions, however, still remain whether chronic intake of arsenic through drinking water is capable of altering miRNA expression and how altered miRNA expression manifests arsenic-induced carcinogenesis.

2.3.2.4 Cocarcinogens

As mentioned earlier, arsenic is generally known not to induce mutation at the levels that humans are exposed to [73]. Thus, diverse modes of actions have been suggested to explain arsenic-induced cancer. A growing body of evidence is, however, indicating that arsenic may act as a cocarcinogen, that is, arsenic itself is not directly mutagenic but can enhance the mutagenicity of other carcinogens like ultraviolet (UV) radiation. Earlier studies reported that when arsenic is administered with several direct mutagens, the incidence of tumors is substantially increased [96]. Since then, two separate studies showed that exposure to arsenicals combined with a physical skin carcinogen, UV irradiation, boosted the occurrence of epidermal cancer in mice [97,98]. They observed that when hairless mice were administered with sodium arsenite through drinking water followed by UV irradiation, the incidence of skin cancer increased in an arsenic dose-dependent manner compared with mice exposed to UV alone. No tumors were observed in any organs in hairless mice given arsenite alone. Motiwale et al. also reported that when mice were coexposed to a carcinogen, dimethylbenzanthracene, on the skin and arsenate (iAsV) through drinking water, the rate of skin cancer development was increased significantly [99].

In addition, arsenic can interact with other environmental contaminants, including tobacco smoke, in inducing lung cancers [100]. The risk of bladder cancer was also increased by interaction of arsenic in drinking water and cigarette smoking [101]. Chen et al. found that interaction between tobacco smoking and arsenic methylation capability was involved in the increased risk of bladder cancer [102]. In a population-based prospective study in 90,378 Japanese, positive trend between arsenic intake and an increased risk of lung cancers was found in men [103]. Interestingly, a recent study reported that coexposure of colon cancer cells to arsenic and ethanol induces angiogenic signal via HIF-1α pathway, suggesting that alcohol consumption may affect arsenic-induced carcinogenesis [104].

2.3.2.5 Low-Dose Issue on Arsenic Carcinogenicity

The current regulation standard for arsenic in drinking water is determined mainly based on the several epidemiological studies from Taiwan, where the arsenic exposure level is very high [43,105,106]. Even if these studies clearly demonstrated that high levels of arsenic (300–1000 µg/L) induce skin, bladder, and lung cancer in human, evidence supporting the association of lower levels of arsenic in drinking water (100 µg/L) with these types of cancer was relatively weak. Furthermore, questions on the Taiwanese studies have been made, contesting about the appraisal of cancer risks of the lower levels of arsenic based upon a simple linearity model [47]. Due to incapability of arsenic to directly induce genetic alteration, several reports suggest that a nonlinear dose–response relationship, that is, a threshold model, is more appropriate for the assessment of arsenic-associated cancer risk to humans [47,107].

This notion was supported by several epidemiological studies that were undertaken to examine arsenic-related cancer in the United States. They reported that less than 100 µg/L of arsenic exposure in drinking water did not increase the risk of bladder, lung, or skin cancer [108,109], signifying the existence of a threshold for arsenic-related cancer, including lung, urinary bladder, and skin cancers. However, it should be noted that in a case–control study in the United States, cigarette smoking potentiated the effect of arsenic (of the range 0.5–160 µg/L; mean 5.0 µg/L) on the risk of bladder cancer [110], suggesting arsenic may act as a cocarcinogen, and threshold model for low dose of arsenic itself may be inappropriate to calculate the cancer risk due to arsenic exposure.

To resolve the issue on cancer risks due to low-dose arsenic contamination in drinking water, many efforts have been made to develop in vivo cancer models in rodents for risk assessment of arsenic. Numerous experiments, however, failed to demonstrate increased incidences of tumors in rodents by arsenic. Recently, using a transplacental model in mice, when pregnant mice were treated with 42.5 and 85 ppm arsenite in drinking water, tumor formation in several tissues (ovary, adrenal, liver, and lungs) of the offspring was observed to increase in a dose-dependent manner [111]. More recently, whole-life exposure of mice (2 years) to arsenic at concentrations of 6, 12, or 24 ppm through drinking water induced dose-related increases in tumors in the tissues of the lungs, liver, gallbladder, adrenal, uterus, and ovary [112]. These target sites of arsenic-associated carcinogenicity were remarkably similar to those observed in the previous transplacental studies. These animal model experiments may contradict the previous theory that rodents are less susceptible to the tumors induced by arsenic in drinking water compared to humans [41]. However, it is still questionable whether these animal models can be applied to assess the cancer risk of low-dose arsenic in drinking water for the following reasons: (1) ovary and adrenal tumors observed in the studies by Waalkes et al. [111] and Tokar et al. [112] have not been reported in humans exposed to arsenic in drinking water, (2) the mode of action for carcinogenesis in these animal models has not been clearly characterized, and (3) a relatively high concentration of arsenic in drinking water was used during the whole-life exposure.

2.3.3 CARDIOVASCULAR DISORDERS

The chronic ingestion of arsenic-contaminated drinking water is closely associated with various cardiovascular complications and peripheral vascular disorders.

2.3.3.1 Endothelial/Smooth Muscle Dysfunction by Arsenic

The vascular endothelium, which covers the inner surface of blood vessels, plays key roles in the regulation of blood vessel tone as well as in the maintenance of vessel integrity. Various active mediators are derived from endothelium, such as nitric oxide (NO), prostanoids, endothelin-1, and cytokines. Among them, NO may be the most important mediator, which induces vasorelaxation and inhibits smooth muscle proliferation/migration, inflammation, and excessive platelet aggregation/activation. Endothelial dysfunction, which is characterized by reduced NO generation, inactivation/uncoupling of endothelial NO synthase (eNOS), and increased oxidative stress, is significantly associated with various CVDs including atherosclerosis and hypertension.

Arsenic is known to initiate endothelial dysfunction both in cells in vitro and in animal models in vivo [113]. In ECs, arsenic induces cell death with increased oxidative stress, disrupts normal cellular signaling such as protein kinase B, and decreases eNOS activity and NO generation [114,115]. Vasorelaxation was significantly impaired by arsenic resulting in vascular dysfunction [115,116]. Interestingly, several recent studies demonstrated that arsenic also causes smooth muscle dysfunction, resulting in abnormal contraction of blood vessels [116,117].

Vascular impairment by arsenic including endothelial and smooth muscle dysfunctions may underlie arsenic-associated CVDs.

2.3.3.2 Hypertension

Many epidemiological studies in arsenic-endemic area have identified a positive association between arsenic exposure and the prevalence of hypertension [118]. The association was also observed in the studies conducted in other areas such as Taiwan, the United States, and Bangladesh [119]. Arsenic-associated hypertension is strongly supported by experimental and mechanistic evidence. Arsenic induces endothelial dysfunction, by increased superoxide accumulation and decreased NO formation in ECs [120], resulting in impaired vasorelaxation and increased vascular tone [115]. Arsenic also promotes inflammation by the stimulation of nuclear factor-κB signaling and expression of cyclooxygenase-2 in ECs [121].

Notably, chronic renal dysfunction as observed in arsenic-exposed population may contribute to the hypertensive effects of arsenic [122]. Although the kidney is not a major target organ for inorganic arsenic toxicity, it is reported that the renal and urinary system may be sensitive targets for dimethylated arsenic metabolites [123].

2.3.3.3 Atherosclerosis

The risk of arsenic-induced atherosclerosis has been evaluated through the epidemiological studies on the prevalence and mortality of PVDs (including BFD), cerebral infarction, and ischemic heart diseases in arsenic-endemic area [124]. The exposure to arsenic is associated with increased risks of all forms of atherosclerotic diseases. The observation of subclinical arterial insufficiency and early microcirculatory defects before the development of PVD and BFD strongly supports progressive cardiovascular deterioration by prolonged arsenic exposure [125]. Notably, a significant increase in atherosclerotic plaques in the innominate artery could be also observed in arsenic-exposed mice [126].

Atherosclerosis has a complicated pathogenesis involving various types of vascular/blood components. Accordingly, the arsenic-induced atherosclerosis can be explained by the multifactorial effects of arsenic on endothelial function, smooth muscle cell proliferation, and inflammation. Endothelial dysfunction following arsenic exposure in drinking water was observed in vivo [127], and alteration in the production of inflammatory mediators was found to be related to arsenic-induced atherosclerosis [126]. In ECs, arsenic induces endothelial dysfunction by increasing oxidative stress and peroxynitrite formation, release of inflammatory cytokines [128], and expression of COX-2 [121]. Arsenic increased the expression of atherosclerosis-related genes that include heme oxygenase-1 (HO-1) and monocyte chemoattractant protein-1 (MCP-1) in endothelial and smooth muscle cells [129].

2.3.3.4 Cardiotoxicity

Epidemiological studies have demonstrated that chronic exposure to arsenic is associated with ischemic heart disease, myocardial injury, cardiac arrhythmias, and cardiomyopathy [130]. Mice exposed to arsenic showed abnormal ultrastructural changes in the cardiac tissue and myocardial injury in vivo [131]. Upregulation of cardiovascular genes that promote neovascularization and tissue remodeling was also observed in mice following chronic arsenic exposure [132]. Although arsenic-induced prolongation of cardiac repolarization was explained by calcium overload and hERG trafficking in ventricular myocytes [133], the mechanisms underlying arsenic-associated cardiotoxicity still remain unclear. The atherogenic potential of arsenic as discussed earlier is known to contribute to cardiac dysfunction through narrowing carotid artery.

2.3.3.5 Thrombosis

Enhanced thrombosis may contribute to arsenic-associated CVDs and atherosclerosis. Alteration in coagulation factors and fibrinolytic pathways can be also related. In platelets, arsenic enhanced the granular secretion resulting in excessive aggregation [134] and externalization of anionic phosphatidylserine on outer cellular membranes, promoting blood coagulation [135,136]. In microvascular ECs, arsenic decreased the expression of tissue-type plasminogen activator (tPA) and increased the expression of plasminogen activator inhibitor type-1 (PAI-1) leading to insufficient fibrinolysis [137]. The prothrombotic effects of arsenic were also observed in

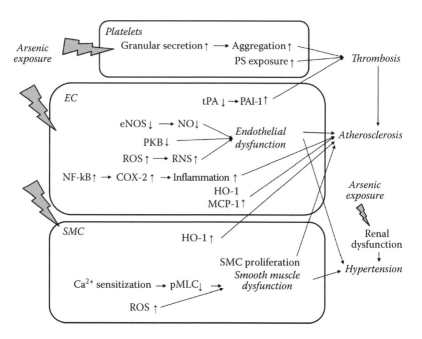

FIGURE 2.2 Mechanisms underlying arsenic-associated cardiovascular complications. COX-2, cyclooxygenase-2; EC, endothelial cells; eNOS, endothelial NO synthase; HO-1, heme oxygenase-1; MCP-1, monocyte chemoattractant protein-1; NF-κB, nuclear factor-κB; NO, nitric oxide; PAI-1, plasminogen activator inhibitor type-1; PKB, protein kinase B; pMLC, phospho-myosin light chain; ROS, reactive oxygen species; RNS, reactive nitrogen species; tPA, tissue-type plasminogen activator; SMC, smooth muscle cells.

animal models, where the extent of venous or arterial thrombosis was increased in rats after arsenic exposure [134,135]. Arsenic treatment in vivo inhibits the synthesis of proteoglycan and reduces the endothelial fibrinolytic activity in vascular ECs. The contributing mechanisms of arsenic toxicity on the cardiovascular systems are shown in Figure 2.2.

2.3.4 OTHER NONCANCER EFFECTS

Besides the most typical arsenic-associated health problems such as cancer, skin disorders, and CVDs, chronic exposure to arsenic is also related to the development of various diseases including diabetes mellitus (DM), liver cirrhosis, and neuropathy. Among them, we will cover arsenic-associated diabetic mellitus, which is the most threatening metabolic disease these days.

2.3.4.1 Diabetes Mellitus

DM, both type 1 and type 2, is one of the major concerns to public health worldwide. DM is a metabolic disease characterized by high levels of blood glucose resulting from defects in pancreatic insulin secretion and/or peripheral insulin resistance in adipocytes, hepatocytes, or myocytes. Many epidemiological studies have indicated a dose–response relationship between arsenic poisoning and the prevalence of DM, which is especially similar to type 2 DM [138]. Studies from Taiwan, Bangladesh, Sweden, and the United States demonstrated the close association of arsenic with DM, suggesting that arsenic serves as a risk factor for DM in arsenic-contaminated environments. Interestingly, recent reviews conclude that there is a strong association between iAs and DM in populations with arsenic in drinking water levels of >500 µg/L, although this association is not sufficiently supported by epidemiological evidence in individuals with low-to-moderate exposure (<150 µg/L) [138–140]. More thorough evaluation with better measure of arsenic exposure and diabetic outcome will be required in future researches in populations with low-to-moderate arsenic exposure.

In arsenic-exposed rats and mice, typical diabetic symptoms such as increased levels of blood glucose, decreased liver glycogen, low insulin sensitivity, and impaired glucose tolerance have been observed. Changes in plasma insulin level are inconsistent between studies [141,142]. The diabetological mode of action by arsenic has been studied in terms of pancreatic beta cell function or peripheral insulin sensitivity. In isolated islets or β cells, arsenic significantly decreased the expression of insulin mRNA and glucose-stimulated insulin secretion [142]. The generation of reactive oxygen species was increased, following disturbance in intracellular antioxidant system, Nrf-2 signaling, and Ca^{2+} homeostasis. The effect of arsenic on insulin resistance has been examined in adipocytes and skeletal muscle cells. In adipocytes, arsenic significantly inhibited insulin-stimulated glucose uptake via impairment of protein kinase B signaling [143]. Arsenic also inhibited adipogenic and myogenic differentiation, interfering with the generation of insulin-sensitive adipocytes and myotubes.

2.4   CONCLUDING REMARKS

Recent advances in the knowledge of arsenic biotransformation and its mode of action for noncarcinogenic and carcinogenic effects may provide clues to understand arsenic poisoning and its health effects. Nevertheless, several issues including unconventional dose–response relationship and low-dose effects still remain unclear. Collective efforts from epidemiology, toxicology, oncology, environmental, analytical and regulatory science would be necessary to find ways to treat or prevent diseases associated with arsenic poisoning, which is a significant threat to public health worldwide.

REFERENCES

1. Nordstrom, D.K., Public health. Worldwide occurrences of arsenic in ground water. *Science*, 2002. **296**(5576): 2143–2145.
2. Oremland, R.S. and J.F. Stolz, The ecology of arsenic. *Science*, 2003. **300**(5621): 939–944.
3. Smedley, P.L. and D.G. Kinniburgh, Sources and behaviour of arsenic in natural water. Chapter 1. In: *United Nations Synthesis Report on Arsenic in Drinking Water*. Geneva, Switzerland: World Health Organization, 2001, pp. 30–46.
4. WHO, World Health Organization. *Guidelines for Drinking Water Quality*. Geneva, Switzerland: World Health Organization, 2006.
5. Bundschuh, J. et al., Arsenic in the human food chain: The Latin American perspective. *Sci Total Environ*, 2012. **429**: 92–106.
6. Watanabe, T. and S. Hirano, Metabolism of arsenic and its toxicological relevance. *Arch Toxicol*, 2013. **87**(6): 969–979.
7. Lin, S. et al., A novel S-adenosyl-L-methionine:arsenic(III) methyltransferase from rat liver cytosol. *J Biol Chem*, 2002. **277**(13): 10795–10803.
8. Kenyon, E.M., L.M. Del Razo, and M.F. Hughes, Tissue distribution and urinary excretion of inorganic arsenic and its methylated metabolites in mice following acute oral administration of arsenate. *Toxicol Sci*, 2005. **85**(1): 468–475.
9. Valenzuela, O.L. et al., Urinary trivalent methylated arsenic species in a population chronically exposed to inorganic arsenic. *Environ Health Perspect*, 2005. **113**(3): 250–254.
10. Calatayud, M. et al., In vitro study of transporters involved in intestinal absorption of inorganic arsenic. *Chem Res Toxicol*, 2012. **25**(2): 446–453.
11. Drobna, Z. et al., Metabolism of arsenic in human liver: The role of membrane transporters. *Arch Toxicol*, 2010. **84**(1): 3–16.
12. Leslie, E.M., A. Haimeur, and M.P. Waalkes, Arsenic transport by the human multidrug resistance protein 1 (MRP1/ABCC1). Evidence that a tri-glutathione conjugate is required. *J Biol Chem*, 2004. **279**(31): 32700–32708.
13. Kojima, C. et al., Chronic exposure to methylated arsenicals stimulates arsenic excretion pathways and induces arsenic tolerance in rat liver cells. *Toxicol Sci*, 2006. **91**(1): 70–81.
14. Cullen, W.R., B.C. McBride, and J. Reglinski, The reaction of methylarsenicals with thiols: Some biological implications. *J Inorg Biochem*, 1984. **21**: 179–194.
15. Hayakawa, T. et al., A new metabolic pathway of arsenite: Arsenic–glutathione complexes are substrates for human arsenic methyltransferase Cyt19. *Arch Toxicol*, 2005. **79**(4): 183–191.
16. Zakharyan, R.A. et al., Interactions of sodium selenite, glutathione, arsenic species, and omega class human glutathione transferase. *Chem Res Toxicol*, 2005. **18**(8): 1287–1295.

17. Raml, R. et al., Thio-dimethylarsinate is a common metabolite in urine samples from arsenic-exposed women in Bangladesh. *Toxicol Appl Pharmacol*, 2007. **222**(3): 374–80.
18. Styblo, M. et al., Comparative toxicity of trivalent and pentavalent inorganic and methylated arsenicals in rat and human cells. *Arch Toxicol*, 2000. **74**(6): 289–299.
19. Eblin, K.E. et al., Arsenite and monomethylarsonous acid generate oxidative stress response in human bladder cell culture. *Toxicol Appl Pharmacol*, 2006. **217**(1): 7–14.
20. Naranmandura, H., K. Ibata, and K.T. Suzuki, Toxicity of dimethylmonothioarsinic acid toward human epidermoid carcinoma A431 cells. *Chem Res Toxicol*, 2007. **20**(8): 1120–1125.
21. Naranmandura, H. et al., Comparative toxicity of arsenic metabolites in human bladder cancer EJ-1 cells. *Chem Res Toxicol*, 2011. **24**(9): 1586–1596.
22. Cohen, S.M. et al., Arsenic-induced bladder cancer in an animal model. *Toxicol Appl Pharmacol*, 2007. **222**(3): 258–263.
23. Tseng, W.P. et al., Prevalence of skin cancer in an endemic area of chronic arsenicism in Taiwan. *J Natl Canc Inst*, 1968. **40**(3): 453–463.
24. Ahsan, H. et al., Arsenic exposure from drinking water and risk of premalignant skin lesions in Bangladesh: Baseline results from the Health Effects of Arsenic Longitudinal Study. *Am J Epidemiol*, 2006. **163**(12): 1138–1148.
25. Hall, M. et al., Blood arsenic as a biomarker of arsenic exposure: Results from a prospective study. *Toxicology*, 2006. **225**(2–3): 225–233.
26. Cuzick, J., R. Harris, and P.S. Mortimer, Palmar keratoses and cancers of the bladder and lung. *Lancet*, 1984. **1**(8376): 530–533.
27. Germolec, D.R. et al., Arsenic can mediate skin neoplasia by chronic stimulation of keratinocyte-derived growth factors. *Mutat Res*, 1997. **386**(3): 209–218.
28. Kachinskas, D.J. et al., Arsenate perturbation of human keratinocyte differentiation. *Cell Growth Differ*, 1994. **5**(11): 1235–1241.
29. Klimecki, W.T. et al., Effects of acute and chronic arsenic exposure of human-derived keratinocytes in an in vitro human skin equivalent system: A novel model of human arsenicism. *Toxicol In Vitro*, 1997. **11**(1–2): 89–98.
30. Yoshida, T., H. Yamauchi, and G. Fan Sun, Chronic health effects in people exposed to arsenic via the drinking water: Dose–response relationships in review. *Toxicol Appl Pharmacol*, 2004. **198**(3): 243–252.
31. Zaldivar, R., L. Prunes, and G.L. Ghai, Arsenic dose in patients with cutaneous carcinomata and hepatic hemangio-endothelioma after environmental and occupational exposure. *Arch Toxicol*, 1981. **47**(2): 145–154.
32. Yu, H.S., W.T. Liao, and C.Y. Chai, Arsenic carcinogenesis in the skin. *J Biomed Sci*, 2006. **13**(5): 657–666.
33. Beane Freeman, L.E. et al., Toenail arsenic content and cutaneous melanoma in Iowa. *Am J Epidemiol*, 2004. **160**(7): 679–687.
34. Lee, C.H., W.T. Liao, and H.S. Yu, Aberrant immune responses in arsenical skin cancers. *Kaohsiung J Med Sci*, 2011. **27**(9): 396–401.
35. Yu, H.S. et al., Arsenic induces tumor necrosis factor alpha release and tumor necrosis factor receptor 1 signaling in T helper cell apoptosis. *J Invest Dermatol*, 2002. **119**(4): 812–819.
36. Yu, H.S. et al., Defective IL-2 receptor expression in lymphocytes of patients with arsenic-induced Bowen's disease. *Arch Dermatol Res*, 1998. **290**(12): 681–687.
37. Tseng, W.P., Blackfoot disease in Taiwan: A 30-year follow-up study. *Angiology*, 1989. **40**(6): 547–558.
38. Yu, H.S. et al., In vitro cytotoxicity of IgG antibodies on vascular endothelial cells from patients with endemic peripheral vascular disease in Taiwan. *Atherosclerosis*, 1998. **137**(1): 141–147.

39. Hong, C.H. et al., Anti-endothelial cell IgG from patients with chronic arsenic poisoning induces endothelial proliferation and VEGF-dependent angiogenesis. *Microvasc Res*, 2008. **76**(3): 194–201.
40. Bustaffa, E. et al., Genotoxic and epigenetic mechanisms in arsenic carcinogenicity. *Arch Toxicol*, 2014. **88**(5): 1043–1067.
41. Cohen, S.M. et al., Evaluation of the carcinogenicity of inorganic arsenic. *Crit Rev Toxicol*, 2013. **43**(9): 711–752.
42. Chen, C.J. et al., Malignant neoplasms among residents of a blackfoot disease-endemic area in Taiwan: High-arsenic artesian well water and cancers. *Canc Res*, 1985. **45**(11 Pt 2): 5895–5899.
43. Chen, C.J., T.L. Kuo, and M.M. Wu, Arsenic and cancers. *Lancet*, 1988. 1(8582): 414–415.
44. Chiou, H.Y. et al., Incidence of transitional cell carcinoma and arsenic in drinking water: A follow-up study of 8,102 residents in an arseniasis-endemic area in northeastern Taiwan. *Am J Epidemiol*, 2001. **153**(5): 411–418.
45. Baastrup, R. et al., Arsenic in drinking-water and risk for cancer in Denmark. *Environ Health Perspect*, 2008. **116**(2): 231–237.
46. Bates, M.N. et al., Case–control study of bladder cancer and exposure to arsenic in Argentina. *Am J Epidemiol*, 2004. **159**(4): 381–389.
47. Schoen, A. et al., Arsenic toxicity at low doses: Epidemiological and mode of action considerations. *Toxicol Appl Pharmacol*, 2004. **198**(3): 253–267.
48. Chen, C.L. et al., Arsenic in drinking water and risk of urinary tract cancer: A follow-up study from northeastern Taiwan. *Canc Epidemiol Biomarkers Prev*, 2010. **19**(1): 101–110.
49. Pou, S.A., A.R. Osella, and P. Diaz Mdel, Bladder cancer mortality trends and patterns in Cordoba, Argentina (1986–2006). *Canc Causes Contr*, 2011. **22**(3): 407–415.
50. Smith, A.H. et al., Increased mortality from lung cancer and bronchiectasis in young adults after exposure to arsenic in utero and in early childhood. *Environ Health Perspect*, 2006. **114**(8): 1293–1296.
51. Mostafa, M.G., J.C. McDonald, and N.M. Cherry, Lung cancer and exposure to arsenic in rural Bangladesh. *Occup Environ Med*, 2008. **65**(11): 765–768.
52. Chen, C.L. et al., Ingested arsenic, characteristics of well water consumption and risk of different histological types of lung cancer in northeastern Taiwan. *Environ Res*, 2010. **110**(5): 455–462.
53. Lamm, S.H. et al., An epidemiologic study of arsenic-related skin disorders and skin cancer and the consumption of arsenic-contaminated well waters in Huhhot, Inner Mongolia, China. *Hum Ecol Risk Assess*, 2007. **13**(4): 713–746.
54. Leonardi, G. et al., Inorganic arsenic and basal cell carcinoma in areas of Hungary, Romania, and Slovakia: A case–control study. *Environ Health Perspect*, 2012. **120**(5): 721–726.
55. Mass, M.J. et al., Methylated trivalent arsenic species are genotoxic. *Chem Res Toxicol*, 2001. **14**(4): 355–361.
56. Chai, C.Y. et al., Arsenic salt-induced DNA damage and expression of mutant p53 and COX-2 proteins in SV-40 immortalized human uroepithelial cells. *Mutagenesis*, 2007. **22**(6): 403–408.
57. Salazar, A.M., M. Sordo, and P. Ostrosky-Wegman, Relationship between micronuclei formation and p53 induction. *Mutat Res—Genet Toxicol Environ Mutagen*, 2009. **672**(2): 124–128.
58. Dopp, E. et al., Forced uptake of trivalent and pentavalent methylated and inorganic arsenic and its cyto-/genotoxicity in fibroblasts and hepatoma cells. *Toxicol Sci*, 2005. **87**(1): 46–56.
59. Akram, Z. et al., Genotoxicity of sodium arsenite and DNA fragmentation in ovarian cells of rat. *Toxicol Lett*, 2009. **190**(1): 81–85.

60. Hernandez, A., A. Sampayo-Reyes, and R. Marcos, Identification of differentially expressed genes in the livers of chronically i-As-treated hamsters. *Mutat Res*, 2011. **713**(1–2): 48–55.
61. Khan, P.K., V.P. Kesari, and A. Kumar, Mouse micronucleus assay as a surrogate to assess genotoxic potential of arsenic at its human reference dose. *Chemosphere*, 2013. **90**(3): 993–997.
62. Kesari, V.P., A. Kumar, and P.K. Khan, Genotoxic potential of arsenic at its reference dose. *Ecotoxicol Environ Saf*, 2012. **80**: 126–131.
63. Basu, A. et al., Enhanced frequency of micronuclei in individuals exposed to arsenic through drinking water in West Bengal, India. *Mutat Res*, 2002. **516**(1–2): 29–40.
64. Ghosh, P. et al., Cytogenetic damage and genetic variants in the individuals susceptible to arsenic-induced cancer through drinking water. *Int J Canc*, 2006. **118**(10): 2470–2478.
65. Banerjee, M. et al., DNA repair deficiency leads to susceptibility to develop arsenic-induced premalignant skin lesions. *Int J Canc*, 2008. **123**(2): 283–287.
66. Lee, C.H. et al., Involvement of mtDNA damage elicited by oxidative stress in the arsenical skin cancers. *J Invest Dermatol*, 2013. **133**(7): 1890–900.
67. Tennant, A.H. and A.D. Kligerman, Superoxide dismutase protects cells from DNA damage induced by trivalent methylated arsenicals. *Environ Mol Mutagen*, 2011. **52**(3): 238–243.
68. Muniz Ortiz, J.G. et al., Catalase has a key role in protecting cells from the genotoxic effects of monomethylarsonous acid: A highly active metabolite of arsenic. *Environ Mol Mutagen*, 2013. **54**(5): 317–326.
69. Ebert, F. et al., Arsenicals affect base excision repair by several mechanisms. *Mutat Res*, 2011. **715**(1–2): 32–41.
70. Lai, Y. et al., Role of DNA polymerase beta in the genotoxicity of arsenic. *Environ Mol Mutagen*, 2011. **52**(6): 460–468.
71. Qin, X.J. et al., Poly(ADP-ribose) polymerase-1 inhibition by arsenite promotes the survival of cells with unrepaired DNA lesions induced by UV exposure. *Toxicol Sci*, 2012. **127**(1): 120–129.
72. Sun, X. et al., Arsenite binding-induced zinc loss from PARP-1 is equivalent to zinc deficiency in reducing PARP-1 activity, leading to inhibition of DNA repair. *Toxicol Appl Pharmacol*, 2014. **274**(2): 313–318.
73. Kitchin, K.T. and K. Wallace, The role of protein binding of trivalent arsenicals in arsenic carcinogenesis and toxicity. *J Inorg Biochem*, 2008. **102**(3): 532–539.
74. Wang, A. et al., Arsenate and dimethylarsinic acid in drinking water did not affect DNA damage repair in urinary bladder transitional cells or micronuclei in bone marrow. *Environ Mol Mutagen*, 2009. **50**(9): 760–770.
75. Clewell, H.J. et al., Concentration- and time-dependent genomic changes in the mouse urinary bladder following exposure to arsenate in drinking water for up to 12 weeks. *Toxicol Sci*, 2011. **123**(2): 421–432.
76. Wnek, S.M. et al., Monomethylarsonous acid produces irreversible events resulting in malignant transformation of a human bladder cell line following 12 weeks of low-level exposure. *Toxicol Sci*, 2010. **116**(1): 44–57.
77. Sun, Y., E.J. Tokar, and M.P. Waalkes, Overabundance of putative cancer stem cells in human skin keratinocyte cells malignantly transformed by arsenic. *Toxicol Sci*, 2012. **125**(1): 20–29.
78. Jiang, R. et al., The acquisition of cancer stem cell-like properties and neoplastic transformation of human keratinocytes induced by arsenite involves epigenetic silencing of let-7c via Ras/NF-kappaB. *Toxicol Lett*, 2014. **227**(2): 91–98.
79. Wang, Z. et al., Epithelial to mesenchymal transition in arsenic-transformed cells promotes angiogenesis through activating beta-catenin-vascular endothelial growth factor pathway. *Toxicol Appl Pharmacol*, 2013. **271**(1): 20–29.

80. Coppin, J.F., W. Qu, and M.P. Waalkes, Interplay between cellular methyl metabolism and adaptive efflux during oncogenic transformation from chronic arsenic exposure in human cells. *J Biol Chem*, 2008. **283**(28): 19342–19350.

81. Martinez, L. et al., Impact of early developmental arsenic exposure on promotor CpG-island methylation of genes involved in neuronal plasticity. *Neurochem Int*, 2011. **58**(5): 574–581.

82. Okoji, R.S. et al., Sodium arsenite administration via drinking water increases genome-wide and Ha-ras DNA hypomethylation in methyl-deficient C57BL/6J mice. *Carcinogenesis*, 2002. **23**(5): 777–785.

83. Chen, H. et al., Genetic events associated with arsenic-induced malignant transformation: Applications of cDNA microarray technology. *Mol Carcinog*, 2001. **30**(2): 79–87.

84. Mass, M.J. and L. Wang, Arsenic alters cytosine methylation patterns of the promoter of the tumor suppressor gene p53 in human lung cells: A model for a mechanism of carcinogenesis. *Mutat Res*, 1997. **386**(3): 263–277.

85. Jensen, T.J. et al., Epigenetic remodeling during arsenical-induced malignant transformation. *Carcinogenesis*, 2008. **29**(8): 1500–1508.

86. Chen, W.T. et al., Urothelial carcinomas arising in arsenic-contaminated areas are associated with hypermethylation of the gene promoter of the death-associated protein kinase. *Histopathology*, 2007. **51**(6): 785–792.

87. Banerjee, N. et al., Epigenetic modifications of DAPK and p16 genes contribute to arsenic-induced skin lesions and nondermatological health effects. *Toxicol Sci*, 2013. **135**(2): 300–308.

88. Zhou, V.W., A. Goren, and B.E. Bernstein, Charting histone modifications and the functional organization of mammalian genomes. *Nat Rev Genet*, 2011. **12**(1): 7–18.

89. Li, J. et al., Tumor promoter arsenite stimulates histone H3 phosphoacetylation of protooncogenes c-fos and c-jun chromatin in human diploid fibroblasts. *J Biol Chem*, 2003. **278**(15): 13183–13191.

90. Jo, W.J. et al., Acetylated H4K16 by MYST1 protects UROtsa cells from arsenic toxicity and is decreased following chronic arsenic exposure. *Toxicol Appl Pharmacol*, 2009. **241**(3): 294–302.

91. Zhou, X. et al., Arsenite alters global histone H3 methylation. *Carcinogenesis*, 2008. **29**(9): 1831–1836.

92. Zhou, X. et al., Effects of nickel, chromate, and arsenite on histone 3 lysine methylation. *Toxicol Appl Pharmacol*, 2009. **236**(1): 78–84.

93. Li, J. et al., Arsenic trioxide promotes histone H3 phosphoacetylation at the chromatin of CASPASE-10 in acute promyelocytic leukemia cells. *J Biol Chem*, 2002. **277**(51): 49504–49510.

94. Lu, J. et al., MicroRNA expression profiles classify human cancers. *Nature*, 2005. **435**(7043): 834–838.

95. Marsit, C.J., K. Eddy, and K.T. Kelsey, MicroRNA responses to cellular stress. *Canc Res*, 2006. **66**(22): 10843–10848.

96. Yamamoto, S. et al., Cancer induction by an organic arsenic compound, dimethylarsinic acid (cacodylic acid), in F344/DuCrj rats after pretreatment with five carcinogens. *Canc Res*, 1995. **55**(6): 1271–1276.

97. Rossman, T.G. et al., Arsenite is a cocarcinogen with solar ultraviolet radiation for mouse skin: An animal model for arsenic carcinogenesis. *Toxicol Appl Pharmacol*, 2001. **176**(1): 64–71.

98. Burns, F.J. et al., Arsenic-induced enhancement of ultraviolet radiation carcinogenesis in mouse skin: A dose–response study. *Environ Health Perspect*, 2004. **112**(5): 599–603.

99. Motiwale, L., A.D. Ingle, and K.V. Rao, Mouse skin tumor promotion by sodium arsenate is associated with enhanced PCNA expression. *Canc Lett*, 2005. **223**(1): 27–35.

100. Ferreccio, C. et al., Lung cancer and arsenic concentrations in drinking water in Chile. *Epidemiology*, 2000. **11**(6): 673–679.

101. Chung, C.J. et al., The effect of cigarette smoke and arsenic exposure on urothelial carcinoma risk is modified by glutathione S-transferase M1 gene null genotype. *Toxicol Appl Pharmacol*, 2013. **266**(2): 254–259.

102. Chen, Y.C. et al., Interaction between environmental tobacco smoke and arsenic methylation ability on the risk of bladder cancer. *Canc Causes Contr*, 2005. **16**(2): 75–81.

103. Sawada, N. et al., Dietary arsenic intake and subsequent risk of cancer: The Japan Public Health Center-based (JPHC) Prospective Study. *Canc Causes Contr*, 2013. **24**(7): 1403–1415.

104. Wang, L. et al., Ethanol enhances tumor angiogenesis in vitro induced by low-dose arsenic in colon cancer cells through hypoxia-inducible factor 1 alpha pathway. *Toxicol Sci*, 2012. **130**(2): 269–280.

105. Wu, M.M. et al., Dose–response relation between arsenic concentration in well water and mortality from cancers and vascular diseases. *Am J Epidemiol*, 1989. **130**(6): 1123–1132.

106. Chen, C.J. et al., Cancer potential in liver, lung, bladder and kidney due to ingested inorganic arsenic in drinking water. *Br J Canc*, 1992. **66**(5): 888–892.

107. Brown, K.G., Inorganic arsenic in drinking water and bladder cancer: A meta-analysis for dose–response assessment. *Int J Environ Res Public Health*, 2007. **4**(2): 193–194; author reply 194.

108. Moore, L.E., M. Lu, and A.H. Smith, Childhood cancer incidence and arsenic exposure in drinking water in Nevada. *Arch Environ Health*, 2002. **57**(3): 201–206.

109. Lamm, S.H. et al., Arsenic in drinking water and bladder cancer mortality in the United States: An analysis based on 133 U.S. counties and 30 years of observation. *J Occup Environ Med*, 2004. **46**(3): 298–306.

110. Bates, M.N., A.H. Smith, and K.P. Cantor, Case–control study of bladder cancer and arsenic in drinking water. *Am J Epidemiol*, 1995. **141**(6): 523–530.

111. Waalkes, M.P. et al., Transplacental carcinogenicity of inorganic arsenic in the drinking water: Induction of hepatic, ovarian, pulmonary, and adrenal tumors in mice. *Toxicol Appl Pharmacol*, 2003. **186**(1): 7–17.

112. Tokar, E.J. et al., Carcinogenic effects of "whole-life" exposure to inorganic arsenic in CD1 mice. *Toxicol Sci*, 2011. **119**(1): 73–83.

113. Balakumar, P. and J. Kaur, Arsenic exposure and cardiovascular disorders: An overview. *Cardiovasc Toxicol*, 2009. **9**(4): 169–176.

114. Tsou, T.C. et al., Arsenite enhances tumor necrosis factor-alpha-induced expression of vascular cell adhesion molecule-1. *Toxicol Appl Pharmacol*, 2005. **209**(1): 10–18.

115. Lee, M.Y. et al., Arsenic-induced dysfunction in relaxation of blood vessels. *Environ Health Perspect*, 2003. **111**(4): 513–517.

116. Bae, O.N. et al., U-shaped dose response in vasomotor tone: A mixed result of heterogenic response of multiple cells to xenobiotics. *Toxicol Sci*, 2008. **103**(1): 181–190.

117. Lee, M.Y. et al., Inorganic arsenite potentiates vasoconstriction through calcium sensitization in vascular smooth muscle. *Environ Health Perspect*, 2005. **113**(10): 1330–1335.

118. Abhyankar, L.N. et al., Arsenic exposure and hypertension: A systematic review. *Environ Health Perspect*, 2012. **120**(4): 494–500.

119. Wang, C.H. et al., A review of the epidemiologic literature on the role of environmental arsenic exposure and cardiovascular diseases. *Toxicol Appl Pharmacol*, 2007. **222**(3): 315–326.

120. Barchowsky, A. et al., Arsenic induces oxidant stress and NF-kappa B activation in cultured aortic endothelial cells. *Free Radic Biol Med*, 1996. **21**(6): 783–790.

121. Tsai, S.H. et al., Arsenite stimulates cyclooxygenase-2 expression through activating IkappaB kinase and nuclear factor kappaB in primary and ECV304 endothelial cells. *J Cell Biochem*, 2002. **84**(4): 750–758.

122. Hsueh, Y.M. et al., Urinary arsenic species and CKD in a Taiwanese population: A case–control study. *Am J Kidney Dis*, 2009. **54**(5): 859–870.

123. Cohen, S.M. et al., Urothelial cytotoxicity and regeneration induced by dimethylarsinic acid in rats. *Toxicol Sci*, 2001. **59**(1): 68–74.

124. Tseng, C.H., Cardiovascular disease in arsenic-exposed subjects living in the arseniasis-hyperendemic areas in Taiwan. *Atherosclerosis*, 2008. **199**(1): 12–18.

125. Chiou, H.Y. et al., Dose–response relationship between prevalence of cerebrovascular disease and ingested inorganic arsenic. *Stroke*, 1997. **28**(9): 1717–1723.

126. Bunderson, M. et al., Arsenic exposure exacerbates atherosclerotic plaque formation and increases nitrotyrosine and leukotriene biosynthesis. *Toxicol Appl Pharmacol*, 2004. **201**(1): 32–39.

127. Kumagai, Y. and J. Pi, Molecular basis for arsenic-induced alteration in nitric oxide production and oxidative stress: Implication of endothelial dysfunction. *Toxicol Appl Pharmacol*, 2004. **198**(3): 450–457.

128. Simeonova, P.P. et al., Arsenic exposure accelerates atherogenesis in apolipoprotein E(−/−) mice. *Environ Health Perspect*, 2003. **111**(14): 1744–1748.

129. Lee, P.C., I.C. Ho, and T.C. Lee, Oxidative stress mediates sodium arsenite-induced expression of heme oxygenase-1, monocyte chemoattractant protein-1, and interleukin-6 in vascular smooth muscle cells. *Toxicol Sci*, 2005. **85**(1): 541–550.

130. Tseng, C.H. et al., Long-term arsenic exposure and ischemic heart disease in arseniasis-hyperendemic villages in Taiwan. *Toxicol Lett*, 2003. **137**(1–2): 15–21.

131. Manna, P., M. Sinha, and P.C. Sil, Arsenic-induced oxidative myocardial injury: Protective role of arjunolic acid. *Arch Toxicol*, 2008. **82**(3): 137–149.

132. Soucy, N.V. et al., Neovascularization and angiogenic gene expression following chronic arsenic exposure in mice. *Cardiovasc Toxicol*, 2005. **5**(1): 29–41.

133. Ficker, E. et al., Mechanisms of arsenic-induced prolongation of cardiac repolarization. *Mol Pharmacol*, 2004. **66**(1): 33–44.

134. Lee, M.Y. et al., Enhancement of platelet aggregation and thrombus formation by arsenic in drinking water: A contributing factor to cardiovascular disease. *Toxicol Appl Pharmacol*, 2002. **179**(2): 83–88.

135. Bae, O.N. et al., Arsenite-enhanced procoagulant activity through phosphatidylserine exposure in platelets. *Chem Res Toxicol*, 2007. **20**(12): 1760–1768.

136. Bae, O.N. et al., Trivalent methylated arsenical-induced phosphatidylserine exposure and apoptosis in platelets may lead to increased thrombus formation. *Toxicol Appl Pharmacol*, 2009. **239**(2): 144–153.

137. Jiang, S.J. et al., Decrease of fibrinolytic activity in human endothelial cells by arsenite. *Thromb Res*, 2002. **105**(1): 55–62.

138. Huang, C.F. et al., Arsenic and diabetes: Current perspectives. *Kaohsiung J Med Sci*, 2011. **27**(9): 402–410.

139. Maull, E.A. et al., Evaluation of the association between arsenic and diabetes: A National Toxicology Program workshop review. *Environ Health Perspect*, 2012. **120**(12): 1658–1670.

140. Wang, S.L. et al., Prevalence of non-insulin-dependent diabetes mellitus and related vascular diseases in southwestern arseniasis-endemic and nonendemic areas in Taiwan. *Environ Health Perspect*, 2003. **111**(2): 155–159.

141. Izquierdo-Vega, J.A. et al., Diabetogenic effects and pancreatic oxidative damage in rats subchronically exposed to arsenite. *Toxicol Lett*, 2006. **160**(2): 135–142.

142. Yen, C.C. et al., The diabetogenic effects of the combination of humic acid and arsenic: In vitro and in vivo studies. *Toxicol Lett*, 2007. **172**(3): 91–105.
143. Paul, D.S. et al., Molecular mechanisms of the diabetogenic effects of arsenic: Inhibition of insulin signaling by arsenite and methylarsonous acid. *Environ Health Perspect*, 2007. **115**(5): 734–742.

3 Arsenic Toxicity
Toxicity, Manifestation, and Geographical Distribution

Yuhu Li and Zhihong Liu

CONTENTS

3.1   INTRODUCTION

Arsenic contamination is a worldwide problem and has become one of the most severe threats to the health of millions of humans and other organisms, even to global sustainability due to its carcinogenicity. Long-term exposure of arsenic may increase the risk of skin, liver, bladder, and lung cancers. It is reported that over 137 million people in more than 70 countries are probably affected by arsenic contamination from drinking water [1]. Arsenicosis is prevalent in certain areas of many countries, including India, Bangladesh, Mongolia, Taiwan, Argentina, and Chile. The two worst affected areas in the world are Bangladesh and West Bengal, India. According to the World Health Organization (WHO), in both of these areas, the groundwater arsenic concentrations were above of 50 µg/L, some even reaching 3400 µg/L [2].

The major cause of human arsenic toxicity is from contamination of drinking water from natural geological sources rather than from mining, smelting, or agricultural sources (pesticides or fertilizers). So, as a result, the WHO set the first International Drinking Water Standard for arsenic concentration at 200 µg/L in 1958 and then lowering the International Drinking Water Standard from 200 to 50 µg/L in 1963. The WHO further revised the guideline for arsenic from 0.05 to 0.01 mg/L in 1993 [3]. The Environmental Protection Agency (EPA) set the permissible level of arsenic in drinking water in the United States to 10 ppb in 1996, while a more vigorous standard of 0.007 mg/L for arsenic level in drinking water was adopted in 2004 [4].

Arsenic is an ubiquitous element that ranks 20th in abundance in the earth's crust. It can be recovered from arsenopyrite or the by-products of the mining and metallurgy of copper, lead, zinc, and gold. Only global resources of copper and lead contained about 11 million tons of arsenic [5]. However, the use of arsenic was rather limited. Most of the arsenic was used for the production of wood preservatives, insecticides, semiconductors, and bronze alloys, and the consumption of arsenic kept decreasing constantly as the environmental legislation for controlling the use of arsenic was improved in many countries [6]. Thus, the demand for arsenic and its compounds is far less than the amount being produced annually; as a result, there is little economic incentive to recover arsenic for marketing as products. The United States had closed domestic production of arsenic since 1985 [7]. On the other hand, there will still be an increase in the emission of arsenic to the environment due to the progressive industrialization of developing countries. Therefore, the best way for the disposal of arsenic is to transform into less hazardous or nonhazardous solids by means of precipitation with other elements, such as iron, or solidification/stabilization process [8]. Without doubt, remarkable progresses and achievements have been made in the safe treatment of arsenic in the last decades, but even to this day, arsenic contamination occurred frequently in many countries; the challenges we faced are still very severe.

3.2 TOXICITY

Most arsenic compounds are especially potent poisons, and arsenic is considered the most potential human carcinogen. Arsenic has been classified as a Group I carcinogen based on human epidemiological data by the International Agency for Research on Cancer [9], and it has also been ranked first in a list of 20 hazardous substances by the Agency for Toxic Substances and Disease Registry and U.S. EPA [10].

Arsenic toxicity is a medical condition caused by the ingestion, absorption, or inhalation of dangerous levels of arsenic. There are two types of arsenic toxicity, acute arsenic toxicity and chronic arsenic toxicity, which have different symptoms due to the ingestion amount and species of arsenic in a certain time.

Once elevated levels of arsenic enter into the human body, the normal processes of the lungs, kidneys, and liver would be disrupted completely. Individuals will suffer from headache, confusion, diarrhea, and convulsions at the beginning of arsenic toxicity. As the poisoning develops, symptoms may include severe diarrhea, vomiting, blood in the urine, stomach pain, and coma, even death [11].

Arsenic can inactivate up to 200 enzymes through interacting with sulfhydryl groups of proteins and enzymes [12], most notably those involved in cellular energy pathways and DNA replication and repair, and is substituted for phosphate in high-energy compounds such as adenosine triphosphate (ATP) due to the similarity between arsenate and phosphate [13]. In addition, arsenic can damage cells through an increase of reactive oxygen species in the cells [14,15].

3.2.1 TOXICITY OF INORGANIC ARSENIC

The toxicity of arsenic compounds depends on its valence state (-3, 0, $+3$, $+5$), its form (inorganic or organic), and the physical and chemical properties, especially the solubility and stability of arsenic compounds. In general, inorganic arsenic is more toxic than organic arsenic, and trivalent arsenite is more toxic than pentavalent and zero-valent arsenic. The trivalent arsenic (As(III)) is considered 60 times more toxic than the pentavalent (As(V)) [16]. Inorganic arsenic compounds are about 100 times more toxic than organic arsenic compounds (DMMA and monomethylarsonic acid [MMAA]) [17]. Thus, it is generally accepted that the toxicity of different arsenic species varies in the order arsine > arsenite > arsenate > monomethylarsonate (MMA) > dimethylarsinate (DMA) > arsenobetaine (AsB) > arsenocholine (AsC) > arsenic metal. However, the distinction of toxicity among these arsenic compounds does not make sense since different forms of arsenic may be interconverted in the environment and microorganisms.

Inorganic arsenic is found throughout the environment; most cases of arsenic toxicity in humans are due to exposure to inorganic arsenic. The main inorganic forms of arsenic relevant for human exposures are arsenate (pentavalent arsenic [As(V)]) and arsenite (trivalent arsenic [As(III)]). Arsenates are stable under

TABLE 3.1

Toxic Mechanisms of Arsenic in Humans

Inorganic Arsenic Species	Reaction Objective	Reaction Results
Arsenite (As(III))	Sulfhydryl groups	The decrease of ATP production
		The inhibition cellular enzymes
		The decrease of glucose uptake
		The decrease of fatty acid oxidation
		The increase of free radical and oxidative damage
Arsenate (As(V))	ADP	The loss of the high-energy phosphate bonds of ATP
Arsine	Red blood cells	The damage of irreversible cell membrane

aerobic or oxidizing conditions, while arsenites mainly exist under anaerobic or mildly reducing conditions [18]. There are several common inorganic arsenic compounds, including arsenic oxides, arsenic sulfides, arsenate, and arsenite.

Although arsenite, arsenate, and arsine belong to the inorganic arsenic, the toxic mechanisms of them in human are quite different. As listed in Table 3.1, once arsenic enters into the human body, it would react with sulfhydryl groups, ADP, or red blood cells, and a series of symptoms would be caused by these biochemical reactions [19–21].

There are many arsenic compounds of environmental importance. Almost all the inorganic arsenic compounds can be found in nature. Representative arsenic-containing compounds are shown in Table 3.2.

The most famous inorganic arsenic compound is arsenic trioxide (As_2O_3), which is also the most common commercial compound of arsenic, mainly used as wood preservatives and herbicides. Arsenic trioxide is recovered from the smelting or roasting of nonferrous metal ores or concentrates, and it also occurs naturally as arsenolite and claudetite. There are no quantitative human data on lethal inhalation exposure to arsenic trioxide, but based on human toxicity data, the toxicity of arsenic trioxide is rather potent. It is estimated that lethal amounts of orally ingested arsenic trioxide were in the range 8–20 g. Another arsenic oxide is arsenic pentoxide, which is similar in appearance with arsenic trioxide, but it is easily soluble in water. In general, the toxicity of arsenic pentoxide is less than arsenic trioxide, but this view is vague and unscientific. There is hardly a single toxicity of arsenic pentoxide, since it is generally converted into the corresponding arsenite species, which may result in similar effects in humans as with the ingestion of arsenic pentoxide.

Metallic arsenic and arsenic sulfides are generally regarded as low-toxicity substances due to their insolubility. However, arsenic oxides are often produced when the surface of metallic arsenic is subject to the erosion of oxygen, resulting in the formation of toxicity for metallic arsenic. The situation of arsenic sulfides is rather complex; it would transform into arsenic oxides, not only by the oxidation of oxygen but also by the effect of microorganisms.

The most famous arsenic salt is potassium arsenite, which is used to prepare Fowler's solution (1% potassium arsenite solution) for the treatment of leukemia.

TABLE 3.2

Common Arsenic-Containing Compounds

Chemical Name	Chemical Formula	CAS Registry	Property	Lethal Dose 50%/Lethal Concentration (LD_{50}/LC_{50})	References
Arsenic acid	H_3AsO_4	7778-39-4 1327-52-2	Appearance, white translucent crystals; melting point, 35.5°C; boiling point, 120°C (decompose); density, 2.2–2.5 g/cm³; solubility, 302 g/L in water at 12.5°C, soluble in alcohol and glycerol	LD_{50}, 48 mg/kg (rat, oral); LD_{Lo}, 10 mg/kg (dog, oral); LD_{50}, 6 mg/kg (rabbit, oral); LD_{Lo}, 5 mg/kg (rabbit, oral)	Milne [22]
Arsenic pentoxide	As_2O_5	1303-28-2	Appearance, white hygroscopic powder; melting point, decomposes at 315°C; boiling point, no data; density, 4.32 g/cm³; odor, no data; solubility, 59.5 and 65.8 g in 100 mL water at 0°C and 20°C, respectively; soluble in alcohol	LD_{50}, 55 mg/kg (mouse, oral) and 110 mg/kg (rat, oral)	Marquardt et al. [23]
Arsenic trioxide	As_2O_3	1327-53-3	Appearance, white powder; melting point, 274°C (arsenolite), 312.2°C (claudetite); boiling point, 465°C; density, 3.74 g/cm³ (cubes) and 4.15 g/cm³ (rhombic crystals); odor, odorless; solubility, 17 and 20 g/L in water at 16°C and 25°C, respectively; 20 g/L in water insoluble in alcohol; soluble in glycerol	LD_{50}, 34.1–52.5 mg/kg (male mouse, oral) and 20–188 mg/kg (rat, oral); LD_{Lo}, 70–180 mg (human, oral)	Marquardt et al., Harrison et al., Kaise et al. [23–25]
Arsenic trisulfide	As_2S_3	1303-33-9	Appearance, orange or yellow crystals; melting point, decomposes at 310°C; boiling point, 707°C; density, 3.43 g/cm³; solubility, 0.0005 g/L in water at 18°C	LD_{50}, 254 mg/kg (mouse, oral), 185 mg/kg (rat, oral), 215 mg/kg (mouse, IP), 85 mg/kg (rat, IP) 936.0 mg/kg (rat, dermal)	

(Continued)

TABLE 3.2 (Continued)
Common Arsenic-Containing Compounds

Chemical Name	Chemical Formula	CAS Registry	Property	Lethal Dose 50%/Lethal Concentration (LD_{50}/LC_{50})	References
Sodium arsenite	$NaAsO_2$	7778-46-5	Appearance, white powder; melting point, decomposes at 550°C; boiling point, no data; density, 1.87 g/cm³; solubility, 156 g in 100 mL water; slightly soluble in alcohol	LD_{50}, 42 mg/kg (rat, oral); LD_{50}, 14 mg/kg (mouse, IP); LD_{50}, 87 mg/kg (mouse, I.M.); lethal dose for humans, approximately 0.1 g	Done and Peart, Tadlock and Aposhian, Franke and Moxon [27–29]
Calcium arsenate	$Ca_3(AsO_4)_2$	7778-44-1	Appearance, white powder; melting point, 1455°C, decomposes on heating; boiling point, no data; density, 3.62 g/cm³; odor, odorless; solubility, 0.13 g/L in water at 25°C; insoluble in organic solvents	LD_{50}, 298 mg/kg (female rat, oral); LD_{50}, >2400 mg/kg (female rat, dermal)	Gaines [30]
Arsenic trichloride	$AsCl_3$	7784-34-1	Appearance, oily, colorless liquid; melting point, −16.2°C (256.9 K); boiling point, 130.2°C (403.3 K); density, 2.163 g/cm³; odor, odorless; solubility, decomposes in water; soluble in alcohol	LC_{Lo}, 100 mg/m³ (cat, inhalative)	Marquardt et al. [23]
Arsine	AsH_3	7784-42-1	Appearance, colorless gas; melting point, −111.2°C; boiling point, −62.5°C; flash point;, −62°C; density;, 4.93 g/L; odor;, Odorless; solubility: 0.07 g in 100 mL water at 25°C	LC_{Lo}, 25 ppm (30 min); 300 ppm (5 min)	RTECS [31]

In addition, arsenic was generally converted into insoluble arsenic salts, such as $Ca_3(AsO_4)_2$ and $FeAsO_4 \cdot 2H_2O$, to lower the environmental risk. These arsenic compounds are considered as low toxicity due to their low solubility or insolubility.

There are abundant evidences that arsenic has carcinogenic effects in humans when ingested orally or inhaled with enough doses, but rarely in animals. Therefore, The LD_{50} data based on animals should not be considered a reliable source to be applied to humans. It has been demonstrated that most laboratory animals appear to have a higher tolerance for arsenic than humans. When exposed orally with inorganic arsenic (0.05–0.1 mg/kg/day), neurological and hematological toxicity may occur in humans, but not in monkeys, dogs, and rats, even when exposed to a higher level of arsenic doses (0.72–2.8 mg/kg/day) [32].

3.2.2 TOXICITY OF ORGANIC ARSENIC

The common organic arsenic compounds are listed in Table 3.3. The major source of organic arsenic comes from the methylation of inorganic arsenic by organisms and microorganisms. Inorganic arsenic can be methylated by certain *bacteria*, fungi, and yeasts, producing methyl arsenic, such as MMAA (V) and dimethylarsinic acid (DMAA (V)). Inorganic arsenic can also undergo a series of reduction and oxidative methylation steps in the human liver and other tissues to form tri- and pentavalent methylated metabolites. Besides biomethylation, anthropogenic activity, such as the use of organic-based pesticides and herbicides, is also identified as the important source of organic arsenic. Although there is a large variety of organic arsenic, the amounts of organic arsenic are usually less than 5% of total arsenics [33]. Moreover, the toxicity of organic arsenic is far less than inorganic arsenic, for example, arsenosugars are even regarded as nontoxic. Therefore, the impact of organic arsenic on the environment is usually neglected. However, it is reported that some organic arsenic compounds, such as the methylated trivalent metabolites, monomethylarsonous acid (MMA(III)) and dimethylarsinous acid (DMA(III)), are significantly more toxic than their pentavalent counterpart and either As(III) or As(V) [34].

Most of the organic arsenic did not participate in further metabolism, and it tends to be eliminated in the urine in unchanged form. Only a small quantity of organic arsenic may metabolize through a series of reduction and oxidation reactions. Trivalent organic arsenic reacts with sulfhydryl groups, as observed with trivalent inorganic arsenic. In vitro binding of trivalent MMA and DMA to protein occurs to a greater extent than with the pentavalent organic forms. Methylated trivalent arsenic is a potent inhibitor of glutathione (GSH) reductase, which may provide a binding site for trivalent arsenic to inactivate the enzyme [35]. Pentavalent organic arsenicals are reduced in vitro by thiols to trivalent organic arsenic, which then bind other thiols.

3.3 ACUTE AND CHRONIC ARSENIC TOXICITY

The toxicology of arsenic is a complex phenomenon as arsenic is considered to be an essential trace element also for some animals, and may be even for humans. According to the symptoms and the onset time, arsenic toxicity can be divided into two types, namely, acute and chronic arsenic toxicity.

TABLE 3.3

Common Organic Arsenic Compounds

Chemical Name	Chemical Formula	CAS Registry	Property (Melting Point/Density/Odor/Solidity)	Lethal Dose 50%/Lethal Concentration (LD_{50}/LC_{50})	References
Arsenobetaine	$C_5H_{11}AsO_2$	64436-13-1	Appearance, solid; melting point, 203°C–210°C (decomposes)	Relatively nontoxic; LD_{50} > 10,000 mg/kg (mouse, oral)	Kaise et al. [25]
Dimethylarsinic acid (cacodylic acid, DMA)	$C_2H_7AsO_2$	75-60-5	Appearance, white crystal or powder; melting point, 192°C–198°C; boiling point, >200°C; density, 1.1 g/cm³; odor, odorless; solubility, 200 g in 100 mL water at 25°C; soluble in alcohol	LD_{50}, 1315 mg/kg (male rat, oral), 644 mg/kg (female rat, oral), 1800 mg/kg (mouse, oral); LC_{50}, 3.9–6.9 mg/L/2 h (rat, inhalation)	Gaines and Linder, Kaise et al., Stevens et al. [36–38]
Arsanilic acid	$C_6H_8AsNO_3$	98-50-0	Appearance, white crystal or powder; melting point, 232°C; density, 1.96 g/cm³; odor, odorless; solubility, slightly soluble in water and alcohol	LD_{50}, 1710 mg/kg (male rat, oral), 1330 mg/kg (female rat)	Confer et al. [39]
Disodium methanearsonate (DSMA)	$CH_3AsO_3Na_2$	144-21-8	Appearance, white solid or powder; melting point, >355°C; density, 1.04 g/cm³; odor, odorless; solubility, 43.2 g in 100 mL water at 25°C; insoluble in most organic solvents	LD_{50}, 928 mg/kg 80.1% MS (male rat, oral), 821 mg/kg 80.1% (female rat), 600 mg/kg 80.1% MS (male rat, oral), 561 80.1% mg/kg (female rat)	Gaines and Linder, Stevens et al. [36,38]
Tetramethylarsonium chloride	$C_4H_{12}NCl$	75-57-0	Appearance, white crystals or powder; melting point, 425°C; density, 1.17 g/cm³; odor, odorless; solubility, soluble	LD_{50}, 580 mg/kg (male mouse; oral), 114 mg/kg (male mouse, IP), 53 mg/kg (mouse, IV)	Shiomi et al. [40]
Monomethylarsonic acid	CH_5AsO_3	124-58-3	Appearance, white solid or powder; melting point, 160.5°C; boiling point, 393.3°C; solubility, 25.6 g in 100 mL water at 20°C; soluble in ethanol	LD_{50}, 961 mg/kg (female rat, oral), 1101 mg/kg (male rat, oral), 1800 mg/kg (mouse, oral); LC_{50}, 3.7 mg/L/2 h (rat, inhalation), 3.1 mg/L/2 h (mouse, inhalation)	Gaines and Linder [36]
Sodium methanearsonate	CH_4AsO_3Na	2163-80-6	Appearance, white solid or powder; melting point, 130°C–140°C; density, 1.55 g/cm³; odor, odorless; solubility, 58 g in 100 mL water at 20°C; insoluble in most organic solvents	LD_{50}, 102 mg/kg (rabbit, oral), 1105 mg/kg 58.4% MSMA (male rat, oral), 1059 mg/kg 58.4% MSMA (female rat, oral)	Jaghabir et al.., Gaines and Linder [36,41]

3.3.1 ACUTE TOXICITY

Most cases of acute arsenic toxicity is related to occupational exposure, such as the inhalation of arsine in the process of zinc electrowinning and copper electrowinning or the inhalation of arsenic trioxide dusts during copper and lead smelting. In addition, it also occurs from accidental ingestion from severely contaminated drinking water or food. Symptoms of acute intoxication after the ingestion of inorganic arsenic compounds usually occur within 30–60 min. The ingestion of small amounts (<5 mg) results in vomiting and diarrhea but recovers in 12 h, and treatment is reported not to be necessary [42]. In the human adult, the lethal range of inorganic arsenic is estimated at a dose of 1–3 mg As/kg [43]. The Risk Assessment Information System database states that "the acute lethal dose of inorganic arsenic to humans has been estimated to be about 0.6 mg/kg/day" [44]. The early symptoms of acute arsenic toxicity are manifested in the gastrointestinal system, which may be characterized by a metallic or garlic-like taste associated with dry mouth, burning lips, dysphagia, violent vomiting, and diarrhea. Following the gastrointestinal phase, multisystem organ damage may occur. If death does not occur in the first 24 h from irreversible circulatory insufficiency, it may result from hepatic or renal failure over the next several days. There may be some delayed effects, such as bone marrow suppression, hemolysis, hepatomegaly, melanosis, and polyneuropathy. In addition, the characterizations of chronic arsenic toxicity may also occur several weeks after ingestion, such as hyperkeratosis and pigmentation [45].

The acute toxicity of arsenic is related to its chemical form, solubility, and valence [18]. The oral LD_{50} for inorganic arsenic ranges from 15–293 mg (As)/kg to 11–150 mg (As)/kg bodyweight in rats and other laboratory animals, respectively [27,46]. Exposure to arsenic trioxide by ingestion of 70–80 mg has been reported to be fatal for humans [47]. Recent studies found that MMA(III) and DMA(III) are more acutely toxic and more genotoxic than their parent compounds [48,49]. These trivalent arsenicals are more toxic than As(V), MMA(V), and DMA(V) in vitro. This may be related to more efficient uptake of trivalent methylated arsenicals than of pentavalent arsenicals by microvessel endothelial cells and Chinese hamster ovary cells [50,51]. Recently, LC50 values were calculated as 571, 843, 5.49, and 2.16 μM for As(V), DMA(V), As(III), and DMA(III), respectively, for human cells [52]. This study also showed that dimethylmonothioarsenic is much more toxic than other pentavalent nonthiolated arsenicals.

3.3.2 CHRONIC ARSENIC TOXICITY

Chronic arsenic toxicity involves multisystem diseases by long-term low-dose arsenic exposure, which may not endanger human life, but is still harmful, especially the potential carcinogenicity for human. The early symptoms of chronic arsenic toxicity are insidious, and it is not easy to distinguish chronic arsenic toxicity from the general diseases due to the similar symptoms, such as abdominal pain, diarrhea, and sore throat. In addition, the clinical features of chronic arsenic toxicity vary between individuals, population groups, and geographic areas. The most representative manifestations of chronic toxicity in humans are skin lesions, which are characterized

TABLE 3.4
Effects Observed in Humans after Chronic Arsenic Exposure

	Affected System or Organ	Effect
1	Dermal	Hyperkeratosis and pigmentation
2	Respiratory tract	Inflammation and tracheobronchitis
3	Vascular	Peripheral vascular disease, myocardial injury
4	Hematological	Bone marrow depression, leucopenia, and anemia
5	Neurological	Peripheral neuropathy, encephalopathy
6	Endocrine	Diabetes mellitus
7	Liver	Hepatomegaly, cirrhosis, altered heme metabolism
8	Kidneys	Proximal tubule degeneration, papillary and cortical nephrosis
9	Gastrointestinal	Diarrhea, vomiting

by hyperpigmentation, hyperkeratosis, and hypopigmentation. In Taiwan, blackfoot disease is also observed in individuals chronically exposed to arsenic in their drinking water [53]. Another manifestation of chronic arsenic exposure is the presence of prominent transverse white lines in the fingernails and toenails called Mees' lines. It is unclear what factors determine the occurrence of a particular clinical manifestation or which body system is targeted. Thus, in persons exposed to chronic arsenic poisoning, a wide range of clinical features are common.

Many different systems within the body are affected by chronic exposure to inorganic arsenic, such as the respiratory, gastrointestinal, cardiovascular, nervous, and hematopoietic systems [54,55]. Some of these systems and their associated toxic effects from chronic arsenic exposure are listed in Table 3.4.

Skin lesions are major indicators of chronic arsenic toxicity following oral or inhalation exposure, but it may take years to manifest. Skin changes may develop and the most serious consequence is skin cancer if the source of arsenic exposure is not cut off. Chronic oral exposure data from studies in humans indicate that the lowest observed adverse effect level for skin lesions is probably about 0.01–0.02 mg As/kg/day (10–20 μg As/kg/day) and that the no observed adverse effect level (NOAEL) is probably between 0.0004 and 0.0009 mg As/kg/day (0.4–0.9 μg As/kg/day) [56,57,58].

Chronic arsenic toxicity produces various systemic manifestations over and above skin lesions, such as lung disease, gastrointestinal disease, liver disease, cardiovascular disease, nervous system diseases, and malignant disease. According to a report provided by the Bangladesh government in 2003, more than 70 million people are infected with tuberculosis and 4 million with diabetes, 7 million suffer from asthma and an equal number from chronic obstructive lung diseases, 10 million have kidney diseases, over 10 million carry the thalassemia gene, and over 80 million people are at risk of arsenic poisoning [59]. An estimated 37% of the populations suffer from heart diseases and 10% from some sort of hearing impairment. Diarrhea is responsible for 21% of child deaths, and pneumonia and other infectious diseases claim the rest.

Lung disease, both restrictive and obstructive lung diseases, was more common in patients with the characteristic skin lesions of chronic arsenic toxicity, which was characterized by chronic bronchitis, chronic obstructive and/or restrictive pulmonary disease, and bronchiectasis. In a hospital-based survey about chronic arsenic toxicity with nonmalignant lung disease in Kolkata, West Bengal, cases of obstructive lung disease, interstitial lung disease, and bronchiectasis were around 58.6%, 31.2%, and 10%, respectively [60].

The adverse effect of low-level arsenic on the gastrointestinal system has been variously reported by investigators, and its characterization symptoms included nausea, diarrhea, anorexia, and abdominal discomfort. Though diarrhea is a major and early-onset symptom in acute arsenic poisoning, in chronic arsenic toxicity, diarrhea occurs in recurrent bouts and may be associated with vomiting.

It is reported that the symptom of liver damage occurred in leukemia patients after taking Fowler's solution as a remedy [61]. All these patients developed features of portal hypertension with signs of liver fibrosis. Hepatomegaly occurred in 62 out of 67 people with evidence of chronic arsenic toxicity who consumed arsenic-contaminated water (200–2000 mg/L) in West Bengal, whereas it was found in only 6 out of 96 people who drank safe water in the same area [62]. There is evidence that chronic arsenic ingestion may cause neurological effects, especially in the peripheral nervous system. The characterization symptoms may include motor paralysis, tingling of the skin of extremities, foot and wrist drop, tremors, severe pain and ataxia. A NOAEL of 0.7 μg/kg/day inorganic arsenic in drinking water has been derived using neurological effects as the reported adverse effect [63].

Of the various genotoxic effects of chronic arsenic toxicity in humans, chromosomal aberration and increased frequency of micronuclei in different cell types have been found to be significant. The results of the study in West Bengal suggest that deficiency in DNA repair capacity, perturbation of methylation of promoter region of p53 and p16 genes, and genomic methylation alteration may be involved in arsenic-induced disease manifestation in humans [64].

3.4 MANIFESTATION OF ARSENIC TOXICITY

Clinical symptoms occurring in the early stage of human arsenic toxicity were unspecific. Early Symptoms of arsenic toxicity included fatigue, headaches, burning sensation of the eyes, dizziness, and palpitations. However, the clinical manifestations of arsenic toxicity are diverse and confusing, because these symptoms are rather similar to many other diseases. Moreover, the manifestation of arsenic toxicity depends on dose, bioavailability, and duration of arsenic exposure. Thus, the manifestations of chronic arsenic toxicity and acute toxicity were completely different.

3.4.1 CHRONIC EFFECT

Manifestations of chronic arsenic toxicity include hyperpigmentation, hyperkeratosis, desquamation, loss of hair, and skin cancer. Whitish lines (Mees' lines) may appear in the fingernails. Both sensory and motor nerve defects can develop. Additionally, liver and kidney functions may be affected. These manifestations have been observed mostly in those who are subjected to chronic arsenic toxicity [54,55,65–67]. Among these symptoms, dermal lesions were most common and dominant. This was maybe because the skin is made up of high content of keratin, which contains several sulf-hydryl groups, resulting in the binding with As(III). However, skin lesions were gen-erally a late manifestation of arsenic toxicity and occurred within a period of about several months, even several years due to the dose of arsenic exposure. It is estimated that hyperpigmentation may appear with 6 months to 3 years at high-dose exposure (e.g., 0.04 mg/kg/day), while it may appear within 5–15 years at lower exposure rates (e.g., 0.01 mg/kg/day) [68]. In addition, hyperkeratoses usually follow the initial appearance of hyperpigmentation within a period of years.

3.4.2   ACUTE EFFECTS

The clinical symptoms of acute arsenic toxicity occurred within 30–60 min. The immediate symptoms of acute arsenic toxicity include vomiting, blood in the urine or dark urine, abdominal pain, hemolysis, diarrhea, and dehydration. These are fol-lowed by numbness and tingling of the extremities, muscle cramping, and death, in extreme cases. With less dramatic cases of intoxication, the main symptoms are gastrointestinal complaints including a metallic taste, dry mouth, burning lips, dysphasia, vomiting attacks, and occasional hematemesis.

The oral LD_{50} values for inorganic arsenic compounds, depending on the arsenic species and the experimental animal, are in the range from 10 to 300 mg/kg body weight [69]. The single oral lethal dose for humans was about 2 mg As/kg. Doses as low as 0.05 mg As/kg/day over longer periods (weeks to months) have caused gas-trointestinal, hematological, hepatic, and dermal diseases, and neurological diseases may occur when oral doses were as low as 0.05 mg As/kg/day over several weeks. Long-term exposure to drinking water at levels as low as 0.001 mg As/kg/day has been related to skin diseases and skin, bladder, kidney, and liver cancers. Long-term inhalation exposure to arsenic has also been associated with lung cancer at air levels as low as 0.05–0.07 mg/m^3 [70].

When compared to acute arsenic toxicity, symptoms of chronic arsenic toxicity were rather secretive, which may be manifested clearly after several months, even years. Therefore, it was too late to diagnosis arsenic toxicity by observing these symptoms unless total arsenic concentrations were checked in the urine, which was widely used as a biomarker for arsenic exposure in humans [71–74].

3.4.3   STAGES OF CLINICAL FEATURES OF ARSENIC TOXICITY

The features of arsenical toxicity have been classified by Dr. Saha, which are now known as Saha's classification of stages, including preclinical, clinical, internal com-plication, and malignancy stage [75].

3.4.3.1   Preclinical (Asymptomatic) Stage

In this stage, the features of arsenic toxicity are absent, and there is no obvious difference between the victim and the normal individual, but arsenic can be found in the urine, blood, or other tissues. After the ingestion of arsenic-contaminated water, the concentration of arsenic in the urine and blood increases rapidly, but it would decrease, even disappear, when the source of arsenic is stopped. The main species of arsenic revealed in urine is dimethylarsonic acid and trimethylarsenic acid.

3.4.3.2 Clinical Stage (Symptomatic or Overt Phase)

Derma changes/signs are the most distinctive feature in the clinical stage. The presence of clinical symptoms is confirmed by the detection of higher arsenic concentration in the nails, hair, and skin scales. The time to show clinical symptoms depends on the arsenic concentration in water, the bioavailability of arsenic, and the physical condition of patients. It may take 6 months to 10 years (average 2 years) to develop clinical features. The major dermatological signs include melanokeratosis, melanosis, spotted melanosis, spotted and diffuse keratoses, leucomelanosis, dorsal keratosis, and combination of melanosis and nodular rough skin. Melanokeratosis is the chief symptom of arsenical dermatosis, which was almost revealed in all the patients. Although melanosis or keratosis may be caused by genetic disorders, the presence of the combination of melanosis and keratosis in the same person is almost definite to arsenical dermatosis. In addition, there are also some uncommon derma signs, such as mucus membrane melanosis, nonpitting edema, and conjunctival congestion. These dermatological features rarely appeared in the patients; the chance of getting conjunctival congestion in a person with arsenic poisoning is likely less than 4% [76].

3.4.3.3 Stage of Internal Complications

Clinical symptoms are associated with biochemical evidence of organ dysfunction as well as histological and histochemical abnormalities and high concentrations of arsenic in the different organs involved. The common symptoms in this stage are obstructive and restrictive lung diseases, gastroenteritis, liver fibrosis, hepatomegaly and splenomegaly, and so on.

3.4.3.4 Stage of Malignancy

Malignancy affecting the skin, lungs, bladder, uterus, or other organs develops if the patient survives the stage of complications. Malignancy does not develop before 10 years of arsenic exposure [77,78]. Usually after 15–20 years from the onset of first symptoms, cancer develops. There are significant associations between these dermatological lesions and the risk of skin cancer. The most common arsenic-induced skin cancers are Bowen's disease (carcinoma in situ), basal cell carcinoma, and squamous cell carcinoma. Further, there is an increased risk of development of urinary bladder cancer and lung cancer due to chronic exposure to arsenic.

3.5 ARSENIC EXPOSURES

3.5.1 Routes of Arsenic Exposure

Human exposure to arsenic can occur in several different routes: ingestion, inhalation, and dermal or skin exposure. The primary routes of arsenic exposure for humans are ingestion and inhalation, whereas dermal exposure has been considered as a minor uptake route.

3.5.1.1 Ingestion

Except for the medicinal ingestion, most of the arsenic is ingested as organic arsenic, and the main source of arsenic exposure is via ingestion of food containing arsenic. Food, with the exception of seafood, contains less than 0.25 mg arsenic/kg [79]. Dietary intake of total arsenic ranges from 10 to 200 µg per person per day [80]. Meat, fish, and poultry account for the majority (60%–90%) of dietary arsenic intake, but most of these arsenic forms in food are generally in the form of organic arsenic, such as AsB and AsC. Various kinds of fish often contain between 1 and 10 mg arsenic/kg, and shellfish has more than 100 mg arsenic/kg. The daily uptake with food is estimated to be between 0.04 (without fish) and 0.19 mg arsenic (with fish). These organic arsenics may cause the rising of arsenic levels in blood but are rapidly excreted unchanged in the urine. Recent studies have shown that one form of seaweed, hijiki, contains high levels of inorganic arsenic [81,82]. Arsenic compounds may enter the plant food chain from agricultural products or from soil irrigated with arsenic-contaminated water. Some vegetables and rice grew in arsenic-contaminated soil with high concentrations of arsenic. In a survey, the total arsenic concentration in rice collected from retail stores in upstate New York and Canada, France, Venezuela, and other countries varied from 5 to 710 µg/kg [83]. In addition, except DMA, inorganic arsenic has also been found as the main arsenic species in rice. It has been estimated that the average daily dietary intake of arsenic by adults in the Spanish diet is 221 µg arsenic per day [84]. The Joint FAO/WHO Expert Committee on Food Additives recommends a provisional tolerable weekly intake of inorganic arsenic of 15 µg/kg body weight, equivalent to about 129 µg/day for a 60 kg person [85]. The amount of organic arsenic intake was not yet set so far.

The use of arsenic-bearing water as drinking water is the other way of arsenic uptake via ingestion. It is estimated that over 57 million people in the world are exposed to drinking water with concentrations of 50 µg/L arsenic [86]. The country worst affected is Bangladesh; over 10,000 people have shown evidence of chronic arsenic toxicity and this number is expected to rise. The species of arsenic ingested from drinking water will vary according to its exposure to air. Deep wells contain predominantly arsenite (As(III)), while surface water will contain predominantly arsenate (As(V)).

Arsenic ingestion from air and soil is usually much less than that from food and drinking water, but this part was not neglected due to the increase of arsenic emission to environment by anthropogenic activities, such as mining and metallurgy and the use of arsenic-based pesticides and insecticides. It is estimated that arsenic concentration in arable surface soil has increased to about 283 kg/km^2 in 2000, which was 30 times than that in 1900 [87]. Certainly, arsenic was not only mobilized to soil but also to air by both natural and anthropogenic processes. Thus, the entry of arsenic into the human food chain due to its mobilization in the air, soil, and water is a serious hazard for humans to face.

3.5.1.2 Inhalation

Major sources of inhaled arsenic may come from air emissions from burning of fossil fuels; smelting of copper, lead, and zinc; manufacturing of glass; and burning of arsenic-bearing materials. In high temperature (>500°C), arsenic can volatilize into fume and condense as arsenic-bearing dusts, in which arsenic generally occurred as arsenic trioxide (As_2O_3) particles. However, some arsenic-bearing dusts may suspend and mobilize in the air due to the ultrafine particle sizes. It is reported that European arsenic levels in air are about 0.2–1.5 ng/m^3 in rural, 0.5–3 ng/m^3 in urban, and 50 ng/m^3 in industrial areas [88]. Except aerosol particles, arsenic may occur as gas state, such as arsine. In addition, organic arsenic can be reduced to volatile arsine under anaerobic conditions, such as monomethylarsine, dimethylarsine, and trimethylarsine [89]. Although arsenic levels in air varied in different areas, the amount of arsenic was rather low. Except occupational exposure, most of us will not be affected by airborne arsenic. Welch et al. measured 8-h averages of airborne arsenic in a U.S. copper smelter from 1943 to 1965; it ranged between 6.9 and 20 mg/m^3 [90]. High-level occupational exposures leading to severe arsenic toxicity may occur through inhalation of airborne arsenic produced from the processing industries of arsenic-bearing materials. Once arsenic enters into the human body via inhalation, some arsenic-bearing particles may deposit in the respiratory tract and finally enter into the lungs, resulting in lesions; others may enter into the gastrointestinal system, resulting in diarrhea or vomiting.

3.5.1.3 Dermal

When arsenic exposures occurred via dermal or skin contact, it may transport from the outer surface to the inner surface of the skin and finally into the systemic circulation. Most of the dermal absorption occurred via occupational exposure, such as exposure to arsenic chemicals, cosmetics, and arsenic-bearing pesticides. Factors such as chemical interactions, solubility, and the presence of other metals can affect the properties of arsenic and its potential for dermal absorption. Current knowledge of percutaneous absorption of arsenic is based on studies of rhesus monkeys using arsenic acid (H_3AsO_4) in aqueous solution and soluble arsenic mixed with soil [91]. These studies produced average dermal absorption rates in the range of 2.0%–6.4% of the applied dose. It is reported that 1.9% was absorbed from water and 0.8% from soil for human skin at a low dose over a 24 h period. Most of us rarely have opportunities contacting high levels of arsenic via dermal route; CCA-treated wood and arsenic-bearing soil and water may be the main sources, but the amount of arsenic via dermal route is generally rather minimal due to the low concentration in these materials.

3.5.2 Sources of Exposure to Arsenic

Exposure to arsenic may come from natural and industrial sources or from administered source, that is, accidental source. The major sources of human exposure to arsenic include food, drinking water, and contaminated air and soil. In addition, tobacco, drugs, and industrial processes may also become significant sources of exposure to arsenic for some people.

3.5.2.1 Food

Food is the principal contributor to the daily intake of arsenic for humans, but it contains usually very low levels, less than 1 mg/kg [92]. Higher levels of arsenic are generally found in sea fish, shellfish, meat, poultry, dairy products, and cereals. However, most of these high levels of arsenic are usually in the organic forms (e.g., AsB) that are of low toxicity. Sea fish may contain arsenic concentrations of up to 5 mg/kg, and concentrations in some crustaceans and bottom-feeding fish may reach several tens of milligrams per kilogram, but the proportion of inorganic arsenic in these fish is very low [93]. After fish, pig and poultry meat are the next most important contributors to the dietary intake of arsenic due to the result of organoarsenical feed additives that may be used as growth promoters.

Rice, as a traditional staple in many countries, may also be contaminated with arsenic. Table 3.5 summarizes the arsenic levels in rice from various countries [94]. In summary, the data presented in Table 3.5 show that the maximum values for inorganic arsenic in rice do not generally exceed 0.2 mg/kg. The fraction of inorganic arsenic from total arsenic showed a wide variation ranging from approximately 10% to 93%. Although brown rice has a higher nutrient content compared to white rice, it also has even higher levels of arsenic. However, it is noted that white rice grown in Arkansas, Louisiana, Missouri, and Texas had higher levels of arsenic than other regions of the world studied, possibly because of past use of arsenic-based pesticides to control cotton weevils [95].

The mean daily intake of arsenic in food for adults has been estimated to range from 16.7 to 129 μg [96,97]. Extreme intakes of arsenic from food depend critically on individuals' dietary habits. There is no doubt that the intake of arsenic for vegetarian is less than that of nonvegetarian due to the high levels of arsenic in fish and meat. Although the levels of arsenic in the food varied widely, the amount of arsenic

TABLE 3.5
Total and Inorganic Arsenic Levels in Rice from Various Countries

Country	Total Arsenic (mg/kg)		Inorganic Arsenic (mg/kg)	
	Content	Mean	Content	Mean
Australia	0.05–1.20	0.29		
China	0.08–5.71	0.29	0.04–0.45	0.13
Japan	0.04–0.43	0.17	0.04–0.37	0.15
EU	0.01–1.98	0.16	0.02–1.88	0.14
United Kingdom	0.12–0.47	0.22	0.06–0.16	0.11
USA	0.04–0.41	0.21	0.025–0.157	0.091
Spain		0.197	0.027–0.253	

was too low to cause any "immediate or short-term adverse health effects." However, it is less clear whether there is a potential for long-term risks.

3.5.2.2   Drinking Water

Background concentrations of arsenic in groundwater in most countries are less than 10 µg/L and sometimes substantially lower. However, the variance of arsenic concentrations in groundwater is even higher, and it may range from 0.5 to 5000 µg/L [98]. Most high levels of arsenic in groundwater are from thermal activity or the dissolution of arsenic mineral. The concentration of arsenic in surface waters was generally less than that of groundwater. Levels of arsenic dissolved in uncontaminated stream waters ranged from 0.1 to 1.7 µg/L, and those in seawaters were 1.5–1.7 µg/L [99]. In contrast, most high levels of arsenic in surface waters are from arsenic pollution by anthropogenic arsenic emission.

For normal populations, assuming consumption of 1.5 L of water daily, intakes of arsenic will be 0.015 mg/day or less; most of this arsenic is likely to be inorganic. Individuals consuming water containing elevated concentrations of arsenic (0.2–0.5 mg/L) will have daily intakes in the range of 0.3–0.75 mg. Certain bottled mineral waters contain up to 0.2 mg/L of arsenic of unidentified species; it is reasonable to suppose that individuals who regularly drink these waters will have daily arsenic intakes from this source of 0.2 mg.

Drinking water poses the greatest threat of arsenic contamination to public health. Inorganic arsenic is naturally present at high levels in the groundwater of a number of countries, such as Argentina, Chile, China, India (West Bengal), Mexico, the United States, and particularly Bangladesh where approximately half of the total population is at risk of drinking arsenic-contaminated water from tube wells. Since it was realized that the high levels of arsenic in drinking water had an adverse effect on human health, the WHO has constantly revised drinking water quality guidelines since 1958, and the present guideline value was 0.01 mg/L, which has been adopted in many countries. Cooking rice with arsenic-contaminated water can actually increase the concentration in rice and further contribute to total dietary arsenic exposure.

3.5.2.3   Air

The major sources of arsenic in air are coal burning and metal smelting where it is emitted as As_2O_3. According to Chilvers and Peterso [100], the atmosphere stores on average 1.74×10^6 kg As, and this mass is unevenly distributed between the hemispheres with 1.48×10^6 kg As in the northern and 0.26×10^6 kg As in the southern hemisphere. Airborne concentrations of arsenic may range from a few nanograms to several hundred nanograms per cubic meter. In addition, the concentrations of arsenic in urban areas are higher than that of suburb areas, and in near point emissions of arsenic, such as smelters and coal-fired power plants, airborne arsenic concentrations have exceeded 100 µg/m^3. Normal intakes of arsenic from air are unlikely to exceed 0.00024 mg/day (0.24 µg/day). In other words, except for occupational exposure, under normal conditions, air contributes only a minute proportion, which may not have an adverse effect on humans. The intakes of arsenic from the air that can be associated with an increased incidence of lung cancer are at least three orders of magnitude greater than normal intakes.

3.5.2.4 Soil

Soils cannot be seen as a homogenous medium and contain highly variable arsenic concentrations. The estimation for a global arsenic concentration in soils is in the range 0.1–55 mg/kg with an average value of 7.2 mg/kg [101]. The high levels of arsenic in soils can be derived mainly from agriculture (the use of pesticides, insecticides, and manure) and from mining and smelting activities. Virgin soils generally contain less than 40 mg/kg arsenic, while contaminated soils may contain up to 500 mg/kg [102]. The resulting arsenic enrichment in the soil may lead to growth defects in plants and the subsequent increase of arsenic in the fruits, seeds, roots, and leaves, and the arsenic may ultimately enter into the food chain and affect human health. Grain harvested from low levels of As in the soils (1.3–11 mg/kg) contained low concentrations of total As (7.7 ± 5.4 µg/kg). In contrast, at one of the trial sites, the As level in the soils was greater (29 mg/kg), and much higher As concentrations (69 ± 17 µg/kg) were present in wheat grain [103]. However, arsenic contaminants in soils generally have a low bioavailability due to the interaction with other soil constituents, such as the adsorption of iron or manganese oxides. A study carried out for evaluation of exposure to arsenic in residential soil shows that correlations between speciated urinary arsenic and arsenic in soil or house dust were not significant for children. Similarly, questionnaire responses indicating soil exposure were not associated with increased urinary arsenic levels [104]. Relatively, low soil arsenic exposure is unlikely to directly affect human health. However, it is noted that high levels of arsenic in the soil may diffuse and enter into the water and food chain, which may have a significant adverse effect on humans.

3.5.2.5 Smoking

Tobacco, like other plants, can uptake arsenic from soil and water, which may lead to smoker exposure. It is found that arsenic concentration in the 1431 tobacco leaves obtained from Africa, Asia, Europe, and South and North America averaged 0.4 ± 0.6 µg/g, which was even higher than some foods [105]. In addition, significant differences in arsenic concentration were found among tobacco types, sampling locations, and crop years.

Depending on the content of arsenic in tobacco, a smoker may inhale between a few micrograms and 20 µg of arsenic daily. When arsenicals, such as lead arsenate, were used to control insect pests, the arsenic content of tobacco was higher. In that case, more than 100 µg might have been inhaled per day. Such a high level of arsenic adequately threatens human health, which may be one of the reasons that lung cancer commonly strikes smokers.

3.5.2.6 Drugs

Arsenic compounds have been used in medicine for many years in many countries, such as China and India. Fowler's solution, a 1% arsenic trioxide preparation, was widely used to treat leukemia, skin diseases (psoriasis, dermatitis herpetiformis, and eczema), stomatitis and gingivitis in infants, and Vincent's angina [106]. Up to now, arsenic trioxide is still widely used to induce remission in patients with acute promyelocytic leukemia. In addition, it was also prescribed as a health tonic. However, it has been confirmed that chronic arsenic intoxication from the long-term use of Fowler's solution occurred in some patients. A study in Singapore identified 17 patients during a 5-year period with cutaneous lesions related to chronic arsenic toxicity, and in 14 (82%) patients toxicity was due to arsenic from Chinese proprietary medicines, while the other three consumed well water contaminated with arsenic [107]. However, the source of arsenic exposure via drugs is valid to specific patients, and most of us rarely have a chance to have contact with these arsenic-bearing drugs.

3.5.2.7 Occupational Exposure

Occupational exposure to arsenic mainly occurs in workplaces that involve treatment of arsenic-bearing materials (ores or by-products) and the production and use of arsenicals, such as copper and lead smelting, coal combustion (especially those using low-grade brown coal), and insecticide and pesticide production. Arsenic trioxide and arsine were the most common arsenic compounds presented in occupational exposure; thus, most of the occupational exposure to arsenic occurs through inhalation. Total arsenic concentration in workplace of gallium arsenide crystal and wafer production air ranged from 2 to 24 $\mu g/m^3$, while it was even more than 1 mg/m^3 in some workplace of metal smelters [108]. Although occupational arsenic exposure is infrequent, and the affected people were very little, arsenic toxicity from occupational exposure often causes severe intoxications, and sometimes death.

3.6 ADSORPTION AND METABOLISM OF ARSENIC

3.6.1 ADSORPTION

Most arsenic entered into the human body via ingestion and inhalation and adsorbed in the gastrointestinal tract and lungs. The major site of arsenic absorption is the gastrointestinal tract, especially the small intestine, which adsorbed arsenic by an electrogenic process involving a proton (H^+) gradient. It is reported that over 90% of soluble inorganic arsenic compounds are rapidly and extensively absorbed from the gastrointestinal tract in humans and most experimental animals, while insoluble arsenic compounds such as arsenic disulfide are poorly absorbed [109]. Except the solubility of arsenicals, the bioavailability depends on other things, such as the pH value of body fluid and the associated elements or ingredients. The optimal pH for arsenic absorption is about 5.0, which is close to the pH value in the milieu of the small bowel [110]. Oral bioavailability of inorganic arsenic was about 54%–80% in humans [111]. Oral bioavailability studies in monkeys indicate that approximately 15% of arsenic from soil and house dust contaminated by smelter emissions was adsorbed [112]. Gastrointestinal absorption of arsenate is inhibited in rats by the presence of phosphate.

After deposition of the particles in the respiratory tract and lungs, the arsenic is desorbed from the deposited particles. Absorption of arsenic in inhaled airborne particles is highly dependent on the solubility and the size of particles; absorption after intratracheal installation in animals is very high (90%) for soluble arsenic compounds [113]. It is reported in an early study that 85%–90% of arsenic deposited in the lungs of human volunteers by smoking arsenic in cigarettes was absorbed within 14 days [114].

After absorption through the lungs or gastrointestinal tract, arsenic is widely distributed by the blood throughout the body. With a half time of 2 h, inorganic arsenic is rapidly eliminated from the blood. The arsenic compounds are distributed in almost all the organs, such as the kidneys, *liver*, spleen, and lungs. Most tissues rapidly eliminate arsenic after several days, and only very little arsenic is found in keratin-rich tissues such as hair, nails, and skin and, to a lesser extent, in bones and teeth [115,116]. It is reported that elimination of arsenic occurs in three phases in humans. About 66% of the administered doses are renally eliminated with a half time of 2.1 days, around 30% with a half time of 9.4 days, and the rest (4%) with a half time of 38.4 days [117].

3.6.2 METABOLISM OF ARSENIC

Methylation of arsenic is a widespread phenomenon in the biosphere. Many organisms, including humans, animals, and even some bacteria and fungi, can methylate arsenic compounds to different extents.

The metabolism of arsenic in humans and other organisms has an important role in its toxic effects. As it is shown in Figure 3.1, arsenic metabolism is characterized by a sequential process involving a two-electron reduction of pentavalent arsenic to

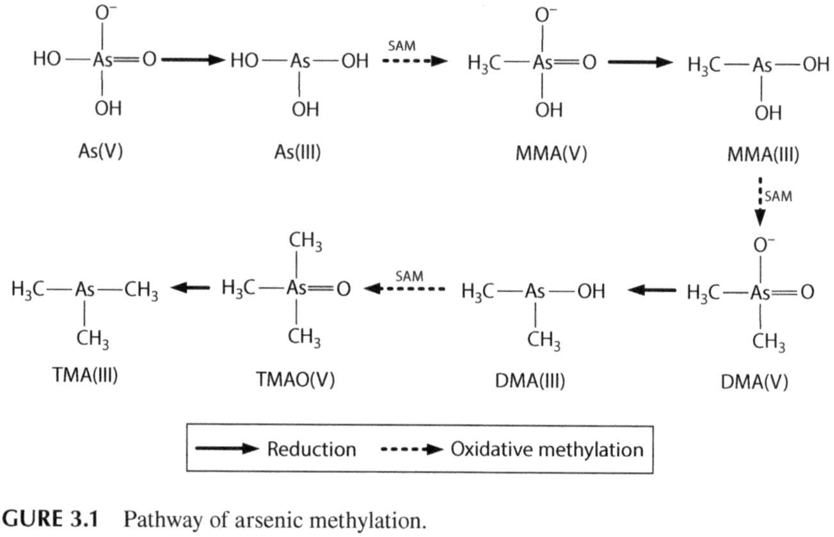

FIGURE 3.1 Pathway of arsenic methylation.

trivalent arsenic, followed by oxidative methylation to form monomethylated and dimethylated arsenic products using S-adenosyl methionine (SAM) as the methyl donor and GSH as an essential cofactor [118,119]. The reduction can occur nonenzymatically in the presence of a thiol such as GSH, while the methylation of arsenic is enzymatic, requiring SAM and a methyltransferase. The predominant metabolite of inorganic arsenic is dimethylarsinic acid (DMA(V), $(CH_3)_2AsVO(OH)$), which is about 60%–70% in all the metabolites [120]. Other metabolites, such as MMAA (MMA(V), $CH_3AsVO(OH)_2$) and trimethylarsine oxide, are found in very low amounts in the urine.

Arsenic metabolites vary between species, such as the metabolites mono-, di-, trimethyl arsenates, but the methylation pathways are very similar, both involving a series of reduction and oxidative methylation reactions. In particular, the final metabolites, the volatile methyl arsines (DMA(-III), TMA(-III)), produced by microorganisms, are totally different from mammals, which may have an environmental significance for bioremediation of polluted soils. The rabbit and mice have similar metabolites as human adults [118,121], suggesting that this may be a good animal model for toxic kinetics in humans. In contrast, the guinea pig, marmoset monkeys, and chimpanzees are not able to methylate inorganic arsenic [122–125].

Organic arsenic is basically not metabolized in vivo, but it is rapidly excreted unchanged in the urine. However, some metabolism of arsenosugars in humans has been found to produce DMA(V) as a major metabolite (67%) in the urine [126]. Diethylarsinoylethanol and TMAO have been found as other minor constituents of arsenic metabolites. In addition, it is reported that arsenosugars can be biotransformed to DMA(V) and inorganic arsenic in microcosm experiments, but the amount of arsenite and arsenate is rather small [127].

The metabolism of inorganic arsenic may be affected by its valence state and exposure doses. Studies in laboratory animals indicate that administration of arsenic trioxide results initially in higher concentrations in most tissues than does the administration of pentavalent arsenic. However, trivalent arsenic is the preferred substrate for methylation reactions; thus, the reduction of arsenic from pentavalent to trivalent may be a critical step in the control of the rate of metabolism of arsenic. Methylation efficiency in humans appears to decrease at high arsenic doses. It is explained as the effect of the methylation threshold. When the dose of arsenic exposure is higher than this value, methylation capacity begins to decline after a certain level, resulting in the increase of arsenic toxic effects.

Methylation of arsenic is considered the primary detoxification mechanism, since it is associated with the conversion of the most potentially toxic inorganic arsenic to the less toxic organic arsenic, followed by accumulation in or excretion from the cell, and MMA(III) and DMA(III) are only considered as intermediates in the metabolism of arsenic. However, MMA(III) and DMA(III) have been detected in the urine of some humans subjected to chronic arsenic toxicity and in the bile of rats administered arsenite intravenously [128,129]. Most alarming of all, these trivalent organic arsenic materials are more toxic than arsenite to the microorganism *Candida humicola* in vitro. It is found that human cells are also more sensitive to the cytotoxic effects of MMA(III) than arsenite [130]. DMA (III) has similar cytotoxic effects in several human cell types as arsenite. Even the LD_{50} of MMA(III) was lower than that of arsenite in the hamster [131]. Therefore, methylation of arsenic cannot be simply considered as the detoxification mechanism due to the greater toxicity of the methylated trivalent intermediates.

3.7   GEOGRAPHICAL DISTRIBUTION OF ARSENIC

Arsenic (As) is widely distributed throughout the earth crusts, soil, sediments, water, air, and living organisms, which occurs from both natural and anthropogenic processes. Although the occurrence of arsenic varies, most of the arsenic existed in the earth crusts, and it generally occurs in over 200 different mineral forms, of which approximately 60% are arsenates and 20% sulfides and sulfosalts, and the remaining 20% includes arsenides, arsenites, oxides, silicates, and elemental arsenic (As) [132]. However, these arsenic-bearing materials are not constant, and they may release arsenic into the water and air due to the effect of microbial activities, oxidation–reduction, dissolution, volcanic activities, and weathering. Arsenic is susceptible to be mobilized under pH conditions typically found in groundwater (pH = 6.5–8.5) and over a wide range of redox (reduction–oxidation) conditions [133].

One of the main sources of arsenic toxicity is groundwater, which was used as the drinking water by millions of people in many countries, especially the Southeast Asian region countries. To date, unacceptably high As levels in groundwater resources have been found in several parts of Bangladesh, Cambodia, China, Chile, Ghana, India, the Lao People's Democratic Republic, Mexico, Mongolia, Myanmar, Nepal, Pakistan, Taiwan, Thailand, the United States, and Vietnam [134,135].

Arsenic contamination of groundwater from natural sources can be attributed to two geochemical processes: reductive dissolution and oxidative dissolution [136].

1. *Reductive dissolution*: Arsenic is assumed to be present in the form of arsenic-rich iron oxyhydroxides or $FeAsO_4 \cdot 2H_2O$ in the sediments, which are derived from weathering or oxidation of base metal sulfides. The organic matter or microbial media deposited with the sediments reduce the arsenic-bearing iron hydroxide and release arsenic into groundwater. According to this hypothesis, the origin of arsenic-rich groundwater is due to a natural process, and it seems that the arsenic in groundwater has been present for thousands of years without being flushed from the delta. This hypothesis (reductive dissolution of arsenic-rich iron oxyhydroxides) is accepted by most researchers. In case of Bangladesh and West Bengal, arsenic-polluted alluvial Ganges aquifers are naturally derived from eroded Himalayan sediments and are believed to become mobile following reductive release from solid phases under anaerobic conditions [137].

2. *Oxidative dissolution*: Arsenic is assumed to be present in certain sulfide minerals (pyrites) that are deposited within the aquifer sediments. During the

dry season, arsenic-rich beds (pyrites/arsenopyrites) are exposed to air and oxidized. During the subsequent rainy season, the recharge period, iron hydroxide releases arsenic into groundwater. For arsenopyrite-rich rocks, aqueous oxidation of arsenic by dissolved oxygen is described by the following equation: $4FeAsS + 14O_2 + 4H_2O = 4Fe^{3+} + 4AsO_4^{3-} + 4SO_4^{2-} + 8H^+$.

However, not all of the lowering of water table below deposits could be attributed to the natural phenomenon, and it may be caused by overexploitation of groundwater; thus, the origin of arsenic-rich groundwater sometimes is man-made, which is a recent phenomenon.

Groundwater is a major source of drinking water in many parts of the world, but it is not always safe for human consumption due to contamination by arsenic. Arsenic toxicity via groundwater has become a worldwide problem with more than 21 countries experiencing arsenic groundwater contamination [138]. Background concentrations of arsenic in groundwater in most countries are less than 10 µg/L; however, in contaminated countries, it shows a very large range from 1 to 5000 µg/L (as seen in Table 3.6). Most high-arsenic groundwater is the result of natural occurrences of arsenic.

Without doubt, widespread groundwater contamination with arsenic has been recognized as a major public health concern in several parts of the world. In the past few decades, a growing body of research has shown that it may cause chronic arsenic toxicity, even when the low level of arsenic was ingested via drinking water. That's why the drinking water standard for arsenic was constantly modified by policy makers. The current WHO provisional guideline was lowered to 10 µg/L in 1993. Most developed countries adopt this recommended limit, but drinking water guidelines in many developing countries remain at the 50 µg/L limit due to the greatly increased water supply costs.

Although major arsenic sources of groundwater were the geochemistry of arsenic minerals in the sediment and rocks, the emission of arsenic by anthropogenic activities must not be neglected, and its risk was even much greater. It is apparent that human activities can increase the rates and amounts of arsenic mobilized and dissolved in groundwater through inputs of organic carbon and from water withdrawals and other changes to natural hydrologic systems. Thus, the key to mitigate arsenic contamination of groundwater is to control and reduce the emission of arsenic into the environment, especially from anthropogenic activities. Its next priority is to restore arsenic-contaminated environments. Certainly, in order to eliminate arsenic toxicity, it is necessary to carry out a systematic screening of groundwater for arsenic contamination. In addition, alternative water supplies should be introduced for those people, who are exposed to contaminated drinking water with concentrations of 50 µg/L or higher. Finally, simple and low-cost water treatment technologies should be developed and generalized to the public. There is no doubt that the more fully we understand how, when, and where arsenic is mobilized from geologic materials, or from anthropogenic releases to the environment, the more effectively we can find solutions to this major contamination problem.

TABLE 3.6

Concentrations of Arsenic in Groundwater of the Arsenic-Affected Countries

Country	Arsenic Concentration in Groundwater (μg/L)	Source	Population at Risk	Reference
Argentina	100–2,000	Natural	2,000,000	[139]
Australia	1–300,000	Both anthropogenic and natural sources	Unknown	[140]
Bangladesh	1–4,700	Natural, derived from geological strata	50,000,000	[141]
Chile, Antofagasta	900–1,040	Natural, associated with quaternary volcanism	437,000	[139]
China	220–2,000	Natural	Over 2,000,000 exposed to high amounts of arsenic; more than 20,000 suffering from arsenicosis	[142]
Hungary	60–4,000	Natural	220,000	[139]
India, West Bengal	10–3,900	Natural, derived from geological strata	Over 5,000,000 exposed to As > 50 μg/L, 300,000 suffering from arsenicosis	[143]
Mexico	10–4,100	Natural	400,000	[139]
Nepal	4–2,620	Natural	550,000 exposed to As > 50 μg/L, 3,190,000 exposed to As >10 μg/L	[144]
Peru	500	Natural	250,000	[139]
Romania	10–176	Natural	36,000	[145]
Taiwan	10–1,820	Natural	140,000	[58]
Thailand, Ronpibool	1–5,000	Mining activities	15,000	[146]
USA	10–48,000	Natural, geothermal, and mining-related sources	Unknown	[147]
Vietnam	1–3,050	Natural	>1,000,000	[148]

3.8 CONCLUSIONS

Arsenic toxicity is a medical condition caused by the ingestion, absorption, or inhalation of dangerous levels of arsenic. There are two types of arsenic toxicity, acute arsenic toxicity and chronic arsenic toxicity, which have different symptoms due to the ingestion amount and species of arsenic in a certain time. Once arsenic enters into the human body, it would react with sulfhydryl groups, ADP, or red blood cells, and a series of symptoms would be caused by these biochemical reactions. In general, inorganic arsenic is more toxic than organic arsenic, and trivalent arsenite is more toxic than pentavalent and zero-valent arsenic.

The acute toxicity of arsenic is related to its chemical form, solubility, and valence [18]. In the human adult, the lethal range of inorganic arsenic is estimated at a dose of 1–3 mg As/kg. Symptoms of acute intoxication after the ingestion of inorganic arsenic compounds usually occur within 30–60 min. Chronic arsenic toxicity involves multisystem diseases by long-term low-dose arsenic exposure, which may not endanger human life, but is still harmful, especially the potential carcinogenicity for human. The most representative manifestations of chronic toxicity in humans are skin lesions, which are characterized by hyperpigmentation, hyperkeratosis, and hypopigmentation.

Clinical symptoms occurring in the early stage of human arsenic toxicity were unspecific. Early symptoms of arsenic toxicity included fatigue, headaches, burning sensation of the eyes, dizziness, and palpitations. However, the clinical manifestations of arsenic toxicity are diverse and confusing, because these symptoms are rather similar to many other diseases. The immediate symptoms of acute arsenic toxicity include vomiting, blood in the urine or dark urine, abdominal pain, hemolysis, diarrhea, and dehydration. When compared to acute arsenic toxicity, symptoms of chronic arsenic toxicity were rather secretive, which may be manifested clearly after several months, even years.

Human exposure to arsenic can occur in several different routes: ingestion, inhalation, and dermal or skin exposure. The primary routes of arsenic exposure for humans are ingestion and inhalation, whereas dermal exposure has been considered as a minor uptake route. The major sources of human exposure to arsenic include food, drinking water, and contaminated air and soil. In addition, tobacco, drugs, and industrial processes may also become significant sources of exposure to arsenic for some people. Food is the principal contributor to the daily intake of arsenic for humans, but it contains usually very low levels, less than 1 mg/kg.

Most arsenic enters into the human body via ingestion and inhalation and is adsorbed in the gastrointestinal tract and lungs. The major site of arsenic absorption is the gastrointestinal tract, especially the small intestine, which adsorbed arsenic by an electrogenic process involving a proton (H^+) gradient. After absorption through the lungs or gastrointestinal tract, arsenic is widely distributed by the blood throughout the body. Most tissues rapidly eliminate arsenic after several days, and only very little arsenic is found in keratin-rich tissues such as hair, nails, and skin and, to a lesser extent, in bones and teeth.

The metabolism of arsenic in humans and other organisms is characterized by a sequential process involving a two-electron reduction of pentavalent arsenic to trivalent arsenic, followed by oxidative methylation to form monomethylated and dimethylated arsenic products using SAM as the methyl donor and GSH as an essential cofactor. Methylation of arsenic cannot be simply considered as the detoxification mechanism due to the greater toxicity of the methylated trivalent intermediates.

Arsenic contamination of groundwater from natural sources can be attributed to two geochemical processes: reductive dissolution and oxidative dissolution. In the reductive dissolution process, arsenic is assumed to be present in the form of arsenic-rich iron oxyhydroxides or $FeAsO_4 \cdot 2H_2O$ in the sediments. The organic matter or microbial media deposited with the sediments reduce the arsenic-bearing iron hydroxide and release arsenic into groundwater. The reductive dissolution of arsenic-rich iron oxyhydroxides is accepted by most researchers. The process of oxidative dissolution occurs when arsenic is assumed to be present in certain sulfide minerals (pyrites) that are deposited within the aquifer sediments. During the dry season, arsenic-rich beds (pyrites/arsenopyrites) are exposed to air and oxidized. During the subsequent rainy season, the recharge period, iron hydroxide releases arsenic into groundwater.

REFERENCES

1. Ravenscroft. Predicting the global distribution of arsenic pollution in groundwater. Paper presented at *Arsenic—The Geography of a Global Problem, Royal Geographic Society Arsenic Conference*, August 29, 2007, London, England.
2. Von Ehrenstein, O., Poddar, S., Mitra, S., Smith, M.H., Ghosh, N., Rohchowdhuroy, P.K., Yuan, Y. et al. Study design to investigate effects of arsenic on reproduction and child development, in West Bengal, India: ISEE-686. *Epidemiology*, 2003, 14(5):S136.
3. WHO (World Health Organization). *Guidelines for Drinking-Water Quality*-Volume 1: *Recommendations*, 3rd edn. WHO, Geneva, Switzerland, 2008.
4. Sundaram, B., Feitz, A.J., De Caritat, P., Plazinska, A., Brodie, R.S., Coram, J., Ransley, T. Groundwater sampling and analysis—A field guide. *Geoscience Australia*, Record 2009/27, 5.
5. U.S. Geological Survey, Mineral Commodity Summaries 2002, arsenic, 25.
6. Leist, M., Casey, R.J., Caridi, D. The management of arsenic wastes: Problems and prospects. *Journal of Hazardous Materials*, 2000, B76:125–138.
7. U.S. Geological Survey, Mineral Commodity Summaries 2004, arsenic, 24.
8. Riveros, P.A., Dutrizac, J.E., Spencer, P. Arsenic disposal practices in the metallurgical industry. *Canadian Metallurgical Quarterly*, 2001, 40(4):395–420.
9. Ng, J.C., Wang, J., Shraim, A. A global health problem caused by arsenic from natural sources. *Chemosphere*, 2003, 52:1353–1359.
10. Guo, H., Stuben, D., Berner, Z. Removal of arsenic from aqueous solution by natural siderite and hematite. *Applied Geochemistry*, 2007, 22:1039–1051.
11. Hughes, M.F. Arsenic toxicity and potential mechanisms of action. *Toxicology Letters*, 2002, 133:1–16.
12. Abernathy, C.O., Lui, Y.P., Longfellow, D., Aposhian, H.V., Beck, B., Fowler, B., Goyer, R. et al. Arsenic: Health effects, mechanisms of actions, and research issues. *Environmental Health Perspectives*, 1999, 107:593–597.
13. Aposhian, H.V. Biochemical toxicology of arsenic. *Reviews in Biochemical Toxicology*, 1989(10):265–269.

14. Li, D., Morimoto, K., Takeshita, T., Lu, Y. Arsenic induces DNA damage via reactive oxygen species in human cells. *Environmental Health and Preventive Medicine*, 2001, 6:27–32.
15. Hei, T.K., Liu, S.X., Waldren, C. Mutagenicity of arsenic in mammalian cells: Role of reactive oxygen species. *Proceedings of the National Academy of Sciences*, 1998, 95:8103–8107.
16. Sullivan, C., Tyrer, M., Cheeseman, C.R. Disposal of water treatment wastes containing arsenic—A review. *Science of the Total Environment*, 2010, 408:1770–1778.
17. Vu, K.B., Kaminski, M.D., Nunes, L. Review of arsenic removal technologies for contaminated groundwaters. ANL-CMT-03/2. Argonne National Laboratories, Lemont, IL, April 2003.
18. Sharma, V.K., Sohn, M. Aquatic arsenic: Toxicity, speciation, transformations, and remediation. *Environment International*, 2009, 35:743–759.
19. Scott, N., Hatlelid, K.M., MacKenzie, N.E., Carter, D.E. Reactions of arsenic (III) and arsenic(V) species with glutathione. *Chemical Research in Toxicology*, 1993, 6:102–106.
20. Delnomdedieu, M., Basti, M.M., Otvos, J.D., Thomas, D.J. Reduction and binding of arsenate and dimethylarsinate by glutathione: A magnetic resonance study. *Chemico-Biological Interactions*, 1994, 90:139–155.
21. Blair, P.C., Thompson, M.B., Bechtold, M., Wilson, R.E., Moorman, M.P., Fowler, B.A. Evidence for oxidative damage to red blood cells in mice induced by arsine gas. *Toxicology*, 1990, 63(1):25–34.
22. Milne, G.W.A. *CRC Handbook of Pesticides*. CRC Press LLC, Boca Raton, FL, 1995.
23. Marquardt, H., Schafer, S.G., McClellan, R.O., Welsch, F. *Toxicology*. Academic Press, London, U.K., 1999.
24. Harrison, J.W.E., Packman, E.W., Abbott, D.D. Acute oral toxicity and chemical and physical 15 properties of arsenic trioxides. *AMA Archives of Industrial Health*, 1985, 17:118–123.
25. Kaise, T., Watanabe, S., Itoh, K. The acute toxicity of arsenobetaine. *Chemosphere*, 1985, 14:1327–1332.
26. Mezhdunarodnaya Kniga. Labor Hygiene and Occupational Diseases. *Gigiena Trudai Professional'nye Zabolevaniya*, 1984, 28(7):53.
27. Done, A.K., Peart, A.J. Acute toxicities of arsenical herbicides. *Clinical Toxicology*, 1971, 4:343–355.
28. Tadlock, C.H., Aposhian, H.V. Protection of mice against the lethal effects of sodium arsenite by, 3-dimercapto-1-propane-sulfonic acid and dimercaptosuccinic acid. *Biochemical and Biophysical Research Communications*, 1980, 94:501–507.
29. Franke, F.W., Moxon, A.L. A comparison of the minimum fatal doses of selenium, tellurium, arsenic and vanadium. *Journal of Pharmacology and Experimental Therapeutics*, 1936, 58:454–459.
30. Gaines, T.B. The acute toxicity of pesticides to rats. *Toxicology and Applied Pharmacology*, 1960, 2:88–99.
31. RTECS (Registry of Toxic Effects of Chemical Substances). Arsine. DHHS (NIOSH) Publ. No. 87-114. National Institute for Occupational Safety and Health, Cincinnati, OH, 1987.
32. Reddy, M.V.B., Sasikala, P. Arsenic induced nephrotoxicity protective role of essential nutrient supplementation with special reference to some selected enzymes in albino rats. *International Journal of Advanced Scientific and Technical Research*, 2013, 3(3):357–369.
33. Cullen, W.R., Reimer, K.J. Arsenic speciation in the environment. *Chemical Reviews*, 1989, 89:713–764.

34. Styblo, M., Del Razo, L.M., Vega, L., Germolec, D.R., LeCluyse, E.L., Hamil, G.A., Reed, W., Wang, C., Cullen, W.R., and Thomas, D.J. Comparative toxicity of trivalent and pentavalent inorganic and methylated arsenicals in rat and human cells. *Archives of Toxicology*, 2000, 74:289–299.

35. Styblo, M., Serves, S.V., Cullen, W.R., Thomas, D.J. Comparative inhibition of yeast glutathione reductase by arsenicals and arsenothiols. *Chemical Research in Toxicology*, 1997, 10:27–33.

36. Gaines, T.B., Linder, R.E. Acute toxicity of pesticides in adult and weanling rats. *Fundamental and Applied Toxicology*, 1986, 7:299–308.

37. Kaise, T., Yamauchi, H., Horiguchi, Y., Tani, T., Watanabe, S., Hirayama, T., Fukui, S. A comparative study on acute toxicity of methylarsonic acid, dimethylarsinic acid and tri-methylarsine oxide in mice. *Applied Organometallic Chemistry*, 1989, 3:273–277.

38. Stevens, J.T., DiPasquale, L.C., Farmer, J.D. The acute inhalation toxicology of the technical grade organoarsenical herbicides, cacodylic acid and disodium methanearsonic acid. A route comparison. *Bulletin of Environmental Contamination and Toxicology*, 1979, 21:304–311.

39. Confer, A.W., Ward, B.C., Hines, F.A. Arsanilic acid toxicity in rabbits. *Laboratory Animal Science*, 1980, 32(2):234–236.

40. Shiomi, K., Horiguchi, Y., Kaise, T. Acute toxicity and rapid excretion in urine of tetramethylarsonium salts found in some marine animals. *Applied Organometallic Chemistry*, 1988, 2:385–389.

41. Jaghabir, M.T.W., Abdelghani, A., Anderson, A.L. Oral and dermal toxicity of MSMA to New Zealand white rabbits. *Oryctolagus cuniculus*. *Bulletin of Environmental Contamination and Toxicology*, 1988, 40:119–122.

42. Kingston, R.L., Hall, S., Sioris, L. Clinical observations and medical outcomes in 149 cases of arsenate ant killer ingestion. *Journal of Clinical Toxicology*, 1993, 31:581–591.

43. Ellenhorn, M.J. *Ellenhorn's Medical Toxicology: Diagnosis and Treatment of Human Poisoning*, 2nd edn. Williams & Wilkins, Baltimore, MD, 1997.

44. Opresko, D.M. Risk Assessment Information System database, Oak Ridge Reservation Environmental Restoration Program, 1992.

45. Saha, J.C., Dikshit, A.K., Bandyopadhyay, M., Saha, K.C. A review of arsenic poisoning and its effects on human health. *Critical Reviews in Environmental Science and Technology*, 1999, 29(3):281–313.

46. Ng, J.C. Environmental contamination of arsenic and its toxicological impact on humans. *Environmental Chemistry*, 2005, 2:146–160.

47. Vallee, B.L., Ulmer, D.D., Wacker, W.E. Arsenic toxicology and biochemistry. *AMA Archives of Industrial Health*, 1960, 21:132–151.

48. Mass, M.J., Tennant, A., Roop, B.C., Cullen, W.R., Styblo, M., Thomas, D.J. Methylated trivalent arsenic species are genotoxic. *Chemical Research in Toxicology*, 2001, 14:355–361.

49. Petrick, J.S., Ayala-Fierro, F., Cullen, W.R., Carter, D.E., Aposhian, H.V. Monomethylarsonous acid (MMAIII) is more toxic than arsenite in Chang human hepatocytes. *Toxicology and Applied Pharmacology*, 2000, 163:203–207.

50. Hirano, S., Cui, X., Li, S., Kanno, S., Kobayashi, Y., Hayakawa, T. Difference in uptake and toxicity of trivalent and pentavalent inorganic arsenic in rat heart microvessel endothelial cells. *Archives of Toxicology*, 2003, 77:305–312.

51. Dopp, E., Hartmann, L.M., Florea, A.M., Von Recklinghausen, U., Pieper, R., Shokouhi, B. Uptake of inorganic and organic derivatives of arsenic associated with induced cytotoxic and genotoxic effects in Chinese hamster ovary (CH) cells. *Toxicology and Applied Pharmacology*, 2004, 201:156–165.

52. Naranmandura, H., Ibata, K., Suzuki, K.T. Toxicity of dimethylmonothioarsenic acid toward human epidermoid carcinoma A431 cells. *Chemical Research in Toxicology*, 2007, 20:1120–1125.

53. Tseng, W.P. Blackfoot disease in Taiwan: A 30-year follow-up study. *Angiology*, 1989, 40:547–558.

54. Ratnaike, R.N. Acute and chronic arsenic toxicity. *Postgraduate Medical Journal*, 2003, 79(933):391–396.

55. Guha Mazumder, D.N. Chronic arsenic toxicity & human health. *Indian Journal of Medical Research*, 2008, 128:436–447.

56. Cebrián, M.E., Albores, A., Aguilar, M. Chronic arsenic poisoning in the north of Mexico. *Human Toxicology*, 1983, 2:121–133.

57. Hindmarsh, J.T., McLetchie, O.R., Heffernan, L.P.M., Hayne, O.A., Ellenberger, H.A.A., Mccurdy, R.R. Electromyographic abnormalities in chronic environmental arsenicalism. *Journal of Analytical Toxicology*, 1977, 1:270–276.

58. Tseng, W.P. Effects and dose-response relationships of skin cancer and Blackfoot disease with arsenic. *Environmental Health Perspectives*, 1977, 19:109–119.

59. Anwar, J., Arsenic levels in Barisal division alarming, 2004. SOS-arsenic.net. Available at: http://www.sos-arsenic.net/english/latest.html#sec0.1.

60. De, B.K., Majumdar, D., Sen, S., Guru, S., Kundu, S. Pulmonary involvement in chronic arsenic poisoning from drinking contaminated ground-water. *Journal of the Association of Physicians of India*, 2004, 52:395–400.

61. Nevens, F., Fevery, J., Van Streenbergen, W., Sciot, R., Desmet, V., De Groote, J. Arsenic and non-cirrhotic portal hypertension: A report of eight cases. *Journal of Hepatology*, 1990, 11:80–85.

62. Guha Mazumder, D.N., Chakraborty, A.K., Ghosh, A., Das Gupta, J., Chakraborty, D.P., Dey, S.B. Chronic arsenic toxicity from drinking tube-well water in rural West Bengal. *Bulletin of the World Health Organization*, 1988, 66:499–506.

63. Persad, A.S., Cooper, G.S. Use of epidemiologic data in Integrated risk information system (IRIS) assessments. *Toxicology and Applied Pharmacology*, 2008, 233(1):137–145.

64. Guha Mazumder, D., Dasgupta, U.B. Chronic arsenic toxicity: Studies in West Bengal, India. *Kaohsiung Journal of Medical Sciences*, 2011, 27:360–370.

65. Rahman, M., Ng, J., Naidu, R., Rahman, M.M., Ng, J.C., Naidu, R. Chronic exposure of arsenic via drinking water and its adverse health impacts on humans. *Environmental Geochemistry and Health*, 31(S1):189–200.

66. Aguilar, E., Parra, M., Cantillo, L., Gómez, A. Chronic arsenic toxicity in El Zapote. *Medicina Cutanea Ibero-Latino-Americana*, 2000, 28(4):168–173.

67. Joshi, S.R., Bhandari, R.P. Chronic arsenic toxicity: Clinical features, epidemiology, and treatment: Experience in Indo-Nepal border. *Epidemiology*, 2008, 19(6):S349–S350.

68. Cayce, K.A., Feldman, S.R., McMichael, A.J. Hyperpigmentation: A review of common treatment options. *Journal of Drugs in Dermatology*, 2004, 3(6):668–673.

69. ATSDR (Agency for Toxic Substances and Disease Registry). Toxicological Profile for Arsenic. Agency for Toxic Substances and Disease Registry, U.S. Public Health Service, Atlanta, GA. ATSDR/TP-88/02, 1989.

70. ATSDR (Agency for Toxic Substances and Disease Registry). Toxicological Profile for Arsenic. U.S. Department of Health & Human Services, Public Health Service 2007. Available at: http://www.atsdr.cdc.gov/ToxProfiles/tp2.pdf.

71. Biggs, M.L., Kalman, D.A., Moore, L.E., Hopenhayn-Rich, C., Smith, M.T., Smith, A.H. Relationship of urinary arsenic to intake estimates and a biomarker of effect, bladder cell micronuclei. *Mutation Research*, 1997, 386:185–195.

72. Hwang, Y.H., Bornschein, R.L., Grote, J., Menrath, W., Roda, S. Urinary arsenic excretion as a biomarker of arsenic exposure in children. *Archives of Environmental Health*, 1997, 52(2):139–147.
73. Calderon, R.L., Hudgens, E., Le, X.C., Schreinemachers, D., Thomas, D.J. Excretion of arsenic in urine as a function of exposure to arsenic in drinking water. *Environmental Health Perspectives*, 1999, 107(8):663–667.
74. Ishinishi, N., Kodama, Y., Kunitake, E., Nobutomo, K., Urabe, M. Symptoms and diagnosis of poisoning: Arsenic and arsenic compounds (metalloid). *Japanese Journal of Clinical Medicine*, 1973, 31(6):1991–1999.
75. Saha, K.C. Arsenic poisoning from groundwater in West Bengal. *Breakthrough*, 1998, 7(4):5–14.
76. Uede, K., Furukawa, F. Skin manifestations in acute arsenic poisoning from the Wakayama curry-poisoning incident. *The British Journal of Dermatology*, 2003, 149(4):757–762.
77. Yoshida, T., Yamauchi, H., Sun, G.F. Chronic health effects in people exposed to arsenic via the drinking water: Dose-response relationships in review. *Toxicology and Applied Pharmacology*, 2004, 198(3):243–252.
78. Nottage, K., Lanctot, J., Li, Z., Neglia, J.P., Bhatia, S., Hammond, S., Leisenring, W. et al. Long-term risk for subsequent leukemia after treatment for childhood cancer: A report from the Childhood Cancer Survivor Study. *Blood*, 2011, 117(23):6315–6318.
79. WHO (World Health Organization). Toxicological evaluation of certain food additives and contaminants. WHO Food Additives Series, No.18, Arsenic, 1983.
80. Xue, J., Zartarian, V., Wang, S.W., Liu, S.V., Georgopoulos, P. Probabilistic modeling of dietary arsenic exposure and dose and evaluation with 2003–2004 NHANES data. *Environmental Health Perspectives*, 2010, 118:345–350.
81. Yokoi, K., Konomi, A. Toxicity of so-called edible hijiki seaweed (*Sargassum fusiforme*) containing inorganic arsenic. *Regulatory Toxicology and Pharmacology*, 2012, 63(2):291–297.
82. Schoof, R.A., Yost, L.J., Eichhoff, J., Crecelius, E.A., Cragin, D.W., Meacher, D.M., Menzel, D.B. A market basket survey of inorganic arsenic in food. *Food and Chemical Toxicology*, 1999, 37:839–846.
83. Zavala, Y.J., Duxbury, J.M., Arsenic in rice: I. Estimating normal levels of total arsenic in rice grain. *Environmental Science and Technology*, 2008, 42(10):3856–3860.
84. Delgado-Andrade, C., Navarro, M., López, H., López, M.C. Determination of total arsenic levels by hydride generation atomic absorption spectrometry in foods from southeast Spain: Estimation of daily dietary intake. *Food Additives and Contaminants*, 2003, 20(10):923–932.
85. Uneyama, C., Toda, M., Yamamoto, M., Morikawa, K. Arsenic in various foods: Cumulative data. *Food Additives & Contaminants*, 2007, 24(5):447–534.
86. Chen, Y., Graziano, J.H., Parvez, F., Liu, M., Slavkovich, V., Kalra, T., Argos, M. et al. Arsenic exposure from drinking water and mortality from cardiovascular disease in Bangladesh: Prospective cohort study. *British Medical Journal* (Clinical research ed.), 2011, 342:2431–2441.
87. Han, F.X., Su, Y., Monts, D.L., Plodinec, M.J., Banin, A., Triplett, G.E. Assessment of global industrial-age anthropogenic arsenic contamination. *Naturwissenschaften*, 2003, 90:395–401.
88. European Commission DG Environment. Ambient air pollution by As, Cd and Ni compounds. Position paper, final version. Brussels, Belgium, 2000.
89. Tamaki, S. Jr., Frankenberger, W.T. Environmental biochemistry of arsenic. *Reviews of Environmental Contamination and Toxicology*, 1992, 124:79–110.
90. Welch, K., Higgins, I., Oh, M., Burchfiel, C. Arsenic exposure, smoking, and respiratory cancer in copper smelter workers. *Archives of Environmental Health*, 1982, 37(6):325–335.

91. Lowney, Y.W., Ruby, M.V., Wester, R.C., Schoof, R.A., Holm, S.E., Hui, X.Y., Barbadillo, S., Maibach, H.I. Percutaneous absorption of arsenic from environmental media. *Toxicology and Industrial Health*, 2005, 21(1–2):1–14.
92. Banejad, H., Olyaie, E. Arsenic toxicity in the irrigation water-soil-plant system: A significant environmental problem. *Journal of American Science*, 2011, 7(1):125–131.
93. DeGieter, M., Leermakers, M., VanRyssen, R., Noyen, J., Goeyens, L., Baeyens, W. Total and toxic arsenic levels in north sea fish. *Archives of Environmental Contamination and Toxicology*, 2002, 43(4):406–417.
94. CODEX Alimentarius Commission. Proposed draft maximum levels for arsenic in rice. *Joint FAO/WHO Food Standards Programme Codex Committee on Contaminants in Foods*, Sixth Session, March 26–30, 2012, Maastricht, the Netherlands.
95. Marin, A.R., Masscheleyn, P.H. Jr., Patrick, W.H. The influence of chemical form and concentration of arsenic on rice growth and tissue arsenic concentration. *Plant and Soil*, 1992, 139(2):175–183.
96. Hazell, T. Minerals in foods: Dietary sources, chemical forms, interactions, bioavailability. *World Review of Nutrition and Dietetics*, 1985, 46:1–123.
97. Dabeka, R.W., McKenzie, A.D., Lacroix, G.M.A. Dietary intakes of lead, cadmium, arsenic and fluoride by Canadian adults: A 24-hour duplicate diet study. *Food Additives and Contaminants*, 1987, 4:89–101.
98. Panagiotaras, D., Panagopoulos, G., Papoulis, D., Avramidis, P. Arsenic geochemistry in groundwater system. In: Panagiotaras, D., ed., *Geochemistry-Earth's System Processes*, pp. 27–38. Intech, Rijeka, Croatia, 2012.
99. Matschullat, J. Arsenic in the geosphere—A review. *Science of the Total Environment*, 2000, 249:297–312.
100. Chilvers, D.C., Peterson, P.J. Global cycling of arsenic. In: Hutchinson, T.C., Meema, K.M., eds., *Lead, Mercury, Cadmium, and Arsenic in the Environment*, Scientific Committee on Problems of the Environment (SCOPE) 31, pp. 279–301. John Wiley & Sons, Chichester, NY, 1987.
101. Allard, B. Groundwater. In: Salbu, B., Steinnes, E., eds., *Trace Elements in Natural Waters*. CRC Press, Boca Raton, FL, 1995.
102. Walsh, L.M., Keeney, D.R. Behaviour and phytotoxicity of inorganic arsenicals in soils. In: Woolson, E.A., ed., *Arsenical Pesticides* (ACS Symp. Ser. No. 7). American Chemical Society, Washington, DC, 1975.
103. Zhao, F.J., Stroud, J.L., Eagling, T., Dunham, S.J., McGrath, S.P., Shewry, P.R. Accumulation, distribution, and speciation of arsenic in wheat grain. *Environmental Science & Technology*, 2010, 44(14):5464–5468.
104. Tsuji, J.S., VanKerkhove, M.D., Kaetzel, R.S., Scrafford, C.G., Mink, P.J., Barraj, L.M., Crecelius, E.A., Goodman, M. Evaluation of exposure to arsenic in residential soil. *Environmental Health Perspectives*, 2005, 113(12):1735–1740.
105. Lugon-Moulin, N., Martin, F., Krauss, M.R., Ramey, P.B., Rossi, L. Arsenic concentration in tobacco leaves: A study on three commercially important tobacco (*Nicotiana tabacum* L.) types. *Water Air and Soil Pollution*, 2008, 192(1):315–319.
106. Jolliffe, D.M. A history of the use of arsenicals in man. *Journal of the Royal Society of Medicine*, 1993, 86(5):287–289.
107. Wong, S.S., Tan, K.C., Goh, C.L. Cutaneous manifestations of chronic arsenicism: Review of seventeen cases. *Journal of the American Academy of Dermatology*, 1998, 38(2 pt 1):179–185.
108. Yamauchi, H., Takahashi, K., Mashiko, M., Yamamura, Y. Biological monitoring of arsenic exposure of gallium arsenide- and inorganic arsenic-exposed workers by determination of inorganic arsenic and its metabolites in urine and hair. *American Industrial Hygiene Association Journal*, 1989, 50:606–612.

109. Vahter, M., Envall, J. In vivo reduction of arsenate in mice and rabbits. *Environmental Research*, 1983, 32:14–24.
110. Silver, S., Misra, T.K. Bacterial transformations of and resistances to heavy metals. *Basic Life Sciences*, 1984, 28:23–46.
111. Buchet, J.P., Lauwerys, R., Roels, H. Comparison of the urinary excretion of arsenic metabolites after a single oral dose of sodium arsenite, monomethylarsonate, or dimethylarsinate in man. *International Archives of Occupational and Environmental Health*, 1981, 48:71–79.
112. Freeman, G.B., Schoof, R.A., Ruby, M.V., Davis, A.O., Dill, J.A., Liao, S.C., Lapin, C.A., Bergstrom, P.D. Bioavailability of arsenic in soil and house dust impacted by smelter activities following oral administration in cynomolgus monkeys. *Fundamental and Applied Toxicology*, 1995, 28:215–222.
113. Marafante, E., Vahter, M. Solubility, retention, and metabolism of intratracheally and orally administered inorganic arsenic compounds in the hamster. *Environmental Research*, 1987, 42:72–82.
114. Holland, R.H., McCall, M.S., Lanz, H.C. A study of inhaled 74As in man. *Cancer Research*, 1959, 19:1154–1156.
115. Yáñez, J., Fierro, V., Mansilla, H., Figueroa, L., Cornejo, L., Barnes, R.M. Arsenic speciation in human hair: A new perspective for epidemiological assessment in chronic arsenicism. *Journal of Environmental Monitoring*, 2006, 7(12):1335–1341.
116. Samanta, G., Sharma, R., Roychowdhury, T., Chakraborti, D. Arsenic and other elements in hair, nails, and skin-scales of arsenic victims in West Bengal, India. *Science of the Total Environment*, 2004, 326(1–3):33–47.
117. Marquardt, H. *Lehrbuch der Toxikologie*. Spektrum alademischer verlag, Heidelberg, Germany, 1997.
118. Vahter, M. Mechanisms of arsenic biotransformation. *Toxicology*, 2002, 181–182:211–217.
119. Roy, P., Saha, A. Metabolism and toxicity of arsenic: A human carcinogen. *Current Science*, 2002, 82(1):38–45.
120. Hopenhayen-Rich, C., Smith, A.H., Goeden, H.M. Human studies do not support the methylation threshold hypothesis of for the toxicity of inorganic arsenic. *Environmental Research*, 1993, 60:161–177.
121. Maiorino, R.M., Aposhina, H.V. Dimercaptan metal binding agents influence the biotransformation of arsenite in the rabbit. *Toxicology and Applied Pharmacology*, 1985, 77:240–250.
122. Healy, S.M., Casarez, E.A., Ayala-Fierro, F., Aposhian, H.V. Enzymatic methylation of arsenic compounds. V. Arsenite methyltransferase activity in tissues of mice. *Toxicology and Applied Pharmacology*, 1998, 148:65–70.
123. Vather, M., Marafante, E. Reduction and binding of arsenate in marmoset monkeys. *Archives of Toxicology*, 1985, 57:119–124.
124. Vahter, M., Marafante, E., Dencker, L. Metabolism of arsenobetaine in mice, rats and rabbits. *Science of the Total Environment*, 1983, 30:197–211.
125. Zakharyan, R.A., Wildfang, E., Aposhian, H.V. Enzymatic methylation of arsenic compounds. *Toxicology and Applied Pharmacology*, 1996, 140(1):77–84.
126. Francesconi, K., Visoottiviseth, P., Sridokchan, W., Goessler, W. Arsenic species in an arsenic hyperaccumulating fern, *Pityrogramma calomelanos*: A potential phytoremediator of arsenic-contaminated soils. *Science of the Total Environment*, 2002, 284:27–35.
127. Francesconi, K.A., Tanggaar, R., McKenzie, C.J., Goessler, W. Arsenic metabolites in human urine after ingestion of an arsenosugar. *Clinical Chemistry*, 2002, 48(1):92–101.
128. Aposhian, H.V., Gurzau, E.S., Le, X.C., Gurzau, A., Healy, S.M., Lu, X., Ma, M. et al. Occurrence of monomethylarsonous acid in urine of humans exposed to inorganic arsenic. *Chemical Research in Toxicology*, 2000, 13(8):693–697.

129. Del Razo, L.M., Styblo, M., Cullen, W.R., Thomas, D.J. Determination of trivalent methylated arsenicals in biological matrices. *Toxicology and Applied Pharmacology*, 2001, 174:282–293.

130. Gregus, Z., Gyurasics, Á., and Csanaky, I. Biliary and urinary excretion of inorganic arsenic: Monomethylarsonous acid as a major biliary metabolite in rats. *Toxicological Sciences*, 2000, 56:18–25.

131. Petrick, J.S., Jagadish, B., Mash, E.A., and Aposhian, H.V. Monomethylarsonous acid (MMAIII) and arsenite: LD50 in hamsters and in vitro inhibition of pyruvate dehydrogenase. *Chemical Research in Toxicology*, 2001, 14:651–656.

132. Onishi, H., Arsenic. In: Wedepohl, K.H., ed., *Handbook of Geochemistry*, Vol. II-2, Chapter 33. Springer-Verlag, New York, 1969.

133. Hossain, M.F. Arsenic contamination in Bangladesh—An overview. *Agriculture, Ecosystem and Environment*, 2006, 113:1–16.

134. Heikens, A., Panaullah, G.M., Meharg, A.A. Arsenic behaviour from groundwater and soil to crops: Impacts on agriculture and food Safety. *Reviews of Environmental Contamination and Toxicology*, 2007, 189:43–87.

135. Mukherjee, A., Sengupta, M.K., Hossain, M.A., Ahamed, S., Das, B., Nayak, B., Lodh, D., Rahman, M.M., Chakraborti, D. Arsenic contamination in groundwater: A global perspective with emphasis on the Asian scenario. *Journal of Health, Population and Nutrition*, 2006, 24(2):142–163.

136. Polizzotto, M.L., Kocar, B.D., Benner, S.G., Sampson, M., Fendorf, S. Near-surface wetland sediments as a source of arsenic release to groundwater in Asia. *Nature*, 2008, 454:505–508.

137. Barringer, J.L., Reilly, P.A. Arsenic in groundwater: A summary of sources and the biogeochemical and hydrogeologic factors affecting arsenic occurrence and mobility. In: Bradley, P.M., ed., *Current Perspectives in Contaminant Hydrology and Water Resources Sustainability*, Chapter 4, pp. 83–116. InTech, Rijeka, Croatia, 2013.

138. Pearson, M., Jones-Hughes, T., Whear, R., Cooper, C., Peters, J., Evans, E.H., Depledge, M. Are interventions to reduce the impact of arsenic contamination of groundwater on human health in developing countries effective: A systematic review protocol. *Environmental Evidence*, 2011, 1:1–7.

139. Sancha, A.M., Castro, M.L. Arsenic in Latin America: Occurrence, exposure, health effects, and remediation. In: Chappell, W.R., Abernathy, C.O., Calderon, R.L., eds., *Arsenic Exposure and Health Effects*, pp. 87–96. Elsevier, Amsterdam, the Netherlands, 2001.

140. Smith, E., Smith, J., Smith, L., Biswas, T., Correll, R., Naidu, R. Arsenic in Australian environment: An overview. *Journal of Environmental Science and Health*, 2003, 38:223–239.

141. Ahmad, K. Report highlights widespread arsenic contamination in Bangladesh. *The Lancet*, 2001, 358:133.

142. Smedley, P.L., Kinniburgh, D.G. A review of the source, behaviour and distribution of arsenic in natural waters. *Applied Geochemistry*, 2002, 17:517–568.

143. Chakraborti, D., Basu, G.K., Biswas, B.K., Chowdhury, U.K., Rahman, M.M., Paul, K., Chowdhury, T.R., Chanda, C.R., Lodh, D., Ray, S.L. Characterisation of arsenic bearing sediments in Gangetic Delta of West Bengal, India. In: Chappell, W.R., Abernathy, C.O., Calderon, R.L., eds., *Arsenic Exposure and Health Effects*, pp. 27–52. Elsevier, Amsterdam, the Netherlands, 2001.

144. Thakur, J.K., Thakur, R.K., Ramanathan, A.L., Kumar, M., Singh, S.K. Arsenic contamination of groundwater in Nepal—An overview. *Water*, 2011, 3:1–20.

145. Gurzau, E.S., Gurzau, A.E. Arsenic in drinking water from groundwater in Transylvania, Romania: An overview. In: Chappell, W.R., Abernathy, C.O., Calderon, R.L., eds., *Arsenic Exposure and Health Effects*, pp. 181–184. Elsevier, Amsterdam, the Netherlands, 2001.

146. Choprapwon, C., Porapakkham, Y. Occurrence of cancer in arsenic contaminated area, Ronpibool District, Nakhon Si thammarat Province, Thailand. In: Chappell, W.R., Abernathy, C.O., Calderon, R.L., eds., *Arsenic Exposure and Health Effects*, pp. 201–206. Elsevier, Amsterdam, the Netherlands, 2001.
147. Welch, A.H., Lico, M.S., Hughes, J.L. Arsenic in groundwater of the western United States. *Ground Water*, 1988, 26:333–347.
148. Berg, M., Tran, H.C., Nguyen, T.C., Pham, H.V., Schertenleib, R., Giger, W. Arsenic contamination of groundwater and drinking water in Vietnam: A human health threat. *Environmental Science & Technology*, 2001, 35:2621–2626.

4 Geochemistry of Arsenic and Toxic Response

Dionisios Panagiotaras, Dimitrios Papoulis, and Elias Stathatos

CONTENTS

4.1   INTRODUCTION

Arsenic (As) is a trace inorganic element with its origin being from mineral phases in their crystal lattice and/or adsorbed onto minerals' surface as inner- or outer-sphere complexes. The interaction between minerals and water is responsible for the presence of arsenic in the aqueous phases (rivers, stream and sea waters, lakes, and groundwaters). For the rate of water, mineral interaction for As mobilization is dependent on the type of mineral and the biogeochemical conditions like redox potential (Eh), pH, temperature, microbial activity, the speciation and the concentration of metals in the fluid, and the ionic strength of the solution. However, once arsenic is released from the solid phases, it can be sorbed onto the mineral phase, precipitate, redissolved, and biointegrate according to the surrounding environmental conditions.

Arsenic can be found as a contaminant in food and water sources. Shellfish and other seafood, as well as fruits, vegetables, and rice, are the foods most commonly contaminated. Arsenic can pose a risk to human health because of its toxicity on the ppb levels [1,2]. It is a contaminant of concern in many industrial products, wastes, and wastewaters and is widely known for its adverse effects on human health, affecting millions of people all over the world [2].

Arsenic is a known neurotoxin and it is one of the most toxic environmental pollutants [2].

Numerous studies linked arsenic exposure to a number of neuropathological disorders. These include production of ß amyloid [3], hyperphosphorylation of tau protein [4], oxidative stress [5], inflammation [6], endothelial cell dysfunction [7], and angiogenesis [8] and are related to cognitive dysfunction and Alzheimer's disease [9]. In addition, a number of morphologic and neurochemical alterations occurring during arsenic exposure in animals are linked with the hippocampus and other memory-related neuronal structures and expected learning and memory deficits [2,7]. Skin diseases and carcinogenesis are effects of arsenic exposure according to epidemiological studies for humans [10]. Enormous populations exposed to arsenic-contaminated water in Taiwan; Japan; Bangladesh; West Bengal, India; Chile, and Argentina have higher cancer risks for skin and other organs, including the lungs, bladder, kidney, and liver [11]. The U.S. Environmental Protection Agency reduced the maximum contaminant level (MCL) for arsenic in drinking water from 50 to 10 µg/L [12]. In addition, drinking water rich in arsenic over a long period leads to arsenic poisoning or arsenicosis under the World Health Organization concluding remarks [13]. A significant exposure response between arsenic concentration and the mortality from cancers has also been reported [14]. In southwestern Taiwan, it has been well documented as one of the major risk factors for blackfoot disease, a peripheral vascular disease that causes severe gangrene in the extremities and cancer. This disease is considered to be correlated with the consumption of arsenic-contaminated groundwater by local inhabitants of this region [15]. Other studies suggest that it might replace the phosphate in the DNA double helix, which might explain the mutagenic, carcinogenic, and teratogenic effects of arsenic [11]. Nevertheless, intracellular reduction of As(+5) into As(+3) may lead to the formation of free radicals responsible for chromosomal and cellular damage [16].

Because As readily changes valence states and reacts to form species with varying toxicity and mobility, effective long-term treatment of arsenic can be difficult [17]. The application of technologies for arsenic removal depends on the characteristics of the contamination site, while precipitation/coprecipitation is frequently used among other methodologies like adsorption, ion exchange, reverse osmosis, and membrane filtration [18]. In addition, natural and modified absorbents like clays, carbonaceous materials, and oxides of iron, aluminum, and manganese are components that may participate in adsorptive reactions with arsenic [19]. Bioprocesses based on bacterial direct or indirect immobilization of arsenic chemical species include reduction (dissimilatory reduction) and oxidation and methylation of As compounds among the four oxidation states and have been shown to occur in both aquatic and terrestrial systems [18,20]. Several studies define passive and active As biotreatments.

The wastewater treatment of the mining industry can be conducted with a passive mine drainage treatment system consisting of a natural or artificial wetland, which functions as a biological filter. In such a system, bacterial metabolism plays an important role, by reducing metals to nonmobile arsenic forms. Sulfate and iron reduction driven by microbes facilitate the removal of metals that precipitate as hydroxides or sulfides [18]. In some wetlands, phytoremediation is an important removal mechanism of metals, and constructed wetlands, planted or unplanted, were highly efficient to remove arsenic, particularly under sulfate-reducing conditions [18,21].

Active treatments are based on the metabolism of arsenic compounds by the microorganisms. For this purpose, sulfate-reducing bacteria oxidize simple organic compounds by utilizing sulfate as an electron acceptor. The effectiveness of the process is based not only on the precipitation of the metalloid as sulfide metal but also on the biosorption of the As(+3) and As(+5) on the sulfate-reducing bacterial cell pellets leading to removal of significant quantities of arsenic ions [22]. On the other hand, the oxidation of As(+3) can be performed biologically with arsenite-oxidizing bacterial strains. Thus, additional consideration must be made for the removal of the As(+5) ions produced [18,23]. In the case of arsenate-reducing or dissimilatory reducing bacteria, As(+5) is reduced to As(+3) and removed by precipitation or complexation with sulfide [38]. Iron- and manganese-oxidizing bacteria have been used for the precipitation of Fe and Mn oxides, which scavenged As(+5) ions in groundwater purification techniques [25]. Iron and manganese oxides have been reported to oxidize arsenite to arsenate; thus, these processes facilitated by the presence of bacteria are used for As(+3) ion extinction [18,26].

However, the effectiveness of the implementation of the aforementioned biotreatment technologies requires the knowledge of the type of pollutants, for example, industrially used water, wastewater, and groundwater. Nevertheless, the availability of materials and the cost and the capacity for the removal of arsenic chemical species are limiting factors for the use of an effective arsenic treatment technology.

This chapter provides valuable information to help scientists to gain the knowledge and understand the biogeochemical processes controlling arsenic distribution and toxicity in environmental systems.

4.2 ARSENIC GEOCHEMISTRY

4.2.1 Occurrence in Nature

The average content of arsenic in the continental crust varies between 1 and 2 mg/kg [27]. Arsenic occurs naturally in water, air, plants, soil, rocks, and animals, whereas forest fires, volcanic eruptions, and rock erosion are activities in nature that can release arsenic into environmental systems. The arsenic concentrations in crustal materials and the major arsenic minerals occurring in nature are presented in Table 4.1 [1,27].

The concentrations of arsenic in the environment vary with atmospheric concentrations being typically low with mean values in air from remote and rural areas that range from 0.02 to 4 ng/m^3 [28]. In urban areas, the atmospheric arsenic concentrations range from 3 ng/m^3 to about 200 ng/m^3 with higher values (>1000 ng/m^3) measured in the vicinity of industrial sources. Concentrations of arsenic in open ocean seawater are typically 1–2 μg/L [28]. Arsenic is widely distributed in surface freshwaters, and concentrations in rivers and lakes are generally below 10 μg/L, although individual samples may range up to 5 mg/L near anthropogenic

TABLE 4.1

Arsenic Concentrations in Crustal Materials, Major and Secondary Arsenic Minerals Occurring in Nature

Materials	Concentration As (mg/kg)	Process	Reference
Igneous material		Cooling and solidification of magma or lava	[27]
Basalt	<1–113		
Ultrabasics	<1–16		
Granites	<1–15		
Sedimentary material		Formed by the deposition of material (organic and/or minerals) at the Earth's surface and within bodies of water	
Shales and clays	<1–500		
Sandstones	<1–120		
Limestones	<1–20		
Phosphorites	3–100		

Major Arsenic Minerals

Mineral	Chemical Formula	Occurrence	[1]
Native arsenic	As	Hydrothermal veins	
Niccolite	NiAs	Vein deposits and norites	
Realgar	AsS	Vein deposits, often associated with orpiment, clays, and limestones, also deposits from hot springs	
Orpiment	As_2S_3	Hydrothermal veins, hot springs, volcanic sublimation products	
Cobaltite	CoAsS	High-temperature deposits, metamorphic rocks	
Arsenopyrite	FeAsS	The most abundant As mineral, dominantly in mineral veins	
Tennantite	$(Cu,Fe)_{12}As_4S_{13}$	Hydrothermal veins	
Enargite	Cu_3AsS_4	Hydrothermal veins	

Secondary Arsenic Minerals

Mineral	Chemical Formula	Occurrence
Arsenolite	As_2O_3	Secondary mineral formed by oxidation of arsenopyrite native arsenic and other As minerals
Clauderite	As_2O_3	Secondary mineral formed by oxidation of realgar arsenopyrite and other As minerals
Scorodite	$FeAsO_4 \cdot 2H_2O$	Secondary mineral
Annabergite	$(Ni,Co)_3(AsO_4)_2 \cdot 8H_2O$	Secondary mineral
Hoernesite	$Mg_3(AsO_4)_2 \cdot 8H_2O$	Secondary mineral, smelter wastes
Hematolite	$(Mn,Mg)_4Al(AsO_4)(OH)_8$	Secondary mineral
Conichalcite	$CaCu(AsO_4)(OH)$	Secondary mineral
Pharmacosiderite	$Fe_3(AsO_4)_2(OH)_3 \cdot 5H_2O$	Oxidation product of arsenopyrite and other As minerals

sources [28]. Arsenic levels in groundwater average about 1–2 μg/L except in areas with volcanic rock and sulfide mineral deposits where arsenic levels can range up to 3 mg/L [28].

However, in polluted waters, for example, in the United States, India, and China, the concentration exceeds the MCL of 10 μg/L for arsenic. In Bangladesh, about 30% of the groundwater sources exceed the 50 μg/L for arsenic concentration, while about 10% of the concentration in the U.S. borehole waters exceed 10 μg/L [29]. In Xinjiang (China), arsenic concentration in groundwaters ranges from 50 to 1860 μg/L [30]. Another disconcerting issue is that high arsenic bioaccumulation rates result to its gathering into the tissues of aquatic biota in a range of 0.007–125.9 μg/g dry weight [31].

The most common sources of arsenic in natural waters are As-containing minerals and pesticides like monosodium methane arsenate, $HAsO_3CH_3Na$; disodium methane arsenate, $Na_2AsO_3CH_3$; dimethylarsinic acid (DMAA, cacodylic acid), $(CH_3)_2AsO_2H$; and arsenic acid, H_3AsO_4 [32].

In sedimentary materials, the mean arsenic concentration range from 1.7 to 400 mg/kg [27]. Background concentrations in soil range from 1 to 40 mg/kg, with mean values often around 5 mg/kg [28]. In soil and sediment located near industrial activity or mining operations, arsenic is observed to be in much higher concentrations [1]. For example, an average arsenic concentration of 903 mg/kg was detected in mine tailings in British Columbia [33]. Marine organisms normally contain arsenic residues ranging from <1 mg/kg to more than 100 mg/kg, predominantly as organic arsenic species such as arsenosugars (macroalgae) and arsenobetaine (invertebrates and fish) [28].

Anthropogenic input of arsenic in the environment includes mining and smelting operations, agricultural applications, burning of fossil fuels and wastes, pulp and paper production, cement manufacturing, and former agricultural uses of arsenic. Other uses of arsenic and its compounds are in wood preservatives, glass manufacture, alloys, electronics, catalysts, feed additives, and veterinary chemicals [34].

A number of biogeochemical processes are responsible for the partitioning of As between solid and dissolved phases, whereas several reactions, including dissolution/precipitation, adsorption/coprecipitation, and reduction/oxidation, control the mobilization of arsenic in environmental systems.

4.2.2 GEOCHEMICAL PROCESSES AND TOXIC RESPONSE

The distribution of arsenic in the natural environment is controlled by a combination of the existing physicochemical conditions, the biological activity, and the type of the primary arsenic source (Figure 4.1). In reducing conditions, arsenite forms insoluble sulfide precipitates such as arsenopyrite (FeAsS), realgar (AsS), and orpiment (As_2S_3) [35]. Arsenic is also found in sedimentary environments adsorbed by Fe(+3) and Mn(+4) oxides—hydroxides after weathering of the sulfide minerals. For example, the electron transfer from the arsenopyrite surface by the ferric hydroxide results to the formation of arsenic oxyanions, which are immobilized by the Fe(+3) hydroxides according to the following equation (4.1),

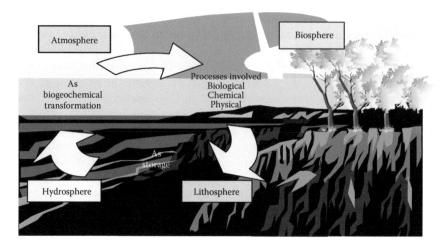

FIGURE 4.1 **(See color insert.)** Simplified geochemical cycle of As in the natural environment.

where x is less than one and the arsenic is adsorbed or coprecipitated with the ferrosoferric hydroxides [36]:

$$FeAsS + Fe(OH)_3 \rightarrow [Fe_x^{2+}Fe^{3+}(OH)_3^{x+}](AsO_4, AsO_3)_x + Products \quad (4.1)$$

The shuttling of electrons between the oxidizing arsenopyrite surface and the ferric hydroxide is facilitated by chemical and biological pathways [36]. Therefore, high levels of arsenic in natural waters come from the reductive dissolution of arsenic-rich iron oxyhydroxides [18,37]. In addition, oxidative dissolution of arsenic-rich pyrite or arsenopyrite is responsible for As existence in natural waters [38]. In this case, the concentration of dissolved oxygen (DO) is the limiting factor for arsenopyrite dissolution in a variety of pH ranges [39]. Arsenic release rates were found to increase with increasing DO concentration and temperature and were similar at low (<7) and high (>10) pH. Oxidation rates were found to pass through a minimum of pH 7–8, which may reflect a switch in the main oxidant species from aqueous Fe(+3) (at low pH) to DO (at high pH). The authors suggested that high activation energies at high pH and DO concentration are consistent with a surface reaction being the rate-determining step [40], whereas the reactions involved in arsenopyrite dissolution reported by Walker et al. [41] are not DO dependent in a neutral pH because of the small range of DO for their experiments lying very close to the experimental error. However, the reaction includes FeAsS dissolution (Equation 4.2) and As(+3) present as H_3AsO_3, while Fe(+2) is further oxidized to $Fe(OH)_3$ (Equation 4.3), with an increase of acidity in the solution [40]:

$$4FeAsS + 11O_2 + 6H_2O \rightarrow 4Fe^{2+} + H_3AsO_3 + 4H_2SO_4^{2-} \quad (4.2)$$

$$4Fe^{2+} + O_2 + 10H_2O \rightarrow 4Fe(OH)_3 + 8H^+ \quad (4.3)$$

In addition, As(+3) further oxidized to As(+5) is presented as $HAsO_4^{-2}$ and $H_2AsO_4^{-2}$ in approximately equal portions (Equations 4.4 and 4.5):

$$2H_3AsO_3 + O_2 \rightarrow 2HAsO_4^{2-} + 4H^+ \tag{4.4}$$

$$2H_3AsO_3 + O_2 \rightarrow 2H_2AsO_4^{2-} + 2H^+ \tag{4.5}$$

The widespread arsenic contamination is thought to be related with As release from iron oxyhydroxides, probably due to the reaction of Fe oxides/hydroxides with organic carbon [1,29]. In such a case, the source of As is adsorbed onto the surface of Fe oxide/hydroxide solid phases, and a parallel release of arsenic during the reductive dissolution of ferric oxides–hydroxides occurs (Equation 4.6) [37]:

$$FeOOH + CH_2O + 7H_2CO_3 \rightarrow 4Fe^{2+} + 8HCO_3^- + 6H_2O \tag{4.6}$$

Oxidation of sulfide minerals such as pyrite is also an important source of arsenic and has been identified as the primary source of arsenic contamination in aquifers [38].

Bioprocess involved in the oxidation/reduction and bioavailability of arsenic chemical species is of great importance in nature and is a result of the metabolism of microorganisms [18]. In the case of reducing bacteria, the process is the reduction of As(+5) to As(+3), which is related to the detoxification of the cells. Arsenate ions enter the cells via phosphate transporters (Pst, high-affinity phosphate-specific transport, or Pit, low-affinity phosphate-inorganic transport), due to structural homologies with phosphate ions. After reaching the cytoplasm, arsenate is reduced into arsenite by the arsenate–reductase enzyme ArsC before being excreted from the cell by transmembrane protein ArsB (which is also known as Acr3 in the context of some eukaryotic microorganisms) or the ArsAB complex. This transformation process followed by the excretion of arsenic is a common occurrence in the living world and is widespread in bacteria [18,42]. Furthermore, phylogenetic groups of bacteria gain metabolic energy utilizing As(+5) as an electron acceptor [18]. Bacterial As(+5) reduction by "breathing" arsenate to arsenite (consisting of two subunits, ArrA and ArrB proteins) has been identified as a membrane-bound heterodimer protein [18,43]. These two bioprocesses for arsenate reduction by microorganisms (the ArsC, for detoxification and the dissimilatory arsenate reductase ArrA) are shown in Figure 4.2.

In microbes, where the ArsC cytoplasmic arsenate reductase (Figure 4.2a) evolves, the arsC gene occurs in ars operons in most bacteria. Through a phosphate uptake channel, Pst or Pit, arsenate ion enters the cell. In the cytoplasm, arsenate is reduced by ArsC and arsenite is chaperoned to an ATPase-activated pump (ArsB). In addition, the total genomes measure 2 Mb or larger as well as in some archaeal genomes [44]. Whereas respiratory arsenate reductase occurs, As(+5) is reduced as a terminal electron acceptor during the anoxic respiration coupled to the oxidation of simple organic substrates (Figure 4.2b). Electron donors known to support dissimilatory arsenate reducing microorganisms are organic acids (acetate, citrate, lactate, malate, etc.), alcohol and sugar (ethanol, glycerol, and glucose), aromatic substances (phenol, toluene, ferulate, etc.), and inorganic constituents like H_2 and H_2S [45]. The reactions involved in the partial and complete oxidation of organic matter (OM)

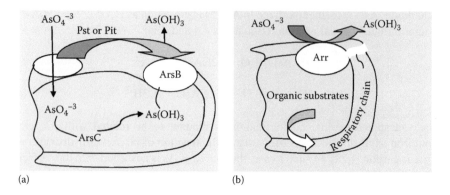

(a)   (b)

FIGURE 4.2 **(See color insert.)** Cellular functions and locations of bacterial (a) cytoplasmic arsenate reductase and (b) respiratory arsenate reductase. (Modified from Silver, S. and Phung, L.T., *J. Ind. Microbiol. Biotechnol.*, 32(11–12), 2005, 587.)

during the microbial arsenate reduction are shown in Equation 4.7 (partial oxidation of OM) [46] and Equation 4.8 (complete oxidation of OM) [47]:

$$CH_3CHOHCOO^- + 2HAsO_4^{-2} + 4H^+ \rightarrow CH_3COO^- + 2HAsO_2 + CO_2\,(g) + 3H_2O \quad (4.7)$$

Lactate ion   As(+5)   Acetate ion   As(+3)

$$CH_2O + 2H_2AsO_4^- \rightarrow HCO_3^- + 2H_2AsO_3^- + H^+ \quad (4.8)$$

As(+5)   As(+3)

By contrast, arsenic-oxidizing bacteria oxidize As(+3) enzymatically to produce arsenite oxidases. Photoautotrophic, heterotrophic, and chemoautotrophic microorganisms reduce oxygen or nitrate using arsenite as the electron donor. In this case, the energy adduced is used for CO_2 production required for carbon generation and the growth of bacterial community [18,48]. The oxidation of the As(+3) species to the less bioavailable As(+5) compounds is crucial for the detoxification process observed in extreme natural environments and could be one of the primary energy resources for the metabolism of chemolithotrophic organisms during the early history of Earth's very first forms of life [18,48]. Microorganisms acting as arsenite oxidizers can facilitate As (+3) oxidation in aerobic environments (O_2 presence) and in anoxic condition using other ions as electron acceptors in the order of $NO_3^- \rightarrow$ Mn oxides $\rightarrow$ Fe(III) oxides $\rightarrow$ sulfate [49]. For example, the microbial oxidation of FeAsS in the presence of oxygen results in the production of arsenious (H_3AsO_3) and arsenic acids (H_3AsO_4) as described by Collinet and Morin [50]. However, the reaction of As(+3) with DO can be interceded by heterotrophic manganese-oxidizing bacteria according to the mechanism (Equations 4.9 and 4.10) proposed by Katsoyiannis et al. [51]:

$$H_3AsO_3 + O_2 + 2H^+ + 2e^- \text{ (catalyzed oxidation)} \rightarrow H_3AsO_4 + H_2O \quad (4.9)$$

$$2(\equiv Mn-OH) + H_3AsO_4 \text{ (sorption)} \rightarrow (\equiv MnO)_2AsO_2 + 2H_2O \quad (4.10)$$

According to Equations 4.9 and 4.10, during the biological oxidation of manganese, the bacteria *Leptothrix ochracea* release a catalase that acts as a peroxidase, catalyzing the formation of hydrogen peroxide. After H_2O_2 formation, oxidation of As(+3) occurs because hydrogen peroxide is a strong oxidant for arsenite ions [51].

Another mechanism refers to the sorption of As(+3) onto Mn oxide/hydroxide phases, which has been reported by Panagopoulos and Panagiotaras [52] in order to delineate controlling geochemical processes in the groundwater pool of the Trifilia karst aquifer, Western Greece. The proposed mechanism is described by the following equation (4.11):

$$\equiv MnOH + As^{3+} \rightarrow \equiv MnOAs^{2+} + H^+ \qquad (4.11)$$

However, adsorbed arsenic species are weak acids and can affect the surface charge due to proton exchange reactions. Whether As adsorbs as a mononuclear or binuclear complex has implications for the level of protonation of the surface species, where this mechanism is elucidated in Equations 4.12 and 4.13 in the case of Fe oxide/hydroxide surfaces [53–55]:

$$\equiv FeOH + H_3AsO_3 \rightarrow \equiv FeH_xAsO_3^{x-2} + H_2O + (2-x)H^+ \text{ where } x = 0\text{--}2 \quad (4.12)$$

$$(\equiv FeOH)_2 + H_3AsO_3 \rightarrow (\equiv Fe)_2H_yAsO_3^{y-1} + 2H_2O + (1-y)H^+ \text{ where } y = 0, 1 \quad (4.13)$$

Therefore, further reductive dissolution of ferric oxides–hydroxides by OM as described in Equation 4.6 contributes to the cycling of arsenic species into the environment [37].

The geochemistry of arsenic is complex and lends itself to energy transformations useful for microbial metabolism. The overall redox pathways, both microbially intervened and chemical, are illustrated in Figure 4.3. The understanding of how the metabolic and geochemical processes potentially influence arsenic mobility, actually generate specific patterns of distribution and speciation including among others Fe and Mn solid phases.

Arsenic can reside in a number of oxidation states and complex ions with negative to neutral charges. This allows a wide array of species that can be utilized during

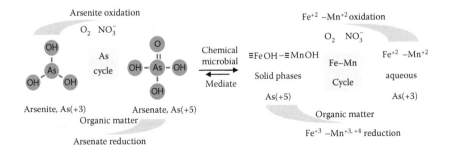

FIGURE 4.3   Geochemical redox arsenic transformation.

redox and complexation reactions so that arsenic undergoes a variety of microbially mediated transformations (Figure 4.3).

The toxicity of arsenic chemical species is considered to be related with the oxidation state of arsenic in both inorganic and methylated forms. Nevertheless, As(+3) compounds are more genotoxic and cytotoxic compared to the As(+5) homologue forms [56]. For example, monomethylarsonous acid [MMA(III)] is a carcinogen for mice [57].

Two mechanisms are involved in arsenic toxicity and are known as cytotoxic and genotoxic. Inactivation of cell enzymes and allosteric inhibition cause cell death during the cytotoxic mechanism [58]. This is related to ester production of arsenate with adenosine diphosphate and the hydrolysis (arsenolysis) of this ester that inhibits energy production and results to cell death [11]. During the genotoxic mechanism, genetic mutations and chromosomal alterations occur [59].

There are many studies that relate the carcinogenicity in humans with arsenic exposure; however, the carcinogenic effects are dependent on the period of exposure and the arsenic concentration and forms [60]. Arsenic has also been shown to interfere with hormonal processes, including estrogen, testosterone, and progesterone biochemical pathways and the immune response [61]. Furthermore, toxic levels lead to impairment of mitochondrial function, which results in optic neuropathy and peripheral neuropathy [62]. Neuropathological disorders are associated with Alzheimer's disease and cognitive dysfunction, with the most conspicuous being angiogenesis, inflammation, hyperphosphorylation of tau protein, oxidative stress, endothelial cell dysfunction, and ß amyloid production [3–5,7,8].

Although arsenic toxicity is a complex issue, the use of As-enriched water may result in skin lesions; chromosomal changes, as well as skin, lung, and bladder cancers; disturbance of the cardiovascular and nervous system functions; and eventually death [63]. Large populations around the world are at risk for toxic As exposure from drinking water and through the consumption of As-rich seafood [64]. In West Bengal, India, people are suffering from arsenicosis and other types of related diseases due to As toxicity. The signs of chronic As toxicity are manifested as melanosis, edema, skin lesions, keratosis, gastrointestinal problems, asthma, bronchial troubles, and cancer of the liver, skin, and limbs [65]. Afterward, the use of contaminated groundwater in aquaculture fish ponds results to the bioaccumulation of arsenic into fishes and shrimps. Some studies have noted concentrations >329 µg/g of total arsenic in cultured fish [66]. Consumption of arsenic-contaminated fish could result in arsenic exposure to humans and lead to adverse health effects [15,67].

Although it is not possible to present all the research on arsenic toxicity and effects, a thorough literature of arsenic toxicities and exposure effects was comprehensively presented in the work of Mudhoo et al. [59] and Schwarzenegger et al. [68].

4.3   CHEMICAL SPECIATION AND THERMODYNAMIC PROPERTIES

Arsenic is apparent as organic and inorganic chemical species with valence states of −3, 0, +3 (arsenite), and +5 (arsenate). The +3 and +5 states are the most common in natural systems. Arsenic in the (−3) oxidation state (arsine) is an extremely toxic compound and can be formed under very reducing conditions, but its occurrence in nature is relatively rare. Inorganic and organic species of As are present in the natural environment, with inorganic forms being typically more abundant in freshwater systems. Mobility and toxicity of arsenic chemical species are dependent on the oxidation state and structure. The inorganic species are more toxic than organic species, and inorganic As(+3) has a higher acute toxicity for mammals than As(+5) [69]. The effects of oxidation state on chronic toxicity are related with the redox conversion of As(+3) and As(+5) within human cells and tissues.

4.3.1 Inorganic Speciation

Inorganic compounds of arsenic include hydrides (e.g., arsine), halides, oxides, acids, and sulfides. Arsenate [As(+5); $HxAsO_4^{x-3}$, $x=0$–3] and arsenite [As(+3); $HxAsO_3^{x-3}$, $x=0$–3] are the two most common inorganic forms of arsenic in freshwater systems. In aqueous systems, arsenic exhibits anionic behavior. As(+5) is thermodynamically stable under oxic conditions, while As(+3) is stable under more reducing conditions. However, As(+5) and As(+3) are often found in both oxic and anoxic waters and sediments [70]. The oxidation of As(+3) by O_2 is slow (on the order of several weeks), while bacterially mediated redox reactions can be much faster [71,72].

Arsenate is an anion at the pH of most natural waters ($H_2AsO_4^-$ and $HAsO_4^{2-}$), while arsenite is a neutral species. The pK_a values for arsenate (H_3AsO_4) are $pK_{a1}=2.19$, $pK_{a2}=6.94$, and $pK_{a3}=11.5$ according to the following equations (4.14 through 4.16):

$$H_3AsO_4 = H_2AsO_4^- + H^+ \tag{4.14}$$

$$H_2AsO_4^- = HAsO_4^{-2} + H^+ \tag{4.15}$$

$$HAsO_4^{-2} = AsO_4^{-3} + H^+ \tag{4.16}$$

$$H_3AsO_3 = H_2AsO_3^- + H^+ \tag{4.17}$$

In oxidative environments, the form $H_2AsO_4^-$ predominates with pH values below 6.9, whereas the $HAsO_4^{-2}$ ions predominate at higher pH levels. Arsenite (H_3AsO_3) pK_a value according to Equation 4.17 is equal to 9.22, while it is the main arsenic chemical species in natural waters with pH<9 and in slightly reducing conditions [18].

In aerobic waters, arsenic acid predominates only at extremely low pH (<2). At the pH range of 2–11, it is replaced by $H_2AsO_4^-$ and $HAsO_4^{2-}$ ions. Arsenious acid appears at low pH and under mildly reduced conditions, but it is replaced by $H_2AsO_3^-$ as the pH increases. Only when the pH exceeds 12 does $HAsO_3^{2-}$ ion appear (Figure 4.4). The $HAsS_2$ arsenic chemical species can form at low pH in the presence of sulfide ions. Arsine derivatives and arsenic metal can occur under extreme reducing conditions [73]. Figure 4.4 shows the speciation of arsenic under varying pH values and redox conditions.

Arsenic changes its valence state and chemical form in the environment. In the pH range of 4–10, As(+5) species are negatively charged in water, and the predominant

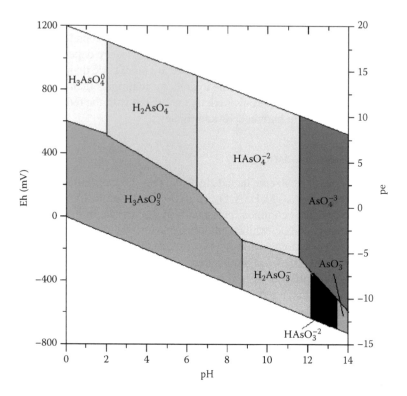

FIGURE 4.4 (See color insert.) Eh–pH diagram of aqueous arsenic species in water at 25°C and 1 bar total pressure. (Modified from Smedley, P. and Kinniburgh, D., *Appl. Geochem.*, 17(5), 2002, 517.)

As(+3) species is neutral in charge (Figure 4.4). Critical parameters for Eh–pH diagram construction and arsenic speciation are the pH and the redox potential. If we take into consideration the half reaction for the reduction of As(+5) to As(+3) in Equation 4.18, then the Nernst equation (4.19) relates the arsenic species concentrations at equilibrium [74]:

$$H_3AsO_4 + 2H^+ + 2e^- = H_3AsO_3 + H_2O \tag{4.18}$$

$$E = E^0 + k \left(Log_{10} \left(\frac{H_3AsO_4}{H_3AsO_3} \right) - 2pH \right) \tag{4.19}$$

In Equation 4.19, E^0 and k are a constant and a collection of constants, respectively. The dependence of the As(+5) fraction with the pH of the studied aquatic system is given by Equation 4.20 as proposed by Stumm and Morgan [49]:

$$a^{+5} = \frac{\left[H_3AsO_4 \right]}{As(+5)} = \frac{1}{1 + K_{a1}10^{pH} + K_{a1}K_{a2}10^{2pH} + K_{a1}K_{a2}K_{a3}10^{3pH}} \tag{4.20}$$

A similar equation can be derived for the fraction of As(+3) in the H_3AsO_3 form (α^{+3}) due to the following equation (4.21):

$$a^{+3} = \frac{[H_3AsO_3]}{As(+3)} = \frac{1}{1 + K_{a1}10^{pH} + K_{a1}K_{a2}10^{2pH} + K_{a1}K_{a2}K_{a3}10^{3pH}} \quad (4.21)$$

Substitution of Equations 4.20 and 4.21 into Equation 4.19 gives Equation 4.22, which relates the equilibrium redox potential to the pH and concentrations of As(+5) and As(+3) measured quantities:

$$E = E^0 + k\left(Log_{10}\left(\frac{As(+5)a^{+5}}{As(+3)a^{+3}} \right) - 2pH \right) \quad (4.22)$$

The major conditions responsible for different arsenic valance and chemical species are redox potential; the presence of complexing ions, such as ions of sulfur, iron, and calcium; and microbial activity. Since arsenic forms anions in solution, it does not form complexes with simple anions like Cl^- and SO_4^{2-} as do cationic metals, whereas anionic arsenic complexes behave like ligands in water.

4.3.2 ORGANIC SPECIATION AND METHYLATION

Arsenic reacts with organic carbon, sulfur, and nitrogen. As(+3) forms bonds with sulfur and sulfhydryl groups such as cystine, organic dithiols, proteins, and enzymes, but it does not react with amine groups or organics with reduced nitrogen constituents [32]. In addition, As(+5) reacts with reduced nitrogen groups, for example, amines, but not sulfhydryl groups. Both the trivalent and pentavalent forms of arsenic react with carbon and produce organoarsenicals. The complexation of arsenic (+3 and +5) by dissolved OM in natural environments prevents sorption and coprecipitation with solid-phase organic and inorganic compounds. Therefore, it increases the mobility of arsenic in aquatic systems and in the soil [75].

Although there are many forms of organic As chemical species, the most common organoarsenic are monomethylated and dimethylated As(+3) and As(+5). Organic As compounds are less acutely toxic than the inorganic species, and methylation of inorganic As is one type of detoxification mechanism for some bacteria, fungi, phytoplankton, and higher level organisms such as humans. Methylation can also occur when organisms are stressed from nutrient limitation [69,71]. Organoarsenicals are typically direct metabolic products and should not be confused with inorganic As complexed with natural OM. Both volatile (e.g., methylarsines) and nonvolatile (e.g., monomethylarsonate and dimethylarsinate) compounds include the redox cycling between As(+3) and As(+5) oxidative states. The biotransformation–biomethylation of arsenic to trimethylarsine was verified in fungi by the work of Challenger et al. [76]. The oxidative addition followed by the reductive expulsion is given in the sequence of equations illustrated in Figure 4.5, whereas substitutions of oxygen atoms by methyl groups occur. This scheme is recognized as enzymatically catalyzed

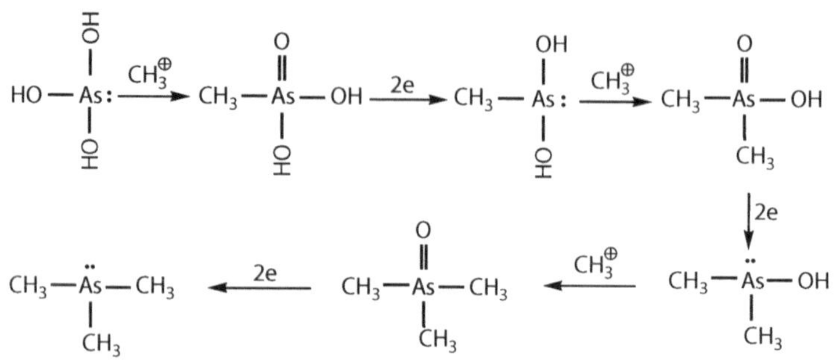

$$HO-As: \xrightarrow{CH_3^{\oplus}} CH_3-As-OH \xrightarrow{2e} CH_3-As: \xrightarrow{CH_3^{\oplus}} CH_3-As-OH$$

$$CH_3-As-CH_3 \xleftarrow{2e} CH_3-As-CH_3 \xleftarrow{CH_3^{\oplus}} CH_3-As-OH$$

FIGURE 4.5 Challenger's sequence of biotransformation—biological methylation of arsenic. (From Challenger, F. et al., *J. Chem. Soc.*, 1, 1933, 95.)

methylation, and the "active methionine," later identified as S-adenosylmethionine, was the methyl group donor.

However, questions about the validity of Challenger's biological pathway for arsenic methylation have prompted speculation about alternative pathways for arsenic biotransformation. Any proposed pathway must be chemically justifiable and this means that each proposed step in the reaction scheme must conform to our erudition of chemical kinetics and thermodynamics [56].

Nevertheless, methylation is thought to be a detoxification mechanism for microbiota and is of great importance for the biogeochemical cycling of arsenic in nature [32]. In accordance to the reductive transformation of arsenate to dimethylarsine by *methanobacterium* in anaerobic conditions, under acidic conditions, the sewage fungi *Candida humicola* transform arsenate to trimethylarsine [32].

Once organic As substances are formed, they can be adsorbed onto mineral surfaces, a reaction that is strongly dependent on pH. In adsorption studies with monomethylarsonic acid (MMAA) and dimethlyarsinic acid (DMAA), the affinity of the organoarsenic species for Fe oxides was observed to be less than that of As(+5) and greater than that of As(+3), that is, below pH=7, but less than that of both inorganic forms of As, that is, above pH=7 [77]. In sediments having circumneutral pH, organoarsenicals may be more mobile than inorganic forms. Significant levels of organoarsenic compounds can be found seasonally in surface waters, but organoarsenicals contribute less than 10% of the total As in interstitial water except in isolated instances [70,71].

Biotransformation and redox reactions of arsenic compounds occurring in nature are of great importance, and their interaction with microorganisms, other biota, and natural products regulates the mobility and toxicity of organoarsenicals in environmental systems.

4.3.3 Stability of Arsenic Minerals

The primary source of arsenic in the natural environment is arsenic minerals. Therefore, the stability of these minerals is a controlling factor for arsenic occurrence in nature. The interaction of these primary minerals with the environmental conditions results either to their dissolution and/or to the formation of secondary As minerals (Table 4.1).

Nowadays, there has been a significant increase in the development of theoretical, computer-based models that enable prediction of the physical and thermodynamic properties of minerals. Thermodynamically, the stability of the primary and secondary As minerals is a function of their free energy (G) as well as the enthalpy (H), entropy (S), and volume (V) in a given temperature and pressure and, in the case of solid solution, the composition. However, the aim is to distinguish the most stable structure at a specific temperature, pressure, and composition.

The standard molal Gibbs free energies and enthalpies of minerals, gases, and aqueous species are represented as apparent standard molal Gibbs free energies ($\Delta G^0_{P,\,T}$) and enthalpies ($\Delta H^0_{P,\,T}$) of formation from the elements at the subscripted pressure (P) and temperature (T). These apparent standard molal properties can be written as Equations 4.23 and 4.24 [78]:

$$\Delta G^0_{P,\,T} = \Delta_f G^0 + (G^0_{P,\,T} - G^0_{Pr,\,Tr}) \tag{4.23}$$

$$\Delta H^0_{P,\,T} = \Delta_f H^0 + (H^0_{P,\,T} - H^0_{Pr,\,Tr}) \tag{4.24}$$

The $\Delta_f G^0$ and $\Delta_f H^0$ indicate the standard molal Gibbs free energy and enthalpy of formation of the species from its elements in their stable phase at the reference pressure ($Pr = 1$ bar) and temperature ($Tr = 298.15$ K). The $G^0_{P,\,T} - G^0_{Pr,\,Tr}$ and the $H^0_{P,\,T} - H^0_{Pr,\,Tr}$ are differences in the standard molal Gibbs free energy and enthalpy of the species that arise from changes in pressure ($P - Pr$) and temperature ($T - Tr$) [78].

According to thermodynamic laws, the most stable structure of a given composition and in a specific pressure and temperature is the one with the lowest Gibbs free energy. However, understanding the macroscopic behavior of rocks and minerals can only be obtained from a detailed knowledge of their microscopic or atomistic nature because of the atomic level response to changes in pressure and temperature. For example, the lattice vibrations of a crystal are linked with heat capacity, enthalpy, etc., while the frequencies of atomic vibrations are determined by the strength and nature of the bonding that holds the atoms in that crystal together. Moreover, these bonding forces can determine not only the vibrational characteristics of a crystal but also its structure and physical properties, such as its elastic and dielectric behavior. In addition, lattice vibrations enable determination of the Gibbs free energy and other minerals' thermodynamic functions, in a given pressure and temperature from which it is possible to determine phase relations between different minerals.

Numerous compilations of thermodynamic data are available for arsenic species (Tables 4.2 through 4.4). However, we report the data from Nordstrom and Archer and references therein [35], whose evaluation has been done by screening and selecting the available data from the literature and who performed a simultaneous weighted multiple least-squares regression on substances of mineral reactions in the As–S–O–H_2O system, with respect to arsenic oxides, sulfides, and their aqueous hydrolysis products.

TABLE 4.2

Thermodynamic Data for Arsenic Oxides and Oxidizing Reactions for Arsenic to Arsenic Trioxides

Species or Reaction Data for Standard State Conditions, 298.15 K and 1 atm	$\Delta_f G^0$(kJ/mol)	$\Delta_f H^0$(kJ/mol)	$\Delta_r S^0$(J/K/mol)	C_p^0(J/K/mol)
As(cr)	0.0	0.0	35.63	24.43
O_2(g)	0.0	0.0	205.029	29.355
H_2(g)	0.0	0.0	130.574	28.824
H_2O(l)	−237.178	−285.830	69.91	75.29
As_2O_3(cubic, arsenolite)	−576.34	−657.27	107.38	96.88
As_2O_3(monoclinic, claudetite)	−576.53	−655.67	113.37	96.98
As_2O_5(cr)	−774.96	−917.59	105.44	115.9
As_2O_3(cubic)$+O_2$(g)$=As_2O_5$(cr)	−198.62	−260.32	−206.97	−10.3
$2As(\alpha,cr)+3H_2O$(l)$=$ As_2O_3(cubic)$ + 3H_2$(g)	135.19	200.22	218.11	−91.38
$2As(\alpha,cr)+3H_2O$(l)$=As_2O_3$ (monoclinic)$+3H_2$(g)	135.00	201.82	224.10	−91.28

Source: Nordstrom, D.K. and Archer, D.G., Arsenic thermodynamic data and environmental geochemistry, in Welch, A.H. and Stollenwerk, K.G., eds., *Arsenic in Ground Water*, Springer, New York, 2003, pp. 1–25.

In Table 4.2, we present elemental arsenic, its simple oxides, and the oxidizing reactions for arsenic to arsenic trioxides. $\Delta_f G^0$, $\Delta_f H^0$, $\Delta_r S^0$, C_p^0, and *Log K* are the standard Gibbs free energy, enthalpy, entropy of formation in their standard reference state, heat capacity at constant pressure, and logarithm of the equilibrium constant, respectively.

Aqua species indicated as (aq), (g) represent gas, (l) is for liquid, (cr) is for crystalline solid phase, and (am) means amorphous phases. The data are available from Nordstrom and Archer [35] and references therein.

In Table 4.3, we tabulate the hydrolysis species for As(+3) and As(+5) in solution, the hydrolysis reactions, and the solubility reactions for the simple oxides from Nordstrom and Archer [35] and references therein.

In Table 4.4, we present the thermodynamic data for the arsenic sulfide minerals and chemical species from Nordstrom and Archer [35] and references therein.

In Table 4.5, we report the secondary arsenic mineral formula together with the crystal system, the $\Delta_f G^0$ values, and solubility data from Drahot and Filippi [79] and references therein.

It is notable that pH and total arsenic solubility values tabulated in Table 4.5 are the highest and lowest levels presented in the reference. There is no relation between the pH and the solubility values in the corresponding column of Table 4.5.

TABLE 4.3

Thermodynamic Data for the Hydrolysis Species and Reactions for As(+3) and As(+5) and Solubility Reactions for Arsenic Oxides

Species or Reaction Data for Standard State Conditions, 298.15 K and 1 atm	$\Delta_f G^0$(kJ/mol)	$\Delta_f H^0$(kJ/mol)	$\Delta_r S^0$(J/K/mol)	Log K
H_3AsO_3(aq)	−640.03	−742.36	195.83	—
$H_2AsO_3^-$(aq)	−587.66	−714.74	112.79	—
$HAsO_3^{2-}$(aq)	−507.4	—	—	—
AsO_3^{3-}(aq)	−421.8	—	—	—
H_3AsO_4(aq)	−766.75	−903.45	183.07	—
$H_2AsO_4^-$(aq)	−753.65	−911.42	112.38	—
$HAsO_4^{2-}$(aq)	−713.73	−908.41	−11.42	—
AsO_4^{3-}(aq)	−646.36	−890.21	−176.31	—
H_3AsO_3(aq) = $H_2AsO_3^-$ (aq) + H^+(aq)	52.37	27.62	−83.04	−9.17
$H_2AsO_3^-$(aq) = $HAsO_3^{2-}$(aq) + H^+(aq)	80.3 ± 1.0	—	—	−14.1
$HAsO_3^{2-}$(aq) = AsO_3^{3-}(aq) + H^+(aq)	85.6 ± 1.5	—	—	−15.0
As_2O_3(cubic) + $3H_2O$(l) = $2H_3AsO_3$(aq)	7.814	30.041	74.55	−1.38
As_2O_3(monoclinic) + $3H_2O$(l) = $2H_3AsO_3$(aq)	8.003	28.443	68.56	−1.34
H_3AsO_4(aq) = $H_2AsO_4^-$(aq) + H^+(aq)	13.10	−7.97	−70.69	−2.30
$H_2AsO_4^-$(aq) = $HAsO_4^{2-}$(aq) + H^+(aq)	39.93	3.02	−123.76	−6.99
$HAsO_4^{2-}$(aq) = AsO_4^{3-}(aq) + H^+(aq)	67.36	18.20	−164.88	−11.80
H_3AsO_4(aq) + H_2(g) = H_3AsO_3(aq) + H_2O(l)	−110.46	−124.74	−47.90	19.35

Source: Nordstrom, D.K. and Archer, D.G., Arsenic thermodynamic data and environmental geochemistry, in Welch, A.H. and Stollenwerk, K.G., eds., *Arsenic in Ground Water*, Springer, New York, 2003, pp. 1–25.

4.3.4 SORPTION MECHANISM ONTO METAL OXIDES

The mobility of arsenic in environmental systems is generally determined by the extent to which it is adsorbed onto mineral surfaces. Arsenic chemical species present in natural, heterogeneous materials are sorbed outer-sphere (physisorption), which describes weak, long-range, attractive forces between the surface and sorbed As, and/or inner-sphere (chemisorptions) sorption that refers to the formation of chemical bonds between the surface and adsorbing As. Stronger adsorption is expected by the formation of a bidentate (two-bond) adsorbed complex rather than a monodentate (one-bond) complex [80,81].

Metal oxides tend to be the primary sorbents of As in the environment, especially the oxides of iron (Fe) and aluminum (Al). In addition, Mn oxides can also sorb arsenic to some extent. Although, metal oxides are generally thought to bind As(+5) readily than As(+3), the extent to which each species is sorbed depends greatly on pH. For example, As(+5) sorption by Fe oxides is greater at acidic pH values, whereas As(+3) sorption by Fe oxides is greater at basic pH values [82,83]. Although As sorption is

TABLE 4.4

Thermodynamic Data for Arsenic Sulfide Minerals and Chemical Species

Species or Reaction Data for Standard State Conditions, 298.15 K and 1 atm	$\Delta_f G^0$(kJ/mol)	$\Delta_f H^0$(kJ/mol)	$\Delta_r S^0$(J/K/mol)	C_p^0(J/K/mol)	Log K
S(cr, rhombic)	0	0	31.8	22.6	—
S_2(g)	79.33	128.37	228.07	32.47	—
H_2S(aq)	−27.87	−40.21	1231	167	—
HS^-(aq)	12.05	−17.64	62.8	−92	—
AsS(α, realgar)	−31.3	−31.8	62.9	47	—
AsS(β, realgar)	−30.9	−31.0	63.5	47	—
As_2S_3(α, orpiment)	−84.9	−85.8	163.8	163	—
As_2S_3(am, orpiment)	−76.8	−66.9	200	—	—
$As_3S_4(SH)_2^-$(aq)	−125.6	—	—	—	—
$AsS(OH)(SH)^-$(aq)	−244.4	—	—	—	—
$4\,AsS(\alpha,cr) + S_2(g) = 2As_2S_3(\alpha,cr)$	−123.9	−168.4	−148.1	106	21.7
$As_2S_3(\alpha,cr) + 6\,H_2O(l) = 2H_3AsO_3(aq) + 3HS^-(aq) + 3H^+(aq)$	264.1	263.1	−3.2	—	−46.3
$As_2S_3(\alpha,cr) + 6H_2O(l) = 2H_3AsO_3(aq) + 3H_2S(aq)$	144.3	195.4	171	—	−25.3
$As_2S_3(am) + 6H_2O(l) = 2H_3AsO_3(aq) + 3H_2S(aq)$	136.2	176.5	136	—	−23.9
$1.5As_2S_3(am) + 1.5H_2S(aq) = As_3S_4(SH)_2^-(aq) + H^+(aq)$	31.4	—	—	—	−5.5
$0.5As_2S_3(am) + 0.5H_2S(aq) + H_2O(l) = AsS(OH)(SH)^-(aq) + H^+(aq)$	45.1	—	—	—	−7.9

Source: Nordstrom, D.K. and Archer, D.G., Arsenic thermodynamic data and environmental geochemistry, in Welch, A.H. and Stollenwerk, K.G., eds., *Arsenic in Ground Water*, Springer, New York, 2003, pp. 1–25.

pH dependent, As(+3) is in general regarded as more mobile than its oxidized As(+5) form under many environmental conditions [80,81].

By contrast, adsorbed arsenic is subject to desorption due to changes in the geochemical conditions, such as phase dissolution resulting from changes in pH and competitive desorption especially by high concentrations of inorganic ligands such as OH^-, PO_4^{-3}, CO_3^{-2}, and SO_4^{-2} ions and by simple organic ligands. However, the availability of coprecipitated arsenic depends more on the bulk mineral solubility, because a big proportion of the arsenic is retained in the interior of the phase rather than at the surface. Furthermore, sulfate, phosphate, and carbonate minerals can assimilate arsenic via adsorption and/or coprecipitation. Results of several studies

TABLE 4.5
Secondary Arsenic Mineral Formula, Crystal System, $\Delta_f G^0$, and Solubility for Standard State Conditions, 298.15 K and 1 atm

Mineral	Formula	Crystal System	Arsenic Solubility Conditions[a]	$\Delta_f G^0$(kJ/mol)
Arsenolite	As_2O_3	Cubic	11.1 g/L; 22°C	−576.34
Claudetite	As_2O_3	Monoclinic	10.1 g/L; 22°C	−576.53
AFA/pitticite	$Fe_x(AsO_4)_y(SO_4)_z \cdot nH_2O$	Amorphous	75–15,370 mg/L at pH 0.5–2.4; 25–130 mg/L at pH 1.82–3.10; 25°C	−1268.72; −1267.1
Arseniosiderite	$Ca_2Fe_3O_2(AsO_4)_3 \cdot 3H_2O$	Monoclinic	3.1–27 mg/L at pH 6.85–8.15; 25°C	—
Kaatialaite	$Fe(H_2AsO_4)_3 \cdot 5(H_2O)$	Monoclinic	—	—
Kaňkite	$Fe_3(AsO_4) \cdot 3.5H_2O$	Monoclinic	—	—
Kolfanite	$Ca_2Fe_3O_2(AsO_4)_3 \cdot 2H_2O$	Monoclinic	1.36 mg/L at pH 8; 20°C	—
Parasymplesite	$Fe_3(AsO_4) \cdot 8H_2O$	Monoclinic	—	—
Pharmacosiderite	$K[Fe_4(OH)_4(AsO_4)_3] \cdot 6.5H_2O$	Cubic	—	—
Scorodite	$FeAsO_4 \cdot 2H_2O$	Orthorhombic	0.33–5.89 mg/L at pH 5.01–6.99; 22°C; 0.11–463 mg/L at pH 0.97–7.92; 23°C; 1.8–10.3 mg/L at pH 5.53–6.36; 25°C	−1282.42; −1285.05; −1279.2; −1263.52
Symplesite	$Fe_3(AsO_4)_2 \cdot 8H_2O$	Triclinic	0.024–7 mg/L at pH 6.0–9.1; 27°C	−3751.02; −3792.01
Yukonite	$Ca_7Fe_{12}(AsO_4)_{10}(OH)_{20} \cdot 15H_2O$	Amorphous	1.16–5.11 at pH 7.56–8.82; 20°C; 6.3–51 mg/L at pH 5.5–6.15; 25°C	—
Beudantite	$PbFe_3(AsO_4)(SO_4)(OH)_6$	Hexagonal	<0.02 mg/L at pH 4.3–4.65; 25°C	−3055.6; −3081.12
Bukovskýite	$Fe_2(AsO_4)(SO_4)(OH) \cdot 7H_2O$	Triclinic	—	−3480
Sarmientite	$Fe_2(AsO_4)(SO_4)(OH) \cdot 5H_2O$	Monoclinic	—	—
Tooeleite	$Fe_6(AsO_3)_4(SO_4)(OH)_4 \cdot 4H_2O$	Monoclinic	—	—

(Continued)

TABLE 4.5 (*Continued*)

Secondary Arsenic Mineral Formula, Crystal System, $\Delta_f G^0$, and Solubility for Standard State Conditions, 298.15 K and 1 atm

Mineral	Formula	Crystal System	Arsenic Solubility Conditions[a]	$\Delta_f G^0$(kJ/mol)
Zýkaite	$Fe_4(AsO_4)_3(SO_4)(OH) \cdot 15H_2O$	Orthorhombic	—	—
Haidingerite	$Ca(AsO_3OH) \cdot H_2O$	Orthorhombic	2050 mg/L at pH 6.22; 23°C; 3120–4360 mg/L at pH 4.93; TCLP test	−1533
Hörnesite	$Mg_3(AsO4)_2 \cdot 8H_2O$	Monoclinic	300–1100 mg/L at pH 6.5–7.4	—
Pharmacolite	$Ca(HAsO_4) \cdot 2H_2O$	Monoclinic	5919 mg/L at pH 6.7; 25°C; 3120–4360 mg/L at pH 4.93; TCLP test	−1808.21
Picropharmacolite	$Ca_4Mg(AsO_4)_2(HAsO_3OH)_2 \cdot 11H_2O$	Triclinic	—	—
Weilite	$CaHAsO_4$	Triclinic	2170–3610 mg/L at pH 4.93; TCLP test; 540–764 mg/L at pH 3–8; 35°C	−1292.48
Adamite	$Zn_2(AsO_4)(OH)$	Triclinic	—	−1252.29
Annabergite	$Ni_3(AsO_4)_2 \cdot 8H_2O$	Orthorhombic	47.8–1449 mg/L at pH 3–9; 22°C	−3488.57; −3482.34
Austinite	$CaZn(AsO_4)(OH)$	Orthorhombic	—	−1651.13
Bayldonite	$PbCu_3(AsO_4)_2(OH)_2$	Triclinic	—	−1810.6
Clinoclase	$Cu_3(AsO_4)(OH)_3$	Monoclinic	—	−1209.48
Conichalcite	$CaCu(AsO_4)(OH)$	Orthorhombic	—	−1470.17
Cornubite	$Cu_5(AsO_4)_2(OH)_4$	Triclinic	—	−2057.9
Duftite	$PbCu(AsO_4)(OH)$	Orthorhombic	—	−959.92
Erythrite	$Co_3(AsO_4)_2 \cdot 8H_2O$	Monoclinic	—	—

(Continued)

TABLE 4.5 (*Continued*)
Secondary Arsenic Mineral Formula, Crystal System, $\Delta_f G^0$, and Solubility for Standard State Conditions, 298.15 K and 1 atm

Mineral	Formula	Crystal System	Arsenic Solubility Conditions[a]	$\Delta_f G^0$(kJ/mol)
Euchroite	$Cu_2(AsO_4)(OH)\cdot3(H_2O)$	Orthorhombic	—	−1552.7
Fornacite	$Pb_2Cu(AsO_4)(CrO_4)(OH)$	Monoclinic	0.03 mg/L at pH 6.96	−1956.86
Köttigite	$Zn_3(AsO_4)_2\cdot8H_2O$	Monoclinic	16 mg/L at pH 4.87	−4030.48
Legrandite	$Zn_2(AsO_4)(OH)\cdot H_2O$	Monoclinic	—	−1488.6
Mansfieldite	$AlAsO_4\cdot2H_2O$	Orthorhombic	—	−1730.78;
				−1720.8
Mimetite	$Pb_5(AsO_4)_3Cl$	Hexagonal	—	−2675.5;
				−2616.8
Olivenite	$Cu_2(AsO_4)(OH)$	Monoclinic	—	−845.52
Schultenite	$Pb(AsO_3OH)$	Monoclinic	8.8 mg/L at pH 4.68; 25°C	−805.66;
				−809.62
Sterlinghillite	$Mn_3(AsO_4)_2\cdot4H_2O$	Monoclinic	—	−4045.17

Source: Drahota, P. and Filippi, M., *Environ. Int.*, 35(8), 1243, 2009.

TCLP, toxicity characteristic leaching procedure.

[a] pH and total arsenic solubility values are the highest and lowest levels presented, and there is no relation between solubility and pH values.

reported by Foster [80] of arsenic species in minerals and synthetic phases where oxyanions are the basic structural unit indicate that arsenic sorption on these minerals primarily occurs by two mechanisms: (1) substitution for other oxyanions (such as carbonate or sulfate) during coprecipitation and (2) bonding through exchange with surface hydroxyl groups to structural metal ions during chemisorption.

Nevertheless, hydrous ferric, manganese, and aluminum oxides (HFOs, HMOs, and HAOs, respectively) are the most important sinks for adsorbed and/or coprecipitated arsenic. These oxides are composed of octahedrally coordinated metal atoms that share edges to form chains in 2D and 3D structures. For example, hydrous ferric oxide phases consist of chains of edge-sharing Fe(+3) octahedra that are linked to opposing chains by face sharing for hematite and feroxyhyte, by edge sharing for lepidocrocite, and by corner sharing for goethite and akaganeite. The common hydrous aluminum oxide phases gibbsite, boehmite, and diaspore contain similar linkages and have structures similar to the hydrous ferric oxides because the Al(+3) ion has the same charge and a nearly identical ionic radius as the Fe(+) ion. However, the hydrous manganese oxide phases found in nature commonly form more complex tunnel and layer structures than hydrous ferric oxides or hydrous aluminum oxides because of the complexity of multiple oxidation states for manganese. The common reactive units for the hydrous ferric, manganese, and aluminum oxides are the surface hydroxyl and/or water groups. The structural arrangement of surface metal atom coordination polyhedra and the pH and ionic strength in the surrounding aqueous milieu delineate the reactivity of these units [80].

However, the affinity of arsenic for surface functional groups on metal oxide surfaces should vary with the distribution and type of surface functional groups available. Depending on the pH, these OH groups can bind or release H^+ ions resulting in the development of a surface charge. In this case, arsenic adsorbs by ligand exchange with OH and OH_2^+ surface functional groups, forming an inner-sphere complex. This type of adsorption requires an incompletely dissociated acid (i.e., $H_2AsO_4^-$) to provide a proton for complexation with the surface OH group, forming H_2O and providing a space for the anion. The reactions (4.25) through (4.30) describe these processes [81]:

$$\equiv SurfOH + H_3AsO_4^{\circ} \rightarrow \equiv SurfH_2AsO_4^{\circ} + H_2O \qquad (4.25)$$

$$\equiv SurfOH + H_2AsO_4^- \rightarrow \equiv SurfHAsO_4^- + H_2O \qquad (4.26)$$

$$\equiv SurfOH + HAsO_4^{-2} \rightarrow \equiv SurfAsO_4^{-2} + H_2O \qquad (4.27)$$

$$\equiv SurfOH + H_3AsO_3^{\circ} \rightarrow \equiv SurfH_2AsO_3^{\circ} + H_2O \qquad (4.28)$$

$$\equiv SurfOH + H_2AsO_3^- \rightarrow \equiv SurfHAsO_3^- + H_2O \qquad (4.29)$$

$$\equiv SurfOH + HAsO_3^{-2} \rightarrow \equiv SurfAsO_3^{-2} + H_2O \qquad (4.30)$$

where
   $\equiv SurfOH$ represents the structural metal atom and associated OH surface functional group
   $\equiv SurfHAsO_4^-$ is the surface arsenic complex

The energy required to dissociate the weak acid at the oxide surface and the amount of arsenic adsorbed varies with pH.

Inner-sphere surface complexation of As has also been inferred from changes in the isoelectric point (*pl*) of the adsorbing solid. The *pl* is the pH at which a particular molecule or surface carries no net electrical charge. At pH values below the *pl*, net surface charge is positive due to adsorption of excess H^+ ions. At pH values greater than the *pl*, net surface charge is negative due to desorption of H^+ ions. Formation of inner-sphere surface complexes of anions has been shown to increase the negative charge of solid surfaces, thereby decreasing the *pl* to lower pH values. The amount of shift in the *pl* depends on the ion and its concentration, as well as the particular solid surface. Adsorption of both As(+3) and As(+5) species has been found to lower the *pl* of various oxides including goethite, ferrihydrite, gibbsite, and amorphous $As(OH)_3$. The arsenic concentration affects the magnitude of this decrease [81].

Notwithstanding, in the case of ferrihydrite, the effect of pH on arsenic adsorption differs between As(+3) and As(+5). Available data from the literature indicate that arsenate adsorption was greatest at low pH values and decreased with increasing pH. Greater adsorption of As(+5) at low pH is attributable to more favorable adsorption energies between the more positively charged surface and negatively charged $H_2AsO_4^-$, the predominate As(+5) species between pH 2.2 and 6.9. As pH increased above 6.9, $HAsO_4^{-2}$ became the predominating aqueous species, surface charge became less positive, and adsorption was less favorable [81].

However, the adsorption of As(+5) as a function of pH is also influenced by arsenic initial concentration in solution. For example, at an initial concentration of As(+5) of 1.33 µM, the As(+5) adsorption was independent of pH > 6. When the concentration of As(+5) was increased to 13.3 µM, available adsorption sites become filled at a lower pH, and As(+5) adsorption began to decrease immediately with any further increase in pH. However, separate adsorption experiments with either As(+3) or As(+5) showed that when As(+5) and As(+3) were present in solution together (2.08 mM each), the As(+3) had little effect on adsorption of As(+5) below pH 6, but caused a decrease in As(+5) adsorption of as much as 25% at pH>6 due to strong adsorption of As(+3). In addition, arsenate resulted in a decrease of 15%–25% in the adsorption of As(+3) between pH 4 and 6; however, by pH=9, As(+5) decreased adsorption of As(+3) by less than 5% [81].

In the case of aluminum oxides, the degree of crystallinity is important for arsenic adsorption. For example, amorphous $Al(OH)_3$ adsorbed more As(+5) per gram than crystalline γ-Al_2O_3 and gibbsite. In a pH range from 5 to 9, experimental data indicated a 17% decrease in the As(+5) adsorption of gibbsite compared to goethite. However, other researchers measured two times more adsorption of As(+5) by goethite (60 m²/g) than by gibbsite (31 m²/g) [81].

Manganese oxides can also adsorb arsenic. The most common Mn oxides formed from weathering processes and have a low *pl* are the birnessite group. The low *pl* value results in a net negative surface charge at a pH range common to that of groundwater. Arsenate adsorption by synthetic birnessite (*pl*=2.5) was reported to be negligible. At the pH=4 to pH=7 values used in these experiments, the negatively charged species, $H_2AsO_4^-$, apparently could not overcome the energy barrier required to displace a functional group from the negatively charged birnessite surface. However, As(+3) was absorbed by birnessite at pH = 7. The neutral charge on $H_3AsO_3°$ results in a lower energy barrier to exchange with surface functional groups. In addition, adsorption of both As(+3) and As(+5) was reported by the manganese oxides cryptomelane ($pl = 2.9$) and pyrolusite ($pl = 6.4$) [81].

The degree of arsenic adsorption by different clay minerals reveals maximum adsorption of As(+5) by kaolinite, montmorillonite, illite, halloysite, and chlorite in up to pH values near 7, while adsorption decreased with further increase of pH. Adsorption of As(+3) by the same clay minerals was a minimum at low pH and increased with increasing pH. Arsenate is adsorbed to a greater extent than As(+3) on all clay minerals at pH < 7. At higher pH values, adsorption of As(+5) and As(+3) was more comparable, and in some cases, As(+3) adsorption exceeded that of As(+5). Thus, only the OH groups associated with Al ions exposed at the edges of clay particles are considered to be proton acceptors and able to complex anionic species of As [81].

In order to describe adsorption/desorption reactions at mineral surfaces, two types of models have been developed. The empirical models that are based on the partitioning relationships of a solute between the aqueous and solid phases and the conceptual models for surface complexation that treat adsorption reactions similar to ion association reactions in solution. All of these models assume that adsorption reactions are at equilibrium.

In the case of empirical adsorption models, the distribution coefficient as in Equation 4.31 is the most known [84]:

$$\bar{C} = K_d C \tag{4.31}$$

In addition, the Langmuir isotherm and the Freundlich isotherm are given in Equations 4.32 and 4.33, respectively [84]:

$$\bar{C} = \frac{K_L S C}{1 + K_L C} \tag{4.32}$$

$$\bar{C} = K_F C^a \tag{4.33}$$

The $\bar{C}$ is the adsorbed concentration, C is the aqueous concentration at equilibrium with the solid phase, S is the adsorption site concentration, K_d is the distribution coefficient, K_L is the Langmuir equilibrium constant, K_F is the Freundlich equilibrium constant, and a is the Freundlich exponent.

The equilibrium constants for empirical models are highly dependent on solution and solid composition, while they can be obtained from measurements of the activity of a solute component in water and the equilibrium adsorbed concentration [84]. To do this, a solid phase is suspended in solution with a known range of solute concentrations, and the equilibrium constants are then derived from various plots of the adsorbed and aqueous concentrations [84]. However, the empirical models are limited to the specific experimental conditions, and in the case of the K_L and K_F

values for arsenic, they are a function of the mineralogy of the adsorbent, the pH of the solution, and the concentration of competing ions [81]. Therefore, modifications to the Langmuir and Freundlich isotherm equations have been made in order to take into consideration the ionic strength of the solution and the pH dependence of the adsorption characteristics. Although these modified equations incorporate the pH and ionic competition variability, they partially simulate to some extent the complexity of the natural environmental conditions.

To overcome this problem, surface complexation models have been developed in order to simulate adsorption in complex geochemical systems. These models provide a rational interpretation of the physical and chemical processes of adsorption, and they consider the chemical reactions at the solid–solution interface as surface complexation reactions analogous to the formation of complexes in solution. In these models, the energy required to penetrate the electrostatic-potential field extending away from the surface is a function of the activities of adsorbing ions, the active mass composition of a reaction mixture at equilibrium and equilibrium constant. All these models are based on the electric double-layer theory, but they differ in their geometric description regarding the oxide–water interface and the treatment of the electrostatic interactions. The more commonly used models include the constant capacitance model, the diffuse double-layer model, and the triple-layer model [81].

By contrast, desorption is related to the type of bond formed between As and the solid surface and also to the pH, the initial concentration of As, and the competitive ions in solution. For example, the slow desorption of As(+5) from the goethite surface is considered as a two-step reaction process. In the first step, a monodentate surface complex was rapidly formed by ligand exchange with OH groups. In the following slower step, a second ligand exchange reaction results in the formation of a bidentate complex [81].

In the case of ferrihydrite, Stollenwerk and references therein [81] reported desorption rates of As(+5) aged for 24 h as a result of the increasing pH. At pH = 8, arsenate was in equilibrium with ferrihydrite for 144 h, and the molar ratio of As(+5)/ Fe = 0.10. However, desorption was initiated by increasing the pH to 9.0, and within a few hours, the molar ratio of As(+5)/Fe = 0.08. At pH = 9 and within 96 h, the concentration of As(+5) still adsorbed was only about 5% greater than the adsorbed concentration of As(+5) determined [81].

In the work of Jain and Loeppert [54], phosphate competition with As(+5) has been observed, and the adsorption of both As(+5) and As(+3) decreased with increasing P(+5) concentration. For As(+5), the decrease was significant over the entire pH range. Competition for adsorption sites between As(+5) and P(+5) involves species with the same charge. Phosphate had the greatest effect on As(+3) adsorption at lower pH values. At pH 9, adsorption of As(+3) decreased by only a few percent, even at the highest P(+5) concentration [81].

The research of Lin and Puls [85], on the effects of aging on desorption of As(+3) and As(+5) by clay minerals, reveals that for halloysite and kaolinite, desorption of As(+3) was almost 100% for the unaged experiments. After aging for 75 days, about 15% As(+3) was desorbed from halloysite and about 30% from kaolinite. Desorption of As(+5) from halloysite decreased from 85% in the unaged experiment to 65% after 75 days of aging. For kaolinite, As(+5) desorption decreased from about 98%

in the unaged experiment to 70% after 75 days of aging. However, arsenic desorption varied with the type of clay mineral, arsenic species, and, in most cases, time of aging [81,85].

4.4 CONCLUSIONS

The geochemistry of arsenic is complex and fits vitality conversions valuable for microbial metabolism. Arsenic can dwell in various oxidation states and complex particles with negative to neutral charges. This permits a wide exhibit of species that might be used throughout redox and complexation reactions so arsenic experiences a variety of microbial intervened changes.

Both soluble and adsorbed As(+5) can serve as an electron acceptor in the bacterial oxidation of OM. Also, As(+5) adsorbed onto hydrous ferric oxides can be released by the respiratory reduction of Fe(+3) by dissimilatory iron-reducing bacteria. The activity of sulfate reducers favors the immobilization of arsenic as metal sulfides. In contrast, oxidation of arsenic sulfide minerals such as arsenopyrite and others will favor arsenic mobilization. In addition, adsorption and coprecipitation of arsenic chemical species onto other solid phases like hydrous aluminum and manganese oxides and clay minerals control the mobility of arsenic in the environment. However, the dissolution of arsenic minerals is responsible for the occurrence of arsenic chemical species in the environment, and in a given temperature and pressure, dissolution of these minerals is a function of the free energy (G), enthalpy (H), entropy (S), and, in the case of solid solutions, the composition. Therefore, mathematical models were established to obtain thermodynamic data of arsenic minerals and chemical species in order to be able to predict their stability in different environmental conditions.

Toxic effects to humans and other organisms are related to the time of exposure, the oxidative stage, and the speciation and the concentration of arsenic compounds. The risk to health for humans during long-term arsenic exposure includes among others reproductive and neurological effects and carcinogenesis.

Detoxification of inorganic arsenic can occur through its reduction and/or methylation. Many transformations that control arsenic flux between various compartments (*sediment/soil* ↔ *aqueous* ↔ *atmosphere*) are driven by and/or accelerated by microbial metabolism.

However, our understanding is presently limited to the potential influence that microbes may exert on these fluxes. The biogeochemical cycling of metals/metalloids has been conducted with a view of these processes occurring on very limited spatial and temporal scales. The influence of these elements on both community succession and on the evolution of microbial species within those communities has generally been ignored.

Conversely, we are only beginning to appreciate the roles that microbial succession and microbial evolution play in transforming the geochemistry and mineralogy of their physical environment. Full understanding of how the abundance metabolic and geochemical processes that conceivably impact arsenic mobility really yields specific patterns of distribution and speciation anticipates examinations that coordinate crosswise over expansive spatial and transient scales.

REFERENCES

1. Smedley, P. and Kinniburgh, D. A review of the source, behaviour and distribution of arsenic in natural waters. *Applied Geochemistry* 17(5) (2002): 517–568.
2. Edwards, M., Johnson, L., Mauer, C., Barber, R., Hall, J., and O'Bryant, S. Regional specific groundwater arsenic levels and neuropsychological functioning: A cross-sectional study. *International Journal of Environmental Health Research* 24(2) (2014): 546–557.
3. Dewji, N.N., Do, C., and Bayney, R.M. Transcriptional activation of Alzheimer's β-amyloid precursor protein gene by stress. *Molecular Brain Research* 33(2) (1995): 245–253.
4. Vahidnia, A., van der Straaten, R.J.H.M., Romijn, F., van Pelt, J., van der Voet, G., and de Wolff, F. Arsenic metabolites affect expression of the neurofilament and tau genes: An in vitro study into the mechanism of arsenic neurotoxicity. *Toxicology In Vitro* 21(6) (2007): 1104–1112.
5. Engström, K., Vahter, M., Johansson, G. et al. Chronic exposure to cadmium and arsenic strongly influences concentrations of 8-oxo-7,8-dihydro-2′-deoxyguanosine in urine. *Free Radical Biology & Medicine* 48(9) (2010): 1211–1217.
6. Vega, L., Styblo, M., Patterson, R., Cullen, W., Wang, C., and Germolec, D. Differential effects of trivalent and pentavalent arsenicals on cell proliferation and cytokine secretion in normal human epidermal keratinocytes. *Toxicology and Applied Pharmacology* 172(3) (2001): 225–232.
7. Luo, J.-H., Qiu, Z.-Q., Shu, W.-Q., Zhang, Y.-Y., Zhang, L., and Chen, J.-A. Effects of arsenic exposure from drinking water on spatial memory, ultra-structures and NMDAR gene expression of hippocampus in rats. *Toxicology Letters* 184(2) (2009): 121–125.
8. Meng, D., Wang, X., Chang, Q. et al. Arsenic promotes angiogenesis in vitro via a heme oxygenase-1-dependent mechanism. *Toxicology and Applied Pharmacology* 244(3) (2010): 291–299.
9. Tan, Z.S., Seshadri, S., Beiser, A. et al. Plasma total cholesterol level as a risk factor for Alzheimer disease: The Framingham Study. *Archives of Internal Medicine* 163(9) (2003): 1053–1057.
10. Chiou, H.Y., Hsueh, Y.M., Liaw, K.F. et al. Incidence of internal cancers and ingested inorganic As: A seven-year follow-up study in Taiwan. *Cancer Research* 55(6) (1995): 1296–1300.
11. Mandal, B.K. and Suzuki, K.T. Arsenic round the world: A review. *Talanta* 58(1) (2002): 201–235.
12. U.S. EPA. National Primary Drinking Water Regulations; Arsenic and Clarifications to Compliance and New Source Contaminants Monitoring; Final Rule. *Federal Register* 66(14) (2001): 6975–7066. http://www.epa.gov/sbrefa/documents/pnl14f.pdf.
13. Smith, A.H., Lingas, E.O., and Rahman, M. Contamination of drinking-water by arsenic in Bangladesh: A public health emergency. *Bulletin of the World Health Organization* 78(9) (2000): 1093–1103.
14. Naujokas, M.F., Anderson, B., Ahsan, H. et al. The broad scope of health effects from chronic arsenic exposure: Update on a worldwide public health problem. *Environmental Health Perspectives* 121(3) (2013): 295–302.
15. Kar, S., Maity, J.P., Jean, J.-S. et al. Health risks for human intake of aquacultural fish: Arsenic bioaccumulation and contamination. *Journal of Environmental Science and Health, Part A: Toxic/Hazardous Substances and Environmental Engineering* 46(11) (2011): 1266–1273.
16. Del Razo, L.M., Quintanilla-Vega, B., Brambila-Colombres, E., Calderon-Aranda, E.S., Manno, M., and Albores, A. Stress proteins induced by arsenic. *Toxicology and Applied Pharmacology* 177(2) (2001): 132–148.

17. U.S. EPA. Arsenic treatment technologies for soil, waste, and water. EPA-542-R-02 004 (2002), http://www.epa.gov/tio/download/remed/542r02004/arsenic_report.pdf.

18. Lièvremont, D., Bertin, P.N., and Lett, M.-C. Arsenic in contaminated waters: Biogeochemical cycle, microbial metabolism and biotreatment processes. *Biochimie*, 91(10) (2009): 1229–1237.

19. Panagiotaras, D., Panagopoulos, G., Papoulis, D., and Avramidis, P. Arsenic geochemistry in groundwater system. In Panagiotaras, D., ed., *Geochemistry-Earth's System Processes*, pp. 27–38. INTECH Publishers, Rijeka, Croatia, 2012.

20. Silver, S. and Phung, L.T. A bacterial view of the periodic table: Genes and proteins for toxic inorganic ions. *Journal of Industrial Microbiology and Biotechnology* 32(11–12) (2005): 587–605.

21. Rahman, K.Z., Wiessner, A., Kuschk, P., Mattusch, J., Kästner, M., and Muller, R.A. Dynamics of arsenic species in laboratory-scale horizontal subsurface-flow constructed wetlands treating an artificial wastewater. *Engineering in Life Sciences* 8(6) (2008): 603–611.

22. Teclu, D., Tivchev, G., Laing, M., and Wallis, M. Bioremoval of arsenic species from contaminated waters by sulphate-reducing bacteria. *Water Research* 42(19) (2008): 4885–4893.

23. Mokashi, S.A. and Paknikar, K.M. Arsenic (III) oxidizing *Microbacterium lacticum* and its use in the treatment of arsenic contaminated groundwater. *Letters in Applied Microbiology* 34(4) (2002): 258–262.

24. Soda, S.O., Yamamura, S., Zhou, H., Ike, M., and Fujita, M. Reduction kinetics of As (V) to As (III) by a dissimilatory arsenate-reducing bacterium, *Bacillus* sp. SF1. *Biotechnology and Bioengineering* 93(4) (2006): 812–815.

25. Katsoyiannis, I.A. and Zouboulis, A.I. Use of iron- and manganese-oxidizing bacteria for the combined removal of iron, manganese and arsenic from contaminated groundwater. *Water Quality Research Journal of Canada* 41(2) (2006): 117–129.

26. Sun, X., Doner, H.E., and Zavarin, M. Spectroscopy study of arsenite [As(III)] oxidation on Mn-substituted goethite. *Clays and Clay Minerals* 47(4) (1999): 474–480.

27. Smith, E., Smith, J., Smith, L., Biswas, T., Correll, R., and Naidu, R. Arsenic in Australian environment: An overview. *Journal of Environmental Science and Health, Part A: Toxic/Hazardous Substances and Environmental Engineering* 38(1) (2003): 223–239.

28. IPCS. Environmental health criteria on arsenic and arsenic compounds. In *Environmental Health Criteria Series: Arsenic and Arsenic Compounds*, no. 224, 521 pp. WHO, Geneva, Switzerland, 2001. http://whqlibdoc.who.int/ehc/WHO_EHC_224.pdf.

29. Welch, A., Westjohn, D.B., Helsel, D.R., and Wanty, R.B. Arsenic in ground water of the United States: Occurrence and geochemistry. *Ground Water* 38(4) (2000): 589–604.

30. Sarkar, B. *Heavy Metals in the Environment*. Marcell Dekker, New York, 2002.

31. Liao, C.-M., Chen, B.-C., Singh, S., Lin, M.-C., Liu, C.-W., and Han, B.-C. Acute toxicity and bioaccumulation of arsenic in tilapia (*Oreochromis mossambicus*) from a black-foot disease area in Taiwan. *Environmental Toxicology* 18(4) (2003): 252–259.

32. Kumaresan, M. and Riyazuddin, P. Overview of speciation chemistry of arsenic. *Current Science* 80(7) (2001): 837–846.

33. Azcue, J.M., Mudroch, A., Rosa, F., Hall, G.E.M., Jackson, T.A., and Reynoldson, T. Trace elements in water, sediments, porewater, and biota polluted by tailings from an abandoned gold mine in British Columbia, Canada. *Journal of Geochemical Exploration* 52(1–2) (1995): 25–34.

34. U.S. EPA. National Primary Drinking Water Regulations; Arsenic and Clarifications to Compliance and New Source Contaminants Monitoring; Proposed Rule. *Federal Register* 65(121) (2000): 38888–38983. http://www.gpo.gov/fdsys/pkg/FR-2000-06-22/pdf/00-13546.pdf.

35. Nordstrom, D.K. and Archer, D.G. Arsenic thermodynamic data and environmental geo-chemistry. In Welch, A.H., Stollenwerk, K.G., eds., *Arsenic in Ground Water*, pp. 1–25. Springer, New York, 2003.
36. Kim, M.-J., Nriagu, J., and Haack, S. Arsenic species and chemistry in groundwater of southeast Michigan. *Environmental Pollution* 120(2) (2002): 379–390.
37. Nickson, R.T., McArthur, J.M., Ravenscroft, P., Burgess, W.G., and Ahmed, K.M. Mechanism of arsenic release to groundwater, Bangladesh and West Bengal. *Applied Geochemistry* 15(4) (2000): 403–413.
38. Chowdhury, T.R., Basu, G.K., Mandal, B.K. et al. Arsenic poisoning in the Ganges delta. *Nature* 401 (1999): 545–546.
39. Yu, Y., Zhu, Y., Gao, Z., Gammons, C.H., and Li, D. Rates of arsenopyrite oxidation by oxygen and Fe(III) at pH 1.8–12.6 and 15–45°C. *Environmental Science and Technology* 41(18) (2007): 6460–6464.
40. Corkhill, C.L. and Vaughan, D.J. Arsenopyrite oxidation—A review. *Applied Geochemistry* 24(12) (2009): 2342–2361.
41. Walker, F.P., Schreiber, M.E., and Rimstidt, J.D. 2006. Kinetics of arsenopyrite oxidative dissolution by oxygen. *Geochimica et Cosmochima Acta* 70: 1668–1676.
42. Paez-Espino, D., Tamames, J., De Lorenzo, V., and Cánovas, D. Microbial responses to environmental arsenic. *Biology of Metals* 22(1) (2009): 117–130.
43. Macy, J.M., Santini, J.M., Pauling, B.V., O'Neill, A.H., and Sly, L.I. Two new arsenate/sulfate-reducing bacteria: Mechanisms of arsenate reduction. *Archives of Microbiology* 173(1) (2000): 49–57.
44. Silver, S. and Phung, L.T. Genes and enzymes involved in bacterial oxidation and reduction of inorganic arsenic. *Applied and Environmental Microbiology* 71(2) (2005): 599–608.
45. Niggemyer, A., Spring, S., Stackebrandt, E., and Rosenzweig, R.F. Isolation and characterization of a novel As(V)-reducing bacterium: Implications for arsenic mobilization and the genus *Desulfitobacterium*. *Applied and Environmental Microbiology* 67(12) (2001): 5568–5580.
46. Newman, D.K., Kennedy, E.K., Coates, J.D. et al. Dissimilatory arsenate and sulfate reduction in *Desulfotomaculum auripigmentum* sp. nov. *Archives of Microbiology* 168(5) (1997): 380–388.
47. Oremland, R.S., Dowdle, P.R., Hoeft, S. et al. Bacterial dissimilatory reduction of arsenate and sulfate in meromictic Mono Lake, California. *Geochimica et Cosmochimica Acta* 64(18) (2000): 3073–3084.
48. Oremland, R.S. and Stolz, J.F. The ecology of arsenic. *Science* 300(5621) (2003): 939–944.
49. Stumm, W. and Morgan, J.J. *Aquatic Chemistry: Chemical Equilibria and Rates in Natural Waters*. John Wiley and Sons, New York, 1996.
50. Collinet, M.-N. and Morin, D. Characterisation of arsenopyrite oxidising *Thiobacillus*. Tolerance to arsenite, arsenate, ferrous and ferric iron. *Antonie van Leeuwenhoek* 57(4) (1990): 237–244.
51. Katsoyiannis, I., Zouboulis, A., and Jekel, M. Kinetics of bacterial As(III) oxidation and subsequent As(V) removal by sorption onto biogenic manganese oxides during groundwater treatment. *Industrial & Engineering Chemistry Research* 43(2) (2004): 486–493.
52. Panagopoulos, G. and Panagiotaras, D. Understanding the extent of geochemical and hydrochemical processes in coastal karst aquifers through ion chemistry and multivariate statistical analysis. *Fresenius Environmental Bulletin* 20(12a) (2011): 3270–3285.
53. Sverjensky, D.A. and Fukushi, K. Anion adsorption on oxide surfaces: Inclusion of the water dipole in modeling the electrostatics of ligand exchange. *Environmental Science and Technology* 40(1) (2006): 263–271.
54. Jain, A. and Loeppert, R.H. Effect of competing anions on the adsorption of arsenate and arsenite by ferrihydrite. *Journal of Environmental Quality* 29(5) (2000): 1422–1430.

55. Goldberg, S. and Johnston, C.T. Mechanisms of arsenic adsorption on amorphous oxides evaluated using macroscopic measurements, vibrational spectroscopy, and surface complexation modeling. *Journal of Colloid and Interface Science* 234(1) (2001): 204–216.

56. Cullen, W.R. Chemical mechanism of arsenic biomethylation. *Chemical Research in Toxicology* 27 (2014): 457–461.

57. Nesnow, S., Roop, B.C., Lambert, G. et al. DNA damage induced by methylated trivalent arsenicals is mediated by reactive oxygen species. *Chemical Research in Toxicology* 15(12) (2002): 1627–1634.

58. Smith, P.G. Arsenic biotransformation in terrestrial organisms—A study of the transport and transformation of arsenic in plants, fungi, fur and feathers, using conventional speciation analysis and X-ray absorption spectroscopy. Doctoral dissertation, Queen's University, Canada, 2007.

59. Mudhoo, A., Sharma, S.K., Garg, V.K., and Tseng, C.-H. Arsenic: An overview of applications, health, and environmental concerns and removal processes. *Critical Reviews in Environmental Science and Technology*, 41(5) (2011): 435–519.

60. ATSDR. Toxicological profile for arsenic. Agency for Toxic Substances and Disease Registry (2007), www.atsdr.cdc.gov/toxprofiles/tp2.pdf.

61. Kaltreider, R.C., Davis, A.M., Lariviere, J.P., and Hamilton, J.W. Arsenic alters the function of the glucocorticoid receptor as a transcription factor. *Environmental Health Perspectives* 109(3) (2001): 245–251.

62. Carelli, V., Ross-Cisneros, F.N., and Sadun, A.A. Optic nerve degeneration and mitochondrial dysfunction: Genetic and acquired optic neuropathies. *Neurochemistry International* 40(6) (2002): 573–584.

63. Ayotte, J.D., Baris, D., Cantor, K.P. et al. Bladder cancer mortality and private well use in New England: An ecological study. *Journal of Epidemiology and Community Health* 60(2) (2006): 168–172.

64. Donohue, J.M. and Abernathy, C.O. Exposure to inorganic arsenic from fish and shellfish. In Chappel, W.R., Abernathy, C.O., Calderon, R.L., eds., *Arsenic Exposure and Health Effects*, pp. 89–98. Elsevier, BV, Amsterdam, the Netherlands, 1999.

65. Samal, A.C., Kar, S., Bhattacharya, P., and Santra, S.C. Human exposure to arsenic through foodstuffs cultivated using arsenic contaminated groundwater in areas of West Bengal, India. *Journal of Environmental Science and Health, Part A: Toxic/Hazardous Substances and Environmental Engineering*, 46(11) (2011): 1259–1265.

66. Huang, Y.-K., Lin, K.-H., Chen, H.-W. et al. Arsenic species contents at aquaculture farm and in farmed mouthbreeder (*Oreochromis mossambicus*) in blackfoot disease hyperendemic areas. *Food and Chemical Toxicology* 41(11) (2003): 1491–1500.

67. Falco, G., Llobet, J.M., Bocio, A., and Domingo, J.L. Daily intake of arsenic, cadmium, mercury, and lead by consumption of edible marine species. *Journal of Agricultural and Food Chemistry* 54(16) (2006): 6106–6112.

68. Schwarzenegger, A., Tamminen, T., and Denton, J.E. Public health goals for chemicals in drinking water arsenic. Office of Environmental Health Hazard Assessment, California Environmental Protection Agency (2004), http://oehha.ca.gov/water/phg/pdf/asfinal.pdf.

69. Ng, J.C., Wang, J., and Shraim, A. A global health problem caused by arsenic from natural sources. *Chemosphere* 52(9) (2003): 1353–1359.

70. Anderson, L.C.D. and Bruland, K.W. Biogeochemistry of arsenic in natural waters: The importance of methylated species. *Environmental Science & Technology*, 25(3) (1991): 420–427.

71. Cullen, W.R. and Reimer, K.J. Arsenic speciation in the environment. *Chemical Reviews* 89(4) (1989): 713–764.

72. Dowdle, P.R., Laverman, A.M., and Oremland, R.S. Bacterial reduction of arsenic(V) to arsenic(III) in anoxic sediments. *Applied and Environmental Microbiology* 62(5) (1996): 1664–1669.

73. Rakhunde, R., Jasudkar, D., Deshpande, L., Juneja, H.D., and Labhasetwar, P. Health effects and significance of arsenic speciation in water. *International Journal of Environmental Sciences and Research* 1(4) (2012): 92–96.

74. Holm, T.R., Kelly, W.R., Wilson, S.D., Roadcap, G.S., Talbott, J.L., and Scott, J.W. Arsenic geochemistry and distribution in the Mahomet Aquifer, Illinois. Illinois Waste Management and Research Center. Illinois Department of Natural Resources (2004), pp. 1–103. http://www.wmrc.uiuc.edu/main_sections/info_services/library_docs/RR/RR-107.pdf.

75. Bauer, M. and Blodau, C. Mobilization of arsenic by dissolved organic matter from iron oxides, soils and sediments. *Science of the Total Environment* 354(2–3) (2006): 179–190.

76. Challenger, F., Higgenbottom, C., and Ellis, L. The formation of organo-metalloid compounds by microorganisms. Part I. Trimethylarsine and dimethylethylarsine. *Journal of the Chemical Society* 1 (1933): 95–101.

77. Bowell, R.J. Sorption of arsenic by iron oxides and oxyhydroxides in soils. *Applied Geochemistry* 9(3) (1994): 279–286.

78. Johnson, J.W., Oelkers, E.H., and Helgeson, H.C. SUPCRT92: A software package for calculating the standard molal thermodynamic properties of minerals, gases, aqueous species, and reactions from 1 to 5000 bar and 0 to 1000°C. *Computers & Geosciences* 18(7) (1992): 899–947.

79. Drahota, P. and Filippi, M. Secondary arsenic minerals in the environment: A review. *Environment International* 35(8) (2009): 1243–1255.

80. Foster, A.L. Spectroscopic investigations of arsenic species in solid phases. In Welch, A.H., Stollenwerk, K.G., eds., *Arsenic in Ground Water*, pp. 27–65. Kluwer Academic Publishers, Boston, MA, 2003.

81. Stollenwerk, K.G. Geochemical processes controlling transport of arsenic in groundwater: A review of adsorption. In Welch, A.H., Stollenwerk, K.G., eds., *Arsenic in Ground Water*, pp. 66–100. Kluwer Academic Publishers, Boston, MA, 2003.

82. Dixit, S. and Hering, J.G. Comparison of arsenic(V) and arsenic(III) sorption onto iron oxide minerals: Implications for arsenic mobility. *Environmental Science & Technology* 37(18) (2003): 4182–4189.

83. Raven, K.P., Jain, A. and Loeppert, R.H. Arsenite and arsenate adsorption on ferrihydrite: Kinetics, equilibrium, and adsorption envelopes. *Environmental Science & Technology* 32(3) (1998): 344–349.

84. Zheng, C. and Bennett, G.D. *Applied Contaminant Transport Modeling, Theory and Practice*. Van Nostrand Reinhold, New York, 2002.

85. Lin, Z. and Puls, R.W. Adsorption, desorption and oxidation of arsenic affected by clay minerals and aging process. *Environmental Geology* 39(7) (2000): 753–759.

Section II

Remediation

5 Introduction to Remediation of Arsenic Toxicity

Application of Biological Treatment Methods for Remediation of Arsenic Toxicity from Groundwaters

Ioannis A. Katsoyiannis, Manassis Mitrakas, and Anastasios I. Zouboulis

CONTENTS

5.1 INTRODUCTION

Arsenic poisoning through drinking water is a matter of worldwide concern. In Southeast Asia (Bangladesh, Vietnam, West Bengal, Nepal, Cambodia, and Mongolia), over 100 million people consume water with arsenic over the WHO, U.S., and European Union (EU) limits of 10 µg/L. In the United States, more than 13 million people, mostly in western states, consume drinking water with more than 10 µg As/L. In Europe, many regions are affected by elevated arsenic concentrations (Hungary, Romania, Greece, Spain, Finland, and Germany). Table 5.1 shows the most severe cases of arsenic contamination in the world and Table 5.2 the most prominent cases of arsenic contamination in European countries [1,2].

In Europe, the situation is particularly severe in some regions of Eastern and Southeastern Europe where smaller communities depend on local groundwater resources that are contaminated with arsenic, and according to the EU directive 98/83, all drinking water sources within the EU should have complied with the new limits by December 2003. However, until to date, there are still areas in Greece, Hungary, and West Romania where the local population depends on drinking water contaminated with arsenic concentrations over 50 µg/L [3,4].

It has been well demonstrated that almost all the maximum contaminant level (MCL) violations of toxic metals (i.e., arsenic, uranium) have been observed in small communities with a population of less than 10,000 people [5]. Large drinking water plants in Northern and Central Europe normally find alternative and arsenic-free water resources, or they treat the water with conventional hi-tech multibarrier treatment plants, including methods such as coagulation–filtration, ion exchange, and nanofiltration. Smaller towns, communities, and individual users in rural areas often rely on local water resources, and removal methods developed for large plants are not applicable because of high operational and capital costs.

TABLE 5.1

Most Prominent Cases of Arsenic Contamination in the World

Country/Region	Maximum Concentration (µg/L)	Potential Population Exposed
Bangladesh	2.5×10^3	30×10^6
West Bengal, India	3.2×10^3	6×10^6
Cambodia	1.3×10^3	320×10^3
Argentina	10×10^3	2×10^6
Chile	1×10^3	400×10^3
Mexico	620	400×10^3

TABLE 5.2
Arsenic Concentrations in Some European Areas

Country/Region	µg As/L	Groundwater Use
Czech Republic/Mokrsko	1.690	Drinking water
Croatia	610	Drinking water
Finland	1040	Drinking water
Germany/Northern Bavaria and Wiesbaden	150	Drinking water
Hungary	800	Drinking water
Iceland	310	No drinking water
Italy/volcanic areas of Ischia, Vesuvius, Etna, Stromboli	1.558	Drinking water in the case of Etna
Romania/Transylvania and Western Plain	200	Drinking water
Serbia/Vojvodina	150	Drinking water
Spain/Duero Basin, Ambles Valley in Avila, Caldes de Malavella	615	Drinking water in Duero basin
Switzerland/Ticino, Wallis	370	Wallis drinking water
Turkey/Kutahya plain	10.700	Drinking water

Consequently, small drinking water systems face this difficult challenge: to provide a safe and sufficient supply of water at a reasonable cost.

In Bangladesh and West Bengal, a number of simple treatment units have been developed (e.g., oxidation of As(III)) with chemical oxidants or with sunlight followed by precipitation with naturally present or with added iron and aluminum salts, removal with zero-valent iron, and sorption in prefabricated filters with iron and aluminum oxide sorbents [6].

Alternative removal units based on biological oxidation of iron and manganese and avoidance of the use of chemical reagents for preoxidation have been examined at the laboratory scale [7,8]. The application of such technologies comprises nowadays the state of the art for arsenic removal in Greece [5] and other countries in Europe, such as in Italy [9] and France [10].

Therefore, in this chapter, we will review the application of biological iron and manganese oxidation for the removal of arsenic from groundwaters, and we will present case studies of the application of this technology in full-scale treatment plants in Greek water treatment units. Besides, a brief overview of arsenic removal methods by macrophytes, bacteria, and algae will be presented.

5.2 BIOLOGICAL IRON REMOVAL

Iron-containing groundwaters have been traditionally treated by chemical oxidation, promoted with the vigorous aeration and/or the addition of chemical oxidizing agents. Although the oxidation of iron by dissolved oxygen at conditions normally found in natural waters (i.e., pH value 6.5–7.5) is in the order of several minutes [11], it was noted that various conventional methods removed iron efficiently, even when the raw water characteristics pointed to poor Fe(II) oxidation. The examination under a microscope of relevant sludge samples revealed that in all similar cases, there has been a massive growth of iron bacteria, that is, *Gallionella ferruginea*, or filamentous ones, such as *Leptothrix ochracea* [12]. It was apparent that iron was removed by biological means.

Since then, several studies have been performed and treatment units for the biological removal of iron have been installed in several European countries, such as in Greece, France, Germany, Denmark, the United Kingdom, and Croatia [13].

The efficient removal of iron in the presence of microbial activity has also been reported during the operation of rapid and slow sand filters, of fluidized bed reactors, and of granular activated carbon filters and during soil percolation [14]. However, the most commonly applied system relies on the presence of a simple sand filter and a limited amount of aeration (i.e., less than 50% of saturation) [7]. Iron-containing groundwater is firstly subjected to aeration and then passes through the sand filter, where Fe(II) is oxidized to Fe(III), and Fe(III) is precipitated in the form of hydrous FeOOH, creating an orange-colored sludge on the sand surface.

The excessive sludge, which contains iron bacteria, and the precipitated iron are removed by backwashing, and these suspensions are left to settle out, for example, in a lagoon or in another sedimentation basin. The necessary iron bacterial inoculum is derived from the groundwater (indigenous) and it is therefore self-seeding [12,15]. The biological removal of iron is very efficient over a long operational period, and residual iron concentrations below 10 μg/L can be constantly achieved. Even when this procedure was cut off for 1 month, the efficient restart of this method and the effective removal of iron were reachieved within only few days (2–3) after restarting the operation [16].

5.3   BIOLOGICAL MANGANESE REMOVAL

Biological oxidation and removal of manganese have also been reported and applied for the removal of dissolved manganese from groundwaters [12,15]. The concentrations of dissolved manganese in anaerobic groundwaters can reach the order of several hundreds of mg/L; however, the usual manganese concentrations fall in the range between 0.1 and 1 mg/L [17].

The removal of dissolved manganese (Mn^{2+}) from groundwaters is generally accomplished by oxidation, followed by precipitation and (sand) filtration for the removal of the oxidized insoluble products [18]. The abiotic oxidation of dissolved manganese by oxygen can be described by the following general equation [17]:

$$-\frac{d[Mn(II)]}{dt} = k_0[Mn(II)] + k_1[Mn(II)][MnO_x] \qquad (5.1)$$

This expression implies that the homogenous manganese oxidation can be accompanied by an autocatalytic action, in the presence of existing manganese solid phases [17]. The abiotic homogenous manganese oxidation by the presence of oxygen is a slow process at pH values below 9 [17]; thus, in the usual pH values encountered in most surface or groundwaters intended for human consumption (i.e., between 6 and 8), it will not be oxidized. Therefore, manganese is very difficult to be removed by application of simple aeration and the subsequent precipitation [19].

TABLE 5.3

Rates of Oxidation of Mn(II): (a) Abiotic Oxidation in the Presence of Goethite and (b) Biological Oxidation

pH	k (min^{-1})	Comment	Literature
8.5	0.0117	In the presence of goethite	[20]
7.2	0.174	Biological oxidation	[15]

It is for these reasons that chemical oxidation is generally required, in order to achieve the precipitation and the effective removal of manganese within reasonable time periods and at pH values relevant to drinking water treatment, that is, between 6.5–8. It has been well established that $KMnO_4$ is an effective oxidant of dissolved manganese over a broad range of pH values, whereas chlorine and ozone can also be applied [18]. To avoid the use of chemicals, biological oxidation of manganese has been considered as a viable alternative for the efficient treatment of groundwaters, which is generally carried out by the presence of microorganisms, which mediate the biotic oxidation of Fe(II), apart from the stalked bacteria of *Gallionella* genus [12].

These bacteria require more stringent conditions to oxidize manganese, than for iron oxidation. Particularly, a completely aerobic environment is required; dissolved oxygen concentration should be higher than 5 mg/L and redox potential over 300 mV (often between 300 and 400 mV), depending on the pH value. In any case, the required redox potential is lower than the value corresponding to the introduction of a chemical oxidant agent, which is stronger than oxygen. Under these conditions, manganese removal is very efficient, and residual manganese concentrations are below 20 μg/L. As in the case of iron, the kinetics of biological manganese oxidation is quite fast (Table 5.3).

When ammonia coexists in the groundwater, biological removal of manganese can take place only after the previous complete nitrification, due to the necessary evolution of redox potential. Low ammonia concentrations (<2 mg/L) did not affect the removal of manganese and could take place simultaneously in the same filter, although for higher ammonia concentrations, the removal of manganese can take place only after nitrification is completed [21].

5.4 ARSENIC OCCURRENCE AND SPECIATION IN GROUNDWATERS

Iron and manganese are often associated with elevated arsenic concentrations of geogenic origin in groundwaters [2,4]. Arsenic is a toxic metalloid element, causing adverse effects on human health. Several studies have been performed to assess arsenic toxicity and its adverse effects on human health [22], demonstrating that arsenic is a human carcinogen.

The distribution of inorganic arsenic species [As(III), As(V)] in natural waters is mainly dependent on redox potential and pH conditions. Under oxidizing conditions, such as those prevailing in surface waters, the predominant species is pentavalent arsenic, which is mainly present with the oxyanionic forms ($H_2AsO_4^-$, $HAsO_4^{2-}$) with $pK_a = 2.19$, $pK_b = 6.94$, respectively. On the other hand, under mildly reducing conditions, such as those existing in anaerobic groundwaters, As(III) is the thermodynamically stable form, which at pH values of most natural waters is mostly present as the nonionic form of arsenious acid (H_3AsO_3, $pK_a = 9.22$) [23].

The sorption of As(V) onto mineral surfaces of soils or sediments is a dominant mechanism for the immobilization and removal of arsenic from aquatic sources [24]. As the respective pH value plays a significant role on the speciation of arsenic in waters and for a given redox potential value, it also comprises the key controlling parameter in arsenic removal processes, when sorption is used for its removal.

Generally, adsorption of As(III) onto iron or manganese oxides is less efficient than of As(V); therefore, the immobilization of As(III) is enhanced by the preliminary oxidation of As(III) to As(V) [24,25]. The higher mobility of As(III) is illustrated clearly by the difference in the respective breakthrough curves of adsorption onto iron oxides, demonstrating the removal efficiencies of both As(III) and As(V) species by their sorption onto hydrous ferric oxides (HFOs), coated on the surface of filters [26].

As(III) oxidation can be performed by a variety of oxidants such as ozone, chlorine, potassium permanganate, or manganese dioxide [27]. However, use of chemical reagents increases the operational costs of the method, while it might introduce undesired by-products in the finished water. Therefore, in iron- or manganese-containing groundwaters, application of biological iron or manganese oxidation can lead to both As(III) oxidation and As(V) removal by sorption onto biogenic iron and manganese oxides.

5.5   ARSENIC REMOVAL DURING BIOLOGICAL OXIDATION OF FE(II)

In several studies over the last decade, biological iron and manganese oxidation has been applied to remove As(III) and As(V) from groundwaters, without the use of chemical oxidizing agents [7,13,16]. It was shown that arsenic can be removed by direct adsorption or coprecipitation on the preformed biogenic iron or manganese oxides, whereas the oxidation of As(III) was induced by the iron-oxidizing bacteria and leads to improved overall removal efficiency of arsenic content.

The main parameter affecting the efficiency of As(III) removal during biological iron oxidation is the applied redox potential [7]. Under specifically applied conditions (i.e., redox potential 280–330 mV, pH value 7.2, dissolved oxygen 2.7–3.2 mg/L), which are optimized for the removal of dissolved iron content, As(III) was found to be oxidized and removed by 95%. It is also worth mentioning that at these conditions the residual arsenic was always below 5 µg/L, when starting from initial arsenic concentrations around 35–40 µg/L. When the same experiments were carried out after disinfecting with NaOCl the filtration columns, it was found that the removal of arsenic was drastically decreased, implying that bacteria contribute significantly to the oxidation of As(III).

These results indicated that during the biological removal of iron, the oxidation of As(III) took place to the respective pentavalent form, which was considered as the cause for the enhanced overall arsenic removal. Oxidation of As(III) by air or oxygen could take days, weeks, or even months, depending on the specific experimental conditions [28]. However, by the application of biological oxidation, the residence time is usually in the range of few minutes (around 7 min) and the dissolved oxygen content far below saturation. Thus, under these conditions, oxygen cannot be the sole (or even the major) oxidative agent for arsenic and iron in this treatment system. Bacteria do play an important role on the sufficient remediation of trivalent arsenic.

Long-term experiments, regarding the removal of As(III), showed that the removal of trivalent arsenic was efficient during the whole operation period. Even when the filters were shut down and the operation was cut off for 1 month, due to a technical problem, the efficient restart of treatment process and the effective removal of arsenic were accomplished within 4 days [16].

As expected, As(V) removal was very efficient with the application of a specific treatment method. From a wide range of initial As(V) concentrations (50–200 µg/L), the residual As concentration was always below 10 µg/L.

It should be also noted that most of physicochemical sorptive treatment methods present the drawback of adsorbate exhaustion (breakthrough). In order to continue the operation, the consumed adsorbing materials need either regeneration or replacement [26]. In this process, the adsorbents were continuously produced by the biological oxidation of dissolved iron, which was also present in groundwater. Therefore, the need for regeneration was eliminated and there was no requirement for monitoring the respective breakthrough point. Once set in effective operation, the only consideration would be to perform the backwashing action periodically, in order to avoid filter clogging and to remove the excessive sludge [7].

5.6 ARSENIC REMOVAL DURING BIOLOGICAL OXIDATION OF MN(II)

The removal of As(III) or As(V) can also be accomplished, even when only manganese is present in the groundwater, but at significant concentrations, that is, higher than 0.5 mg/L. Around 80% removal of arsenic can be achieved for both inorganic species, when the initial concentrations are in the range of 30–45 µg/L [8]. The results are less efficient than those obtained during the biological oxidation of iron from groundwaters [7]. During iron oxidation, the percentage removal of arsenic accounts for more than 90% and the MCL (10 µg/L) can be easily achieved, even starting from rather high initial arsenic concentrations (230 µg/L).

The difference in the removal efficiency can be attributed to the fact that iron oxides are efficient adsorbents for arsenic, presenting a strong tendency to create surface complexes with arsenic ions, whereas the use of manganese oxides is not equally efficient. Manganese oxides usually present point of zero charge values between 2 and 4.5 [17]. Therefore, at the pH values of most natural waters (6–8), the net surface charge of manganese oxides is negative. As a result, the sorption of arsenic onto manganese oxides can be attributed to chemical interactions through specific adsorption, because the respective coulombic interactions do not favor the sorption.

This technology appears to suit well treatment of manganese-containing waters, which are slightly contaminated with arsenic and can be applied without the

TABLE 5.4

Half-Life Values of As(III) Oxidation, by Biological Oxidation of Manganese, by Manganese Dioxide, by Dissolved Oxygen, and by Ozone

As(III) Oxidation	Half-Life, $t_{1/2}$	References
Biological oxidation of manganese	3 min	[8]
Oxidation with MnO_2	0.17 h	[29]
	0.33 h	[30]
Oxidation with dissolved oxygen (abiotic)	2.2 days	[28]
Oxidation with ozone	4.5 min	[28]

additional use of chemical reagents for the oxidation of Mn(II) or As(III). This renders the method cost-effective, while the fact that these filtration systems can be also used for efficient long-term operational periods, without replacing the filter medium, consolidates further this statement.

As with the case of arsenic removal using iron oxides, the oxidation of arsenic was catalyzed by the presence of bacteria. During this process, the depletion of As(III) concentration is faster than the depletion of total arsenic concentration, indicating that As(III) is oxidized prior to removal. As(III) oxidation using biological manganese oxidation is faster, as compared with the respective half-life values for arsenic oxidation applying other oxidants (Table 5.4).

It can be observed that the oxidation of As(III) in the specific treatment system is much faster than the other physicochemical methods usually applied. The faster rates of As(III) oxidation support the conclusion that the oxidation of As(III) is catalyzed by the bacteria, most likely by the same microorganisms that mediate the oxidation of Mn(II). Furthermore, the oxidation of As(III) (presenting a half-life $t_{1/2}=3$ min) was found to proceed almost in parallel with the bacterial manganese oxidation for this specific treatment unit, because the kinetic analysis of Mn(II) oxidation yielded a relevant half-life value ($t_{1/2}=3.98$ min).

5.7   EFFECT OF PHOSPHATE ON REMOVAL OF AS(III)

Among the anions, phosphate may be present in aquatic sources and can therefore inhibit the sorption of arsenic. The effect of phosphate on the removal of arsenic by sorption onto iron oxides has been previously documented. Meng et al. [31] have reported that in order to achieve the efficient removal of arsenic by coagulation, when phosphate was present, a ratio of Fe/As=40:1 was needed, whereas in the absence of phosphate, the respective ratio was only 12:1.

The effect of phosphate on the removal of arsenic by sorption onto biogenic manganese oxides is strong. Phosphate inhibits arsenic(III) removal, which from 80% removal efficiency (in the absence of phosphate) drops down to around 30%. Phosphate competes with arsenic for the available sorption sites on the surface of manganese oxides. The sorption of phosphate onto manganese oxides is a very efficient and fast procedure. Manganese oxides present high affinity for creating surface complexes with phosphates, with the respective value of log K equal to 29 [32]. However, the presence of phosphate did not exercise any impact on the oxidation of As(III), even if the overall arsenic removal is inhibited. Katsoyiannis et al. [8] showed that although the effluent concentration of total arsenic was around 20 µg/L, the respective concentration of As(III) accounted only for 1–2 µg/L.

5.8   USE OF PLUG FLOW REACTORS COMBINED WITH MICROFILTRATION

A modification of the traditional biological iron and manganese oxidation taking place in fixed bed bioreactors is the use of plug flow reactors followed by membrane microfiltration [33]. The application of such experimental unit was examined for the removal of As(III) and As(V) from groundwaters. The examined groundwater (Berlin, Marienfelde) contained average Fe(II) and Mn(II) concentrations of 2.9 and 0.6 mg/L, respectively. Oxidation of these metals provided sufficient adsorption sites, and therefore, arsenic species were removed from groundwater very efficiently. From initial arsenic concentrations in the range 20–250 µg/L, the residual concentrations were in all cases below 10 µg/L. A significant advantage of this technology is the uptake of oxidized iron and manganese onto recirculated suspended solids that flocculated in the pipe reactor, thus eliminating the need for mechanical cleaning of the membrane, while keeping the transmembrane pressure constantly low. Furthermore, As(V) removal capacity—in terms of µg As(V) removed per mg Fe(III)—of this hybrid plug flow reactor-microfiltration (PR-MF) unit was found to be significantly higher than that achieved by conventional coagulation—filtration with Fe(III). The presence of phosphate at concentrations up to 1.6 mg/L did not exercise any negative impact on As(III) removal, from initial As(III) concentrations up to 100 µg/L. The PR-MF process efficiently removed iron, manganese, and arsenic without the use of chemical reagents for oxidation or pH adjustment, and without the need for regular regeneration or backwashing, following the principles of green chemistry.

5.9   ARSENIC REMOVAL UNITS IN GREECE

5.9.1   Arsenic Occurrence in Greek Groundwater Sources

Elevated arsenic concentrations in groundwater have been reported for many areas in Greece. The regions where arsenic is found in Greek groundwater sources are classified in three categories: the geothermal regions, such as in Chalkidiki and in Aridaia regions; the rivers' alluvial deposits such as those in the basins of Aksios, Nestos, and Strymon rivers; and aquifers that are influenced by mineralization, resulting in arsenic mobilization over the centuries. The fourth case is that of the lake water and its influence in the nearby aquifers, like the case of Lake Volvi [4,5]. Table 5.5 contains the most remarkable arsenic cases found in Greek groundwaters under the described classifications. It must be noted that arsenic concentration in all bottled waters was found well below the drinking water regulation limit.

TABLE 5.5

Major Cases of Arsenic Contamination in Greek Groundwaters

Area	μgAs/L	Conditions
Chalkidiki/Agia Paraskevi	4500	Geothermal
Chalkidiki/Petralona	1500–2000	Geothermal
Chalkidiki/Triglia	200–400	Geothermal
Kavala/Eleftheres	800	Geothermal
Loutraki Aridaias	350–450	Geothermal
Thermopylae	200–300	Geothermal
Island of Kos/Kefalos	5–35	Geothermal
Aksios delta/Malgara	20–27	Alluvial deposits
Nestos Delta/Keramoti	15–20	Alluvial deposits
Eastern Thessaly/Agia	20–35	Mineralization
Mpourmpoulithra	40–60	Mineralization
Kavala/Nikisiani	25–30	Mineralization

Source: Data are obtained from Katsoyiannis, I.A. et al., *Desalin. Water Treat.*, 54(8), 2100, 2015.

5.9.2 Applied Treatment Technologies in Full-Scale Plants in Greece

In Greece, treatment measures have been taken in order to provide the population with drinking water containing arsenic concentration below 10 μg/L. The applied treatment technologies were greatly dependent on the environmental conditions and on arsenic speciation in the affected groundwaters [5,34].

In the deltaic area of Aksios, arsenic is mainly present in the trivalent form, due to the reducing conditions prevailing in the groundwaters. Treatment units have been constructed and placed in operation in the cities of Malgara, Kumina, and Vrachia. These units comprised biological oxidation for the removal of iron, manganese, and ammonia by practicing oxygen, resulting also in As(III) oxidation. Then, As(V) is removed by chemical precipitation with $FeClSO_4$ for the cases of Kumina and Malgara and by adsorption on granular ferric hydroxide (GFH) for the case of Vrachia. Similar to Vrachia treatment plant, biological oxidation followed by adsorption on GFH was applied in Mitrousi in Serres region (Strymon river basin) with initial arsenic concentration ranging between 10 and 20 μg/L, with As(III) to be the predominant species.

In the village of Triglia, where iron, manganese, and ammonia are in negligible concentrations, due to relatively oxidative conditions, oxidation of arsenic was performed by ozonation, and arsenic is effectively removed by chemical precipitation with $FeClSO_4$. Arsenic is also removed by chemical precipitation with $FeClSO_4$ from drinking water of Daidalos in Kos Island after biological treatment for iron, nitrite, and As(III) oxidation. In contrast, in the area of Pagaion Mountain, close to ancient Macedonian's mines, spring water, serving as drinking water for the municipality of

TABLE 5.6
Water Treatment Plants Installed in Greece for Treatment of Arsenic-Contaminated Groundwaters

Area	Oxidation Method	Treatment	As(tot) (μg/L) (Inlet)	As(III) (μg/L) (Inlet)
Malgara	Biological	Chemical precipitation	18	13
Kumina	Biological	Chemical precipitation	44	26
Vrachia	Biological	GFH/adsorption	7–11	4–6
Triglia	Ozone	Chemical precipitation	208	21
Nikisiani	No treatment	AquAszero/adsorption	27	4
Daidalos	Biological	Chemical precipitation	33	25
Melivoia	No treatment	Bayoxide/adsorption	41	4

Nikisiani, contains arsenic that is directly removed by adsorption onto AquAszero, since arsenic is mainly in the pentavalent form [34]. Table 5.6 presents a summary of the results from the various treatment units, installed in Greece.

Based on the residual arsenic concentrations mentioned in Table 5.6 and the iron concentration used for the removal of arsenic in each case, the specific arsenic removal q (mgAs/gFe) can be calculated for each treatment plant. Highest specific arsenic removal was observed at the treatment plant of Triglia with 41 mgAs/gFe, where chemical precipitation was applied and the pH was 7.5. In general, application of chemical precipitation results in much higher q values than adsorption of arsenic onto specific adsorption media, such as GFH or Bayoxide. Another important observation is that the pH value of groundwater significantly affects the efficiency of chemical precipitation as reflected by the values of specific arsenic removal. In Triglia plant, the q value was 41 and the pH of water was 7.5, whereas in Daidalos plant, the q value was 26 and the pH of groundwater was 7.7. In general, we observed that by increasing the pH by 0.5 units, the q value was decreased by approximately 50%.

5.10 CASE STUDY: ARSENIC, IRON, MANGANESE, AND AMMONIA REMOVAL BY BIOLOGICAL OXIDATION FOLLOWED BY COAGULATION–FILTRATION

5.10.1 DESCRIPTION OF TREATMENT UNIT

In the city of Malgara in Northern Greece, arsenic is present in concentrations between 15 and 20 μg/L, from which almost 70% is in the form of As(III). A treatment unit was constructed for the removal of iron, manganese, ammonia, and arsenic [21]. The detailed groundwater composition is shown in Table 5.7.

The applied method is based on the oxidation of Fe(II) by aeration, followed by biological oxidation of Mn(II), ammonium, and As(III) in upflow filters and subsequent coagulation–direct filtration in downflow filters.

TABLE 5.7
Groundwater Constituents in the Influent and
Treated Groundwater

Components	In Groundwater	In Treated Water	EC Limit
As(tot) (μg/L)	20	4	10
As(III) (μg/L)	14	1.4	—
Fe (μg/L)	165	50	200
Mn (μg/L)	235	5	50
NO_3–N (mg/L)	0.25	4.3	50
NH_4–N (mg/L)	1.2	<0.05	0.5
PO_4 (μg/L)	550	30	—

5.10.2 GROUNDWATER CHARACTERIZATION AND DECISION ON THE TREATMENT METHOD

The chemical composition of the groundwater is presented in Table 5.7, together with the water composition after the treatment [21].

The groundwater, directly after pumping, has a pH of 7.9, is relatively anoxic (DO < 2 mg/L), and has a redox potential slightly over 0 mV, indicating the prevalence of reduced conditions. The presence of Fe(II), Mn(II), and NH_4^+ is typical of reducing conditions. Manganese and ammonium concentrations (0.235 and 1.2 mg/L, respectively) were over the EC parametric values (0.05 and 0.5 mg/L, respectively), whereas Fe(II) concentration (165 μg/L) was below the current EC parametric value.

The total arsenic concentration of 20 μg/L is higher than the EC parametric value of 10 μg/L and requires remedial action. Arsenic is mostly present in its trivalent form (As(III)/As(tot)=70%), in consistency with the observed groundwater reducing conditions. Arsenic speciation is of great importance in the decision of treatment, because As(III) is generally less efficiently removed by conventional methods, and therefore, for efficient As(III) removal, oxidation to As(V) is necessary.

While the concentration of phosphate at 550 μg/L does not constitute a problem for the drinking water quality, it can adversely affect arsenic removal efficiency, since phosphate competes with arsenic for the available sorption sites on iron and manganese oxides.

These factors are important in deciding the treatment method to use. Arsenic, manganese, and ammonium must be removed from the water before it will be distributed for consumption. Iron and manganese removal can take place by biological oxidation mediated by indigenous iron- and manganese-oxidizing bacteria. In addition, it is also well established that ammonium can be removed by biological oxidation by indigenous nitrobacteria, and the end product will be nitrate through the process of nitrification.

The application of biological oxidation can also lead to As(III) oxidation, according to the aforementioned description of the treatment method. However, because the concentrations of iron and manganese are not sufficiently high to produce excessive iron and manganese oxides, it is expected that phosphate will outcompete arsenic sorption and hinder its removal. For these reasons, the subsequent coagulation and filtration step is applied.

5.10.3 FE(II), NH$_4$, MN(II), AND PO$_4$ REMOVAL

During the first treatment stage of aeration, which is essential for the subsequent biological oxidation (the bacteria require oxygen to mediate the reactions), only the concentration of dissolved iron in the filtered samples changes. This is expected since the kinetics of iron oxidation at pH 7.9 is very fast ($t_{1/2}$, roughly 2–3 min), and therefore, by simple aeration, Fe(II) is oxidized to Fe(III), which at pH 7.9 forms HFOs. Therefore, the Fe(II) concentration in the filtered samples after aeration was negligible. On the other hand, oxidation of Mn(II) by dissolved oxygen results only in a low decrease (i.e., 5%–10%) of the dissolved manganese concentration, attributable to the fact that Mn(II) oxidation by dissolved oxygen does not take place significantly at pH values below 9. Ammonium concentration remained practically constant as well [21].

Significant decrease in dissolved phosphate concentration (27% or 1.6 μM) occurs during the stage of aeration. This can be attributed to phosphate sorption onto the particulate iron oxides, which are formed during Fe(II) oxidation. Indeed, during aeration, we observed 2.7 μM of Fe(II) oxidation. Considering that 1 mol of phosphate can formally react with 1 mol of Fe(III) to form FePO$_4$ and in practice Fe/P ratios of 1.5–2.0 are needed to precipitate phosphate, the formation of 2.7 μM Fe(III) resulted in the expected sorption of phosphate (1.6 μM) and removal in the filters [21].

Manganese and ammonium are removed during the next treatment stage, which involves upflow filtration and biological oxidation. Ammonium oxidation leads to its conversion to nitrate through the process of nitrification, which is confirmed by the increase in the nitrate concentration during treatment (Table 5.7). From 1.2 mg/L of ammonium, the residual concentration is ≤0.05 mg/L, which means that roughly 64 μM of ammonium is converted to nitrate during the biological treatment stage. Indeed, the nitrate concentration increases from 0.25 to 4.3 mg/L by biological oxidation, which corresponds to the formation of approximately 65 μM of nitrate, implying a 1:1 ammonium conversion to nitrate [21].

Mn(II) is efficiently oxidized and removed in the second stage, due to the presence of manganese-oxidizing bacteria and contact with manganese oxide surfaces, which accumulate in the filter during the filtration of manganese oxides on the sand and anthracite, as also reported in the literature. Iron removal occurs mainly by filtration of the HFO formed in the first (aeration) stage. During biological filtration and oxidation, roughly 68% phosphate removal (4 μM removal) was also observed, which is most likely attributable to the retention of phosphate sorbed on the HFO formed during aeration and additional sorption on biogenic manganese oxides formed in the filter [21].

Additional phosphate removal was recorded during the coagulation–direct filtration step. The use of FeClSO$_4$ as a coagulant adsorbed the rest of the phosphate, which was removed during the subsequent filtration step.

5.10.4   As(III) Oxidation and As Removal

In the first treatment stage, aeration of the groundwater takes place. As already discussed, this process promoted the oxidation of ferrous iron but not of manganese. Similarly to manganese, As(III) oxidation was insignificant. The As(III) concentration was reduced by roughly 10%, while the total arsenic concentration decreased by 3%. This is attributed to the fact that abiotic oxidation of As(III) by dissolved oxygen is very slow ($t_{1/2}$, 2–5 days), depending on the concentrations of dissolved ferrous iron. Hug and Leupin [35] showed also that As(III) oxidation by dissolved oxygen was not measurable in the absence of Fe(II) and took place only in the presence of 20 µM or higher Fe(II). In the present case study, the concentration of dissolved ferrous iron is relatively low (roughly 3 µM), and although the water was air saturated, oxidation of As(III) on the timescale of minutes was negligible [21].

However, As(III) was almost completely oxidized in the following treatment stage, in which manganese and ammonium were oxidized as well. Nonetheless, the removal of As(tot) in this stage was low (roughly 22% in the filtered samples), and therefore, after this treatment stage, the arsenic concentration was still higher than 10 µg/L.

In the present case, low Fe(II) and Mn(II) concentrations in addition to the relatively high phosphate concentration are the reasons for the ineffectiveness of As(tot) removal in this treatment stage. The low concentrations of Fe(II) and Mn(II) result in only low amounts of oxides acting as adsorbents for arsenic removal. The limited adsorption sites on the iron and manganese oxides are occupied by phosphate, since roughly 68% phosphate removal was recorded. These results are in agreement with previous studies, where As(III) removal was examined in Fe(II)- and Mn(II)-containing groundwaters (150 and 600 µg/L, respectively) in the presence of phosphate in the range of 0.6–1 mg/L, and it was shown that although arsenic removal was adversely affected by the presence of phosphates, As(III) oxidation was practically unaffected [8].

The removal of arsenic takes place in the next stages of coagulation and the subsequent downflow filtration. The water, after biological oxidation, is treated with the coagulant (FeClSO$_4$). As(V) is sorbed on the HFO colloids formed by hydrolysis of the added Fe(III), which stay in suspension in the coagulant tank until they are removed in the next stage of downflow filtration. This is clearly shown by arsenic analysis, which depicted that the As(tot)-unfiltered concentration was still over 12 µg/L, while the As(tot)-filtered concentration was below 4 µg/L and did not change significantly in the last filtration stage.

5.11   PHYTOREMEDIATION FOR ARSENIC REMOVAL BY AQUATIC MACROPHYTES

Phytoremediation of toxic contaminants can be readily achieved by aquatic macrophytes or by other floating plants since the process involves biosorption and bioaccumulation of the soluble and bioavailable contaminants from water. In aquatic phytoremediation systems, aquatic plants can be either floating on the water surface or submerged into the water [36,37]. In early studies, it was found that the toxic elements can be accumulated in the plants by an order of magnitude higher than in the water bodies. These findings indicated the potential use of macrophytes for remediation of toxic metals from waters. A large number of aquatic macrophytes have been studied for the phytoremediation of toxic metals from waters, such as *Microspora* and *Lemna minor* and *Typha latifolia*. Some species of aquatic macrophytes have been proven to be able to accumulate high amounts of arsenic, such as water hyacinth, duckweed, water fern, butterfly fern (*Salvinia*), water lettuce, watercress, waterweed, and esthwaite waterweed. *Eichhornia crassipes*, belonging to the water hyacinth family, showed a removal rate of 600 mg arsenic/ha/day under field conditions and a removal efficiency of 80%. However, water hyacinth presents water management problems, because of its huge vegetation and high growth rate.

Duckweeds are small free-floating aquatic plants, which do not have distinct stems and leaves. The duckweed family comprises four genera and 34 species, from which *Lemna*, *Spirodela*, and *Wolffia* have been reported to accumulate arsenic from water. Alvarado et al. [37] found that the removal rate of *L. minor* was 140 mg arsenic/ha/day with a removal recovery rate of 5%. *Wolffia globosa* was reported to accumulate more than 1000 mg arsenic kg^{-1} dry weight and can tolerate up to 400 mg arsenic/kg dry weight. Species of water fern showed arsenic sorption capacity ranging from 29 to 397 mg/kg dry weight upon exposure to As(V) concentration of 50 mM. Several species of watercresses have shown significant arsenic uptake. An average of 29 and 16 mg/kg of arsenic has been found in the leaves and stems of *Lepidium sativum* grown in Waikato River in New Zealand, and higher concentrations have been reported in *Nasturtium microphyllum* (up to 138 mg/kg fresh weight).

From the family of esthwaite waterweed, *Hydrilla verticillata* accumulates higher arsenic upon exposure to As(III) than to As(V). Maximum arsenic accumulation was 231 mg/kg dry weight, observed for As(III) concentration of 10 μM for 7 days, whereas when exposed to the same concentration of As(V), the sorption capacity was 121 mg/kg dry weight.

Uptake mechanisms for arsenic by aquatic macrophytes have been investigated in several studies. Three mechanisms have been proposed for arsenic uptake: (a) active uptake through phosphate uptake transporters, (b) passive uptake through aquaglyceroporins, and (c) physicochemical adsorption on root surfaces. Through the first mechanism, As(V) is usually removed, most likely by substitution of phosphate by arsenates. As(V) is also removed by physicochemical adsorption on root surfaces. As(III) and organoarsenicals such as DMAA and MMAA get into the plant through the mechanism of aquaglyceroporin.

Based on the obtained results, phytoremediation of arsenic-contaminated water would be an alternative, good, option in the long term. However, up to now, large-scale implementation of this technology has not been reported yet. The application of large-scale plants will raise the issue of management (treatment and disposal) of the huge amount of phytoremediating plants with high content of loaded arsenic.

5.12 ARSENIC REMOVAL BY BACTERIA AND ALGAE

The use of bacteria for the reduction of As(V) to As(III) has been traditionally used for decontamination of soils and sediments, since reduction of As(V) to As(III) increases the mobility of arsenic, which can be more easily leached out from the soil. However, in water treatment, the critical step is the oxidation of As(III) to As(V), because As(V) is more efficiently removed by traditional methods, such as coagulation with iron and aluminum salts, ion exchange, lime softening, and adsorption on specific media. Therefore, in water treatment, the identification of bacteria that can oxidize As(III) is of high importance. The main bacteria that are able to oxidize As(III) are iron-oxidizing bacteria, such as *L. ochracea* or *G. ferruginea*, which work well at pH values relevant to groundwater treatment and therefore have found wide application in water treatment plants, as described earlier in the text. These bacteria belong to the aerobic oxidizers of As(III). Bacterial oxidation of As(III) was first reported in 1918. In 1949, Turner isolated 15 strains of heterotrophic As(III)-oxidizing bacteria [38]. Currently, physiologically diverse As(III) oxidizers are found in various groups of the domains Bacteria and Archaea and include both heterotrophic As(III) oxidizers and chemolithoautotrophic As(III) oxidizers (CAOs) [38]. Heterotrophic As(III) oxidation is considered as a detoxification mechanism that converts As(III) into less toxic As(V). In contrast, CAOs use As(III) as an electron donor during fixation of CO_2 coupled with reduction of oxygen. In recent studies, As(III) oxidizers isolated from a mining environment with high oxidizing activity have been used for water treatment [39].

With regard to arsenic removal by algae, literature studies indicate that species of algae are able to uptake heavy metals, which occurs when algae release a protein called metallothionein [40]. When the protein is released, the alga starts to bind the metal to itself, as a mechanism of defense. The results of a study performed by Suhendrayatna et al. [41] indicated that when *Chlorella* sp. is exposed to concentrations of arsenite up to 100 µg/L, cell growth started to become affected at concentrations higher than 50 µg/L, and at another study [42], it was concluded that *Chlorella* sp. can retain almost 50% of arsenite from solution.

In general, arsenic removal either by macrophytes or by bacteria or algae does not comprise technology applied in full scale for removing arsenic from groundwater, either because the technology is not so efficient or because it is not so convenient in application and operation. The biological oxidation of As(III), however, is finding wider applications but needs to be combined with a physicochemical method, such as adsorption or coagulation for removing arsenic from water.

5.13 SUMMARY

This chapter summarized the development of water treatment methods for arsenic removal from groundwaters based on the biological oxidation of iron and manganese. Biological iron and manganese oxidation, mediated by iron- and manganese-oxidizing bacteria, leads to the formation of iron and manganese oxides and, if As(III) is present in the groundwater, leads also to the oxidation of As(III) and to some extent to the removal of the oxidized arsenic by sorption onto biogenic iron and manganese oxides. The effect of phosphate was also described and the reasons for reduced arsenic removal in the presence of phosphate were analyzed. This chapter describes the application of such treatment methods in full-scale plants in Greece and presents in detail the operation of a treatment plant in the area of Malgara in Northern Greece. In this unit, As(III) is simultaneously present in groundwater with Fe(II), Mn(II), NH_3, and phosphate. The application of biological oxidation leads to the oxidation of iron, manganese, ammonia, and As(III). In the next stage, iron and manganese are removed and ammonia is transformed to nitrate and As(III) to As(V). However, because of the presence of phosphate in water, sorption of As(III) onto iron and manganese oxides is not so efficient, and therefore, coagulation with iron salts is additionally applied to lead to overall arsenic removal of more than 95% and final arsenic concentrations of less than 10 µg/L. Overall, the use of biological oxidation for the oxidation of As(III) comprises nowadays a state-of-the-art technology, especially for the treatment of groundwaters containing iron or manganese.

REFERENCES

1. Nordstrom, D.K., 2002. Worldwide occurrences of arsenic in groundwater. *Science* 296, 2143–2145.
2. Smedley, P.L., Kinniburgh, D.G., 2002. A review of the source, behaviour and distribution of arsenic in natural waters. *Applied Geochemistry* 17, 517–568.
3. Rowland, H.A.L., Omoregie, E.O., Millot, R., Jimenez, C., Mertens, J., Baciu, C., Hug, S.J., Berg, M., 2011. Geochemistry and arsenic behaviour in groundwater resources of the Pannonian Basin (Hungary and Romania). *Applied Geochemistry* 26, 11–17.
4. Katsoyiannis, I.A., Hug, S.J., Amman, A., Zikoudi, A., Hatziliontos, C., 2007. Arsenic speciation and uranium concentrations in drinking water supply wells in Northern Greece: Correlations with redox indicative parameters and implications for groundwater treatment. *Science of the Total Environment* 383, 128–140.
5. Katsoyiannis, I.A., Mitrakas, M., Zouboulis, A.I., 2015. Arsenic occurrence in Europe: Emphasis in Greece and description of the applied full scale treatment plants. *Desalination Water Treatment* 54(8), 2100–2107.
6. Hussam, A., Sad, A., Munir, A.K.M., 2008. Arsenic filters for groundwater in Bangladesh. Toward a sustainable solution. *The Bridge* 38(3), 14–23.
7. Katsoyiannis, I.A., Zouboulis, A.I., 2004. Application of biological processes for the removal of arsenic from groundwaters. *Water Research* 38, 17–26.
8. Katsoyiannis, I.A., Zouboulis, A.I., Jekel, M., 2004. Kinetics of bacterial As(III) oxidation and subsequent As(V) removal by sorption onto biogenic manganese oxides during groundwater treatment. *Industrial & Engineering Chemistry Research* 43, 486–493.
9. Sorlini, S., Gialdini, F., Collivignarelli, M.C., 2014. Survey on full scale drinking water treatment plants for arsenic removal in Italy. *Water Practice & Technology* 9(1), 42–51.
10. Guezenec, A.G., Michel, C., Touze, D., Breeze, D., Joulian, C., Coulon, S., Dictor, M.C., Klein, J., Deluchat, V., Dagot, C., 2010. Arsenic removal from a drinking water supply: Biological oxidation of arsenite. Arsenic in geosphere and human disease. *As 2010, 3rd International Congress: Arsenic in the Environment*, Kuang-Fu Campus, National Cheng Kung University (NCKU), Taiwan, pp. 441–442.
11. Stumm, W., Lee, G.F., 1961. Oxygenation of ferrous iron. *Industrial & Engineering Chemistry* 53(2), 143–146.
12. Mouchet, P., 1992. From conventional to biological removal of iron and manganese in France. *Journal of American Water Works Association* 84(4), 158–167.

13. Katsoyiannis, I.A., Zouboulis, A.I., 2006. Use of iron- and manganese-oxidizing bacteria for the combined removal of iron, manganese and arsenic from contaminated groundwater. *Water Quality Research Journal of Canada* 41, 117–129.
14. Bouwer, E.J., Crowe, P., 1988. Biological processes in drinking water treatment. *Journal of American Water Works Association* 80(9), 82–91.
15. Katsoyiannis, I.A., Zouboulis, A.I., 2004. Biological treatment of Mn(II) and Fe(II) containing groundwaters: Kinetic consideration and product characterization. *Water Research* 38, 1922–1932.
16. Zouboulis, A.I., Katsoyiannis, I.A., 2005. Recent advances in the bioremediation of arsenic contaminated groundwaters. *Environment International* 31, 213–219.
17. Stumm, W., Morgan, J.J., 1996. *Aquatic Chemistry: Chemical Equilibria and Rates in Natural Waters.* Wiley Interscience, New York.
18. Knocke, W.R., van Benschoten, J.E., Keanny, M.J., Soborski, A.W., Reckhow, D.A., 1991. Kinetics of manganese and iron oxidation by potassium permanganate and chlorine dioxide. *Journal of American Water Works Association* 83(7), 80–87.
19. Diem, D., Stumm, W., 1984. Is dissolved Mn^{2+} being oxidized by O_2 in the absence of Mn—Bacteria or surface catalysts? *Geochimica Cosmochimica Acta* 48, 1571–1573.
20. Davies, S.H.R., Morgan, J.J., 1989. Manganese(II) oxidation kinetics on metal oxide surfaces. *Journal of Colloid and Interface Science* 129, 63–77.
21. Katsoyiannis, I.A., Zikoudi, A., Hug, S.J., 2008. Arsenic removal from groundwaters containing iron, ammonium, manganese and phosphate: A case study from a treatment unit in northern Greece. *Desalination* 224, 330–339.
22. Hughes, M.F., 2002. Arsenic toxicity and potential mechanisms of action. *Toxicology Letters* 133, 1–16.
23. Cullen, W.R., Reimer, K.J., 1989. Arsenic speciation in the environment. *Chemical Reviews* 89, 713–764.
24. Katsoyiannis, I.A., Zouboulis, A.I., 2006. Comparative evaluation of conventional and alternative treatment methods for the removal of arsenic from contaminated groundwaters. *Reviews on Environmental Health* 21, 25–41.
25. Jekel, M.R., 1994. Removal of arsenic in drinking water treatment. In: Nriangu, J.O., ed. *Arsenic in the Environment. Part 1: Cycling and Characterization.* John Wiley & Sons, New York, pp. 119–130.
26. Katsoyiannis, I.A., Zouboulis, A.I., 2002. Removal of arsenic by iron oxide coated polymeric materials. *Water Research* 36, 5141–5155.
27. Dodd, M.C., Vu, N.D., Amman, A., Le, V.C., Kissner, R., Pham, H.V., Cao, T.H., Berg, M., von Gunten, U., 2006. Kinetics and mechanistic aspects of As(III) oxidation of aqueous chlorine, chloramines and ozone: Relevance to drinking water treatment. *Environmental Science and Technology* 40, 3285–3292.
28. Kim, M.J., Nriangu, J., 2000. Oxidation of arsenite in groundwater using ozone and oxygen. *Science of the Total Environment* 247, 71–79.
29. Scott, M.J., Morgan, J.J., 1995. Reactions at oxide surfaces. 1. Oxidation of As(III) by synthetic birnessite. *Environmental Science and Technology* 29, 1898–1905.
30. Oscarson, D.W., Huang, P.M., Liaw, W.K., Hammer, U.T., 1983. Kinetics of oxidation of arsenite by various manganese dioxides. *Soil Science Society American Journal* 47, 644–648.
31. Meng, X., Korfiatis, G., Christodoulatos, C., Bang, S., 2001. Treatment of arsenic in Bangladesh well water using a household co-precipitation and filtration system. *Water Research* 35, 2805–2810.
32. Yao, W., Millero, F.J., 1996. Adsorption of phosphate on manganese dioxide in seawater. *Environmental Science and Technology* 30, 536–541.

33. Katsoyiannis, I.A., Zouboulis, A.I., Mitrakas, M., Althoff, H.-W., Bartel, H., 2013. A hybrid system incorporating a pipe reactor and microfiltration for biological iron, manganese and arsenic removal from anaerobic groundwater. *Fresenius Environmental Bulletin* 22(12), 3848–3853.

34. Tresintsi, S., Simeonidis, K., Zouboulis, A., Mitrakas, M., 2013. Comparative study of As(V) removal by ferric coagulation and oxy-hydroxides adsorption. Laboratory and full scale case studies. *Desalination and Water Treatment* 51, 2872–2880.

35. Hug, S.J., Leupin, O., 2003. Iron-catalyzed oxidation of As(III) by oxygen and by hydrogen peroxide: pH dependent formation of oxidants in the Fenton reactions. *Environmental Science and Technology* 37, 2734–2742.

36. Rahman, M.A., Hasegawa, H., 2011. Aquatic arsenic: Phytoremediation using floating macrophytes. *Chemosphere* 83, 633–646.

37. Alvarado, S., Guedez, M., Lue-Meru, M., Nelson, G., Alvaro, A., Jesus, A.C., Gyula, Z., 2008. Arsenic removal from waters by bioremediation with the aquatic plants Water Hyacinth (*Eichhornia crassipes*) and Lesser Duckweed (*Lemna minor*). *Bioresource Technology* 99, 8436–8440.

38. Yamamura, S., Amachi, S., 2014. Microbiology of inorganic arsenic: From metabolism to bioremediation. *Journal of Bioscience and Bioengineering* 118, 1–9.

39. Battaglia-Brunet, F., Dictor, M.-C., Garrido, F., Crouzet, C., Morin, D., Dekeyser, K., Clarens, M., Baranger, P., 2002. An arsenic(III)-oxidizing bacterial population: Selection, characterization and performance in reactors. *Journal of Applied Microbiology* 93, 656–667.

40. Kauser, J., Mosto, P., Mattson, C., Frey, E., Derchak, L., 2006. Microbial removal of arsenic. *Water, Air, Soil Pollution: Focus* 6, 71–82.

41. Suhendrayatna Ohki, A., Kuroiwa, T., Maeda, S., 1999. Arsenic compounds in the freshwater green microalga *Chlorella vulgaris* after exposure to arsenite. *Applied Organometallic Chemistry* 13, 127–133.

42. Beceiro-Gonzalez, E., Taboada-de la Calzada, A., Alonso-Rodriguez, E., Lopez-Mahia, P., Muniategui-Lorenzo, S., Prafa-Rodriguez, D., 2000. Interaction between metallic species and biological substrates: Approximation to possible interaction mechanisms between the alga *Chlorella vulgaris* and arsenic(III). *TrAC Trends in Analytical Chemistry* 19, 475–480.

6 Remediation of Arsenic Toxicity

Sudeshna Mukherjee, Sanjay Kumar Verma, and Rajdeep Chowdhury

CONTENTS

6.1 INTRODUCTION

The presence of high level of arsenic in groundwater is a major concern around the world, especially in South Asia. Only in Bangladesh, an estimated 60 million people are exposed to unsafe levels of naturally occurring arsenic in their drinking water, radically raising the risk for cancer and other chronic diseases [1,2]. In areas where drinking water contains unsafe levels of arsenic, the pressing need is finding a safe source of drinking water. Two main options are adopted to solve the problem: finding a new safe source or removing arsenic from the contaminated source. Since most of the contaminated water is near the surface, many people now in Bangladesh have installed deep wells to tap the groundwater that is relatively free of arsenic—a practice that is essentially compromising on access to clean drinking water across the country, according to a report in the journal *Science* in 2010 [3]. Not only that, people in Bangladesh have started pumping clean water from the deep aquifers for irrigation. This can be a major cause of concern, as when irrigation wells pump high enough volumes, it can simultaneously pull down arsenic-contaminated water from the surface and jeopardize the quality of the groundwater below, which is used for drinking [3]. So, if deep wells are not a rational solution to the dreadful menace, a short-term goal can be to reduce arsenic levels through filtration techniques from the contaminated water. A wide range of technologies have been developed for the removal of high concentrations of arsenic from drinking water [4]. To our understanding, arsenic water filtration is therefore vital to many communities inside and outside Bangladesh where arsenic has been detected at dangerous levels in public drinking water supplies. In this context, this chapter shall discuss the various methods currently in use to filter arsenic from contaminated water. In this regard, it is worthwhile to mention that the most common valence states of arsenic in raw water sources are As(V) or arsenate and As(III) or As(III). In the pH range of 4–10, the As(III) species are neutral in charge, while As(V) compounds are negatively charged [5,6]. Therefore, the removal efficiency of As(III) is less compared to that of As(V) by any of the conventional technologies discussed later; hence, for effective and complete removal of arsenic from water, an oxidation of As(III) to As(V) is a necessary prerequisite. As(III) can generally be oxidized by the use of any one of the following chemicals: oxygen, hypochlorite, permanganate, and hydrogen peroxide. Additionally, to complicate the issue, all the arsenic treatment technologies discussed later ultimately concentrate arsenic in the adsorbing media, the residual sludge, or in a liquid media, and, therefore, disposal or treatment of these wastes is as equally important as arsenic water filtration itself. Therefore, the requirements for a suitable technique for removal of arsenic from water must include the following: high efficiency, safe technology to ensure the maintenance of the maximum contaminant level, simple operation, and minimum residual mass. As of now, there are numerous arsenic removal technologies developed by several sectors starting from universities to government organizations, groups, and private sectors. Some of the widely used technologies classified into categories based on the dominant removal process (although at times, a given technology may use multiple treatment processes) are discussed next.

6.2 CONVENTIONAL COAGULATION/FILTRATION

Coagulation process encompasses reactions used to remove suspended and dissolved solids from water which results in particle growth (floc formation); particle aggregation, including in situ coagulant formation; chemical particle destabilization; and interparticle collision within the source water that is being treated. Coagulation ideally involves the elimination of colloidal (0.001–100 μm) and settleable (>100 μm) particles; however, the term is also often applied to the expulsion of dissolved ions (<0.001 μm), which is actually precipitation. It can therefore be defined as a process by which dissolved ions in a solution form an insoluble solid through chemical reaction. Coagulation has been one of the most commonly used water treatment processes. Coagulation-reaction-generated aggregated suspended solids or flocs are thereafter generally removed by sedimentation and/or filtration. Alum and iron salts (e.g., ferric chloride or ferric sulfate) are few of the most commonly used coagulants for drinking water treatment [7]. Coagulation/filtration with the use of metal salts independently or with lime followed by subsequent filtration is the most profoundly documented method of arsenic removal from arsenic-laden water [8,9]. During the treatment process, the alum or ferric chloride is added into arsenic-laden water and is allowed to dissolve with efficient stirring for one to few minutes. During the flocculation process, microparticles and negatively charged ions attach to the flocs by electrostatic attraction including the arsenic species. The adsorbed arsenic can then be removed partially by sedimentation, while further filtration may be required to guarantee complete removal of all flocs [10]. The filtering medium may be of various types that include, for example, paper, sand, coal, activated carbon, cloth, and others that can retain the insoluble solid formed on its surface and let the water pass through. The pH of the solution, coagulation dose, and the nature of the used coagulation–precipitation agent are critical determinants for arsenic removal by this process; coagulation using ferric chloride works best at pH around 8, while alum has an effective range from pH 6 to pH 8 [11]. In ferric coagulation method, arsenic removal was found to decrease rapidly with increasing pH above pH 8; however, removal was considerably insensitive to changes in pH below 8. Likewise, at low coagulant doses, arsenic removal tends to increase with escalating iron concentrations; however, added increase in dose produced diminishing results. It is presently possible to reduce arsenic load even 40 times of initial concentration, assuming pH, oxidizing, and coagulation agents are stringently controlled [10]. However, in source water with large amounts of As(III), preoxidation of As(III) to As(V) should be considered as the latter is removed more readily than As(III). While the coagulation process followed by filtration has been considerably successful in removing arsenic, there is still need for further research in this direction to optimize the process, such as standardizing the exact dose of iron to be used, the optimal pH to be set in order to achieve a specific reduced arsenic concentration. Such endeavors shall not only save money, as the socioeconomic condition of majority of people suffering from arsenic burden in water demands low-cost treatment units that can be implemented in rural areas, but will also result in the production of less iron (e.g., ferric hydroxide) as waste reducing the pain of disposal.

6.2.1 BUCKET TREATMENT UNIT

The bucket treatment unit (BTU) was developed by the DPHE-DANIDA project in Bangladesh and is based on the aforementioned coagulation and filtration process [12]. It consists of two buckets, 20 L each, placed on top of the other. Chemicals like powdered alum (aluminum sulfate) and powdered potassium permanganate are mixed manually in raw arsenic-contaminated water in the upper bucket and stirred with a wooden spoon briskly for 30–60 s. Arsenic-infested water is then allowed to stand for around 3 h until the coagulated flocs settle down in the red bucket. The supernatant water from the top red bucket is then allowed to flow into the lower green bucket having a coarse sand filter box installed in it. Flocs, if present in water, can thus be removed by filtration in the lower bucket. Water to be treated is generally preoxidized before the alum is added to have predominantly As(V) species. One of the major concerns with the use of BTU is the application of aluminum and manganese above WHO-recommended concentrations [12]. Aluminum is suspected to be a neurotoxic agent, and many studies confirm its positive correlation with Alzheimer's disease [13]. Manganese can also have synergistic effect on aluminum toxicity, and an excess of manganese ion can impart undesirable taste to beverages [14]. Alternatively, iron salts can be used instead of alum to avoid such issues.

6.2.2 STEVENS INSTITUTE TECHNOLOGY PROCESS

This process is quite similar to that of BTU. The Stevens Institute Technology (SIT) process also uses two buckets, one to mix chemicals (iron sulfate and calcium hypochlorite) and the other to separate flocs produced by sedimentation and filtration [4]. The second bucket in SIT has an inner third bucket with slits to facilitate sedimentation process and to keep the underneath filter sand bed in place. However, the coarse sand bed used for filtration is quickly clogged by flocs produced and requires frequent washing in the SIT unit [4]. A plastic pipe is attached to the sand bed to deliver the clean water outside the unit.

6.2.3 FILL AND DRAW UNITS

These are community-type water treatment units developed under the DANIDA arsenic mitigation project. The tank has a high (600 L) water storing capacity with tapered bottom for effective disposal of accumulated sludge. The tank is also equipped with manually operated mixer with flat blade impeller. Arsenic-laden water along with oxidants and coagulants is poured in the tank, mixed by rotating the device, and left for few hours for sedimentation. Flocs are formed by the hydraulic gradient of rotating water inside the tank; the water is thereafter allowed to settle that facilitates sedimentation. The settled water is withdrawn through a pipe that is fitted at a few inches above the level where sedimentation occurs; the water is then passed through sand filter followed by ultimate collection for drinking [4].

Similar arsenic removal units based on the principle of coagulation/filtration is also often attached to the tube wells in Bangladesh or India. They have been found to effectively remove 90% of arsenic from contaminated water when initial concentration ranges around 300 µg/L. The treatment procedure generally includes addition of sodium hypochlorite and aluminum alum, mixing, flocculation, sedimentation, and subsequent filtration in a compact unit.

6.2.4 COAGULATION WITH LIME

Water treatment by quick lime or hydrated lime also is known to remove arsenic [4]. The principle followed in lime treatment process is quite similar to what is done in coagulation with metal salt. The precipitated calcium hydroxide here acts as an adsorbing flocculent for arsenic [15]. According to bench scale studies by Sorg and Logsdon, arsenic removal by lime is pH dependent. Removal of both arsenic species is significantly low below pH 10, while the adsorbing capacity increases considerably—close to 100% for As(V) and 75% for As(III) at pH above 10.5. The highest removal efficiency is achieved between pH 10.6 and 11.4 [16]. Also, another study by McNeill and Edwards showed that As(V) removal was highest if pH was high enough to precipitate magnesium hydroxide (around pH 11); coprecipitation of arsenic with magnesium hydroxide and calcium carbonate seems to be the predominant mechanism for arsenic removal by lime. Also, addition of iron was found to increase arsenic removal by lime; however, competition of arsenic and carbon on sorption sites reduced efficiency to some extent [17]. Ideally, lime softening may be used as a pretreatment to reduce hardness of water and enhance clarification before coagulation by alum or iron and subsequent filtration.

Arsenic removal from drinking water has been extensively studied at the laboratory or pilot-scale level; however, translation of this process to sustained full-scale applications is still awaited. There is a pressing need for determination of effectiveness of coagulation/filtration process in producing drinking water on a long-term basis under different operational and seasonal conditions. Also gap exists in information on the amounts and chemical composition of the disposable residue generated by the aforementioned processes and the environmentally accepted procedure for their disposal.

6.3 ADSORPTIVE FILTRATION

Arsenic-infested water can also be efficiently removed by passing it though adsorptive granular media enclosed in a pressure vessel. When the contaminated water passes through the media, the negatively charged arsenic ions are adsorbed onto the surfaces of the positively charged media granular particles. The efficiency of adsorptive media depends on the use of oxidizing agents as aids to augment the sorption of arsenic on the media [18]. In addition to strong arsenic removal efficiency, low-cost, minimal operator attention during runs and simple operating mechanism make adsorptive media treatment a very attractive option for arsenic water filtration [9]. There are presently several adsorption media available: activated alumina (AA), iron-based sorbents, titanium-based media, zirconium-based media, activated carbon, manganese-coated sand, silicium oxide, kaolinite clay, and activated bauxite [19]. The media that are used most widespread include modified

AA and iron-based materials. Here, in this chapter, we discuss the pros and cons of few of the most prevalently used adsorbants.

6.3.1 ACTIVATED ALUMINA

Activated alumina (AA), for example, has a good adsorptive surface in the range of 200–300 m^2/g. AA, with its high surface area, is a perfect choice for the adsorption of various unwanted minerals in water, with proven results for arsenic [20]. When arsenic-contaminated water passes through a packed column of AA, the impurities with arsenic present in water are efficiently adsorbed on the surface of AA grains. In due course, the column becomes saturated, and then, the regeneration of saturated alumina is carried out by exposing the medium in the column to 4% NaOH resulting in generation of highly arsenic-contaminated caustic wastewater for disposal. Arsenic removal by AA is dependent on pH, and there is a drop in efficiency as the point of zero net charge is approached; at pH 8.2 where the surface is negatively charged, the removal capacity is only 2%–5% of the capacity at optimum pH [21]. Also, the presence of phosphate, sulfate, fluoride, and chloride in raw water may reduce the adsorption capacity of AA substantially. Chemical handling requirements can make this technique too complex and dangerous for many small systems; also, AA may not be efficient in long-term applications, as it loses significant adsorptive capacity with each regeneration cycle; further, generation of highly concentrated waste may be a point of concern. Some examples of AA-based media include the *Actiguard AA400G Alumina*, *BUET AA*, *Alcan Enhanced AA*, and *Apyron Arsenic Treatment Unit*.

6.3.2 GRANULAR FERRIC HYDROXIDE

Granular ferric hydroxide (GFH) is another widely used adsorptive medium for removal of As(V) and As(III) from water [22]. The GFH adsorbent was first developed at the Technical University of Berlin, Germany, Department of Water Quality Control, especially for selective removal of arsenic from natural water. GFH reactors are fixed bed adsorbers that operate like a conventional filter with a downward flow of water [4]. Of the known removal systems, the adsorption of arsenic by GFH is the simplest, safe, and the most effective method for elimination of arsenic from contaminated groundwater. GFH is poorly crystallized FeOOH prepared from ferric chloride solution by neutralization and precipitation with sodium hydroxide. As no drying procedure is included while it is manufactured, all the pores are entirely filled with water, leading to a high density of accessible adsorption sites and hence a high adsorption capacity. The water containing high dissolved iron and suspended matters are aerated and filtered through a gravel/sand bed as a pretreatment to avoid clogging of the adsorption bed. The granulation of the bed does not lead to a considerable decrease in adsorption capacity. The particle free water passes through an adsorption tower, where the arsenic content in water is brought down as water containing As(V) and As(III) binds on the surface of ferric hydroxide building inner sphere complexes (chemisorptions). The adsorbent can be effectively applied in the pH range from 5.5 to 9; however, As(V) adsorption drops to some extent with pH, which is quite typical for anion adsorption. At high pH values, GFH outperforms AA. Below a pH of 7.6, the performance is quite comparable. The competition of sulfate on arsenic adsorption was not robust; phosphate, however, competed strongly with As(V), which reduces As(V) removal efficiency through GFH. The plants are very simple to install, compact, and practically almost maintenance free. The adsorption capacity of GFH is 5–10 times better than that of AA. Also, as the equipment involves fixed bed adsorption, the adsorbent is better utilized; the quantity of solid residue at the end of the adsorption process is less and needs no further dehydration. The used material is a nontoxic and nonhazardous solid waste, and its volume being small, its disposal is comparatively less challenging. The typical residual mass of the spent GFH is in the range of only 5–25 g/m^3 treated water, whereas that of the spent AA is approximately about 10 times higher. Also, under regular conditions, no leaching of arsenic takes place out of spent GFH, while in other systems the disposal of the spent material (sludge) creates big environment issue; the spent adsorbent can be profitably utilized as a useful component for manufacturing bricks [22]. To our knowledge, GFH is one of the safest and most effective systems for removal of arsenic from groundwater. The only Achilles' heel of this technology appears to be its expenditure; currently, GFH media costs around $4000 per ton. However, if a GFH fixed bed can be used several times longer than an AA bed, it may be more cost effective.

6.3.3   IRON OXIDE-COATED SAND

Adsorption of arsenic and its removal with AA or fixed bed GFH is effective but the total cost incurred on arsenic elimination from drinking water was on the higher side. An effective yet economical technique was therefore desired to reduce arsenic load in the rural areas of South Asia [19]. The use of iron oxide-coated sand (IOCS) consisting of sand grains coated with ferric hydroxide used in fixed bed showed some tendency for arsenic removal and also proved to be cost effective. The metal ions were exchanged with the surface hydroxides on the IOCS and studies confirmed that it can be effectively used to achieve a low level of arsenic in drinking water supplies for small communities or for home treatment units. However, pH was found to have an effect on arsenic adsorption by IOCS; pH increase from 5.5 to 8.5 decreased the sorption of As(V) by about 30%. Like other processes discussed previously, the arsenic oxidation state did play a role in its removal: As(V) was more easily removed than As(III). The effect of other ions like sulfate and chloride had little impact on IOCS-mediated As(V) removal [19]. Recent studies confirm that the use of limestone as a pH controller or a buffer along with IOCS can increase its arsenic removal efficiency [23]. Regeneration of IOCS is performed similar to that of AA and is accomplished using a strong base, typically NaOH, and subsequent neutralization by strong acid, typically H_2SO_4. However, the removal efficiency was found to decline slightly after two total regeneration procedures. With low cost and simple operation, IOCS can be a promising novel medium for arsenic removal in drinking water at least for small-scale or household purposes.

6.3.4 SULFUR-MODIFIED IRON

Sulfur-modified iron (SMI) is a kind of reactive media that is often used for the removal of contaminants like arsenic and selenium from water [24,25]. The media is ideally prepared from finely divided metallic iron and powdered elemental sulfur that are hydrothermally prereacted and granulated for use. The powdered iron and powdered sulfur are thoroughly mixed with an oxidizing agent and then added to the arsenic-contaminated water to be treated. The SMI process converts raw iron from a porous particle to a particle that is smoother and more homogeneous structure; arsenic is removed from water through sorption onto the iron and sulfur matrix. With the SMI technology, high adsorptive capacities were obtained with final arsenic concentration of 0.050 mg/L. However, arsenic removal was found to be influenced by pH. Interestingly, significant removal of both As(V) and As(III) was obtained with SMI. The problems that are often encountered with SMI processes are rusting of the media, cementing, and clogging; formation of separated preferential flow paths and better arsenic removal were observed with reduced flow rates. A higher iron-to-sulfur ratio showed improved column flow dynamics and As(V) removal efficiency; an iron-to-sulfur ratio of 10:1 produced the best removal potential in a study. The operating cost for SMI, when the process is operated below pH 8, is considerably lower than alternative arsenic exclusion technologies such as ferric chloride addition and AA. However, cost increases proportionally with increased flow rates and increased arsenic concentrations [24,25]. Packed bed and fluidized bed reactors appear to be the most promising for successful arsenic removal in pilot-scale and full-scale treatment systems based on present knowledge of the SMI process.

6.3.5 USE OF NATURALLY OCCURRING IRON

The cost of arsenic removal by the aforementioned methods is unfortunately unaffordable to a large population of people residing in the interior rural areas of Bangladesh or Mongolia, where the average household income is very meager. To have a low-cost process, naturally occurring iron oxides are often used in Inner Mongolia for effective removal of arsenic from groundwater. This process can be used affordably and on a relatively small scale, allowing for rapid propagation into households in rural interiors. Magnetic particles of naturally occurring iron oxides are generally used as adsorbents to remove arsenic. The magnetic iron oxide particles are produced by grinding iron ores that are available at low prices and segregated from nonmagnetic materials by the use of a magnet drum separator. The magnetic particles are then dispersed in water in the form of powder or sludge to adsorb arsenic. The particles are again recovered by a magnet drum separator after the treatment is over. Also the magnet drum separator is designed to be able to operate without electricity to minimize the cost. The use of naturally occurring iron definitely holds promise for small-scale household arsenic removal. However, the efficiency depends on arsenic contents of water. Also, it has been observed that in about 50% and above of the hand tube wells installed in Bangladesh where water has iron in excess, the arsenic concentration satisfies the standard required, while the ones that have low iron in water show arsenic above recommended level [4]. The iron precipitates formed by oxidation of dissolved iron present in groundwater probably have the affinity for adsorption of arsenic.

6.3.6 CERIUM-BASED ADSORBERS

Other adsorptive medium like hydrous cerium oxide (HCO) is also found to be a good adsorbent [26]. It was found that cerium hydroxide, one of the rare earth elements, has an elevated selective adsorption against negative ions, such as arsenic, fluorine, and boron. HCO was introduced into practical use as an adsorbent called cerium adsorbent READ-As series by Nihonkaisui Co., Ltd. HCO possesses the least solubility against acid among the rare elements on earth, and since the adsorbent does not elute when harmful ions in water are removed, it is often safely used for the adsorption of arsenic. One of the advantages of HCO-based technology is that in laboratory-based experiments, it shows a high adsorption for both trivalent and pentavalent states of arsenic, and hence no pretreatment like oxidation or pH control is required [26]. Also, this methodology can be used repeatedly by regenerating it using an alkali. Overall, HCO can be an effective alternative strategy to AA or GFH-based technology in use.

6.3.7 LOW-COST INDIGENOUS FILTERS IN USE

The conventional large-scale methods described previously for the removal of arsenic from water through adsorption and filtration often cannot meet the requirements mandated by the region's numerous remote locations and limited fiscal resources. Such a recognized need forced the scientists to develop affordable technology, which can be applied in remote areas, can be operated without electricity, requires no chemical addition, and yet is simple to operate and maintain. In answer to this challenge and to address the magnitude of this problem, various low-cost technologies for purifying arsenic-contaminated groundwater at the household level have been developed mostly centered around adsorption principle discussed earlier as the mechanism of functioning. In the following section, we discuss some of these low-cost methods.

6.3.7.1 Apyron Arsenic Treatment Units

Apyron Technologies, Inc. developed an integrated treatment system that is easily adaptable to the rural setting, is effective under a wide range of water chemistry applications, and is extremely cost effective [27]. The key component to the system is Apyron's Aqua-Bind media, an alumina-based adsorption media confined in a cylindrical adsorber vessel that can selectively remove the two most common forms of arsenic, arsenite and arsenate, from water. This media consists of nonhazardous aluminum oxide and manganese oxide. The selective capabilities of the Aqua-Bind media permit high adsorption capacities, even in the presence of competing ions, which significantly extend the shelf life of the media and reduce maintenance cost. The column is set to receive water under slight positive pressure and water is allowed to pass downward through two chambers capable of capturing particulate iron and adsorbing arsenic. This system can operate for 6 months before requiring a media change and is designed to treat 1000 L of water/day with influent levels of arsenic around 250 µg/L [27].

6.3.7.2 Shapla Filter

The *Shapla arsenic filter* is an earthen household arsenic removal technology developed by International Development Enterprises, Bangladesh. The functioning of the Shapla filter is based on the same adsorption principle of arsenic to iron coated on brick chips, which works as good as iron-coated sand. The bricks are generally coated with a ferrous sulfate solution [28]. The filter is known to hold up to 3 L of water. As contaminated water with arsenic passes through the filter, the latter is rapidly adsorbed by iron on the brick chips. The media can filter up to 400 L of arsenic water. Also the effluent water levels after purification through these units were consistently below 0.050 mg/L arsenic, meeting the treatment standards.

6.3.7.3 Sono Filter

Another innovative, locally designed arsenic filter, known as the Sono filter, offers hope for millions of people in Bangladesh who lack access to safe drinking water. Invented by U.S. National Academy of Engineering in 2006, the Sono filter is a simple device that uses a *composite iron matrix* that is manufactured locally from cast iron turnings, along with river sand, wood charcoal, wet brick chips, and two buckets [29]. The top bucket is filled with coarse river sand that filters coarse particles and controls the flow of water; it also contains composite iron matrix that removes inorganic arsenic. The water from the first bucket then flows into a second bucket where it is again filtered through coarse river sand and wood charcoal to remove other contaminants. Finally, water passes through fine river sand and wet brick chips to remove fine particles and stabilize water flow. The filter's humble housing in a pile of two buckets has the potential to change lives in arsenic-infested areas. It is said to remove 98% of the arsenic in water, as well as other organic and mineral impurities, but may harbor bacterial growth [30].

6.3.7.4 Magc-Alcan Filter

The Magc-Alcan is also a two-bucket filter. The buckets are in series and both are filled with enhanced AA media. The media is developed by MAGC Technologies and Alcan of the United States; and it is produced by thermal dehydration of aluminum hydroxide at 250°C–1150°C. The Magc-Alcan filter removes arsenic by adsorption of arsenic to the enhanced AA which is porous and has a high surface area, as discussed previously [31]. A similarly working filter called *Nirmal filter* exists in India. The only difference is that it uses arsenic adsorption on an Indian-made AA and followed by filtration through a ceramic candle. It is comparatively less expensive than a Magc-Alcan filter ($10–$15 capital cost), but it needs to be reproduced every 6 months.

6.3.7.5 SAFI Filter

Another filter called the SAFI filter is a type of household ceramic candle filter made of composite porous materials like kaolinite and iron oxide on which hydrated ferric oxide is deposited by various treatments [32]. It works by the same principles of adsorption and filtration on the chemically treated active porous composites materials of the candle. The SAFI filter was initially reported to have good arsenic removal efficiency, but one of its major hindrances is the clogging of the filter media; the unit also suffers rapid erosion if mechanical cleaning is attempted.

6.3.7.6   Kanchan™ Filter

The Kanchan filter (KAF) developed by MIT, United States, along with the Environment and Public Health Organization (ENPHO) of Nepal, combines slow sand filtration and adsorption on iron hydroxide and is quite efficient in removing arsenic from drinking water [33]. The top part of the filter contains iron nails installed in it. These nails when exposed to air and water rust quickly producing ferric hydroxides particles, which is, as discussed before, a first-rate absorbent of arsenic. Thus, when arsenic-containing water is poured, surface reaction occurs, and arsenic is rapidly adsorbed onto the surface of the GFH particles. The lower part of the filter is made up of a concrete box that is filled with sand and gravel. The arsenic-loaded iron particles as resultant of surface adsorption are then flushed into the sand layer below. Because of the small pore space in the fine sand layer, the arsenic-loaded iron particles are trapped in the top of the fine sand layer. Hence, arsenic is effectively eliminated from the water. Overall, study statistics portray that the household arsenic filtering units discussed previously are effective and appropriate, but with regular clogging problems which need to be addressed in future. However, alternatively, it is also true that they have been partly successful in removing this slow but steady assassin from the midst.

In summary, there are many treatment technologies available for arsenic removal, but one of the simplest and most widely used methods is probably the arsenic selective adsorbent media. This system is well suited not only for residential systems but for larger commercial purposes as well, requiring little or no maintenance and only intermittent monitoring. Majority of these selective media, as discussed previously, are based on adsorptive principles where arsenic is captured onto iron oxide. At a pH level of around 7, these media have exceptionally high throughput capacity for arsenic predominantly in the form of As(V) and low to moderate for As(III). A speciated water from As(III) to As(V) as inlet adds to the efficiency of the system. As the pH level rises in the input water, there are constituents that can interfere with arsenic adsorption which include silica, phosphate, sulfur, selenium, etc. The effect of pH higher than 8.5 can be drastic, often reducing the capacity of an adsorbent media significantly. An economical solution for treating arsenic-laden water when the pH levels are at the higher end is therefore the need of the hour.

6.4   ION EXCHANGE

Ion exchange (IE) is a frequently used treatment technology for arsenic removal. As arsenic-laden water is passed through the resin, a kind of synthetic media with better-defined IE capacity, arsenic anions are exchanged for other ions in the resin. The synthetic resin is generally based on a cross-linked polymer skeleton, known as the matrix. The charged functional groups are attached to the matrix through covalent bonding [21,34]. Importantly, the IE process is least dependent on the pH of water. The two major types of anion exchange resins commonly used today are Type 1 and Type 2 strongly basic resins. Both are used to remove arsenic and alkalinity. Type 1 resin contains the trimethylamine group, while the Type 2 resin derives its functionality from dimethylethanolamine group [21,34]. The capacities of the two resins are virtually identical in arsenic removal systems. However, the relative order of affinity of these basic anion exchangers for some familiar ions in drinking water is sulfate > arsenate > nitrate > chloride > bicarbonate. Therefore, based on the affinity relationships, we can assume why the standard resins are limited by the amount of sulfate in water. IE is hence often preceded by treatment contaminants that can foul the resins and reduce their efficacy. As(III), being uncharged, is ideally not eliminated by the IE process; thus, a preoxidation of As(III) to As(V) is required for complete removal of arsenic. However, an excess of oxidant can have a negative impact on the sensitive resins. Also, as the resins become exhausted, they need to be regenerated which is often achieved by washing with a NaCl solution; alternatively, single-use ion nonregenerable resins can also be used. Some of the other disadvantages of IE process include the generation of highly concentrated waste by-product; sulfate levels in water are also found to affect run length.

6.4.1 GREENSAND FILTRATION

Greensand filtration (GSF) is a basic technology based on the chemical theories of oxidation and reduction where manganese greensand is developed from the mineral glauconite. This oxidation filtration process has demonstrated effectiveness for removal of arsenic [35]. The GSF medium is manufactured by treating glauconite sand with potassium permanganate until the granular sand is coated and firmly attached with a layer of shiny, hard, finite thickness manganese oxides. Arsenic in contaminated-water displaces species from the manganese oxide and gets bound to the greensand surface—actually an exchange of ions [35]. The GSF has a high buffering or oxidation–reduction potential due to the well-defined manganese oxide coating. This oxidative nature of the manganese surface converts As(III) to As(V), and eventually the latter is adsorbed onto the surface. The grains of manganese greensand are of perfect size and shape to capture the fine precipitates of arsenic during normal conditions. A pH of 5 was found to be the optimal pH for arsenic extrusion by this technology [36]. Also, the greensand columns were found to be most effective only after the media had been pretreated with dilute acid; a solution of dilute HCl when passed through the media until the influent and effluent pH reach steady state gives the best result. This technology is considered to be a small system technology; however, approximate cost for operation of larger system sizes is unavailable [36]. There are a number of simple technologies for arsenic removal discussed earlier; however, when only arsenic is to be removed, this technology is probably not that cost effective compared to others, but in a situation where manganese and iron are to be removed with arsenic, this procedure probably holds great promise.

6.5   ARSENIC REMOVAL WITH MEMBRANE PROCESSES

If the total arsenic burden of source water is under 50 µg/L, then regardless of the turbidity of source water, arsenic concentration can be efficiently lowered below the WHO guideline of 10 µg/L by conventional coagulation–filtration. But, if the arsenic concentration in source water exceeds 50 µg/L, then more coagulant dosage for enhanced removal is required. Therefore, a constant monitoring of raw water arsenic concentration is essential to adopt an optimum coagulant dosage for arsenic removal; however, it is difficult to manage that in the rural settings and arsenic measurement is time consuming too. Additionally, if the source water contains exceeding amount of As(III), it is hard for coagulation or filtration system to meet an arsenic maximum contaminant level of 2 µg/L [37]. To resolve this dilemma, membrane filtration technology has evolved over time. Membrane filtration procedure is a promising technique to remove many contaminants from water, including bacteria, salts, and heavy metals. Membranes are ideally synthetic materials with millions of pores in them that act as a selective barrier; the structure of the membrane allows some components to pass through, while others are rejected. The movement of molecules across the membranes typically needs a driving force, for example, pressure difference between the two sides of the membrane. For arsenic removal, similar synthetic membranes are used. Membrane processes are efficient at removing arsenic through the following methods: filtration, electric repulsion, and adsorption of arsenic-containing compounds. If arsenic compounds are larger than membrane pore size, they will be rejected due to size exclusion. Size, however, is not the only determining factor that influences exclusion. There are membranes that can reject arsenic of two orders of magnitude smaller than the membrane pore size, indicating other removal mechanisms in function [38]. In this context, the shape and chemical characteristics of arsenic compounds play a key role in arsenic exclusion. A considerably higher amount of As(III) could be removed using membranes without the use of any chemical additives, whereas, trivalent arsenic species could not be removed effectively by filtration system without preoxidation of As(III) to As(V). Also, chemical characteristics particularly charge, and hydrophobicity of both the membrane and the water constituents can also have a significant impact on removal dynamics. Membrane-mediated arsenic removal thus depends on a number of factors extending from pore size distribution, membrane morphology, surface charge of membrane, membrane material, configuration of the module and operating conditions, including solution concentration, operating pressure, solution pH, and solution temperature. Some of the membrane filtration processes are discussed in detail next.

6.5.1   MICROFILTRATION

Microfiltration (MF) along with ultrafiltration falls under the category of low-pressure membrane filtration. Low-pressure membranes have larger pore sizes and are operated at pressures of around 10–30 psi. Membranes in MF have the largest pore size, extending from 0.1 to 10 µm. A wide variety of membranes are used for MF extending from polyethersulfone (PES), polysulfone (PS), cellulose acetate (CA), polypropylene, and polyvinylidene fluoride (PVDF). MF's efficacy as a technique for arsenic removal is primarily dependent on the size of arsenic-bearing particles in the water source. Mostly, MF pore size is too large to considerably remove dissolved or colloidal arsenic. Although, MF can eliminate particulate forms of arsenic, this method alone is not significantly efficient unless a large proportion of arsenic is found in particulate form. Typically, less than 10% of particulate arsenic is found in groundwater [38,39]. Hence, recently, to increase efficiency, MF systems are often used where contaminants are converted to particulate form in a pretreatment step. Also to amplify removal efficiency in water with a low particulate arsenic content, MF can be joined with coagulation processes discussed earlier. Essentially, the removal efficiency of arsenic by MF is increased by increasing the particle size of arsenic-bearing particles. Membrane MF processes, which simultaneously use properties of iron salts, have shown to significantly reduce arsenic in drinking water to below detectable limits. Ferric chloride and ferric sulfate have been predominantly used in many studies as a standard pretreatment of arsenic-contaminated water before MF. The presence of sulfate groups though was found to yield better results as they generated larger flocs that are more easily rejected by the membrane, compared to chloride ions. The negatively charged arsenic is adsorbed onto positively charged ferric particles, which are thereafter removed by microfiltration. The amount of ferric compound required to remove arsenic is strongly dictated by pH. The most effective removal takes place at pH values below 8, where the positive species generally dominates. However, for utilities using MF alone, removal would mostly depend on the percentage of particulate arsenic since the MF rejection mechanism is ideally mechanical sieving [40,41]. The effectiveness of MF arsenic rejection is therefore inversely proportional to the pore size. Microfiltration is thus a process with a marginal ability to eliminate arsenic due to its comparatively large pore size in comparison to other more efficient membrane processes. To improve efficiency, modulating the chemical properties, particularly charge and hydrophobicity of the membrane was found to pay dividends; for instance, a membrane with a fixed negative charge was more effective in removing anionic As(V) than a membrane which is uncharged. However, though MF is an effective method to remove arsenic, a major impediment in its application is the deposition of materials on the membrane surface and in pores causes membrane fouling, and therefore MF treatment requires periodic backwashing. Recently, it has been observed that iron hydroxide and manganese oxide particles are the primary reason for membrane fouling, and their rejection by different membranes should be evaluated before selection of membranes for arsenic removal by MF.

6.5.2 ULTRAFILTRATION

Ultrafiltration (UF) is also a low pressure driven membrane operation that is generally capable of removing colloidal and particulate constituents with molecular weight higher than few thousand Daltons. UF membranes generally have pore sizes ranging from 10 to 10,000 Å and are capable of retaining species in the range of 1,000 to 500,000 Da. Considering this fact, UF alone, like MF, may not be an independently viable technique for arsenic removal from groundwater; however, UF may be an appropriate choice for surface water with high colloidal and particulate arsenic content [40,41]. However, recent studies have shown that UF with electric repulsion may have a better arsenic removal efficacy compared to UF functioning only as a pore-size-dependent sieve. AWWARF (1998) performed a series of bench-scale studies to elucidate the effect of membrane charge on removal of arsenic by using uncharged FV2540F and negatively charged GM2540F UF membrane [38]. Results of this study showed that for the negatively charged GM2540F membrane, As(V) rejection was higher compared to the uncharged one; however, at neutral pH, rejection was very low but better results were obtained at basic pH probably due to electrostatic interaction between arsenic ions and the negatively charged surface of the membrane. The rejection of As(V) increased with increase of pH from approximately 13% at pH 2 to above 80% at pH 10. Again, As(III) rejection was considerably lower than that of As(V) ions. The rejection of As(III) was reported to be about 15% in pH range of 4–10 and 40% with an increase in pH from 10 to 11. In a separate study, a negatively charged, thin film, composite sulphonated polysulphone UF membrane was used to investigate its arsenic removal potential and also the effect of cooccurring divalent ions and natural organic matter (NOM) on arsenic rejection by the charged membrane. The presence of divalent ions (e.g., Ca, Mg) depicted a significant reduction in arsenic removal capacity of the membrane probably due to interaction of the solutes with the membrane. These interactions probably locally neutralize the charge of the membrane, thus reducing its arsenic rejection capacity. However, when NOM was present the arsenic rejection potential of the membrane increased, probably because NOM complexes with the divalent ions in solution that hinders with arsenic removal [38]. Also, the NOM is adsorbed on the membrane surface forming a layer that increases rejection. Adsorption of NOM actually reduces the net surface charge of the membrane and, in effect, increases the repulsion toward negatively charged arsenic compounds. Hence, the charge density of the layer is directly proportional to the concentration of the NOM in the solution, and a higher NOM concentration leads to higher arsenic removal [38]. Interestingly, the presence of cooccurring inorganic solutes (e.g., HCO_3^-, HPO_4^{2-}) did not alter the arsenic removal efficiency by a flat sheet hydrophilic PES-UF membrane; however, the presence of a surfactant cetylpyridinium chloride (CPC) had significant positive impact on arsenic removal [42]. In a different study conducted by Lin-Han Chiang Hsieh in Taiwan in 2006, it was observed that upon application of an electrical voltage of 25 V to the UF system, it was able to reduce the total arsenic concentrations from groundwater significantly. The theory behind the high rejection rate of charged UF is the electrostatic interaction between negatively charged membrane surface and the arsenic ions. This is again pH dependent since the anionic As(V) and the nonionic As(III) will be charged (protonated/deprotonated) at different pH levels. Thus, membrane charge and pH play a key role in arsenic rejection by UF. However, neither MF nor UF can remove dissolved substances efficiently unless they are pretreated (with activated carbon) or coagulated (with alum or iron salts).

6.5.3 NANOFILTRATION

Nanofiltration (NF) membranes are usually used to segregate out multivalent ions from the monovalent ones. They are competent enough in removing significant amount of dissolved arsenic, both As(V) and As(III), from water attributed to their small pore size. This makes NF a very dependable arsenic removal process for groundwater that has around 90% of dissolved arsenic [38]. The minimal pore size of the membranes in NF is typically about 1 nm. The NF membranes are usually negatively charged at neutral and alkaline media but tend to lose their charge in acidic pH; hence, the separation of arsenic anions is based not only on different rates of their diffusion through the membrane but also on repulsion (Donnan exclusion) between anions in solution and the surface groups. Thus, a high ion rejection can be achieved but at a higher water flux through the membrane. Based on several studies, it is now understood that NF can be an effective arsenic removal process. According to Brandhuber and Amy, the negatively charged NF membranes showed significant As(V) rejection; however, reduced As(III) removal was obtained may be because As(III) is small and can more easily diffuse through very small NF pores [43,44]. They studied rejection of arsenic by three NF membranes—NF70 4040-B (Film Tec), HL-4040F1550 (DESAL), and 4040-UHA-ESNA (Hydranautics)—where the As(V) rejection was more than 95%, whereas the As(III) rejection was 20%–53% for all of the three membranes. Other studies further corroborated the same fact, where Levenstein et al. used a commercial loose, porous polyamide thin film composite membrane, NF-45 which showed 90% removal of As(V), whereas the removal of As(III) was between 10% to 20% [45]; Urase et al. used low pressure aromatic polyamide NF membrane, ES-10 which showed 50%–89% rejection of As(III) but around 87%–93% rejection of As(V) [46]. That sustains the ever-existing issue of removing As(III) species from water efficiently. An oxidization process converting As(III) to As(V) may thus yield better results. The differential rejection for As(III) and As(V) could be because of the fact that As(V) persists as an anion at pH (5–8) in natural water, while As(III) remains as a neutral molecule at the same pH. Thus, probably because of electrostatic repulsion between As(V) molecules and the charged NF membrane at neutral and alkaline pH, the rejection of As(V) was considerably higher compared to As(III). Vrijenhoek and Waypa reported in 2000 that rejection of As(V) increased significantly from 25% at pH 4 to above 80% at pH 9 probably due to change of As(V) species from monovalent ions to divalent with a rise in pH; the fact that divalent ions have larger hydrated radii compared to monovalent ions could be one possible reason for increased rejection of As(V) species [47]. Also, the charged sites of membranes at pH 4–8 are negative; hence, a higher Donnan exclusion was achieved resulting in higher rejection. In another study by Seidel et al., it was found that considerably higher arsenic removal efficiency was obtained in the presence of sodium chloride, especially at low arsenic content in source water [48]. Saitúa et al. further showed with NF membranes that cooccurrence of dissolved inorganic salts does not hinder arsenic removal process [49]. A recent study by Zhao et al. with a simulated aqueous solution of arsenic salt ($Na_2HAsO_4 \cdot 7H_2O$) and using a self-made PMIA (poly m-phenylene isophthalamide), NF membrane showed more than 90% rejection of As(V) which increased from 83% at pH 3 to 99% at pH 9. The presence of NaCl increased As(V) rejection; however, the existence of Na_2SO_4 reduced rejection by about 8% [50]. Surprisingly, arsenic rejection rates in NF filters were found to decline over time; the underlying reason behind it was reduction of As(V) to the As(III) form reducing overall arsenic removal rate. This decline in rejection with time indicates that a negatively charged membrane cannot keep high As(V) rejection rates for long durations. The membranes used in NF are also subject to scaling and fouling and often modifiers such as antiscalants are required for proper use. The small pore size of NF membranes renders them more prone to clogging; hence, the application of NF for water treatment is ideally not completely accomplished without extensive pretreatment for particle removal. Removal though depends on operating conditions, membrane properties, and also arsenic speciation. Typically, high pressure, high pH, low temperature, and groundwater preoxidation before NF treatment favor more efficient removal of arsenic through NF. NF has better arsenic removal potential than low-pressure membrane processes; however, additional long-term testing studies are further required to use NF in large-scale operations. Furthermore, for large treatment plants, a large water body would likely be needed to dispose the contaminated brine stream from the NF technologies. Also, the operational and maintenance cost for NF membranes are usually considerably greater than equivalent cost for low-pressure processes.

6.6   REVERSE OSMOSIS

Reverse osmosis (RO) is an old membrane technology that utilizes a semipermeable membrane to purify water. In RO, a pressure is applied to overcome osmotic pressure; it can eliminate low molecular mass compounds and ions from solutions. It is generally used in both industrial processes and for the production of potable water. The net result of maintaining a pressure gradient across the membrane is that the solute with arsenic is retained on the pressurized side of the RO membrane, and the filtered solvent is allowed to pass to the other side. A high osmotic pressure is achieved in RO systems compared to other membrane processes due to concentration of salts on the pressure side of the membrane. The RO units are generally installed either at the point-of-entry or at the point-of-use; however, the RO at the point-of-use will be more economical. Around 70% of the raw water supplied to an RO system will be passed through the membrane; the rest is wastewater, which is flushed out to the drainage system; so water rejection can be an issue in use of RO systems [40]. Similar to NF, RO also requires extensive pretreatment for particle removal, and oxidizing As(III) to As(V) may be required to improve efficiency; however, the pretreatment process can make RO processes expensive. RO has very high rejection rate for As(V) but very low for As(III) at neutral pH; thus, RO systems may need pH manipulation for better performance [51]. Also, water treated with RO may not have beneficial minerals such as calcium and magnesium which may be a concern as people suffering from arsenic are mostly already undernourished. It is therefore important to consume a reasonably well-balanced diet to offset the removal of these minerals if arsenic extruded RO water is used for drinking. RO-treated water is also found to be corrosive; hence, extensive corrosion control could be required. RO is however a very effective arsenic removal technology, and since groundwater typically contains 80%–90% of dissolved arsenic, RO is very effectual in removing dissolved arsenic from water. Existing studies indicate that RO is a very useful process for arsenic removal, compared to low-pressure membrane processes, but it is typically associated with higher operational and maintenance cost if performed at large scale.

6.7   ELECTRODIALYSIS REVERSAL

Electrodialysis reversal (EDR) is conventionally a water desalination membrane process that is also used for the removal of arsenic from water. In EDR an electric current helps in transfer of ions through membranes that are selectively permeable to cations or anions. Dissolved ions typically migrate from a lesser to a higher concentrated solution through an electrodialysis stack consisting of alternating layers of cationic and anionic IE membranes, and this results in alternating sets of compartments having water with low and high concentrations of particular ions. Periodically, the direction of flow of ions is reversed by reversing the polarity of the applied current; this leads to decreased potential for fouling of EDR membranes, which thus negates pretreatment requirement and also cost [52]. Also, other advantages of the EDR system are that it is completely automated, needs little operator attention, and does not require any chemical addition; it is, however, not as much competitive with respect to costs compared to NF or RO systems [52]. Additionally, use of EDR in water-scarce regions is an issue due to high water rejection (about 20%–25% of influent).

In summary, considering the available options, arsenic level can be reduced effectively by the use of coagulation or membrane technology. However, the efficiency of removal varies greatly depending upon the technology and the process parameters used. The operating pH, temperature, other contaminants in water, membrane characteristics, and initial concentration of arsenic are certain factors that were found to have a significant effect on arsenic removal efficiency. In majority of removal methods, the pentavalent form of arsenic was found to be more effectively removed than the trivalent form. Probably because of the presence of ionic charge, As(V) is more readily eliminated from source water than As(III). Unfortunately, arsenic in groundwater exists mostly as As(III). For most of the arsenic removal system based on conventional processes, a secondary treatment system is required in order to reduce the arsenic concentration to meet the standard. Also, large amount of chemical reagents such as alum or ferric chloride/sulfate is typically required for conventional coagulation. Hence, these processes produce a large volume of sludge, which requires even further treatment before ultimate disposal. Also, when the sludge is piled up, a huge amount of leaching with elevated arsenic concentration is produced. From this aspect, membrane technologies like MF and UF require relatively less quantity of coagulants or flocculants to increase particle size of arsenic, and hence the sludge produced is not so much in volume in comparison to conventional methods. The removal of arsenic, especially by NF and RO membranes, essentially requires no chemical reagent, and hence no sludge is therefore produced in these processes. Furthermore, water obtained from conventional technologies may contain fine particles which cannot be removed by sedimentation processes or through gravity settling. However, in the membrane technology, not only arsenic is removed but also some other dissolved minerals and microorganisms are efficiently rejected, thus reducing chances of infectious diseases. Disinfection is hence not a separate step in membrane-based approaches, whereas conventional coagulation/filtration technologies cannot remove pathogens and therefore require disinfection of treated water. Among the most prevalently used membrane technologies, high pressure membrane processes, for example, NF and RO, are more efficient in removing arsenic species compared to UF and MF. Use of negatively charged UF membranes and separation of anionic species of arsenic through Donnan exclusion were also found to be very effective in rejecting arsenic species. Further added advantages of membrane systems are that, typically, membrane-technology-based plants are highly compact, while the conventional treatment technology requires large area for setting up of a plant; the fouled membranes can be replaced with little fuss by new membranes and the membrane-based process can be scaled up readily only by adding more membrane modules. Also, the performance of the membranes can be controlled by controlling parameters of the feed solution [53]. Hence, maintenance of membrane technology is simpler, and if handled with care and proper planning, this technology can satisfy environmental requirements. However, a coalesion of membrane technologies like RO, NF, UF with coagulation–filtration has also provided good results and has been quite effective in removing arsenic from water and also meets the standard of arsenic level in water. The use of hybrid treatment technology has proved to be efficient not only from the performance perspective but it is also less energy consuming and a low-cost technology if efficiency level is taken into consideration. In future, may be the use of solar-driven direct contact membrane modalities can be pursued as an effective alternative arsenic removal strategy for the vast arsenic-affected rural areas of Southeast Asian countries.

Finally, the most prevalent issues with existing arsenic filtration technologies have been the differential filtering potential for predominantly As(III) and waste disposal. Disposal of arsenic-laden coagulation sludge from coagulation/filtration technologies needs to be addressed before their full-scale use in rural areas of Southeast Asia. A large amount of water would most likely be needed in large-scale plants to discharge the contaminated brine from RO or NF technologies; additionally a waste of water is on the cards if RO/NF/EDR technologies are used on a large scale. Installation of inland treatment plants that would pretreat the wastewater prior to its discharge would be helpful. Also, the water quality and its constituents should be premonitored before deciding on the type of treatment unit to be installed in a particular area, rather than the random use of any reportedly highly efficient filtering system. Thus, though quite a good number of arsenic removal methods are presently widely used, a more cost-effective, eco-friendly technology is yet to be developed so as to further improve the efficiency of arsenic removal as well as to resolve sludge and arsenic concentrated waste management problems. Hence, an unremitting investigation of the available arsenic removal technologies and its further development taking into consideration the pros and cons of the existing methods is absolutely essential to develop an economical yet effective arsenic removal system.

REFERENCES

1. Khan, M.M., Magnitude of arsenic toxicity in tube-well drinking water in Bangladesh and its adverse effects on human health including cancer: Evidence from a review of the literature. *Asian Pac J Cancer Prev*, 2003. 4(1): 7–14.
2. Smith, A.H., Lingas, E.O., Rahman, M., Contamination of drinking-water by arsenic in Bangladesh: A public health emergency. *Bull World Health Organ*, 2000. 78(9): 1093–1103.

3. Fendorf, S., Michael, H.A., Geen, A.V., Spatial and temporal variations of groundwater arsenic in South and Southeast Asia. *Science*, 2010. 328(5982): 1123–1127.

4. Ahmed, M.F., An overview of arsenic removal technologies in Bangladesh and India. *Proceedings of BUET-UNU International Workshop on Technologies for Arsenic Removal from Drinking Water*, Dhaka, Bangladesh, 2001, pp. 251–269.

5. Charles, F., Harvey, K.N.A., Winston, Y., Badruzzaman, A.B.M., Ashraf, A.M., Oates, P.M., Michael, A.H., Neumann, R.B., Beckie, R., Islam, S., Ahmed, M.F., Groundwater dynamics and arsenic contamination in Bangladesh. *Chem Geol*, 2006. 228(1): 112–136.

6. Terlecka, E., Arsenic speciation analysis in water samples: A review of the hyphenated techniques. *Environ Monit Assess*, 2005. 107(1–3): 259–284.

7. Amirtharajah, A., O'Melia, C.R., Coagulation processes: Destabilization, mixing, and flocculation. In American Water Works Association (ed.), *Water Quality and Treatment: A Handbook of Community Water Supplies*. New York: McGraw-Hill, 1990, pp. 269–365.

8. Edwards, M., Chemistry of arsenic removal during coagulation and Fe–Mn oxidation. *J AWWA*, 1994 (September) 86(9): 64–78.

9. Katsoyiannis, I.A., Zouboulis, A.I., Comparative evaluation of conventional and alternative methods for the removal of arsenic from contaminated groundwaters. *Rev Environ Health*, 2006. 21(1): 25–41.

10. Sancha, A.M., Review of coagulation technology for removal of arsenic: Case of Chile. *J Health Popul Nutr*, 2006. 24(3): 267–272.

11. Johnston, R., Heijnen, H., Safe water technology for arsenic removal. *Report: World Health Organization (WHO)*, Geneva, Switzerland, 2002.

12. Tahura, S., Shaidullah, S.M., Rahman, T., Milton, A.H., Evaluation of an arsenic removal household device: Bucket treatment unit (BTU). *BUET-UNU International Workshop on Technologies for Arsenic Removal from Drinking Water*, Dhaka, Bangladesh, 2001 (May 5–7).

13. Tomljenovic, L., Aluminum and Alzheimer's disease: After a century of controversy, is there a plausible link? *J Alzheimers Dis*, 2011. 23(4): 567–598.

14. Zatta, P., The role of metals in neurodegenerative processes: Aluminum, manganese, and zinc. *Brain Res Bull*, 2003. 62(1): 15–28.

15. EPA, Arsenic removal from drinking water by coagulation/filtration and lime softening plants. Prepared by Battelle under contract 68-C7-0008 for EPA ORD, 2000 (June).

16. Sorg, T.J., Csanady, M., Logsdon, G.S., Treatment technology to meet the interim primary drinking water regulations for inorganics: Part 2. *J AWWA*, 1978. 7: 379–392.

17. McNeill, L., Edwards, M., Soluble arsenic removal at water treatment plants. *J AWWA*, 1995 (April). 87: 105–113.

18. Gupta, S.K., Chen, K.Y., Arsenic removal by adsorption. *J WPCF*, 1978. 3(3): 493.

19. Benjamin, M.M., Sletten, R.S., Bailey, R.P., Bennett, T., Sorption of arsenic by various adsorbents. *AWWA Inorganic Contaminants Workshop*, San Antonio, TX, 1998 (February).

20. EPA, Arsenic removal from drinking water by ion exchange and activated alumina plants. Prepared by Battelle under contract 68-C7-0008 for EPA ORD, 2000 (October).

21. Clifford, D., Ion exchange, activated alumina, and membrane processes for arsenic removal from groundwater. *Proceedings of the 45th Annual Environmental Engineering Conference*, University of Kansas, Lawrence, KS, 1995 (1st February).

22. Driehaus, W., Jekel, M., Hildebrandt, U., Granular ferric hydroxide—A new adsorbent for the removal of arsenic from natural water. *J Water SRT—Aqua*, 1998. 47(1): 30–35.

23. Devi, R., Das, B., Borah, K., Thakur, A.J., Raul, P.K., Banerjee, S., Singh, L., Removal of iron and arsenic (III) from drinking water using iron oxide-coated sand and limestone. *Appl Water Sci*, 2014. 4: 175–182.

24. Hydrometrics, I., Summary report on the sulfur-modified iron (SMI) process, Hydrometrics, Inc., Helena, MT, 1997.

25. Hydrometrics, I., Second interim report on the sulfur-modified iron (SMI) process for arsenic removal, Hydrometrics, Inc., Helena, MT, 1998.
26. Shimoto, T., Arsenic removal technology—Cerium adsorbent Virtual Center for Environmental Technology Exchange, Apec Virtual Center, Nakagaito, Japan, 2007.
27. Senapati, K., Alam, I., Apyron arsenic treatment unit—Reliable technology for arsenic safe water. *Technologies for Arsenic Removal from Drinking Water*, 2001, pp. 146–157.
28. The World Bank, Towards a more effective operational response: Arsenic contamination of groundwater in South and East Asia countries. Water and Sanitation Program. I & II, 2012.
29. Khan, A.H., Rasul, S.B., Munir, A.K.M., Alauddin, M., Habibuddowlah, M., Hussam, A., On two simple arsenic removal methods for groundwater of Bangladesh. *Bangladesh Environment-2000*, Bangladesh Poribesh Andolon, Dhaka, Bangladesh, 2000, pp. 151–173.
30. WS Atkins International-Epsom, BAMWSP, Rapid assessment of household level arsenic removal technologies, Phase-I and Phase-II, Final report. WS Atkins International Limited, Surrey, U.K., 2001.
31. Hanchett, S., Johsnton, R., Khan, M.H., Arsenic removal filters in Bangladesh: A technical and social assessment. *UNC Water and Health Conference*, Chapel Hill, NC, 2011.
32. Visoottiviseth, P., Ahmed, F., Technology for remediation and disposal of arsenic. *Rev Environ Contam Toxicol*, 2008. 197: 77–128.
33. Sutherland, D., Rapid assessment of household level arsenic removal technologies: Phase II Executive Summary. Water Aid Bangladesh, Bangladesh Arsenic Mitigation Water Supply Project, 2001.
34. Clifford, D., Ceber, L., Chow, S., As (III) and As (V) separation by chloride-form ion exchange resins. *Proceedings AWWA WQTC*, Norfolk, VA, 1983.
35. Subramanian, K.S., Viraraghavan, T., Phommavong, T., Tanjore, S., Manganese greensand for removal of arsenic in drinking water. *Water Qual Res J Canada*, 1997. 32(3): 551–561.
36. Hanson, A., Bates, J., Heil, D., Bristol, A., Arsenic removal from water using manganese greensand: Laboratory scale batch and column studies. Water Treatment Technology Program Report No. 41, New Mexico State University, Las Cruces, NM, 1999.
37. Sato, Y., Performance of nanofiltration for arsenic removal. *Water Res*, 2002. 36(13): 3371–3377.
38. Amy, G.L., Edwards, M., Benjamin, M., Carlson, K., Chwirka, J., Brandhuber, P., McNeill, L., Vagliasindi, F., Arsenic treatability options and evaluation of residuals management issues. Draft Report. AWWARF, Denver, CO, 1998.
39. McNeill, L.S., Edwards, M., Arsenic removal during precipitative softening. *J Environ Eng—ASCE*, 1997. 125(5): 453.
40. EPA, Technologies and costs for removal of arsenic from drinking water. Under Contract with the USEPA No. 68-C6-0039, Washington, DC, 2000.
41. AWWARF, *Water Treatment Membrane Processes*. McGraw-Hill Publishing Company, New York, 1996.
42. Ergican, E., Gecol, H., Fuchs, A., The effect of co-occurring inorganic solutes on the removal of arsenic (V) from water using cationic surfactant micelles and an ultrafiltration membrane. *Desalination*, 2005. 181: 9–26.
43. Brandhuber, P., Amy, G., Alternative methods for membrane filtration of arsenic from drinking water. *Desalination*, 1998. 117(1–3): 1–10.
44. Amy, G., Arsenic rejection by nanofiltration and ultrafiltration membranes: Arsenic forms vs. membrane properties. *AWWA Inorganic Contaminants Workshop*, San Antonio, TX, 1998.
45. Levenstein, R., Hasson, D., Semiat, R., Utilization of the Donnan effect for improving electrolyte separation with nanofiltration membranes. *J Membr Sci*, 1996. 116: 77–92.

46. Urase, T., Oh, J., Yamamoto, K., Effect of pH on rejection of different species of arsenic by nanofiltration. *Desalination*, 1998. 117(1–3): 11–18.
47. Vrijenhoek, E.M., Waypa, J.J., Arsenic removal from drinking water by a 'Loose' nanofiltration membrane. *Desalination*, 2000. 130: 265–277.
48. Seidel, A., Waypa, J.J., Elimelech, M., Role of charge (Donnan) exclusion in removal of arsenic from water by a negatively charged porous nanofiltration membrane. *Environ Eng Sci*, 2001. 18(2): 105–113.
49. Saitúa, H., Campderos, M., Cerutti, S., Padilla, A.P., Effect of operating conditions in removal of arsenic from water by nanofiltration membrane. *Desalination*, 2005. 172(2): 173–180.
50. Zhao, C., Du, S., Wang, T., Zhang, J., Luan, Z., Arsenic removal from drinking water by self-made PMIA nanofiltration membrane. *Adv Chem Eng Sci*, 2012. 2: 366–371.
51. Waypa, J., Menachem, E., Janet, G.H., Arsenic removal by RO and NF membranes. *AWWA*, 1997. 89: 102–114.
52. EPA, Review of the draft drinking water criteria document on inorganic arsenic. EPA SABDWC-94-004, Washington, DC, 1994.
53. Uddin, M.T., Mozumder, M.S.I., Figoli, A., Islam, M.A., Drioli, E., Arsenic removal by conventional and membrane technology: An overview. *Indian J Chem Technol*, 2007. 14: 441–450.

7 Bioremediation of Arsenic Toxicity

Nguyen Van Hue

CONTENTS

7.1 CHEMICAL AND ENVIRONMENTAL PROPERTIES OF ARSENIC

Arsenic (As) is a highly toxic element that has poisoned many people by being added to their food and drink, mistakenly, unknowingly, or deliberately [1]. Arsenic occurs in trace quantities in all rock, soil, water, and air [2,3]. Naturally, total As is about 1–2 mg/kg in rock, 5–10 mg/kg in soil, and 1–3 μg/L in seawater [4]. Volcanoes and microbial activities can release As into the atmosphere as arsine gas (AsH_3) or methylated Arsine species. The atmospheric residence time of As species is relatively short and the As concentration is generally low (0.02 μg/m^3), except in the vicinity of these sources [5]. However, anthropogenic sources such as fossil fuel combustion, mining, smelting of sulfide ores, pesticide application, timber preservation, and the application of sludge and manure have elevated As levels and may cause As contamination in the food chain and drinking water [6]. As an example, past use of arsenical herbicides in sugarcane fields resulted in total As levels ranging from 50 to 950 mg/kg in some Hawaiian soils [7,8].

Although As can have −3, 0, +3, and + 5 oxidation states, the two oxidation states of As (+3) (arsenite, with ionic radius, r = 0.58 Å) and As (+5) (arsenate, r = 0.46 Å) are most important environmentally and biologically. Because of its size and polarizability, As (+3) has a relatively high affinity for softer S- and N-donating species. Thus, As (+3) can react strongly with thiols of cysteine residues and/or imidazolium nitrogens of histidine residues from cellular proteins, inactivating many enzymes [9]. In aqueous solution, it is found mainly as the neutral [$As(OH)_3$] species whose first pK_a = 9.2. Arsenate (+5) with its greater charge and smaller ionic radius leads to a higher stability with harder O-donating species, and its prevalence in aqueous

155

solution as $H_2AsO_4^-$, $HAsO_4^{2-}$ depending on pH (pK_a's = 2.3, 7.0, and 11.5), chemically, AsO_4^{3-} and PO_4^{3-} whose pK_a's 2.1, 7.2, and 12.7 are very similar. Thus, As (+5) can replace phosphate in energy transfer phosphorylation and block protein synthesis. On the other hand, like phosphate, arsenate can be adsorbed strongly on sesquioxides, particularly amorphous $Fe(OH)_3$, rendering it less bioavailable and less toxic [7,10].

The As (+3)/As (+5) two-electron redox potential ($E^{0'} = +140$ mV at pH 7.0, 25°C vs. normal hydrogen electrode) is significantly higher than that of phosphate ($E^{0'} = -690$ mV), resulting in the existence of both As (+3) and As (+5) in environmental and biological conditions [5]. In addition, As can form stable bonds with carbon, yielding compounds such as mono-$[CH_3AsH_2]$, di-$[(CH_3)_2AsH]$, tri-methyl arsines $[(CH_3)_3As]$, arsenobetain $[(CH_3)_3–As^+–CH_2COO^-]$, arsenocholine $[(CH_3)_3–As^+–CH_2–CH_2–OH]$, and arsenosugars [5,6].

7.2 BIOLOGICAL PROPERTIES OF ARSENIC AND ITS TOXICITY

Due to its chemical similarity to phosphate, As (+5) is taken up by microorganisms, plant roots, and animal (intestinal) cells by two pathways used for phosphate [11]. The low-affinity P inorganic transport (Pit) pathway uses energy from the transmembrane proton gradient, while the high-affinity P specific transport (Pst) pathway has certain selectivity for phosphate over arsenate with a periplasmic phosphate binding protein and an ATP-hydrolyzing membrane transporter [12,13]. Neutral arsenite $[As(OH)_3]$, on the other hand, diffuses through membrane-spanning channels created by aquaglyceroporin proteins, which allow the diffusion of water, glycerol, $Si(OH)_4$, and other neutral species [12,14]. The properties of organoarsenic species are modulated by their organic substituent(s), and this affects their uptake by these or other pathways.

Normal human blood levels of As are 0.3–2 μg/L but can be one to two orders of magnitude higher when elevated levels of As are consumed with drinking water or food [5]. The overall half-life of As in humans is about 10 h, and 50%–80% of absorbed As is excreted in about 3 days [15]. Most excretory As is in urine [16].

Ingestion of inorganic As (60–120 mg as As_2O_3) would result in acute toxicity characterized by vomiting, abdominal pain, bloody diarrhea, which lead to dehydration, convulsion, coma, and death [17]. Chronic As exposure leads to skin lesions, hyperpigmentation, keratosis, diabetes, and cardiovascular disease [18]. "Black foot" disease, which shows a discoloration and blackening of the extremities, especially the feet, in Southwestern Taiwan was caused by drinking water from deep artesian wells high in As [19]. Millions of people in Bangladesh have been poisoned by consuming groundwater contaminated with high levels of As, sometimes as high as 800 μg/L [18,20]. The World Health Organization (WHO) and the United States (U.S.) Environmental Protection Agency have set 10 μg/L As as the maximum concentration for drinking water [21]. Rice grown in the southeastern United States is of concern to human health because of measured As levels of over 300 μg/kg due to past use of Ca-arsenate as a cotton defoliant [22]. Although currently there are no national standards for As in food, previous WHO guidelines established a provisional tolerable weekly intake for inorganic As of 15 μg/kg body weight, but these are currently being reconsidered [21]. Inorganic As species are of most concern to human health, with As (+3) being more toxic than As (+5) [23–25]. Organic As species, particularly arsenobetain and arsenocholine are generally considered less or nontoxic, though some debates remain [26].

7.3   BIOREMEDIATION OF ARSENIC

Bioremediation is the use of microorganisms or plants to detoxify an environment (mainly soil or water) by transforming or degrading pollutants. In case of diffused pollution, *in situ* bioremediation is better adapted for treatments of large areas. Such treated land becomes available for less risky uses at economically acceptable cost.

Arsenic being a metalloid, unlike many organic pollutants, cannot be converted to CO_2 and dissipated into the atmosphere (although the release and capture of the toxic arsine species are possible). The likely strategy would be oxidation of the more toxic As (+3) to the less toxic As (+5), methylation of inorganic species, and/or extraction (uptake) by plants and then disposing of the high-As plant biomass [27,28].

7.3.1 ARSENITE OXIDATION

Although pH and concentration dependent, the redox potential of As (+3)/As (+5) is about +140 mV at pH 7, making both species exist in environments that support many microbial growth and activities [29]. Arsenite itself can serve as an electron (e^-) donor for microbial respiration processes, oxidizing to As (+5) with e^- being passed to suitable e^- acceptors, such as oxygen (O_2) or nitrate (NO_3^-) under aerobic conditions. A wide array of microorganisms have evolved an energy requiring detoxification process catalyzed by the *ars* operon, linked to the intracellular reduction of As (+5) by the ArsC protein and its efflux as As (+3), as shown in Figure 7.1.

The ArsC is an As (+3)-specific exporter, which removes As from the cell. This process can be passive or active, which the latter case involves an associated ATPase. The ArsC is a cytoplasmic protein of 13–15 kDA related to tyrosine phosphate phosphatases facilitates the reduction of As (+5) when a suitable e^- donor, such as reduced thioredoxin or glutaredoxin, is provided. The genes involved are clustered in an *ars* operon that is located on plasmids or chromosomes of a diverse group of organisms, including Archaea, Bacteria, and yeasts [9,30,31].

The microbial oxidation of As (+3), a recognized detoxification process, involves two enzymes: Aso (also called Aox) and Arx [9,32].

$$As(+3)O_2^- + 2H_2O \xrightarrow{\text{Aso}} As(+5)O_4^{3-} + 4H^+ + 2e^-$$

The reaction sequence is completed once the electrons generated are passed to a physiological e^- acceptor, such as a *c*-type cytochrome [33,34]. Aso, normally located in the periplasm (Figure 7.1), has been isolated from a variety of organisms,

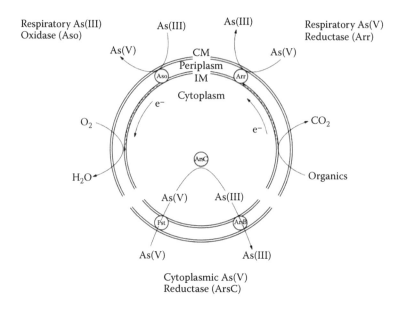

FIGURE 7.1 Biochemical transformation of inorganic As species by microbial cells. (Adapted from Lloyd, J.R. et al., Microbial transformations of arsenic in the subsurface, in: *Microbial Metal and Metalloid Metabolism: Advances and Applications*, ASM Press, Washington, DC, pp. 77–90, 2011.)

including *Rhizobium* sp. Str. NT-26, *Hydrogenophaga* sp. Str. NT-14, *Alcaligenes faecalis*, *Arthrobacter* sp. Str. 15b, and *Ralstonia* sp. Str. NT-14 [34]. Aox consists of two heterologous subunits: AoxA and AoxB. The large catalytic AoxA contains a molybdenum (Mo) atom coordinated by two pterin molecules and a [3Fe-4S] cluster; the smaller AoxB subunit contains a [2Fe-2S] cluster, as illustrated in Figure 7.2 [32]. The As (+3) oxidase, Arx, usually operates under anaerobic conditions and is distantly related to Aso, but has not been fully characterized.

Oremland et al. [35] isolated a facultative chemoautotrophic bacterium, strain MLHE-1, from arsenite-enriched bottom water from Mono Lake, California, that oxidized As (+3) anaerobically to As (+5) using nitrate as terminal e⁻ acceptor [35]. This organism was also able to grow heterotrophically with acetate as carbon and energy sources and oxygen (aerobic growth) or nitrate (anaerobic growth) as terminal e⁻ acceptors. Phylogenetic analysis based on its 16S rDNA places this organism with the haloalkaliphilic *Ectothiorhodospira* of the γ-*Proteobacteria*.

Mateos et al. [9] proposed that *Corynebacterium glutamicum* (a member of the genera *Corynebacterium* of biotechnological importance for the large-scale production of amino acids, such as L-glutamate and L-lysine), which is gram-positive with a thick cell wall, be used to accumulate/sorb As (3+) (and As (+5) after it is reduced to As (+3)) as a means to clean up (bioremediation) As in water [9]. The authors showed that *C. glutamicum* can tolerate up to 12 mM As (+3) and more than 400 mM As (+5) [9].

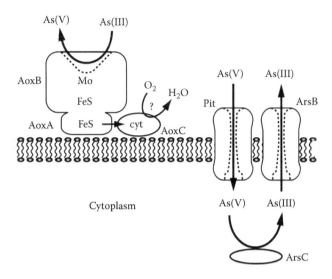

FIGURE 7.2 Model of arsenite oxidase (AoxA and AoxB) and arsenate reductase (ArsC). (Adapted from Saltikov, C.W., Regulation of arsenic metabolic pathways in prokaryotes, in: *Microbial Metal and Metalloid Metabolism*, ASM Press, Washington DC, pp. 195–210, 2011.)

7.3.2 METHYLATION OF INORGANIC AS SPECIES

Methylation is another mechanism that can confer As resistance and detoxification. The methylated As species include monomethyl arsonate [MMA (+5)], monomethylarsonite [MMA (+3)], dimethylarsinate [DMA (+5)], dimethylarsenite [DMA (+3)], and trimethylarsine oxide (TMAO), as well as several volatile arsines, including mono-, di-, and tri-methyl arsines (TAMs). Gosio (1897, as cited in Cullen and Reimer [36]) was the first to establish that fungi could generate methylated As. Challenger proposed a scheme in which As (+5) was eventually transformed to TMAs [37]. In this scheme, As (+5) is first reduced to As (+3) then methylated, and each methylation step results in the reoxidation of the As, thus requiring a reductive step to As (+3) prior to further methylation as shown in the equation [36]:

$$As(+3)O_3^{3-} \xrightarrow{RCH3} \underset{\text{methylarsonic acid}}{CH_3As(+5)O(OH)_2} \xrightarrow{2e^-} \underset{\text{methylarsinous acid}}{CH_3As(+3)(OH)_2} \xrightarrow{RCH3}$$

$$\rightarrow \underset{\text{dimethylarsinic acid}}{(CH_3)_2As(+5)O(OH)} \xrightarrow{2e^-} (CH_3)_2As(+3)(OH) \xrightarrow{RCH3}$$

$$\rightarrow \underset{\text{TMAO}}{(CH_3)_3As(+5)O} \xrightarrow{2e^-} \underset{\text{Trimethylarsine}}{(CH_3)_3As(+3)}$$

The methyl donors (RCH$_3$) in these reactions can be a form of methionine [36]. Several different enzymes have been identified with the methylase activity, such as S-adenosine methyltransferase in *Rhodobacter sphaeroides* [38,39].

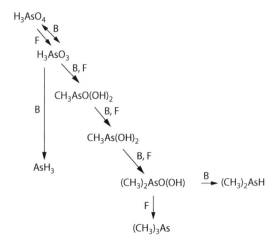

FIGURE 7.3 Summary of observed microbial interactions with arsenic species performed by (B) bacteria and (F) fungi.

Fungi, such as *Scopulariopsis brevicaulis*, *Aspergillus*, *Mucor*, *Fusarium*, *Peacilomyces*, and *Candida humicola*, have been found to be active in such As methylation [27,40,41]. Huysmans and Frankenberger [42] isolated a *Penicillium* sp. from an agricultural evaporation pond in California capable of producing trimethylarsine from methylarsonic acid and dimethylarsinic acid. The transformations of arsenic species by bacteria and fungi are summarized in Figure 7.3 [42].

The methylation of arylarsonic acids (e.g., Roxarsone) is important because their wide use as food supplements for swine, turkeys, and poultry. Methylphenylarsinic acid and dimethylphenylarsine oxide are reduced to dimethylphenylarsine by *C. humicola*. These arsines species are volatile and can be captured by activated carbon traps as illustrated in Figure 7.4 for As-contaminated water [27].

7.3.3   ARSENIC EXTRACTION BY PLANTS (PHYTOREMEDIATION)

There are two basic strategies by which higher plants can tolerate elevated levels of toxic metals, including As [43]: (1) exclusion, whereby transport of As is restricted, and low, relatively constant As concentrations are maintained in the shoot or grain over a wide range of soil concentrations, and (2) accumulation, whereby As is accumulated in nontoxic form(s) in upper plant parts at both high and low soil concentrations.

Most plants do not accumulate As, their As concentrations in leaves or seeds are often below 1 mg/kg [44]. The As hyperaccumulator fern, *Pteris vittata*, was discovered by Ma et al. [45] by screening many plant species growing at an As-contaminated site in Florida. Its fronds can contain in excess of 1% As (or 10,000 mg/kg dry weight) [45,46]. The ability of this fern to translocate As from the roots to the fronds and accumulate it was due in part to its ability to maintain high phosphate in its roots [47]. In fact, once entering the roots, through P transporters (Pit and Pst proteins), As (+5) is reduced to As (+3) before being expelled to cell vacuoles.

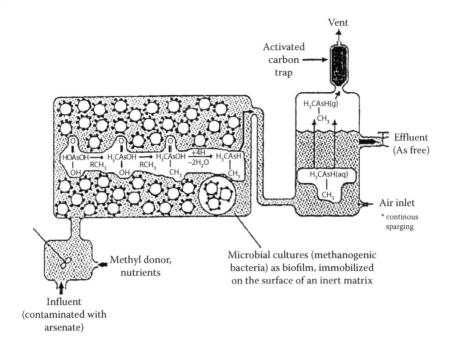

FIGURE 7.4 Bioreactor design for treatment of As-contaminated water. (Adapted from Frankenberger, W.T. and Losi, M.E., Applications of bioremediation in the clean up of heavy metals and metalloids, in: *Bioremediation: Science and Applications*, American Society of Agronomy, Madison, WI, Soil Science of Society of America Special Publication No. 43, pp. 173–210, 1995.)

Cytosolic arsenite, whether as a product of arsenate reductase or from uptake via an aquaglyceroporin, is detoxified by removal from the cytosol [48]. It is rather counterintuitive that As (+5), which is less toxic, should be converted to the more toxic As (+3) before removal from the cell interior. Perhaps, arsenate efflux would have caused phosphate efflux as well, a detrimental consequence that no living cells can afford, or perhaps it is an accident of evolution as speculated by Rosen [12]:

> Since the primordial atmosphere was not oxidizing, most As would have been in the form of As (+3), and early organisms would have evolved detoxification mechanisms to cope with As (+3), not As (+5). [*Furthermore, As (+3) can be chelated/detoxified by phytochelatins and other SH-containing proteins*] [49,50]. Once the atmosphere became oxidizing, most As (+3) in the environment would have been oxidized to As (+5). The mechanisms to cope with As (+5) make use of existing As (+3) extrusion systems. Besides, the conversion of a phosphatase to a reductase is relatively facile (at least easy in laboratory conditions), its evolution during the formation of an oxygenic world would have been rapid.

In addition to *P. vittata*, a few other fern species, such as *Pteris cretica*, *Pteris longifolia*, and *Pteris umbrosa*, also hyperaccumulate As [51]. Except for *Pityrogramma calomelanos*, all known As hyperaccumulators are ferns in the *Pteris* genus.

Brake fern (*Pteris vittata*)

FIGURE 7.5 **(See color insert.)** Chinese brake fern (*P. vittata* L.) grown in the greenhouse (left) and in the field. (NV Hue's personal images.)

However, not all *Pteris* ferns hyperaccumulate As [47,51,52]. Wang et al. [53] examined variation of As accumulation by ferns collected at different locations in south China (Guangxi Province) and found genotypic variations within *P. vittata* that could be useful in breeding improved cultivars. Ma et al. [45] found that fronds of *P. vittata* were able to accumulate as much as 15,000 mg/kg As in 2 weeks of growth in a sandy soil spiked with 1500 mg/kg As. Kertulis-Tartar et al. [54] implemented a field trial in a copper–chromium–arsenate (CCA)-contaminated soil in Florida, with soil As averaging 278 mg/kg in the 15–30 cm depth. After 2 years of cropping, soil at that depth averaged 158 mg/kg, a 43% depletion. The bioconcentration factor (BF), the ratio of As concentration in the plant to the total As concentration in soil, ranged from 10 to 100, depending on total As levels, soil characteristics, and growing environments [7,55]. McGrath and Zhao [56] calculated that a harvest of 10 tons biomass per hectare (ha) with a BF of 20 could reduce soil As in the top 20 cm depth by 50% after 10 harvests (*P. vittata* could be harvested every 3 months; Figure 7.5).

While *P. vittata* is capable of producing substantial biomass under favorable conditions as discussed in detail by Cai and Ma [57] and others [45,56], field results have been suboptimal [55]. Cai and Ma [57] could obtain only about 1 ton of frond biomass/ha/year, regardless of harvesting procedure and frequency (although better soil and plant management could surely improve the fern biomass). It also should be noted that *P. vittata* does not tolerate frost, is hard to germinate from spores, and only grows best in the tropics and subtropics. Its introduction into nonnative locations may also pose ecological dangers.

Genetic manipulation has the potential to transfer the ability to hyperaccumulate As to desired species. Chen et al. [58] demonstrated the principle by transferring the PvARC3 gene, a key As (+3) antiporter in *P. vittata* to *Arabidopsis thaliana*. The resulting transgenic plants had an increased ability to tolerate and accumulate As. Further work is needed to develop As hyperaccumulating capacities in other plant species that could produce more biomass, be more easily harvested, or more ecologically appropriate.

7.4 FUTURE TREND IN BIOREMEDIATION

The biological treatments, using microorganisms and plants, of As-contaminated systems are making significant progress toward practical technologies that are applicable to different agroecological regions. Recent developments in genomics and proteomics have led to the identification and characterization of As-resistance/tolerance genes and proteins, such as *ars* operon, *aio* gene in As (+3) oxidation and ArsC protein in As (+5) reduction. Such As involving genes could be engineered into other bacteria and plants that are adapting well to local environmental conditions, having high As tolerance, or increasing As uptake capacity, thus improving the efficiency of bioremediation. Applying the knowledge of many interdisciplinary sciences (e.g., microbiology, chemistry, hydrology, geology, engineering, soil and plant sciences, as well as biotechnology), bioremediation will probably become more effective, eco-friendly, socially acceptable, and profitable: the sustainable technology it should be.

REFERENCES

1. Emsley, J. *The Elements of Murder*, pp. 93–193. Oxford, U.K.: Oxford University Press, 2005.
2. WHO. *Environmental Health Criteria: Arsenic and Arsenic Compounds*, p. 224. Geneva, Switzerland: World Health Organization, 2001.
3. Ortiz-Escobar, M.E., Hue, N.V., and Cutler, W.G. Recent developments on arsenic: Contamination and remediation. In *Recent Research Developments in Bioenergetics*, Vol. 4. Kerala, India: Transworld Research Network, 2006, pp. 1–32.
4. Matschullat, J. Arsenic in the geosphere—A review. *Sci. Total Environ.* 249 (2000): 297–312.
5. Wilcox, D.E. 2013. Arsenic: Can this toxic metalloid sustain life. In *Metal Ions and Life Sciences*, pp. 475–498. Berlin, Germany: Springer, 2013.
6. Frankenberger, W.T. *Environmental Chemistry of Arsenic*. New York: Marcel Dekker, 2002.
7. Hue, N.V. Arsenic chemistry and remediation in Hawaiian soils. *Int. J. Phytoremed.* 15 (2013): 105–116.
8. Cutler, W.G., Brewer, R.C., El-Kadi, A. et al. Bioaccessible arsenic in soils of former sugar cane plantations, Island of Hawaii. *Sci. Total Environ.* 442 (2013): 177–188.
9. Mateos, L.M., Ordonez, E., Letek, M., and Gil, J.A. *Corynebacterium glutamicum* as a model bacteria for the bioremediation of arsenic. *Int. Microbiol.* 9 (2006): 2007–2015.
10. Cutler, W.G., El-Kadi, A., Hue, N.V. et al. Iron amendments to reduce bioaccessible arsenic. *J. Hazard. Mater.* 279 (2014): 554–581.
11. Willsky, G.R. and Malamy, M.H. Effects of arsenate on inorganic phosphate transport in *Escherichia coli. J. Bacteriol.* 144 (1980): 366–374.
12. Rosen, B.P. 2002. Biochemistry of arsenic detoxification. *FEBS Lett.* 529: 86–92.
13. Rao, N.N. and Torriani, A. Molecular aspects of phosphate transport in *Escherichia coli. Mol. Microbiol.* 4 (1990): 1083–1090.
14. Liu, Z., Shen, J., Carbrey, J.M. et al. Arsenite transport by mammalian aquaglyceroporins AQP7 and AQP 9. *Proc. Natl Acad. Sci.* 99 (2002): 6053–6058.
15. Liu, J., Goyer, R.A., and Waalkes, M.P. Toxic effects of metals. In *The Basic Science of Poisons*, pp. 931–979. New York: McGraw-Hill, 2008.
16. Basta, N.T., Rodriguez, R.R., and Casteel, S.W. Bioavailability and risk of arsenic exposure by the soil ingestion pathway. In *Environmental Chemistry of Arsenic*, pp. 117–139. New York: Marcel Dekker, 2002.
17. Ravenscroft, P., Brammer, H., and Richards, K. *Arsenic Pollution: A Global Synthesis*. New York: John Wiley & Sons, 2009.

18. Smith, A.H., Lingas, E.O., and Rahman, M. Contamination of drinking water by arsenic in Bangladesh: A public health emergency. *Bull. WHO* 78 (2000): 1093–1103.
19. Chiou, H.-Y., Chiou, S.T., Hsu, Y.-H. et al. Incidence of transitional cell carcinoma and arsenic in drinking water: A follow-up study of 8,102 residents in an arseniasis-endemic area in Northeast Taiwan. *Am. J. Epidemiol.* 153 (2001): 411–418.
20. Anawar, H.M., Akai, J., Mihaljevic, M. et al. Arsenic contamination in groundwater of Bangladesh: Perspective on geochemical, microbial and anthropogenic issues. *Water* 3 (2011): 1050–1076.
21. WHO. *Exposure to Arsenic: A Major Public Health Concern.* Geneva, Switzerland: World Health Organization, 2010.
22. Consumer Reports. 2012. Arsenic in your food, November issue, http://www.consumerreports.org/cro/magazine/2012/11/arsenic-in-your-food/index.htm. Accessed on March 24, 2013.
23. Hughes, M.F. Arsenic toxicity and potential mechanisms of action. *Toxicol. Lett.* 133 (2002): 1–16.
24. Cohen, S.M., Arnold, L.L., Beck, B.D., Lewis, A.S., and Eldan, M. Evaluation of the carcinogenicity of inorganic arsenic. *Crit. Rev. Toxicol.* 43 (2013): 711–752.
25. Contreras-Acuna, M., Garcia-Barrera, T., Garcia-Sevillano, M.A., and Gomez-Arina, J.L. Arsenic metabolites in human serum and urine after seafood (*Anemonia sulcata*) consumption and bioaccessibility assessment using liquid chromatography coupled to inorganic and organic mass spectrometry. *Microchem. J.* 112 (2014): 56–64.
26. Watanabe, T. and Hirano, S. Metabolism of arsenic and its toxicological relevance. *Arch. Toxicol.* 87 (2013): 969–979.
27. Frankenberger, W.T. and Losi, M.E. Applications of bioremediation in the clean up of heavy metals and metalloids. In *Bioremediation: Science and Applications*, pp. 173–210. Madison, WI: American Society of Agronomy, Soil Science Society of America Special Publication No. 43, 1995.
28. Chaney, R.L., Broadhurst, L.C., and Centofanti, T. Phytoremediation of soil trace elements. In *Trace Elements in Soils*. Oxford, U.K.: John Wiley & Sons, 2010.
29. Lloyd, J.R., Gault, A.G., Hery, M., and Mackae, J.D. 2011. Microbial transformations of arsenic in the subsurface. In *Microbial Metal and Metalloid Metabolism: Advances and Applications*, pp. 77–90. Washington, DC: ASM Press.
30. Stolz, J.F. Overview of microbial arsenic metabolism and resistance. In *The Metabolism of Arsenite*, pp. 55–60. Leiden, the Netherlands: CRC Press, 2012.
31. Saltikov, C.W. Regulation of arsenic metabolic pathways in prokaryotes. In *Microbial Metal and Metalloid Metabolism*, pp. 195–210. Washington, DC: ASM Press, 2011.
32. Heath, M.D., Schoepp-Cothenet, B., Osborne, T.H., and Santini, J.M. Arsenite oxidase. In *The Metabolism of Arsenite*, pp. 81–97. Leiden, the Netherlands: CRC Press, 2012.
33. Santini, J.M., Kappler, U., Ward, S.A. et al. The NT-26 cytochrome C552 and its role in arsenite oxidation. *Biochim. Biophys. Acta* 1767 (2007): 189–196.
34. Vanden Hoven, R.N. and Santini, J.M. Arsenite oxidation by the heterotrophy *Hydrogenophaga* sp. Str. NT-14: The arsenite oxidase and its physiological electron acceptor. *Biochim. Biophys. Acta* 1656 (2004): 148–155.
35. Oremland, R.S., Hoeft, S.E., Santini, J.M. et al. Anaerobic oxidation of arsenite in Mono Lake water and by a facultative, arsenite-oxidizing chemoautotroph, strain MLHE-1. *Appl. Environ. Microbiol.* 68 (2002): 4795–4802.
36. Cullen, W.R. and Reimer, K.J. Arsenic speciation in the environment. *Chem. Rev.* 29 (1989): 713–764.
37. Challenger, F. Biological methylation. *Adv. Enzymol.* 12 (1951): 429–491.
38. Qin, J., Rosen, B.P., Zhang, Y. et al. Arsenic detoxification and evolution of trimethylarsine gas by a microbial arsenite S-adenosylmethionine methyltransferase. *Proc. Natl. Acad. Sci.* 103 (2006): 2075–2080.

39. Jie, Q., Lehr, C.R., Yuan, C. et al. Biotransformation of arsenic by a Yellowstone thermoacidophilic eukaryotic alga. *Proc. Natl. Acad. Sci.* 106 (2009): 5213–5217.
40. Martin, A. *Introduction to Soil Microbiology.* New York: John Wiley & Sons, 1977.
41. Pickett, A.W., McBride, B.C., Cullen, W.R., and Manji, H. The reduction of trimethylarsine oxide by *Candida humicola. Can. J. Microbiol.* 27 (1981): 773–778.
42. Huysmans, D.K. and Frankenberger, W.T. Evolution of trimethylarsine by a *Penicillium* sp. Isolated from agricultural evaporation pond water. *Sci. Total Environ.* 105 (1991): 13–28.
43. Baker, A.J., McGrath, S.P., Reeves, R.D., and Smith, J.A.C. Metal hyperaccumulator plants: A review of the ecology and physiology of a biological resource for phytoremediation of metal-polluted soils. In *Phytoremediation of Contaminated Soil and Water*, pp. 85–107. Boca Raton, FL: Lewis Publishers, 2000.
44. Schat, H., Llugany, M., and Berhard, R. 2002. Metal-specific patterns of tolerance, uptake, and transport of heavy metals in hyperaccumulating and non hyperaccumulating metallophytes. In *Phytoremediation of Contaminated Soil and Water*, pp. 171–188. Boca Raton, FL: Lewis Publishers.
45. Ma, L., Komar, K.M., Tu, C. et al. A fern that hyperaccumulates arsenic—A hardy, versatile, fast-growing plant helps to remove arsenic from contaminated soils. *Nature* 409 (2001): 579.
46. Tu, C. and Ma, L. Effects of arsenate and phosphate on their accumulation by an arsenic hyperaccumulator *Pteris vittata* L. *Plant Soil* 249 (2003): 373–382.
47. Luongo, T. and Ma, L. Characteristics of arsenic accumulation by *Pteris* and non-*Pteris* ferns. *Plant Soil* 277 (2005): 117–126.
48. Rensing, C., Ghosh, M., and Rosen, B.P. Families of soft-metal-ion-transporting ATPases. *J. Bacteriol.* 181 (1999): 5891–5897.
49. Zhao, F.J., Wang, J.R., Barker, J.M.A. et al. The role of phytochelatins in arsenic tolerance in the hyperaccumulator *Pteris vittata. New Phytol.* 159 (2003): 403–410.
50. Yong, C., Su, J., and Ma, L.Q. Low molecular weight thiols in arsenic hyperaccumulator *Pteris vittata* upon exposure to arsenic and other trace elements. *Environ. Pollut.* 129 (2004): 69–78.
51. Zhao, F.J., Dunham, S.J., and McGrath, S.P. Arsenic hyperaccumulator by different fern species. *New Phytol.* 156 (2002): 27–31.
52. Meharg, A.A. Variation in arsenic accumulation-hyperaccumulation in ferns and their allies. *New Phytol.* 157 (2003): 25–31.
53. Wang, H.B., Wong, M.H., Lan, C.Y. et al. Uptake and accumulation of arsenic by 11 *Pteris* taxa from Southern China. *Environ. Pollut.* 145 (2007): 225–233.
54. Kertulis-Tartar, G.M., Ma, L., Tu, C., and Chirenje, T. Phytoremediation of an arsenic-contaminated site using *Pteris vittata* L.: A two-year study. *Int. J. Phytoremed.* 8 (2006): 311–322.
55. Zhao, F.J., McGrath, S.P., and Meharg, A.A. Arsenic as a food chain contaminant: Mechanisms of plant uptake and metabolism and mitigation strategies. *Annu. Rev. Plant Biol.* 61 (2010): 535–559.
56. McGrath, S.P. and Zhao, F.J. Phytoextraction of metals and metalloids from contaminated soils. *Curr. Opin. Biotechnol.* 14 (2003): 277–282.
57. Cai, Y. and Ma, L.Q. Metal tolerance, accumulation, and detoxification in plants with emphasis on arsenic in terrestrial plants. In *Biogeochemistry of Environmentally Important Trace Elements*, pp. 95–114. Washington, DC: American Chemical Society, 2003.
58. Chen, Y.S., Xu, W.S., Shen, H.L. et al. Engineering arsenic tolerance and hyperaccumulation in plants for phytoremediation by a PvACR3 transgenic approach. *Environ. Sci. Technol.* 47 (2013): 9355–9362.

8 Bioremediation for the Removal of Arsenic

Dayene C. Carvalho, Luciana M. Coelho,
Luciene M. Coelho, Helen C. de Rezende,
Nívia M.M. Coelho, and Thais S. Neri

CONTENTS

8.1   INTRODUCTION

Arsenic is a metalloid that is widely distributed in the earth's crust, and thus it is an element of concern given its toxicological significance, even at low concentrations. Sources of arsenic in the environment can be natural or anthropogenic, since it occurs as a trace element in most rocks as well as in soil, water, and atmospheric dust. Once released into the environment, arsenic compounds reach water sources, such as rivers and groundwater systems, and subsequently food sources. Thus, it is important to develop technologies for the remediation of arsenic-contaminated sites. Recent years have seen a growth in the research on the use bioremediation for the removal of heavy metal ions from aqueous solutions. Bioremediation is based on the ability of certain biomolecules or types of biomass to bind to and concentrate selected ions or other molecules from contaminated media (soil, sediment, air, and water). The major advantages of the bioremediation technology are its effectiveness in reducing the concentration of heavy metal ions to very low levels and its use of inexpensive biosorbent materials. The main purpose of this chapter is to provide an update on the recent literature concerning the strategies available for arsenic bioremediation and to discuss critically their main advantages and weaknesses. The discussions herein are focused on the main biosorption processes involved in this type of treatment.

8.2 GENERAL OVERVIEW

Arsenic (As) has been an element of great interest since its discovery in the thirteenth century, mainly due to its potential toxicity. According to the U.S. Agency for Toxic Substances and Disease Registry (ATSDR), arsenic is considered to be the most dangerous chemical element on the Priority List of Hazardous Substances, which makes it an element worthy of considerable attention [1].

Until the early twentieth century, arsenic was used as a poisoning agent due to specific characteristics that contributed to its relatively wide popularity, such as its harmless appearance and slightly sweet flavor allowing it to be easily mixed with food [2].

Arsenic contamination in the environment occurs from both anthropogenic and natural sources. Its presence in a variety of types of mineral deposits has been reported, primarily in the form of arsenopyrite (FeAsS), and it can change to arsenates and sulfoarsenate at the surface. Consequently, arsenic can be partially released into water bodies and later immobilized via adsorption onto oxides and hydroxides of iron, aluminum and manganese or onto clay minerals. Its use in industry, especially in the casting of lead, gold, silver, copper, zinc, and cobalt, is another potential source of arsenic exposure. Other sources include the production of glass, paint, fabric, leather, agricultural products, such as insecticides, pesticides and herbicides, and wood preservatives [3].

Arsenic-contaminated soil, sediment, and sludge are the major sources of arsenic contamination in the food chain, surface water, groundwater and drinking water [4]. Arsenic concentrations typically vary from below 10 mg/kg in noncontaminated soils to as high as 30,000 mg/kg in contaminated soils [5]. Environmental studies related to the exposure of this element have verified that the contamination of surface and groundwater by arsenic is mainly the result of mining and other industrial activities as well as agricultural practices.

In water, the most common forms of arsenic are oxyanions of As(V), under conditions of moderate to high redox potential, and in more reducing conditions As(III) is present. The concentration of arsenic in drinking water, according to the World Health Organization, should not exceed 10 mg/L [6]. Arsenic is a carcinogenic substance and the inorganic form is the most harmful to humans. The toxicity of the species As(III) is considered to be several times higher than that of As(V). The most common route of human exposure is through the consumption of contaminated water, but the inhalation of gases and dust intake may also be important sources of chronic arsenic exposure, which can cause serious problems related to metabolism, including hyperkeratosis; gangrene of the extremities, known as "black foot disease"; skin cancer; lung cancer; disorders of the nervous system; increased frequency of miscarriages and other serious health problems. Arsenic acts on the inactivation of approximately 200 enzymes, especially those involved in cellular energy production and related synthesis and DNA repair.

Arsenic has several oxidation states (+III, +V, 0, −III) and a variety of inorganic and organic forms. Extensive toxicity studies on arsenic have provided information on the different toxicities of the different arsenic forms [7]. The more reduced forms of arsenic are the most toxic, arsine (AsH_3) > arsenite (As(III)) > arsenate (As(V)) [8,9], while the methylated organic species monomethylarsonic acid (MMA) and dimethylarsinic acid (DMA) are less toxic, and organoarsenicals, arsenobetaine (AsB), and arsenocholine (AsC) are generally considered to be nontoxic [10]. Tri- and pentavalent inorganic arsenicals are apparently of comparable bioavailability, but differ in terms of their biochemistry, in part due to the increased preference of the trivalent forms for binding with thiols [8,9]. Thus, the oxidation state of this element profoundly affects its toxicity.

Various physicochemical and biological processes are commonly employed for the removal of arsenic from contaminated waters, such as coagulation/filtration, ion exchange, lime softening, adsorption onto iron oxides or activated alumina, and reverse osmosis [11,12]. Most of these technologies are applied only for the removal of As(V). In the case of As(III) contamination, a preoxidation step is usually required to transform the trivalent form to the pentavalent form. The oxidation procedure is mainly performed with the addition of chemical reagents, such as potassium permanganate, chlorine, ozone, hydrogen peroxide, or manganese oxide [11,13]. Although these reagents are effective for the oxidation of trivalent arsenic, they may also cause several secondary problems, arising mainly from the presence of residues or from the formation of by-products, incurring a significant increase in the cost of the respective methods [14].

Thus, it is important to develop alternative technologies to minimize the negative environmental impacts of arsenic contamination. These technologies should offer effective decontamination, simplicity of implementation, relatively short processing time and reduced costs. In this context, bioremediation is a technique aimed at reducing/removing contaminants in soil, water, sediment, wastewater, sewage, agricultural wastes, and groundwater using living organisms, such as microorganisms, plants and algae, or dead biomass. Various materials used in the removal of arsenic through a biosorption process are shown in Figure 8.1.

Several aspects of the biosorption process are relevant to the removal of pollutants [15–17]. Biosorption can involve physicochemical processes (e.g. ion exchange, complex formation, and physical adsorption), intracellular uptake, and the (bio) chemically mediated conversion of the species (e.g., methylation, demethylation, and redox processes). Materials with and without biological activity can be used in the process, including vegetal materials (e.g., plant leaves, seeds, and fruit bagasse), algae, and microorganisms (e.g., bacteria, fungi, and yeast) in their natural state or after appropriate immobilization.

For a number of years it has been reported that biosorption is a promising biotechnology for pollutant removal and/or recovery from solution, due to its simplicity,

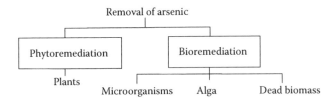

FIGURE 8.1 Materials used for arsenic removal through biosorption.

its operation being analogous to conventional ion exchange technology, its apparent efficiency, and the availability of biomass and waste bioproducts [15,16,18–23].

Bioremediation using microorganisms shows great potential for future development due to its environmental compatibility and potential cost effectiveness. Algae and microorganisms, including bacteria, fungi, and yeasts, can act as biologically active methylators that are able to at least modify toxic species. Many microbial detoxification processes involve efflux or the exclusion of metal ions from the cell, which in some cases can result in high local concentrations of metals at the cell surface where they may react with biogenic ligands and precipitate. Although microorganisms cannot destroy arsenic, they can alter its chemical properties via a surprising array of mechanisms.

Phytoremediation, which is a bioremediation process, involves the use of various types of plants that can accumulate, immobilize, dissipate, and/or degrade organic and inorganic contaminants. To ensure the success of this technology, the choice of plant is important since it must have certain characteristic, such as a deep root system and high resistance to pollutants, since the physicochemical properties of land fill capping soils, for instance, can interfere with the ability of the plant to perform the bioremediation [24,25]. These processes can involve the absorption or biotransformation of contaminants by roots, steams, and leaves or the production of biochemicals by the plants, which can be released into the soil or groundwater in the immediate vicinity of the roots and precipitate, or otherwise immobilize, contaminants.

The main purpose of this chapter is to provide an update on the recent literature concerning the strategies available for arsenic remediation (microorganisms, phytoremediation, and so on) and to discuss critically their main advantages and weaknesses. This chapter summarizes existing knowledge on various aspects of the fundamentals and applications of biosorption, reviews the obstacles hindering its commercial success, and discusses the future perspectives.

8.3 PHYTOREMEDIATION

Several different techniques are commonly used for the removal of arsenic and a notable example is phytoremediation. This approach seems attractive due to the nondestructive technology involved, which leaves the soil intact and biologically productive [26]. The appropriateness of phytoremediation techniques is dependent on their applicability and the type of contaminant, and they can involve different processes including phytodegradation, rhizodegradation, phytoextraction, rhizofiltration, phytovolatilization, phytostabilization, and hydraulic control. Table 8.1 shows the sites at which these phytoremediation processes occur and the possible mechanisms involved.

Figure 8.2 shows the increase in recent years in the amount of research studies on the proposed use of phytoremediation for the removal of arsenic. This highlights the great potential for the continued development of this process, due to its environmental compatibility and potential cost effectiveness.

According to Wang et al. [22], the first reported arsenic hyper accumulator was the brake fern *Pteris vittata*, which is able to remove 12–64 mg arsenic/kg in its fronds from uncontaminated soils containing 0.5–7.5 mg arsenic/kg and up to 22,630 mg arsenic/kg from a soil amended with 1500 mg arsenic/kg.

TABLE 8.1
Main Processes Associated with Phytoremediation

Main Characteristics	Type of Phytoremediation	Site at Which Phytoremediation Occurs	Mechanism
Degradation	Rhizodegradation	Enhancement of biodegradation in the below-ground root zone by microorganisms	Breakdown of contaminants due to the presence of proteins and enzymes produced by the plants or by soil organisms, such as bacteria, yeast, and fungi.
	Phytodegradation	Contaminant uptake and metabolism above or below ground, within the root, stem, and leaves	Absorption and metabolization for the transformation to more stable, less toxic, or less mobile forms.
Accumulation	Phytoextraction	Accumulation of contaminants in the roots and aboveground shoots or leaves	A mass of plants and contaminants (usually metals) can be collected for disposal or recycling.
	Rhizofiltration	Accumulation of contaminants in the roots	The plant roots absorb, concentrate, and precipitate toxic metals from contaminated groundwater.
Dissipation	Phytovolatilization	Contaminant uptake and volatilization	The contaminant, present in the water and soil, is taken up by the plant, passes through it or is modified by it, and is released to the atmosphere.
Immobilization	Hydraulic control	Control of groundwater flow through plant uptake of water	Use of vegetation to influence the movement of groundwater and soil water, through the uptake and consumption of large volumes of water. Root penetration throughout the soil can help counteract the slow flow of water in low-conductivity soils.
	Phytostabilization	Contaminant immobilization in the soil	The plants reduce the mobility and migration of contaminated soil.

Source: Pivetz, P.E., Phytoremediation of contaminated soil and ground water at hazardous waste sites, *Ground Water Issue*, 1–36, EPA/540/S-011/500 February 2001.

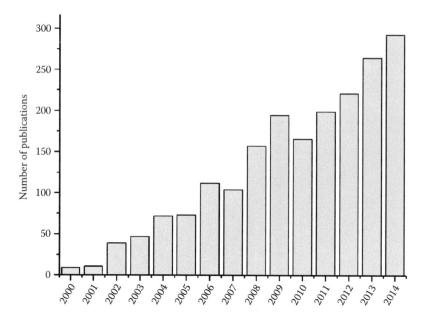

FIGURE 8.2 Published studies related to phytoremediation of arsenic. (From http://www. sciencedirect.com/, accessed on December 20, 2014.)

In a more recent study carried out by Titah et al. [27] on the phytoremediation of soil samples, the experiment included control *Ludwigia octovalvis* plants and three phytoremediation treatments with *L. octovalvis* plants with an arsenic (As) concentration of 39 mg/kg only, with the addition of a six rhizobacterial consortium at 2% (v/v) with an As concentration of 39 mg/kg, and with the addition of NPK fertilizer at 0.02% (w/w) with an As concentration of 39 mg/kg. The effectiveness of the phytoremediation in terms of As uptake at day 42 reached 49.8% in the phytoremediation treatment with 0.02% the NPK fertilizer addition. When phytoremediation is applied to groundwater samples, there is more efficient adsorption compared to its use with solid samples. Rahman et al. [28] performed long-term experiments with *Juncus effusus* to investigate the removal of As from domestic sewage containing 200 µg/L of As and obtained a metal retention >90%.

The inorganic forms (As(V) and As(III)) and the methylated forms (MMAA(V) and DMAA(V)) are the main species of arsenic in natural waters [29]. Plants show different mechanisms of retaining the various As species, and different adaptive mechanisms can be used to accumulate or exclude metals, thus maintaining their growth. This is considered to be a complex phenomenon and few studies have been carried out to propose the mechanisms involved in the application of phytoremediation for the removal of arsenic. The movement of metals across the root membrane, loading and translocation of metals through the xylem, and sequestration and detoxification of metals at the cellular and whole plant levels are important mechanisms adopted by accumulator plants [30].

According to Rahman et al. [30], there are three types of mechanisms of phytore-mediation involved in the removal of arsenic: (1) active uptake through phosphate transporters, (2) passive absorption through aquaglyceroporins, and (3) physical–chemical adsorption onto the root surface. Inorganic and methylated As species enter the plants via a passive mechanism through the aquaglyceroporin channels [30,31]. However, other mechanisms of phytoremediation involving physicochemical adsorption on root surfaces are also proposed for the phytoremediation of the arsenic species [32,33].

8.4 BIOSORPTION

8.4.1 PLANT MATERIALS

Living plants or dead biomass can be used for biosorption. Biomass wastes represent an interesting alternative for the adsorption of metallic ions since they are a cheap and an environmentally safe source of adsorbent material. There have been numer-ous studies carried out on the adsorption of heavy metals from water by lignocel-lulosic materials [34].

Ghimire et al. [35] used orange juice residue to prepare an adsorption gel by means of simple chemical modification by phosphorylation followed by loading with iron(III). The adsorption characteristics aimed at the adsorption of arsenate and arsenite from various aqueous media were then investigated. The performance of pure cellulose for the adsorption of arsenic was evaluated using the same treat-ment process for comparison purposes. The loading capacity for iron(III) in the gel prepared from orange waste (POW) was 1.21 mmol/g compared with 0.96 mmol/g for the gel prepared from cellulose (PC). Arsenite removal was favored under alka-line conditions for both PC and POW gels; however, the POW gel showed some removal capacity even at neutral pH. In contrast, arsenate removal took place under acidic conditions, at pH = 2–3 and 2–6 for the PC and POW gels, respectively. Since the iron(III) loading is higher for the POW gel than for the PC gel, greater arsenic removal is achieved applying the POW gel compared with the PC gel.

Blue pine shavings, walnut shells (*Juglans regia*), and chickpeas (*Cicer arieti-num*, Indian variety), which are widely available biomaterials, were investigated in terms of their potential for arsenic removal from drinking water. The conditions that can affect the adsorption were evaluated and under optimal conditions the removal efficiencies were 97%, 88%, and 33% for blue pine, walnut shells, and chickpeas, respectively. Thus, the blue pine biomass represents a potential low-cost biosorbent for arsenic removal from water samples, and after treatment with this biosorbent (blue pine) most samples were found to contain As levels under the permissible limit set by WHO, 10 µg/L [36].

The performance of three adsorbents (rice husk [RH], tea leaves, and newspaper) was evaluated for the removal of As(III) and As(V) from aqueous solutions. In the initial studies it was observed that only RH allowed the complete removal of As(III) and As(V) and therefore the authors selected this adsorbent for subsequent studies. The RH material was used without pretreatment, and the adsorption of As(III) and As(V) was performed using column methods, with the following initial conditions:

As concentration of 100 μg/L, adsorbent amount of 6 g, average particle sizes of 780 and 510 μm, treatment flow rates of 6.7 and 1.7 mL/min, and pH values of 6.5 and 6.0, respectively. The adsorbent was recovered after the treatment and a 1 mol/L potassium hydroxide solution was found to be appropriate for the desorption of arsenic from the RH surface [37].

Coconut coir pith (CP), a lignocellulosic residue, was used for the preparation of an anion exchanger through reaction with epichlorohydrin and dimethylamine followed by treatment with hydrochloric acid. The material produced was evaluated in relation to the extraction of As(V) from aqueous solution and groundwater, through batch experiments. A maximum removal of 99.2% was obtained for an initial concentration of 1 mg/L As(V) at pH 7.0 with an adsorbent dose of 2 g/L. The authors performed adsorbent regeneration studies and 0.1 N HCl showed potential for the regeneration of the material [34].

Hassan et al. [38] rated three different adsorbent materials—potassium hydroxide activated carbon from apricot stone (C), calcium alginate beads (G), and calcium alginate/activated carbon composite beads (GC)—for use in the removal of arsenic. Scanning electron microscopy (SEM), Fourier transform infrared spectroscopy (FTIR), N_2-adsorption at $-196°C$, and the determination of the point of zero charge were carried out to characterize the materials. The experimental results revealed that the adsorption of arsenic by the three different adsorbent materials is dependent on the textural properties of the adsorbate and the pH has a strong effect on the adsorption efficiency. GC exhibited the maximum As(V) adsorption (66.7 mg/g at 30°C). The adsorption of arsenic ions followed a pseudo-second-order mechanism and the thermodynamic parameters also confirmed an endothermic spontaneous process and a physisorption process.

An agricultural residue, known as "rice polish," was employed for the removal of arsenic from aqueous solutions. The parameters that influence the adsorption process were optimized. The sorption of arsenic onto rice polish was found to be highly pH dependent. Also, it increased with an increase in the initial metal ion concentration and decreased with an increase in the temperature. The system followed Langmuir, Freundlich, and Dubinin–Radushkevich (D–R) isotherm models, and the maximum sorption capacities calculated using the Langmuir model were 138.88 μg/g at 20°C and pH 7.0 for As(III) and 147.05 μg/g at 20°C and pH 4.0 for As(V). The mean sorption energy (E) calculated from the D–R model indicated that the nature of the sorption was chemisorption. A study on the thermodynamic parameters revealed the exothermic and spontaneous nature as well as the feasibility of the sorption process for both As(III) and As(V). The agricultural residue was found to be a potential biosorbent for the removal of both arsenic species (As(III) and As(V)) [39].

The removal of arsenic from water samples of different origins (lake, canal, and river) was studied using biomass derived from stems of *Acacia nilotica*. The main parameters affecting arsenic removal using the biosorbent studied, such as pH, biosorbent dosage, arsenic concentration, contact time, and temperature, were investigated. The Langmuir and Freundlich isotherm models presented good results, with the biosorbent material showing an adsorbent capacity of 667 μmol/g (50.8 mg/g). The study of the thermodynamic parameters revealed the endothermic and spontaneous nature as well as the feasibility of the biosorption process. The pseudo-second-order rate equation best described the arsenic biosorption kinetics and the thermodynamic calculations, also indicating the endothermic and spontaneous nature as well as the feasibility of the biosorption process at 298–318 K [40].

The sorption of Cr(III), Cr(VI), and As(V) onto biochars derived from RH and onto other materials such as the organic fraction of municipal solid wastes and sewage sludge and sandy loam soil has also been investigated. A kinetic study showed that the sorption process can be well described by the pseudo-second-order kinetic model, while the best fit in the simulation of the sorption isotherms was obtained for the Freundlich model. The materials tested removed more than 95% of the initial Cr(III). However, the removal rates for As(V) and Cr(VI) anions were considerably lower [41]. Table 8.2 summarizes the application of vegetal materials in arsenic bioremediation.

8.4.2 Algae

Algae have been extensively used for the biosorption of heavy metals for several reasons: they are (1) available in large quantities, (2) abundant in seawater and fresh water, (3) widely cultivated worldwide, (4) effective for the removal of metallic elements in aqueous solutions, and (5) of low cost [42].

Figure 8.3 shows the number of published papers related to studies on algae and arsenic, indexed in ScienceDirect for the period 2000–2014. The relatively high number of papers published each year highlights the importance of this issue. Also, it can be observed that the trend over the past 15 years is generally toward an increase in such publications, reaching a total of 248, although the numbers were mostly lower in 2004–2009.

In a more general context, several authors have reported the use of algae for the sorption of metals, including arsenic [43] and other elements such as Al [44]; Au [45]; Cd, Cu, Cr [46]; Co, Ni [47]; Fe, Hg [48]; Pb [49]; Se [50]; Zn [51]; and U [52].

Algae range in size from unicellular, such as the genus *Chlorella*, to multicellular forms, such as the giant kelp. One description of algae is that they have chlorophyll and carry out oxygenic photosynthesis [53]. Biosorption by algae has mainly been attributed to the cell wall structure containing functional groups such as amino, hydroxyl, carboxyl, and sulfate groups, which can act as metal binding sites through electrostatic attraction, ion exchange, or complexation. Pawlik-Skowrońska et al. [53] reported that arsenic can be present in algal cells in various complexes with nonprotein single-bond sulfhydryl groups. Some of these complexes dissociate under acidic conditions, but others are able to dissociate only at alkaline pH. According to Pacheco et al. [54], the cell walls of brown algae contain alginic acid (10%–40%), fucoidan (5%–20%), and cellulose (2%–20%), the carboxylic groups being the most abundant acidic functional group. The cell walls of red algae contain agar, xylans, lectin, and cellulose, while the cell walls of green algae contain mainly pectic substances and cellulose.

Bioremediation employing algae has been widely used to remove arsenic from freshwater [55], contaminated wastewaters [56], and river water [57]. Reports in the

TABLE 8.2
Vegetal Materials Used in Arsenic Bioremediation

Analyte	Vegetal Materials	Modification	Isotherm Model	Kinetics Model	Sorption Capacity/ Removal Efficiency	Sample	Reference
As(V)	Rice hush	Biochar (pyrolysis)	Freundlich	Pseudo-second order	25%	Aqueous solutions	[41]
As(III) As(V)	Rice hush	—	—		96% 96%	Groundwater	[37]
As(V)	Coconut coir pith	Epichlorohydrin and dimethylamine and treatment with HCl	Langmuir	Pseudo-second order	99.2%	Groundwater, industrial effluents	[34]
As	*A. nilotica*	—	Langmuir Freundlich	Pseudo-second order	50.8 mg/g; 95%	Lake, canal, and river water	[40]
Arsenate arsenite	Orange waste	Phosphorylation followed by loading with iron(III)	Langmuir	—	1.21 mmol/g	Wastewater	[35]
As(V)	Apricot stone	Carbonization	Langmuir	Pseudo-second order	26.6 mg/g (20°C)	—	[38]
As(III) As(V)	Rice polish	—	Langmuir Freundlich Dubinin–Radushkevich	Pseudo-second order	138.88 µg/g (20°C) 147.05 µg/g (20°C)	Aqueous solutions	[39]
As	Pine wood shavings Walnut shell Chickpea	—	Langmuir Freundlich	—	97% 88% 33%	Water samples	[36]

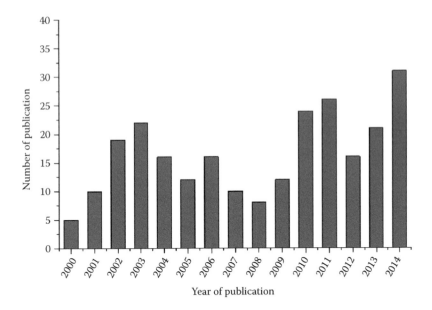

FIGURE 8.3 Evolution of the number of published papers related to algae and arsenic per year from 2000 to 2014. (From http://www.sciencedirect.com/, accessed on December 20, 2014.)

literature describe the application of several different algal species to arsenic adsorption, including *Caulerpa fastigiata*, *Ceramium* sp., *Chlorella vulgaris*, *Cladophora*, *Cystoseira barbata*, *Enteromorpha* sp., *Fucus virsoides*, *Gelidium*, *Maugeotia genuflexa*, *Oedogonium* sp., *Padina pavonica*, and *Polysiphonia* sp. Blue-green algae (Cyanobacteria), of which *Dunaliella, Spirulina (Arthrospira), Nostoc, Anabaena*, and *Synechococcus* are typical genera, have shown potential as biosorbents for the efficient removal of heavy metals from wastewaters. Table 8.3 summarizes the various types of algae used in arsenic bioremediation.

Different conditions can be employed to study arsenic removal by bioremediation using algae, varying parameters such as the initial metal ion concentration, pH, temperature, biosorbent dosage, biomass particle size, and contact time [28].

The bioremediation of both water resources and soils using algae has been tested and found to have great potential. A method has been developed for the removal of arsenic from wastewater samples using living algae employing HPLC coupled with ICP. Primarily, the authors investigated the growth capacity of six different types of algae (*Cladophora*, *Chlorodesmis* (hair algae), *Chorella*, *Spirogyra*, *Spirulina*, and *Dunaliella*) in an arsenic-rich medium. Out of the six different algal species studied, only *Cladophora* grew efficiently in the medium. It was found that *Cladophora* species can survive in a medium with an arsenic concentration of up to 6 mg/L in water and at a concentration of As of 80 g/L, close to 100% biosorption being achieved with a contact time of up to 9–10 days. Also, the presence of arsenide, arsenate, arsenosugar, MMA and DMA in *Cladophora* species has been reported [58].

TABLE 8.3

Papers Involving the Biosorption of Arsenic Using Algae

Algae (Biosorbent)	Analyte	Sample	Detection	Reference
Spirodela polyrhiza L	Arsenate and DMAA	River water	ET AAS	[57]
Cladophora	As(III), As(V), DMA, MMA, arsenosugar	Wastewater	HPLC-ICPMS	[58]
J. effuses	As (V)	Artificial wastewater	HG AAS	[28]
Scenedesmus quadricauda	As(III)	Standard solution	ETAAS	[59]
M. genuflexa	As (III)	Aqueous solution	HG AAS	[60]
Tolypothrix tenuis, Nostoc muscorum and Nostoc minutum	As(III), As(V)	Standard solution		[61]
Microcystis aeruginosa	As(III), As(V)	Freshwater system		[62]
Green algae	As (III)	Freshwater	ET AAS	[63]

The algal species *Lessonia nigrescens* has been evaluated for the biosorption of As(V) and its maximum adsorption capacities were estimated at 45.2 mg/g (pH = 2.5), 33.3 mg/g (pH = 4.5), and 28.2 mg/g (pH = 6.5), indicating better adsorption at lower pH. These values are high in comparison with other arsenic adsorbents reported in the literature [64].

Cyanobacteria have some advantages over other microorganisms, including their greater mucilage volume with high binding affinity, large surface area, and simple nutrient requirements. Cyanobacteria are easily cultivated in large scale in laboratory cultures, providing a low-cost biomass for the biosorption process [28,65,66].

Seaweed has been the focus of many studies where attention has been directed toward identifying and quantifying the arsenic species present in algal samples. Typically, this is carried out using HPLC with detection techniques such as ICP-AES and HG-ICP-AES. Although the focus of these studies is not bioremediation, they indirectly demonstrate the ability of algae to retain arsenic species. Notable in this regard are the studies by Salgado's research group [67,68], who studied the arsenic speciation in algal species such as *Sargassum fulvellum, C. vulgaris, Hizikia fusiformis*, and *Laminaria digitata*. The species As(V) was identified in *Hizikia* (46 ± 2 µg/g), *Sargassum* (38 ± 2 µg/g), and *Chlorella* (9 ± 1 µg/g) algae and DMA was also detected in *Chlorella* (13 ± 1 µg/g). However, for *Laminaria*, only an unknown arsenic species was detected, which was eluted from the dead volume.

Slejkovec et al. [69] reported a study on ten different marine algae, collected from a coastal zone, employing high performance liquid chromatography (anion and cation exchange) UV photochemical digestion hydride generation atomic fluorescence spectrometry (HPLC-UV-HG AFS). *Ceramium* sp., *Cy. barbata, Enteromorpha* sp., *F. virsoides*, two species of *Gelidium, P. pavonica, Polysiphonia* sp., and *Ulva rigida* were collected along the Adriatic Sea coast of Slovenia. The speciation results show that arsenosugars are the most abundant arsenic compounds found in most algae. In all red and green algae studied, with the exception of *Ceramium* sp. and *Polysiphonia* sp., arsenosugar was found to be the major arsenic compound. The brown algae analyzed contained, in general, arsenosugars (except *Cy. barbata*). High fractions of inorganic arsenic, especially As(V), were present in the red algae *Ceramium* sp. and *Polysiphonia* sp. and in the brown alga *Cy. barbata*.

Thus, based on these studies, the potential for the use of algae in As removal has become widely recognized. Biosorption employing algae is an efficient and inexpensive method of remediating contaminated sites.

8.4.2.1 Microorganisms

The study of the interaction between toxic metals and microorganisms (fungi and bacteria) has long been of scientific interest. The common occurrence of microorganisms in soils and sediments contaminated with toxic elements and heavy metals, including arsenic, has been extensively reported. The mechanisms of contaminant removal by microorganisms can include (1) extracellular accumulation/precipitation, (2) cell surface adsorption or complexation, and (3) intracellular accumulation [70].

Some microorganisms are able to use either the oxidized form of inorganic arsenic (arsenate) or the reduced form (arsenite) for metabolism, and an even greater number are resistant to arsenic toxicity due to the ars genetic system [71,72].

Fungi have a high growth capacity, they have the potential to produce a large number of enzymes and also they are good accumulators of various metals, which can be screened as promising bioremediation agents [73]. The application of fungi to remediate land and water contaminated with toxic compounds has received increasing interest because fungi are ubiquitous in the natural environment and are the dominant organisms in many soils, particularly those with low pH values [74]. The ubiquitous presence of As(V)-reducing microorganisms and the ease with which they can be stimulated provide the basis for the development of *in situ* remediation techniques involving the controlled mobilization of contaminants. The effective recovery of mobilized As(III) is the most important factor for the successful implementation of this type of remedial option. Fungi may be able to accumulate these metal(loid)s in their cells. The intracellular uptake of metal ions from a substrate into living cells, otherwise known as bioaccumulation, may lead to the biological removal of metals by fungi. As fungi play fundamental roles in the natural environment, especially regarding decomposition, transformation, and nutrient cycling, knowledge of their response to high metal concentrations is particularly relevant [75].

Many fungal species are capable of transforming inorganic arsenic compounds, arsenite and arsenate, through biomethylation into methylated arsenic species such as MMA, DMA, trimethylarsine (TMA), and trimethylarsine-oxide (TMAO). Three of these, MMA, DMA and TMAO, are less toxic than arsenate, arsenite, and TMA. Several fungal species (*Aspergillus glaucus, Candida humicola, Scopulariopsis brevicaulis, Gliocladium roseum, Penicillium gladioli*, and *Fusarium* spp.) are reportedly capable of transforming arsenic from its inorganic and methylated forms into volatile trimethylarsine, via the reductive methylation process [70].

The microbial biotransformation of arsenic plays a significant role in the fate and toxicity of this metalloid in the environment. Several different types of arsenic biotransformation pathways (mainly involving oxidation or reduction reactions) in a variety of microorganisms have been proposed [76].

A facultative marine fungus *Aspergillus candidus* showed profuse growth in different concentrations (25 and 50 mg/L) of the trivalent and pentavalent forms of arsenic. The highest level of arsenic removal (mg/g) was recorded on day 3. Hence, the test fungus *As. candidus* is a promising candidate for arsenic remediation [73].

Su et al. [77] studied the As-resistant fungi *Trichoderma asperellum* SM-12F1, *Penicillium janthinellum* SM-12F4, and *Fusarium oxysporum* CZ-8F1 on exposure to 50 mg/L of As(V). The biotransformation of arsenic and the concomitant change in the Eh and pH of the medium were evaluated. After cultivation for 15 days, the total As levels in the culture systems of these strains were all notably lower than the originally spiked As levels of 2500 µg. This was largely attributed to arsenic volatilization in these fungi and also possible interference during the detection due to the precipitation and/or chelation of arsenic, although intermediate biomethylation species of As were not found in the case of *Penicillium* and *Fusarium* strains.

In addition, resistance to arsenic has been shown in laboratory strains of bacteria such as *Escherichia coli* [78], *Staphylococcus aureus* [79], and *Pseudomonas aeruginosa* [80]. Since arsenic is known to act as an electron donor or acceptor, it may be an active component of the electron transport chain in some bacteria [81]. As(III) and As(V) have similar structures and their uptake tends to occur via the glycerol and phosphate transporters, respectively [82]. GlpF homologs have been identified as a member of the glycerol channels in *Leishmania major* [83] and *Pseudomonas putida*; all of them are expected to facilitate the transport of As(III) across the cell membrane [81].

In another study, Tian et al. [84] analyzed stress due to the presence of As(V) on bacteria using the SERS (surface enhanced Raman scattering) technique to explore the As-resistance mechanism. The notable changes in the SERS spectra for the cell wall in the presence of As(V) indicate that As(V) mainly acts upon the cell wall, except in the case of *Acinetobacter* sp. On the other hand, As(V) was found to have a negligible effect on protoplasts. This suggests that bacteria may be resistant to elevated concentrations of As(V) because this species is mostly sequestered by the cell wall. The changes in the SERS spectra upon As(V) attack suggest that As(V) mainly interacts with functional groups, such as polysaccharides, flavin derivates, C\N, COO−, and CH_2 in the cell walls.

According to Joshi et al. [85], amino acid sequencing data for arsenic binding proteins showed only one cysteine residue and unusually high contents of aspartic acid and glutamic acid, indicative of a higher degree of ionic interaction compared with that of the thiol groups of protein with arsenite. An earlier report suggests that multiple physiological responses occur in *Pseudomonas stutzeri* in response to arsenate-induced stress [86]. Considering the modest number of studies carried out on metal binding proteins in bacteria, it is possible that similar proteins are present in other bacterial species.

Liu et al. [87] investigated whether the expression of the *arsM* gene in bacteria significantly increased the ability of organism to methylate arsenic in aqueous and soil systems. In aqueous systems, compared with the wild-type cells without the expression of *arsM*, the levels of gaseous arsenic generated by

Sphingomonas desiccabilis and *Bacillus idriensis* strains (with the expression of *arsM*) were around 10-fold higher. Also, the biovolatilization by these strains can result in around 2.2%–4.5% of arsenic removal from soil after 30 days of incubation. It has been proposed that genetically engineering bacteria with the *arsM* gene could be a promising strategy for bioremediation processes aimed at remediating arsenic-contaminated environments.

It has been postulated that the physiological function of TtArsC (gene encoding arsenate reductase) in the bacterium *Thermus thermophilus* HB27 could be to significantly contribute to arsenic resistance. According to this hypothesis, once in the cell, arsenate would increase the TtarsC expression, probably through an as yet unidentified arsenic-dependent transcription regulator. As the arsenate is reduced by the enzyme, the even more toxic arsenite produced could be extruded by up to three arsenite permeases, two encoded by genes located on the chromosome and one encoded by a gene carried on the plasmid, which have not yet been functionally characterized. Even though TtarsC expression levels are quite low, microbial cells could benefit from both an efficiently encoded enzyme and finely regulated export to the outside of the cell. The contribution of the other proteins to arsenic tolerance remains to be determined and will be investigated in the near future [88].

Comprehensive knowledge on the molecular and genetic aspects of arsenic metabolism and detoxification, besides being of interest from an evolutionary point of view, also represents an important starting point for developing efficient and selective arsenic bioremediation strategies, an environmentally friendly approach to metal pollutant removal. The results of the studies described in these articles open new perspectives on bioremediation technology for the removal of arsenic from highly contaminated soils and groundwater. The accumulation of arsenic in the biomass of the test microorganisms suggests the role for microorganisms, through their bioaccumulation capability, as agents for the bioremediation of arsenic-contaminated environments.

8.5 CONCLUSIONS

Biosorption has been widely used in the removal of arsenic from different environments, including contaminated water resources. Biosorption is the most economical and eco-friendly method for the removal of arsenic of both anthropogenic and natural sources. Notable advantages are (1) low cost of the biosorbent, (2) high efficiency for metal removal at low concentration, (3) potential for biosorbent regeneration and metal valorization, (4) high sorption and desorption rates, (5) limited generation of secondary residues, and (6) relatively environmentally friendly life cycle of the material (e.g., easy to eliminate compared with conventional resins). Even though biosorbents represent an attractive alternative to synthetic materials, the seasonal variability in their composition and the required pretreatments can be important drawbacks. Furthermore, after the metal removal from aqueous solutions by the biomass, the recovery of the metal is an important issue. This can be achieved through a metal desorption process, aimed at weakening the metal-biomass binding. Thus, studies to evaluate the reversibility of the adsorption reactions involved in the biosorption of arsenic are of great importance. In this way, problems associated with the disposal of the adsorbent can be solved, since arsenic is potentially toxic and its disposal even after appropriate treatment has to be carefully evaluated. In relation to biosorption and desorption processes, another important aspect is the biosorbent reuse in successive biosorption–desorption cycles, the viability of which is determined by the cost–benefit relationship between the loss in biosorption capacity during the desorption steps and the operational yield in terms of metal recovery.

The use of nanotechnology has contributed to continually improving this approach. However, a new scientific challenge has emerged in relation to applying bioremediation to speciation analysis, since the mechanisms associated with the removal of arsenic in its different oxidation states have not been fully elucidated, particularly in relation to the food chain and biogeochemical cycles. Thus, further studies are needed that focus on the development of new clean environmentally acceptable technologies with commercial feasibility.

ACKNOWLEDGMENTS

The authors are grateful for the financial support from the Brazilian governmental agencies Conselho Nacional de Desenvolvimento Científico e Tecnológico (CNPq) and Coordenação de Aperfeiçoamento de Pessoal de Nível Superior (CAPES), the Minas Gerais state government agency Fundação de Amparo à Pesquisa do Estado de Minas Gerais (FAPEMIG), and the Goiás state government agency Fundação de Amparo á Pesquisa do Estado de Goiás (FAPEG).

REFERENCES

1. ATSDR (Agency for Toxic Substances and Disease Registry). Comprehensive Environmental Response, Compensation and Liability Act (CERCLA). Priority list of hazardous substances. U.S., http://www.atsdr.cdc.gov/SPL/index.html, accessed on 28 March 2015.
2. Gontijo, B., Bittencourt, F. Arsenic—A historical review. *Brazilian Annals of Dermatology* 80(1) (2005): 91–95.
3. Azcue, J., Nriagu, J. Arsenic: Historical perspectives. In: *Arsenic in the Environment*, J. O. Nriagu (ed.) (New York: John Wiley & Sons, 1994, pp. 1–15).
4. Frankenberger Jr., W., Arshad, M. Volatilisation of arsenic. In: *Environmental Chemistry of Arsenic*, W. T. Frankenberger Jr. (ed.) (New York: Marcel Dekker, 2002, pp. 363–380).
5. Adriano, D. C. *Trace Elements in the Terrestrial Environments* (New York: Springer, 1986, pp. 219–262).
6. World Health Organization (WHO). *Arsenic and Arsenic Compounds*. Geneva, Switzerland: International Programme on Chemical Safety, 2001.
7. Leermakers, M., Baeyens, W., De Gieter, M. et al. Toxic arsenic compounds in environmental samples: Speciation and validation. *Trends in Analytical Chemistry* 25(1) (2006): 1–10.
8. Hindmarsh, J., McCurdy, R. Clinical and environmental aspects of arsenic toxicity. *Critical Reviews in Clinical Laboratory Sciences* 23(4) (1986): 315–347.
9. Templeton, D., Ariese, F., Cornelis, R. et al. Guidelines for terms related to chemical speciation and fractionation of elements. Definitions, structural aspects, and methodological approaches. *Pure and Applied Chemistry* 72(8) (2000): 1453–1470.
10. Pizarro, I., Milagros, G., Cámara, C. et al. Arsenic speciation in environmental and biological samples: Extraction and stability studies. *Analytica Chimica Acta* 495(1–2) (2003): 85–98.

11. Jekel, M. Removal of arsenic in drinking water treatment. In: *Arsenic in the Environment: Part 1 Cycling and Characterization*, J. O. Nriagu (ed.) (New York: Wiley-Interscience, 1994, pp. 119–130).

12. Zouboulis, A., Katsoyiannis, I. Removal of arsenates from contaminated water by coagulation–Direct filtration. *Separation Science and Technology* 37(12) (2002): 2859–2873.

13. Kim, M.-J., Nriangu, J. Oxidation of arsenite in groundwater using ozone and oxygen. *Science of the Total Environment* 247(1) (2000): 71–79.

14. Zouboulis, A., Katsoyiannis, I. Recent advances in the bioremediation of arsenic-contaminated groundwaters. *Environment International* 31(2) (2005): 213–219.

15. Tsezos, M., Volesky, B. Biosorption of uranium and thorium. *Biotechnology and Bioengineering* 23(3) (1981): 583–604.

16. Gadd, M., White, C. Microbial treatment of metal pollution—A working biotechnology? *Trends in Biotechnology* 11(8) (1993): 353–359.

17. Texier, A., Yves, A., Faur-Brasquet, C. et al. Fixed-bed study for lanthanide (La, Eu, Yb) ions removal from aqueous solutions by immobilized *Pseudomonas aeruginosa*: Experimental data and modelization. *Chemosphere* 47(3) (2002): 333–342.

18. Volesky, B., Holan, Z. Biosorption of heavy metals. *Biotechnology Progress* 11(3) (1990): 235–250.

19. Volesky, B. Detoxification of metal-bearing effluents: Biosorption for the next century. *Hydrometallurgy* 59(2–3) (2001): 203–216.

20. Volesky, B. Biosorption and me. *Water Research* 41(18) (2007): 4017–4029.

21. Veglio, F., Beolchini, F. Removal of metals by biosorption: A review. *Hydrometallurgy* 44(3) (1997): 301–316.

22. Wang, J., Chen, C. Biosorption of heavy metals by *Saccharomyces cerevisiae*: A review. *Biotechnology Advances* 24(5) (2006): 427–451.

23. Mack, C., Wilhelmi, B., Duncan, J. et al. Biosorption of precious metals. *Biotechnology Advances* 25(3) (2007): 264–271.

24. Coutinho, D., Barbosa, A. Fitorremediação: Considerações Gerais E Características De Utilização. *Silva Lusitana* 15(1) (2007): 103–117.

25. Kim, K.-R., Owens, G. Potential for enhanced phytoremediation of landfills using biosolids—A review. *Journal of Environmental Management* 91(4) (2010): 791–797.

26. Sabir, M., Waraich, E., Hakeem, K. et al. Phytoremediation: Mechanisms and adaptations. In *Soil Remediation and Plants: Prospects and Challenges*, K. Hakeem, M. Sabir, M. Ozturk and A. Murmet (eds.) (London, U.K.: Elsevier, 2015, pp. 85–105).

27. Titah, H. S., Abdullah, S. R. S., Mushrifah, I. et al. Effect of applying rhizobacteria and fertilizer on the growth of *Ludwigia octovalvis* for arsenic uptake and accumulation in phytoremediation. *Ecological Engineering* 58 (2013): 303–313.

28. Rahman, A., Hassler, C. Is arsenic biotransformation a detoxification mechanism for microorganisms? *Aquatic Toxicology* 146 (2014): 212–219.

29. Cullen, W., Kenneth, R. Arsenic speciation in the environment. *Chemical Reviews* 89(4) (1989): 713–764.

30. Rahman, A., Kehtarnavaz, N. Aquatic arsenic: Phytoremediation using floating macrophytes. *Chemosphere* 83(5) (2011): 633–646.

31. Zhao, F., Ma, J., Meharg, A. et al. Arsenic uptake and metabolism in plants. *New Phytologist* 181(4) (2009): 777–794.

32. Robinson, B., Kim, N., Marchetti, M. et al. Arsenic hyper accumulation by aquatic macrophytes in the Taupo Volcanic Zone, New Zealand. *Environmental and Experimental Botany* 58(1–3) (2006): 206–215.

33. Rahman, M., Hasegawa, H., Ueda, K. et al. Arsenic uptake by aquatic macrophyte *Spirodela polyrhiza* L.: Interactions with phosphate and iron. *Journal of Hazardous Materials* 160(2–3) (2008): 356–361.

34. Anirudhan, T. S., Rijith, S., Suchithra, P. S. Preparation and characterization of iron(III) complex of an amino-functionalized polyacrylamide-grafted lignocellulosics and its application as adsorbent for chromium(VI) removal from aqueous media. *Journal of Applied Polymer Science* 115(4) (2010): 2069–2083.

35. Ghimire, K. N., Inoue, K., Makino, K., Miyajima, T. Adsorptive removal of arsenic using orange juice residue. *Separation Science and Technology* 37 (2002): 2785–2799.

36. Saqib, A. N. S., Waseem, A., Khan, A. F. et al. Arsenic bioremediation by low cost materials derived from Blue Pine (*Pinus wallichiana*) and walnut (*Juglans regia*). *Ecological Engineering* 51 (2013): 88–94.

37. Amin, Md. N., Kaneco, S., Kitagawa, T. et al. Removal of arsenic in aqueous solutions by adsorption onto waste rice husk. *Industrial & Engineering Chemistry Research* 45(24) (2006): 8105–8110.

38. Hassan, A. F., Abdel-Mohsen, A. M., Elhadidy, H. Adsorption of arsenic by activated carbon, calcium alginate and their composite beads. *International Journal of Biological Macromolecules* 68 (2014): 125–130.

39. Hasan, S. H., Ranjan, D., Talat, M. "Rice polish" for the removal of arsenic from aqueous solution: Optimization of process variables. *Industrial & Engineering Chemistry Research* 48(9) (2009): 4194–4201.

40. Baig, J. A., Kazi, T. G., Shah, A. Q. et al. Biosorption studies on powder of stem of *Acacia nilotica*: Removal of arsenic from surface water. *Journal of Hazardous Materials* 178(1–3) (2010): 941–948.

41. Agrafioti, E., Kalderis, D., Diamadopoulos, E. Ca and Fe modified biochars as adsorbents of arsenic and chromium in aqueous solutions. *Journal of Environmental Management* 146(15) (2014): 444–450.

42. Ibrahim, W. M. Biosorption of heavy metal ions from aqueous solution by red macroalgae. *Journal of Hazardous Materials* 191 (2011): 1827–1835.

43. Alvarado, S., Guédez, M., Lué Meru, M. et al. Arsenic removal from waters by bioremediation with the aquatic plants water hyacinth (*Eichhornia crassipes*) and lesser duckweed (*Lemna minor*). *Bioresource Technology* 99(17) (2008): 8436–8440.

44. Sari, A., Mustafa, T. Equilibrium, thermodynamic and kinetic studies on aluminum biosorption from aqueous solution by brown algae (*Padina pavonica*) biomass. *Journal of Hazardous Materials* 171(1–3) (2009): 973–979.

45. Mata, Y., Torres, E., Blázquez, M. L. et al. Gold(III) biosorption and bioreduction with the brown alga *Fucus vesiculosus*. *Journal of Hazardous Materials* 166(2–3) (2009): 612–618.

46. Mishra, S., Srivastava, S., Tripathi, R. et al. Thiol metabolism and antioxidant systems complement each other during arsenate detoxification in *Ceratophyllum demersum* L. *Aquatic Toxicology* 86(2) (2008): 205–215.

47. Chen, Z., Ma, W., Han, H. Biosorption of nickel and copper onto treated alga (*Undaria pinnatifida*): Application of isotherm and kinetic models. *Journal of Hazardous Materials* 155(1–2) (2008): 327–333.

48. Bakatula, E., Cukrowska, E., Weiersbye, I. et al. Biosorption of trace elements from aqueous systems in gold mining sites by the filamentous green algae (*Oedogonium* sp.). *Journal of Geochemical Exploration* 144 Part C (2014): 492–503.

49. Bulgariu, D., Bulgariu, L. Sorption of Pb(II) onto a mixture of algae waste biomass and anion exchanger resin in a packed-bed column. *Bioresource Technology* 129 (2013): 374–438.

50. Tuzen, M., Sari, A. Biosorption of selenium from aqueous solution by green algae (*Cladophora hutchinsiae*) biomass: Equilibrium, thermodynamic and kinetic studies. *Chemical Engineering Journal* 158(2) (2010): 200–206.

51. Małgorzata, R., Andrzej, K., Wacławek, M. Sorption properties of algae *Spirogyra* sp. and their use for determination of heavy metal ions concentrations in surface water. *Bioelectrochemistry* 80 (2010): 81–86.

52. Ghasemi, M., Keshtkar, R., Dabbagh, R. et al. Biosorption of uranium(VI) from aqueous solutions by Ca-pretreated *Cystoseira indica* alga: Breakthrough curves studies and modeling. *Journal of Hazardous Materials* 189(1–2) (2011): 141–149.

53. Pawlik-Skowrońka, B. Arsenic availability, toxicity and direct role of GSH and phytochelatins in as detoxification in the green alga *Stichococcus bacillaris. Aquatic Toxicology* 70 (2005): 201–212.

54. Pacheco, P. H., Gil, R. A., Cerutti, S. et al. Biosorption: A new rise for elemental solid phase extraction methods. *Talanta* 85(5) (2011): 2290–2300.

55. Magellan, K., Barral-Fraga, L., Rovira, M. et al. Behavioural and physical effects of arsenic exposure in fish are aggravated by aquatic algae. *Aquatic Toxicology* 156 (2014): 116–124.

56. Rubio, R., Ruiz-Chancho, M., López-Sánchez, F. et al. Sample pre-treatment and extraction methods that are crucial to arsenic speciation in algae and aquatic plants. *Trends in Analytical Chemistry* 29(1) (2010): 53–69.

57. Rahman, A., Hasegawa, H., Ueda, K. et al. Arsenic accumulation in duckweed (*Spirodela polyrhiza* L.): A good option for phytoremediation. *Chemosphere* 69 (2007) 493–499.

58. Jasrotia, S., Kansal, A., Kishore, V. Arsenic phyco-remediation by *Cladophora* algae and measurement of arsenic speciation and location of active absorption site using electron microscopy. *Microchemical Journal* 144 (2014): 197–202.

59. Zhang, J., Ding, T., Zhang, C. Biosorption and toxicity responses to arsenite (As[III]) in *Scenedesmus quadricauda. Chemosphere* 92 (2013): 1077–1084.

60. Sari, A., Uluozlu, O. D., Tuzen, M. Equilibrium, thermodynamic and kinetic investigations on biosorption of arsenic from aqueous solution by algae (*Maugeotia genuflexa*) biomass. *Chemical Engineering Journal* 167 (2011): 155–161.

61. Ferrari, S., Silva, P., González, D. et al. Arsenic tolerance of cyanobacterial strains with potential use in biotechnology. *Revista Argentina de Microbiología* 45(3) (2013): 174–179.

62. Wang, Z., Luo, Z., Yan, C. et al. Arsenic uptake and depuration kinetics in *Microcystis aeruginosa* under different phosphate regimes. *Journal of Hazardous Materials* 276 (2014): 393–399.

63. Wang, N., Li, Y., Deng, X.-H. et al. Toxicity and bioaccumulation kinetics of arsenate in two freshwater green algae under different phosphate regimes. *Water Research* 47(7) (2013): 2497–2506.

64. Henrik, H., Ribeiro, A., Mateus, E. Biosorption of arsenic (V) with *Lessonia nigrescens. Minerals Engineering* 19(5) (2006): 486–490.

65. Rahman, A., Hasegawa, H., Lim, R. Bioaccumulation, biotransformation and trophic transfer of arsenic in the aquatic food chain. *Environmental Research* 116 (2012): 118–135.

66. Sharma, V., Sohn, M. Aquatic arsenic: Toxicity, speciation, transformations, and remediation. *Environment International* 35(5) (2009): 743–759.

67. Salgado, S., Quijano, A., Simon, N. B. Optimisation of sample treatment for arsenic speciation in alga samples by focussed sonication and ultrafiltration. *Talanta* 68(5) (2006) 1522–1527.

68. Salgado, S., Quijano, A., Simon, N. B. Assessment of total arsenic and arsenic species stability in alga samples and their aqueous extracts. *Talanta* 75(4) (2008): 897–903.

69. Slejkovec, Z., Kápolna, E., Ipolyi, I. et al. Arsenosugars and other arsenic compounds in littoral zone algae from the Adriatic Sea. *Chemosphere* 63 (2006): 1098–1105.

70. Srivastava, K., Vaish, A., Dwivedi, S. et al. Biological removal of arsenic pollution by soil fungi. *Science of the Total Environment* 409(12) (2011): 2430–2442.
71. Jackson, C., Jackson, E., Dugas, S. et al. Microbial transformations of arsenite and arsenate in natural environments. *Recent Research Developments in Microbiology* 7 (2003): 103–118.
72. Oremland, R., Stolz, J. The ecology of arsenic. *Science* 300(5621) (2003): 939–944.
73. Vala, A., Sutariya, V. Trivalent arsenic tolerance and accumulation in two facultative marine fungi. *Jundishapur Journal of Microbiology* 5(4) (2012): 542–545.
74. Visoottiviseth, P., Panviroj, N. Selection of fungi capable of removing toxic arsenic compounds from liquid medium. *Science Asia* 27 (2001): 83–92.
75. Adeyemi, A. Bioaccumulation of arsenic by fungi. *American Journal of Environmental Sciences* 5(3) (2009): 364–370.
76. Rahman, A., Hogan, B., Duncan, E. et al. Toxicity of arsenic species to three freshwater organisms and biotransformation of inorganic arsenic by freshwater phytoplankton (*Chlorella* sp. CE-35). *Ecotoxicology and Environmental Safety* 106 (2014): 126–135.
77. Su, S., Zeng, X., Bai, L. et al. Arsenic biotransformation by arsenic-resistant fungi *Trichoderma asperellum* SM-12F1, *Penicillium janthinellum* SM-12F4, and *Fusarium oxysporum* CZ-8F1. *Science of the Total Environment* 409(23) (2011): 5057–5062.
78. Carlin, A., Shi, W., Dey, S. et al. The ars operon of *Escherichia coli* confers arsenical and antimonial resistance. *Journal of Bacteriology* 177 (1995): 981–986.
79. Ji, G., Silver, S. Regulation and expression of the arsenic resistance operon from *Staphylococcus aureus* plasmid pI258. *Journal of Bacteriology* 174 (1992): 3684–3694.
80. Cai, J., Salmon, K., DuBow, M. A chromosomal ars operon homologue of *Pseudomonas aeruginosa* confers increased resistance to arsenic and antimony in *Escherichia coli*. *Microbiology* 144 (1998): 2705–2713.
81. Tsai, S.-L., Singh, S., Chen, W. Arsenic metabolism by microbes in nature and the impact on arsenic remediation. *Current Opinion in Biotechnology* 20(6) (2009): 659–667.
82. Rosen, B. P., Liu, Z. Transport pathways for arsenic and selenium: A mini review. *Environment International* 35(3) (2009): 512–515.
83. Gourbal, B., Sonuc, N., Bhattacharjee, H. et al. Drug uptake and modulation of drug resistance in *Leishmania* by an aquaglyceroporin. *Journal of Biological Chemistry* 279(30) (2004): 31010–31017.
84. Tian, H., Zhuang, G., Ma, A. et al. Arsenic interception by cell wall of bacteria observed with surface-enhanced Raman scattering. *Journal of Microbiological Methods* 89(3) (2012): 153–158.
85. Joshi, D. N., Flora, S. J., Kalia, K. *Bacillus* sp. strain DJ-1, potent arsenic hyper tolerant bacterium isolated from the industrial effluent of India. *Journal of Hazardous Materials* 166(2–3) (2009): 1500–1505.
86. Patel, P., Florence, G., Christopher, B. et al. Arsenate detoxification in a *Pseudomonad* hyper tolerant to arsenic. *Archives of Microbiology* 187 (2007): 171–183.
87. Liu, S., Zhang, F., Chen, J. et al. Arsenic removal from contaminated soil via biovolatilization by genetically engineered bacteria under laboratory conditions. *Journal of Environmental Sciences* 23(9) (2011): 1544–1550.
88. Del Giudice, I., Limauro, D., Pedone, E. M. et al. A novel arsenate reductase from the bacterium *Thermus thermophilus* HB27: Its role in arsenic detoxification. *Biochimica et Biophysica Acta* 1834 (2013): 2071–2079.
89. Pivetz, P. E. Phytoremediation of contaminated soil and ground water at hazardous waste sites. *Ground Water Issue* 1–36, EPA/540/S-01/500 February 2001.

9 Remediation in Food

Domingo A. Román-Silva

CONTENTS

9.1 INTRODUCTION

Alarming quantities of arsenic in rice, vegetables, sea foods as well as beer and soft drinks render a safeguard to have these consumables. Generally, these problems are concomitant with the enrichment of soil with heavy metal, the quality of water sources used for irrigation, the supply of drinkable water plants, and the intentional use of arsenic compounds for commercial purposes like mineral process and the use of arsenic additives for animal feed to produce more animals in less time at lower cost; generally, the additives are antimicrobial drugs, including arsenicals, antibiotics, etc. Time has demonstrated that mitigation actions must be made. The amount of information about these paradigms, except in the United States and some Asian countries, is scarce.

Arsenic causes cancer even at the low levels currently found in our environment. The Evidence also suggests that it contributes to others diseases, including heart disease [1–5], diabetes [6], and declines brain's intellectual functioning. Some human exposure to arsenic stems directly from its natural occurrence in the earth's crust. In some cases arsenic is mined and then used intentionally for commercial purposes and in other cases they come from the mining wastes and metallurgical emissions.

As advised by multiple bodies of scientific experts, the United States Environmental Protection Agency (US-EPA) finally lowered its long-outdated drinking water standard in 2001 to that of the World Health Organization's (WHO) 10 µg/L, that is to say, dropping by fivefold the amount of legally allowed arsenic in tap water. Various Latin American countries like Chile are daily exposed to cancer-causing arsenic that comes from copper mining, drinking water, foods, playground equipment, and from a variety of other sources [7]. Regulatory actions have reduced some of this exposure. However, arsenic also contaminates many of the favorite foods including fish, rice, baby and children foods, animal foods such as that of chicken and swine, vegetables, fruit juices, soft drinks, wine, beer, etc. It is regrettable that some food contamination stems from intentional uses of arsenic.

9.2   ARSENIC CONTAMINATION IN CHICKEN, SWINE, GOAT, AND CATTLE

Clearly the presence of arsenic residues in animals that are food for human beings concern, principally, with certain geomedical areas of the earth enriched with arsenic, anthropic contamination of soil, irrigation water, and/or animal feed, but also with the intentional practice of adding arsenic in chicken and swine feed [8–12]. Adding arsenic in chicken feed means exposing more people to arsenic. It is estimated that 1.7–2.2 million pounds of roxarsone, a single arsenic organometallic compound feed additive, are given each year to chickens. Arsenic is an natural endogen element, that is, it does not degrade or disappear and is a persistent pollutant. Arsenic subsequently contaminates much of the 26–55 billion pounds of litter or waste generated each year by the U.S. broiler chicken industry [9]. The average total arsenic in uncooked chicken products purchased in California was in the range of 1.6–21.2 µg/kg, and it appears to be apparent that giving arsenic to chickens further adds to an already elevated arsenic burden in our environment from other intentional uses.

Roxarsone, arsanilic acid, nitarsone, and a fourth arsenical called carbarsone are US-FDA approved as feed additives for chickens and turkeys; arsanilic acid is also approved for use in swine feeds. Industry and other sources estimates that at least 1.7–2.2 million pounds of arsenic compounds like roxarsone, that is, almost 1000 tons, are given to broiler chickens in the United States each year. Roxarsone-like compounds can be degraded to other higher toxic metabolites when the excretions of these animals and birds are used as manure, enhancing the uptake of As metabolites by vegetables [13]. Hence, the FDA has set arsenic-tolerance levels in various foods; see Table 9.1. Notice that in chicken livers four times as more arsenic is allowed than breasts, thighs, or other muscle tissues.

Most poultry feed additives are not used to treat sickness, rather they are given to healthy birds to promote faster growth on less feed, or prevent diseases. Conventional chicken meat had higher inorganic arsenic concentrations than conventional antibiotic-free and organic chicken meat samples. Cessation of the use of

TABLE 9.1
U.S. Tolerance Levels for Total Arsenic Residuals in Food

In eggs and edible tissues of chicken and turkey	500 µg/kg in uncooked muscle tissue
	2000 µg/kg in uncooked edible by-products
	500 µg/kg in eggs
In edible tissues of swine	2000 µg/kg in uncooked liver and kidney
	500 µg/kg in uncooked muscle tissue and by-products other than liver and kidney

Source: Wallinga, D., *Playing Chicken: Avoiding Arsenic in Your Meat*, Institute for Agriculture and Trade Policy Food and Health Program, Minneapolis, MN, 2006, pp. 1–33.

arsenical drugs could reduce the exposure and burden of arsenic-related diseases in chicken consumers [14]. Finally, on September 30, 2013, under the threat of a lawsuit, the US-FDA responded to a nearly 4-year-old petition, calling for the immediate withdrawal of a vast majority of arsenic-containing compounds used as feed additives for chickens, turkeys, and hogs.

9.3   ARSENIC CONTAMINATION IN RICE, CEREALS, FRUITS, AND OTHER VEGETABLE PRODUCTS

U.S.-grown rice contains 1.4–5 times more arsenic on average than rice from Europe, India, and Bangladesh. Scientists, think the likely culprit is the American practice of growing rice on former cotton fields contaminated with long-banned arsenic pesticides. Besides, for decades, Americans were exposed intentionally to arsenic from the use of lumbers pressure-treated with chromated copper arsenate (CCA), a pesticide mixture that has 22% arsenic by weight and which is still in use. The United States Environmental Protection Agency (US-EPA) ended its manufacture in 2004.

Contamination of shallow groundwater aquifers with arsenic has been reported in over 70 countries around the world [15]. In addition to drinking water health risk, Food and Agriculture Organization (FAO) was concerned about the potential levels of arsenic entering the food chain via absorption by crops from irrigated water. Because rice is the staple food in some Asian countries and is consumed in large quantities, arsenic-contaminated rice could aggravate human health risk when consumed along with drinking water enriched in arsenic. An international symposium held in Dhaka in January 2005 confirmed the presence of high levels of arsenic in irrigated rice and vegetables. Widespread use of As-contaminated irrigation water leads to issues of food security, food safety, and degradation of the environment [16] and to negative impacts such as the following:

1. Reduced agricultural productivity
2. Constraints on land use due to arsenic presence in soils, toxicity of rice, and/or unacceptable quality of agricultural products

3. Creation of spatial variability in soil As, Fe, Mn, and P levels that makes agricultural management of land difficult
4. Enhanced exposure of humans to As through agricultural products containing elevated levels of arsenic

Rice productivity is affected by soil arsenic, As toxicity interferes with translocation of As from vegetative tissues to grain, that is to say, the grain and straw As concentrations for rice ground in spots and the field yield (ton/ha); wheat yields were negatively correlated with soil As, although it seems unlikely that As was the cause of reduced wheat yield as the concentration of As is low under aerobic soil conditions in which wheat is grown. Irrigation water deposits both As and P in soils, consequently, wheat yield was also negatively correlated with the available P (Olsen method). Also, maize growth and productivity in pots reduced with increasing soil arsenic and grain; arsenic levels were very low in the stem and higher concentrations were found in the leaf than the stem.

In Bangladesh, arsenic concentrations in deep tube wells were almost twice (181 µg/L) than the average levels allowed in surface tube wells (0.097 µg/L). In spite of P was also higher in the water of deep tube wells, the more high level of As would accelerate As contamination of soils and exacerbate the food chain and environment As contamination. Deep tube well water, on an average, was also higher in P but lower in Fe and Mn than surface tube well water [16]. Conditions like these are proper of some As enriched ground waters.

9.4 ARSENIC-ENRICHED ENVIRONMENTS

Although microorganisms play an essential role in the environmental fate of arsenic in relation to mechanisms of arsenic transformations (e.g., soluble and insoluble forms and toxic and nontoxic forms), human activities have exacerbated arsenic contamination in the environment. Adverse environmental impacts include mining, waste disposal, indiscriminate use of fertilizers, pesticides, herbicides, and manufacturing and chemical spillage. Many incidents of arsenic contamination of the environments have been reported in several countries around the world. Adverse effects on health due to arsenic uptake through water and food is especially high in developing and rural populations that depend on local sources of food and water. Therefore, any arsenic geochemical anomaly may impact negatively on health. Some examples of arsenic-enriched environments are [17] as follows.

9.5 AGRICULTURAL ENVIRONMENTS

The dominant source of arsenic in soils is the parent rock, but pesticides and phosphates can enhance arsenic concentration in soils. Arsenic has been used and is still used as pesticides, insecticides, and in cattle and sheep dips and for control of moth in fruit crops. Arsenate portrays certain characteristics of phosphate through its absorption by ligand exchange on hydrous iron and aluminum oxide. Hence, arsenic accumulates in soil, contaminates both surface and groundwater, and is taken up by plants and is then entrenched in mammalian–insectivore food chain. Irrigation, especially with wastewater, can cause a problem of buildup of mobile and potentially toxic metals including arsenic in soils and in surface runoff.

9.6 MINING-RELATED ENVIRONMENTS

Mining activities cause arsenic to be released in high concentration from oxidized sulfide minerals. This has resulted in high concentrations of arsenic in surface water, groundwater, soil, and vegetation. In the region of Antofagasta in Chile, it is an serious problem due to volcanism and also because Chilean copper minerals are enriched in arsenic and that copper production in Chile has increased from 700,000 metric tons in 1970 to more than 5,700 metric tons in 2013 [18,19].

9.7 RIVERINE AND VOLCANIC ENVIRONMENTS

Volcanoes are important natural sources of arsenic especially in the Southern Hemisphere. Under high temperatures, arsenic is very mobile in the fluid phase and may also be present in fumaroles as sublimates, incrustations, and volcanic ashes increasing arsenic concentrations in water, generating low pH in surface- and groundwater. The coastal–Andean Mountain–Upper Highlands ecosystem of the Antofagasta region in Chile is an important area of the Atacama Desert, of which the River Loa basin is a part. This particular ecosystem suffers from the chronic impact of endogenous arsenic due to volcanism in the area and anthropogenic delivery of arsenic and other heavy metals due to mining activity, which transports trace elements more rapidly into the ecosystem in comparison to the normal geological process, thus spreading the heavy metals to human beings, through the biogeochemical cycles [17,20].

The Loa river suffered five disasters from contamination during the years 1996 and 1998 and in that the most important happened in 1997; the causes were different, but the consequences were similar. Small agricultural towns such as Chíu Chíu, Lasana, and Quillagua were affected majorly and negatively by the contamination and the mining tasks of the proximities were mainly responsible [21,22].

The recent estimation of arsenic in global soils was at 5.0 mg/kg with a mean concentration of 7 mg/kg in surface soils in the United States. Due to the release of arsenic from industrial sources and in mining areas, its concentration in soils can reach 20,000 mg/kg; soil with arsenic levels about 5 mg/kg are considered not contaminated [23], but some actual soil As enrichment depend from their geogenic and anthropogenic As fraction distribution.

Without considering the direct discharge of materials, the contamination of soils takes place dominantly due to the emission of inorganic and organic pollutants into the atmosphere and its later immission. A technical difference of the terms pollutant of contaminant is that pollutant is, in general, an environmental stressor that converts it to the category of contaminant when they equal or they overcome primary guidelines of quality and/or secondary or emission guidelines. Pollutants of diverse origins mix in the atmosphere, taking place then the immission, that is to say, the

descent for graveness and the later soil deposit of these substances or their chemical artifacts, which enter to the biogeochemical cycles of the elements under gases
and particulate forms, affecting adversely on men, animals and the environment in
general. As particulate material or powder understands each other in the group of
materials dispersed in the atmosphere and condensed in solid or liquid form and
whose sizes oscillates between 0.05 and 500 μm. A size under 10 μm defines the
fractions of breathable particulate material, that is, PM10 and PM2.5, respectively.
Usually the particles with environmental connotation are classified, assisting the
formation process, into primary that are the direct result of physical or chemical
peculiar processes of the source from where they are emitted and secondary that
are those that take place starting from chemical reactions in the atmosphere. The
classification of sizes has special importance in knowing the dispersion capacity or
transport. According to this approach they are distinguished into two categories of
particles [24]:

- Setting particles, that is to say, those that have a diameter around 5–100 μm
  at the most, and they reach the soil far from the emission source according
  to their size.
- Particles in suspension or aerosols present inferior diameters less than 5 μm.

The particles of small size are controlled by the atmospheric turbulences and wind;
their fall speed is very low, since the Brownian movement is worthless, due to which
they can be transported to long distances. On the contrary, the particles of large size
tend to be similar to the local conditions of the soil, that is to say, with the characteristic emissions of the industrial area feinted by them [24]. Of particular importance
for the paradigm of this chapter are the emission-immission tandems for essential
elements, heavy metals, hydrocarbons, SO_2 and its gas, aerosol and particulate matter artefacts. Recent works carried out in countries like Iran and Jordan that has similar climatic and geographic conditions of desert as that of the area of Antofagasta
illustrate with clarity the type of environmental impact that smelting and cement
plants produce [25–27].

The gas emissions like NO_x and SO_2 from the industrial plants can travel long distances and become their corresponding acids, HNO_3 and H_2SO_4, due to the reaction
with environmental water (sea breeze, fogs, Camanchaca, rain, etc.) humidity and
a phenomenon that is known as *acid rain* [28]; however, in less humid atmosphere,
these gases can also be adsorbed for inorganic particulate materials, also emitted
as powders and be fixed directly on the terrestrial surface, and thus precipitating
into the soil. The main gas emitted by the smelters is sulfur dioxide (SO_2), which is
mitigated with added value producing sulfuric acid in plants dedicated for this end.
However, a fraction of this gas escapes to the environment, which is transformed into
acid fog after chemical transformations or it is adsorbed in finely divided phase minerals that will form salts of sulfates in the soil. On the other hand, SO_2 is a powerful
reduction agent, but weaker than the sulfhydric acid, with which it reacts and gets
oxidized to elementary sulfur.

As in Bangladesh, India, Pakistan, Taiwan, China, Hungary, and U.S. Midwest
states such as New England, California, and Oklahoma, in many Latin American

countries such as Argentina, Chile, Bolivia, Peru, and Mexico, at least 14 million people depend on drinking and irrigation water with toxic As concentrations. Also, in Bolivia, Ecuador, Costa Rica, El Salvador, Guatemala, and Nicaragua, As in drinking water has been detected at toxic levels over the last 10–15 years, but the extent of the problem and the number of affected people are still unknown. Sustainable land use and agricultural practices in Latin American countries are threatened by the use of irrigation water with high arsenic contents. However, only Chile, Peru, and Mexico have a few available but incomplete studies on the contamination of soils, crops, and animal fodder in the region. Despite the relative lack of availability of detailed data, it is estimated that at least 4.5 million people in Latin America are currently drinking arsenic-contaminated water ranged 50–2000 µg/L, that is to say, 200 times higher than the current WHO standard (10 µg/L) for drinking water [29].

In Tables 9.2 and 9.3, details of metal contents in corn in San Salvador River area and other vegetables in Chiu-Chiu, Calama, Chile, is presented [30]. Arsenic speciation in vegetables and fruits samples are dominated by inorganic arsenic, that is to say, Arsenate (As^{5+}) and Arsenite (As^{3+}) [31].

Arsenic was first documented by Albertus Magnus in AD 1250; but at present we cannot forget "we are what we drink, we eat, and we breathe"; the chronic human exposure to arsenic causes a number of adverse health effects, including among others cardiovascular and neurologic pathologies and several types of cancer [32,33]. Therefore, the WHO standard for drinking water is a critical parameter and presently the world needs to take urgent measures to limit the values of arsenic, termed as *a silent global* poison, in foods and alcoholic drinks, in particular the wines, soft drinks, and fruit juices, to safely allowed limits and mitigate its impact on the quality of healthy life. In the twenty-first century, arsenic continues being a global calamity [34,35], one of which few governments speak, but only the scientists concerned about healthy human life doing it.

TABLE 9.2

Mean[a] of the Total Residual Heavy Metal Concentrations in Teeth and Cores of Corns from Sector San Salvador River in Calama, Chile

Element	Core	Teeth
Cu	2.02	4.16
Zn	13.4	13.5
Cd	0.087	0.086
Pb	0.17	0.22
Hg	0.018	0.027
As	2.58	3.31
Se	0.79	0.72

Source: Román, D.A. and Román, H.A., *Agreement of Collaboration between University of Antofagasta Technical Attendance*, Agricultural and Cattle Service, Antofagasta, Chile, 2004.

[a] µg/g, wet weight.

TABLE 9.3

Mean and Concentration Factors (Chilean/Spain) of Total Residual Heavy Metal Concentrations in Lyophilized Carrots, Beets, and Quinoa from Chiu-Chiu, Calama, Chile

Vegetables	As (µg/g)	Cd (µg/g)	Pb (µg/g)	Cu (µg/g)	Mn (µg/g)
Carrots					
Flesh—C[a]	0.52±0.04	0.05±0.01	0.12±0.02	7.75±0.91	4.75±0.51
Flesh—C[b]	0.54±0.06				
Flesh—S[a]	0.02±0.01	0.07±0.01	0.09±0.01	8.82±0.59	2.28±0.18
Factor	24	0.7	1.3	0.9	2
Peel—C[a]	1.62±0.05	0.08±0.01	0.21±0.02	16.3±1.50	9.95±0.20
Peel—S[a]	0.15±0.01	0.12±0.02	0.23±0.03	17.30±0.9	13.70±0.9
Factor	11	0.7	0.9	0.9	0.7
Beets					
Flesh—C[a]	0.62±0.05	0.09±0.00	0.36±0.01	6.71±0.95	8.38±0.16
Flesh—C[b]	0.64±0.07				
Flesh—S[a]	0.02±0.00	0.05±0.00	<LD(0.006)	3.82±2.10	3.84±0.16
Factor	26	1.8	—	1.7	2.2
Peel—C[a]	3.20±0.06	0.11±0.01	0.31±0.05	28.6±0.50	21.80±0.20
Peel—S[a]	0.22±0.01	0.03±0.01	<LD(0.008)	3.80±1.2	11.10±0.3
Factor	14	3.6	—	7	2
Quinoa—C[a]	0.20±0.02	0.38±0.07	0.04±0.02	7.89±0.73	13.50 ± 2.40
Quinoa—C[b]	0.21±0.04	<LD	0.09±0.01	1.14±0.16	4.11±0.11
Quinoa—C [30]	0.01±0.01				
Factor	20	—	0.4	7	3

Source: Palacios, M. A. et al., Personal communication, 2014.

C, Chile; S, Spain; Factor = As concentration in Chilean/Spanish samples.

[a] ICP-MS.

[b] HGAAS.

9.8   ARSENIC CONTAMINATION IN BEER AND WINE

Arsenic has been a constituent of traditional Chinese medicine for at least 3000 years; Egyptians used arsenic to harden copper at least 3000 year ago and also added arsenic to their embalming fluids; arsenic minerals such as realgar (As_4S_4) and orpiment (As_2S_8) have, for several thousand years, been used in Hindu medicine. Hippocrates (469–377 BC) recommended arsenic as a tonic while Dioscorides (AD 54–68) recommended it for asthma. The Roman Pliny Secundus (AD 23–89) refers to arsenic and Nero (AD 37–68) using it to kill Britanicus in AD 55 [36]. One of the more popular activities in China during the Dragon Boat Festival is the consumption of wine mixed with the arsenic mineral realgar; a practice that stems from realgar being a mineral reputed to possess magical properties. This is an ancient festival and celebration of traditional Chinese culture and folklore that dates back to 300 BC, which has permeated to many countries within Asia such as Korea, Singapore, Malaysia, and Indonesia. Furthermore, its use in Chinese Medicine has been prevalent, where it is given to prevent sickness as it is a general cure-all, however, the consumption of this wine in the practice indicates that a considerable amount of arsenic is absorbed by the body and acute renal failure after drinking realgar wine have occurred [37,38].

Arsenic was officially introduced as a pesticide in viniculture in 1925. Its purpose was to protect the wine plants. But it was banned in 1942 and in Germany it was used until the mid-1950s. Consumption of the so-called wine-grower's house drink led to severe symptoms and illness, particularly liver damage. This homemade wine was produced by watering down the wine obtained from a second pressing of the grape skin it had low alcohol content (3%–5%), but high arsenic content, leading to cirrhosis and angiosarcoma among farmers exposed to arsenic insecticides [38]. Arsenic is usually present in wine as a consequence of herbicides and insecticides used for grape production, soil type, and the kind of process for wine production and wine storage conditions. For instance, sodium arsenite ($NaAsO_2$) has been employed as a fungicide in viticulture against the *esca* plant disease (*Eutypa lata*). Considerable efforts are made by the wine producing countries to improve the image of the product. As a consequence, the product characteristics and origin have been well defined. The amounts of arsenic present in wine may not produce acute intoxication but might impose an excessive intake on wine drinkers. The maximum total arsenic concentration permissible is 10 µg/L, but the presence of a few ng/L of arsenic has been accepted as uncontaminated wines. However, there is no legislation with respect to the maximum allowable concentration of specific arsenic species in wine. It is known that arsenic species exhibit different toxicities and the information given by the total arsenic concentration may be insufficient. The toxicity of arsenic species varies, ranging from relatively harmless organometallic arsenicals compounds (e.g., arsenobetaine) to more toxic species (i.e., monomethylarsonic acid [MMA] and dimethylarsinic acid [DMA]), as well as the inorganic arsenic species (i.e., arsenite and arsenate) [39].

Many studies about arsenic speciation in wine have been published, but most of them are from European and North American wines; there is little information about arsenic speciation in wines of Argentina, Brazil, and Chile that are the major wine producers in South America [40]. The total arsenic concentrations in wines of Argentina, Brazil, and Chile ranges from 13 to 18.5, 1.8 to 17.2, and 10.1 to 35 µg/L, respectively [41]. These levels are below the limit of 200 µg/L suggested by the International Organization of Vine and Wine [42]. Arsenic, cadmium, and lead in alcoholic beverages are considered carcinogenic in humans [43] and alcoholism it is a serious illness in these countries, particularly in young people. According to the Food and Agriculture Organization of the United Nations (UN-FAO) and the WHO, the provisional tolerable weekly intake of As is equal to 15 µg/kg of body weight, which represents a maximum tolerable daily ingestion of 2.1 µg/kg of body weight, corresponding to 147 µg/person/day, considering an average body weight of 70 kg for an adult. Such a dose would be ingested by eating more than 20 kg of grape berries/day, considering the grapes are with the highest As concentrations, that is to say, 36.8 µg/kg dry weight, which is approximately 7 µg/kg fresh weight; more recent literature has reported As concentrations in white and red wines ranging from <0.5 to 17 µg/L [42]. Wines contributed with higher As quantities compared with the arsenic intake through the consumption of beer (0.47 µg/person/day) and sherry brandies (0.28 µg/person/day) [44].

9.9   ARSENIC CONTAMINATION IN JUICE FRUITS AND SOFT DRINKS

In additions to seafood, rice, beer, wine, chickens, and other farm animals eaten by human beings, some fruits like apples and other juice fruits have also been assayed for arsenic. A 2012 U.S.-consumed report study revealed total arsenic levels that exceeded federal drinking-water standards set by US-EPA and followed by US-FDA (10 µg/L). Finally, in July 2013, US-FDA published the Draft Guidance for arsenic in apple juice, which provided information to manufacturers on the action level for inorganic arsenic in apple juice that US-FDA considers safe for human health and achievable with good manufacturing practices. US-FDA considers 10 µg/L of arsenic as the action level for inorganic arsenic in apple juice to be safe for public health [45].

The use of canned and noncanned beverages like natural fruit juices or water with flavors and coloring substitutes, and soft drinks are deeply embedded in the lifestyle of people. However, for health reasons, water used in soft drinks and canned fruit juices must be soft and free from any appreciable amount of toxic trace metals and organic matter, particularly arsenic higher than 10 µg/L. The trace element levels of fruit juices may be expected to be influenced by the nature of the fruit, the mineral composition of the soil from which it originated, the composition of the irrigation water, the weather conditions, and agricultural practices such as the types and amounts of fertilizers used, etc. [46]. The number of countries that control the levels of As and other trace elements in fruit juices and soft drinks has increased [47–51].

In fact, poultry is not the only animal industry utilizing organic arsenicals in animal diets, the swine and pig industry also use them in a lesser extent [52–54]. Nevertheless, today the main use of arsenic is as pesticides, veterinary drugs, herbicides, and silvicides, and smaller amounts are used in the glass and ceramics industries as feed additives [55]. However, the arsenicosis in goat, sheep, and cattle is due principally to the arsenic ecosystem contamination of groundwater and soils, affecting the food chain through irrigation water–soil–crop pathway and the livestock reared in areas such as Bangladesh, West Bengal in India, Iran, and other parts with similar conditions like Rio Loa Basin in the region of Antofagasta, Chile [11,29,56,57]. Animals and persons ingest large amount of arsenic-contaminated straw that comes out in milk and its derivatives, meat, egg, etc. These animal produces are consumed by human beings and maybe those most affected ones are the infants and children [54]. However, environmental pollution is the principal cause of heavy metal contamination in the food chain, in particular, arsenic in food chain is a community health risk [58] as it is a highly toxic element that is of concern to the human well-being.

9.10   ARSENIC CONTAMINATION IN HERBAL MEDICINES AND SPICES

Use of herbal medicines to relieve and treat many human diseases is increasing around the world due to their mild features and low side effects. An emerging public health concern relates to hazards posed by the exposure to toxicants through contaminated everyday merchandise, including natural health products (NPHs). Most individuals in the Orient and Western world consume some form of NPHs [59]. However, herbs may be contaminated easily during growing and processing, so it is important to have a good quality control for herbal medicines in order to protect consumers from contamination with dangerous heavy metals like Pb, Cd, Cr, Ni, Hg, Tl, and Arsenic, which are widely considered as potential contaminants in our ecosystems due to their human toxicities [60].

Ayurvedic practices stem from the Vedic culture of Southern Asia and date back over 5000 years. Within Ayurvedic tradition it is thought that metals and metalloids should be included in the formulations to maintain a proper balance for health. Findings revealed that some Ayurvedic and traditional Chinese herbal medicines (CHMs) have been found to contain significant amounts of Pb, Hg, and arsenic [61–65]. Thus, metal-content supplements in these herbal preparations may result from intentional additives, rather than from contamination. Around 60% of Americans are now using NPHs, 50% of Europeans, and approximately 71% of the population in Canada uses NPHs [59]. Because, inorganic arsenic is considered to be more toxic than methylated species of As(V), the assessment of human health risk associated with As in edible plants or foodstuffs mainly depend on the concentration of inorganic arsenic. In a recent work inorganic arsenic was the predominant species detected in all of CHMs investigated [65]. Inorganic arsenic levels in CHMs from fields and pharmacies ranged from 67.2 to 550 ng/g and 94.0 to 8638 ng/g, respectively. Therefore, the human risk due to inorganic arsenic in CHMs it is high. Now Ayurvedic medicine realize bioassay for obtain information about toxicity or nontoxicity of their herbo-metallic formulations [66].

Spices can be defined as vegetable products used for flavoring, seasoning, and imparting aroma in food [67]. Many spice and culinary herb plants are regarded as having medicinal properties also, and so there is some overlap between them and medicinal aromatic plants (MAPs). Around 50 spices are of global trade importance, and many other spices and herb crops are used in traditional cooking and health care in the countries, which are proper of the folklore. World market for dried and fresh spices and medicinal herbs are in expansion and can offer good return. However, spices and herbs must be free of contaminants, like arsenic, others heavy metals, and persistent xenobiotics. African, Asian, and developed countries are now working toward the quality control and traceability of arsenic and other heavy metals in medicinal herbs and spices [68–75]. The maximum levels for heavy metals tolerated in spices and herbs under the European Spice Association (ESA) are 5.00, 20.0, 10.0, and 50.0 mg/kg for arsenic, copper, lead and zinc, respectively. At the actual level of knowledge, we can say that ESA maximum value threshold for arsenic is high and does not protect the health of human beings.

Governmental and other organizations in the United States, and particularly in California, and several international venues, have provided information relevant to limits on daily consumption of arsenic, cadmium, lead, and mercury. Some of these have provided levels for total daily consumption from all sources, while others have focused on the intake of these heavy metals from a single source. Only Health Canada has specified limits for individual finished NPHs [76]:

	Stated Limit	Calculated Daily Limit (Adult, 70 kg)
Arsenic	0.14 μg arsenic and its salts and derivatives/kg body weight	10 μg

Arsenic occurs naturally in the environment as an element of the earth's crust and is found in rocks, soil, water, air, plants, and animals. Arsenic can be released into the environment through natural activities such as volcanic action, erosion of rocks, and forest fires, or through human actions, and can appear in inorganic and organic forms; elemental arsenic combines with other elements such as oxygen, chlorine, and sulfur to form inorganic arsenic compounds. At one time organic arsenic was considered less toxic than inorganic arsenic and safe at low levels, but its toxicity is now well-documented. Recent studies show that organic arsenic can easily be converted to inorganic in the environment and in the body when ingested by human and animals. Some organic forms of arsenic synthesized by the body's metabolism appear to be more toxic than inorganic arsenic.

Arsenic chronic exposures causes a multitarget disease named arsenicism that involves a variety of adverse health effects to human such as dermal changes and respiratory, pulmonary, cardiovascular, gastrointestinal, hematological, hepatic, renal, neurological, developmental, reproductive, immunological, genotoxic, mutagenic, and carcinogenic effects [77]. Arsenic also can contribute to declines in intellectual function and can decrease a body's ability to respond to viruses [78,79].

Inorganic arsenic is a known human carcinogen [3,80]. As early as 1870, high rates of lung cancer in Saxony miners were attributed in part to inhaled arsenic. By 1992, the combination of evidence from Taiwan and elsewhere was sufficient to conclude that the ingested inorganic arsenic, such as that in found in contaminated drinking water and food, was likely to increase the incidence of several internal cancers. In additions to being a carcinogen, arsenic can cause diabetes and cardiovascular diseases [1]. In February 2012, one editor of *The Lancet Journal* said that bladder cancer in Antofagasta, Chile, has reached the epidemiological conditions of cluster [81]. The basic cause would be the contamination due to a synergistic combination of heavy metals in which arsenic acting it as a pivots element, deteriorating the environmental health of the region of Antofagasta.

Arsenic contamination has become a major environmental concern because it not only adversely affects humans but also causes highly toxic effects on the metabolic process of plants such as mitotic abnormalities, leaf chlorosis, growth inhibitions, reduced photosynthesis, DNA replication, and inhibition or activation of enzymatic activities. Arsenic toxicity in soil and water is an increasing menace across the world and it is causing significant health damage to people living in developing and third-world countries. Such a situation demands an increase in and strengthening of the investigation at proteomic and metabolomics level of arsenic [82], considering the arsenic stress proteins [83] and the arsenic metabolites formed by the cells in association with its metabolism, aiming to better know the homeostasis of toxic species, to develop low-cost less-invasive animal and human treatments, mitigations technologies, and find treatments for arsenicosis.

9.11 REMEDIATION/MITIGATION OF ARSENIC TOXICITY IMPACTS

People must be careful if living in an known medical geology zone [8] or in/near actual and old mining area where arsenic and other heavy metals exposes their health to risks due to the consumption of vegetables grown in these soils [84,85] and eating farm animals as cattle, goat, bovine, pig, sheep, and rabbits that has been fed with products cultivated in the risk area and also the side products such as eggs, milk, and cheese.

Arsenic is a heavy metal widely spread through earth's crust, omnipresent in all environmental compartments, natural processes, and miscellanea anthropogenic activities. Arsenic form part of insecticides, wood preservatives, livestock dips, herbicides, food additives, drugs, and other vast amounts of products. A seriously problem has been the undiscriminating use of arsenic pesticides worldwide from the early to mid-1900s that resulted in the extensive deterioration of soils [86]; problem is also the level of heavy metals in chemical fertilizers that lead to water and soil pollution, in the latter case the effects are not immediately obvious because soils have strong buffering power [87]. In addition, smelting and mining process impart arsenic contamination since arsenic is a natural constituent of zinc, lead, gold, and copper ores. In Chile an 80:20 mixture of $Na_2S-As_2O_3$ named anamol, was for years employed in the separation of molybdenite from copper concentrates [88]. In 1997, in Antofagasta, Chile, during the annual inundated phase of the dryland Loa River, the sediment was removed and the contamination by arsenic residues and metal xanthate stabilized by adsorption on particular matter that affected river and marine ecosystems was proven [20]. Before the onset of the twenty-first century, ground water arsenic contamination was already reported in 20 countries, out of which 4 major incidents were from Asia. The most severe occurrence includes Bangladesh, West Bengal (India), Mongolia (PR China), and Taiwan. Documented groundwater arsenic incidents around the world have been reported in Poland; Ontario, Canada; New Zealand; Spain; Hungary; Cordoba, Argentina; Lagunera, Mexico; Taiwan; Antofagasta, Chile; Lassen County, California; Sri Lanka; Nova Scotia, Canada; Fairbanks, Alaska; Fallon, Nevada; inner Mongolia, China; Xingjang Utghur, China; Bangladesh; and West Bengal, India [89].

There is growing awareness about relationships between ecosystem and human health and the distribution of chemicals in the environment. The problem of chronic exposure to potentially toxic chemicals in the environment also is one of the priorities for international organization like the European Environment Agency (EEA), United Nations Environment Programme (UNEP), WHO, UNICEF, etc. However, among Organisation for Economic Co-operation and Development (OECD) countries, Japan appears to be the only nation that applies environmental monitoring data to evaluate the risk of chemicals in the environment [90]. Arsenic toxicity is an increasing menace across the world and it is causing significant health damage to people living in developing and third world countries. It is being considered as a global hazard.

9.12 MITIGATION/REMEDIATION STRATEGIES

Water and soil arsenic enrichment is the keystone of the multiphase disease named arsenicism in animals and human beings. Therefore, both matrices need to be mitigated or remediated. The water arsenic removal technologies and the mitigation technology to be applied are strongly dependent on the socioeconomic conditions of the countries [91], and ranges from developing filtration systems employing arsenic-removal filters packed with adsorbent media with a highly affinity for arsenic (typically iron or aluminum oxides) to desalting reversed osmosis (OR membranes) technology. The basic principles of arsenic removal from water are based on conventional techniques of oxidation, coprecipitation and adsorption on coagulated flocs, adsorption onto sorptive media, ion exchange, and membrane filtration. Oxidation of As(III) to As(V) is needed for the effective removal of arsenic from groundwater by most treatment methods. Arsenic removal depends highly on the composition and chemistry of arsenic-contaminated water. The most common arsenic removal technologies can be grouped into [92]

- Oxidation and sedimentation
- Coagulation and filtration
- Sorptive filtration
- Membrane filtration

Elemental pollutants are difficult to be remediated from soil, water, and air because, unlike organic pollutants that can be degraded to harmless small molecules, heavy metals such as mercury, arsenic, cadmium, lead, copper, and zinc are immutable by biochemical reactions. Nevertheless, plants have evolved several properties that give them selective advantages for use in environmental remediation approach [93], with the additional advantage of moving into the possibilities of the open field of the environmental biotechnology [94] in relation with DNA codes for RNA, which code proteins, produce metabolites, and lead to the physiology of the cell, the consortia, the community, and finally the ecosystem. The remediation of soils contaminated with heavy metals is a challenging task because metals cannot be degraded and the danger they pose are aggravated by their persistence in the environment.

Soil phytoremediation is a suitable option for developing countries that are in economic crisis and thus cannot afford technologically sophisticated solutions for their huge populations. Many plants species, aquatic macrophytes, and some wetland plants have shown promising ability to uptake arsenic from contaminated environments [95,96]. Phytoremediation is an emerging and ecologically benign technology for decontamination of soils, which can be microbial assisted via the interface between microbes and plant roots (rhizosphere) favoring the growth and survival of plants.

To acquire sufficient iron, bacteria of rhizosphere have had to develop strategies to solubilize this metal for its efficient uptake. One of the most common strategies evolved by bacteria is the production of siderophores, low-molecular-mass iron chelators with high conditional constants for iron [97]. So, siderophores act as solubilizing agents for iron from minerals or organic compounds under conditions of iron limitations. Besides to iron, siderophores can also form stable complexes with other metals that are of environmental concern, such as Al, Cd, Cu, Ga, In, Pb, and Zn and some radionuclides; binding of the siderophores to a metal increases the soluble metal concentration. Although they differ widely in their overall structure, the functional groups that coordinate the iron atom are not as diverse. In their metal binding sited, siderophores have α-hydroxicarboxilic acid, catechol, or hydroxamic acid moieties and thus can be classified as hydrocarboxilate-, catecholate-, or hydroxamate-type siderophores [98].

Phytoremediation exploits plant's intrinsic biological mechanism for human benefit. Four modes of phytoremediation technology are applicable to toxic metal remediation from soil and water [99]:

1. *Phytoextraction*: The use of metal-accumulating plants to remove toxic metals from soil.
2. *Phytovolatilization*: Evaporation of certain metals from aerial parts of the plants.
3. *Phytostabilization*: The use of plants to eliminate the bioavailability of toxic metals in soils.
4. *Rhizofiltration*: The use of plant roots to remove toxic metals from polluted waters.

Halophyte plants are those that grow naturally in saline environment. Grown in many arid and semiarid regions around the world, they are distributed from coastal areas to mountains and deserts. Halophytic plants have many uses; they can be used as animal feed, vegetables, and drugs, as well as wind shelter, soil cover, for cultivation of swampy saline lands, laundry detergents, paper production, for sand dune fixation, and so on [100], and could be a plant resource of interest in biotechnology as As-hyperaccumulating halophyte species. In additions, the discovery of two ABC type phytochelatin (PCs) transporters required for arsenic detoxification in the As-hyperaccumulating plant *Arabidopsis thaliana* [101] and the discovering of As-hyperaccumulating fern species [102] has attracted the attention and further research focusing on understanding the mechanism behind this phenomenon and evaluation of the phytoremediation potential of various As-hyperaccumulator plants.

Arsenate is the main arsenic species in aerobic soils and its concentration in soil solutions are usually low due to their strong affinity for iron oxides/hydroxides in soil. Physiological and electrophysiological studies have shown that arsenate and phosphate share the same transport pathway in higher plants, with the transporter having a higher affinity for phosphate than arsenate. Arsenite is the dominant As-species in reducing environments such as flooded paddy soils. Thermodynamically, reduction of arsenate to arsenite can occur quite readily at intermediate redox potentials.

Flooding of paddy soils leads to mobilization of arsenite into the soil solution and enhanced arsenic bioavailability to rice plants. Plant roots are capable of rapidly taking up arsenite from the external medium [103].

Plants have mechanisms for metal homeostasis that allow uptake and distribution of metals to tissues while maintaining metal within cells or subcellular compartments below levels that cause toxic symptoms. Given that the more toxic forms of essential metals ions like Fe, Zn, Mn, and Cu are the free metal ion species, their cellular concentration is expected be very low. No protein metal ions are expected to be bound to low-molecular-weight metal ligands in order to fulfill as intracellular chelators, which may be used either for mobilization from the soil or for translocation within the plant. In order to fulfill these roles, transport mechanisms are required for the secretion or uptake of metal ligands by cells or for the movement of ligands between subcellular compartments [104].

Plants also have developed mechanisms to combat adverse environmental heavy-metal-toxicity problems or increase tolerance against abiotic stress. These include using nonprotein and protein thiols as scavengers for toxic metals. In the first defensive line, they use sulfur-containing amino acids, cysteine and methionine, followed by the most abundant low-molecular-weight thiol, the tripeptide glutathione (γ-glutamyl-cysteinyl-glycine; GSH). Glutathione is a key water-soluble antioxidant and play an important role in reactive oxygen species (ROS)-scavenging through GSH-ascorbate cycle and as an electron donor to glutathione peroxidase (GPx). GSH is a substrate for PC synthesis, which is crucial for the detoxification of heavy metals like cadmium, nickel, and arsenic. PCs are small, heavy metal binding, cysteine-rich polypeptides with the general structure $((\gamma\text{-Glu-Cys})_n\text{-Gly}$, whereby n varies between 2 and 11); the enzyme responsible for their synthesis is PC synthase (EC 2.3.2.15) that is activated in the presence of metal ion. The PC complexes with ions are captured in the vacuole, thus ameliorating the toxic effects of metals; PCs most effectively chelate cadmium, followed by arsenic [105,106].

9.13   MITIGATION OF AS CONSUMPTION THROUGH FOOD

The possible mitigation options for reducing arsenic consumption directly through food and beverages coming from cultivable products can be classified in two categories [92]:

1. Changes in agriculture practices
2. Changes in food process

9.14   AGRICULTURAL PRACTICES

Possible mitigation options under irrigation and soil management are reducing arsenic in irrigation water change by free irrigation arsenic water, apply intermittent irrigation, periodic removal of top soil, fixation of arsenic in more unavailable forms and addition of ions like phosphate and silicates. Intermittent irrigation can decrease the bioavailability of arsenic during early plant growth and this may be a promising means of reducing arsenic input in paddy soils. Removal of arsenic as a mitigation option is not suitable for irrigation purposes except when Fe is present naturally, which could favor precipitations of oxides/hydroxides by removing arsenic from water under aerobic conditions using open irrigation channels. Under anaerobic conditions iron oxides will release the adsorbed arsenic and thus increase the fraction of bioavailable arsenic. The plant uptake of arsenic can be reduced by adding iron to aerated and sandy soils low in iron. Iron must be added as ferrous sulfate, iron grit, or other Fe(III) materials suitable for aerated soils. Under aerobic conditions, As(III) is oxidized and bound to iron oxides and is not readily available to dry land crops except when application of phosphate fertilizers may desorb the As(V) making it bioavailable. The alternative is change the soil redox conditions and further developing and extending the suitable irrigation and soil management practices to reduce the risk of arsenic contamination to human health and to secure food safety and production. The toxic and redox-sensitive arsenic trace element can be present in large concentrations in floodplain soils along the drainage catchment of rivers where As-containing ores have been mined and processed. In this case, risk assessment of the transfer of arsenic into the food chain is thus poorly predicted by total arsenic concentrations. Operationally invasive, more or less destructive soil fractionation arsenic approach and arsenic speciation into soil solution of soil fractions are then necessary to obtain knowledge about possible changes in the speciation and mobility of arsenic with changing redox (E_H) conditions. In particular, the destructive or invasive nature of the applied techniques for soil fractionation approach, do not allow obtaining the necessary and required certainty that is required [107].

It is recommended that prior to food processing and consumption the edible roots and tubers (potatoes, carrots, radishes, and turnips) must be washed, because arsenic accumulates mainly in the root peel. When food processing and cooking practices are formulated, it must also be considered that the total arsenic content and arsenic speciation are affected by the cooking methods used during the preparation of meals [108,109]. In summary, arsenic is a dangerous food chain contaminant [110]. Following a call of EFESA (European Food Safety Authority), 15 European countries submitted more than 100,000 results on arsenic concentrations in various food commodities. Two thirds of the samples were below the limit of detection. The highest total arsenic levels were measured in fish and seafood, food products or supplements based on algae, especially *hijiki*, and cereal and cereal products, with high concentrations in rice [111].

The knowledge about the means of human As exposure mitigation actions, particularly in geomedical areas and in arsenic polluted zones of the world, are related with the "human being quality of life". The problem not alone implicate soil mitigation, free As human consumption water and irrigation water according national quality guides, and carried out sustainable agriculture. At same time also involves advance in the agricultural reconversion without food production for human consumption.

Five concepts must be very clear, that is, mitigation, remediation, remedial action plan, remediation criteria, and prevention.

1. *Mitigation*: To minimize or to intervene an anthropic action or with the purpose of making less invasive the pollution that otherwise affects the eco-systems, animal and vegetable health, biodiversity of the organisms, quality of life, and the human life. The action is making the environmental negative impacts friendlier, repairing the provoked damages less partially, without considering the *environmental compensations.*

2. *Remediation*: It is the improvement of a contaminated site to prevent, minimize, or mitigate damages to the environment and human health. Remediation involves the development and application of a planned approach that removes, destroys, contains, or otherwise reduces the availability of contaminants to receptors of concern.

3. *Remediation action plan*: A report that identifies site-specific remedial scopes for a site, identifies remedial options and outlines their feasibility, and recommends and describes a preferred conceptual remediation plan, a performance monitoring plan, and, if appropriate, requirements for ongoing site management.

4. *Remediation criteria*: The numerical limits or narrative statements pertaining to individual variables or substances in environmental compartments and products (water, sediment, soil, terrestrial and marine foods, wine, beer, soft drinks, fruit juice, herbal medicines, etc.), which are recommended to protect and maintain the specified uses of contaminated sites.

5. *Prevention*: The essential and basic effort for the reduction of effects of chronic arsenic toxicity is prevention and according to the precatory principle, prevention is better than cure [112].

    a. *Primary prevention*: Due to low socioeconomic status of the large population of the affected area, it is not possible to eliminate total arsenic and provide arsenic-free drinking water to everyone. Thus, it is suggested to use alternative water source such as rainwater or to remove the arsenic from contaminated water. The most important remedial action for the person who suffered from arsenicosis is preventing the use of arsenic-contaminated drinking water or at least consuming drinking water with arsenic below 10 μg/L (MCL of WHO and USEPA).

    b. *Secondary prevention*: For the reduction of toxicity and elimination of heavy metals from the body, chelating treatment is used. In acute arsenic toxicity, this treatment is a good remedial action; the most common chelating reagent used for arsenic and other heavy metals is British anti-Lewisite (BAL) (2,3-dimercaptopropanol); however, this also has toxic activity. Other treatments using thiol chelators like meso 2,3-dimeercaptosuccinic acid (DMSA), sodium 2,3-dimercaptopropane-1-sulfonate (DMPS), and monoisoamyl DMSA are used, in both acute and chronic arsenic toxicity. Regrettably, the lack of selectivity of these chelant agents is a problem, since they also remove essential trace elements and present residual toxicity [113].

    Available scientific evidence does not support claims that chelation therapy is a safe treatment. Chelation therapy may produce toxic effects, including kidney damage, irregular heartbeat, and swelling of the veins. Chelation products, even when used under medical supervision, can cause serious harm, including dehydration, kidney failure, and death. The drugs may also cause nausea, vomiting, diarrhea, and temporary lowering of blood pressure. Since the therapy removes minerals from the body, there is a risk of developing low calcium levels (hypocalcemia) and bone damage. Chelation therapy may also impair the immune system and decrease the body's ability to produce insulin. Each individual chelating drug can cause its own side effects, such as allergic reactions, coma, seizures, low blood pressure, and infections, and should only be used under close medical supervision. Chelation therapy may be dangerous in people with kidney disease, liver disease, or bleeding disorders. Women who are pregnant or breastfeeding should not use this method. Chelation therapy is often given along with large doses of vitamins and other minerals, which may actually contribute to the processes that produce dangerous free radicals in the body [114–116]. Biochelation treatments with lyophilized microalgae biomass selected for some heavy metals are known to pose less danger [117,118].

c. *Tertiary prevention*: Safe drinking water and well-nourished food free of arsenic is essential for the prevention of chronic arsenic toxicity. The diet with low protein, fats, vitamins, and minerals may increase the risk of arsenic skin lesions and other malignant diseases. Deficiency in proteins, folate, and vitamin B in diet affected the biomethylation or biotransformation mechanism of arsenic by which it is excreted, and will be accumulated in the body causing dangerous adverse health effects. Organic forms of calcium, iron(II), and zinc reduces arsenic toxicity, however, their inorganic free ions are chemically very active and can be a problem more than help; deficiency of vitamins, antioxidants, and selenium(IV) (cofactor of enzyme gluthathion peroxidase) increases the ROS that causes tissue damage and cancer; arsenic and selenium presents bioinorganic antagonism [7]. In human beings, the nutritional functions of selenium are achieved by 25 selenoproteins that have selenocysteine at their active centre. In contrast to many to other micronutrients, the intake of selenium varies hugely worldwide, ranging from deficient to toxic concentrations, the plasma baseline concentration of 122 μg/L appear be a critical value and the range for minimal mortality is between 122 and 150 μg/L [119].

9.15   REGULATORY NEED FOR ARSENIC MITIGATION

Mitigating arsenic pollution is the bridge between knowledge and practice [120]. As in others, the case of rice illustrate the problem and throw light on the solution; given the lack of regulations in *developing countries* and the lack of effective regulations in *most developed countries*, highly As-contaminated rice and others foods may be sold and unknowingly consumed [121]. Regulations related to arsenic toxicity in ecosystem matrices, biological samples, domestic water, solid and liquid wastes, food, beverages such as wine, beer, fruit juice, soft drink, natural products as herbal medicines, and spices and emissions to the atmosphere are often very complex, vary from nation to nation, and change in time. Considering the complexity and the number of regulations and how often modifications occur, a comprehensive and up-to-date summary of the regulations is not possible. Readers that need more current or additional regulations or guides should access government, international, and private organizations' websites such as WHO guidelines, Environmental Protection Agency of countries such as Australia–New Zealand (EPA), Food Standards Australia New Zealand (FSANZ), Bangladesh, Canada, China, European Union, Germany, India (Ministry of Health and Family Welfare), Italy, Japan, Norway, Russia, Sri Lanka (ESA, European Spice Association; The Spice Council of Sri Lanka), United States of America (US-EPA, US-FDA; USGS; NOAA; AOAC; AHPA [American Herbal Products Association]), South Korea, Switzerland, Taiwan, Thailand, or search on the Internet for additional information.

Although sources of arsenic in Latin America are mainly geogenic, anthropogenic activities such as mining have contributed substantially to the irrigation and drinking water arsenic problems. Geographical zones with high arsenic concentrations in surface- and groundwater show high bioaccumulation of arsenic in plants, fish/shellfish, livestock meat, milk, and cheese. Therefore, these facts and the lack of arsenic regulations and faulty inspections by the environmental and public health authorities have increased the arsenic impacts on ecosystem and people of Latin America, resulting in much higher health risk to local inhabitants including indigenous populations living in Andes and Chaco plains and the residents living near mines or metallurgical industries [29,30]. In the case of the region of Antofagasta in Chile in particular, the environmental impact has affected human population badly. Recently, Antofagasta region has been epidemiologically signed in a World Report as "cluster cancer due to arsenic contamination" [81].

In spite of this discussion, it is still a long road to travel—from what began in June 5, 1851, in England, when *anno decimo quarto of the Victoria Reginae*, an act in which the sale of arsenic was regulated [122]—before we can maintain arsenic under safe levels for human beings. Currently, in the case of pollutants that have ended up being a global contamination problem, the regulations related with the environmental health should also be globally implemented.

9.16 PREVENTIVE TREATMENTS FOR HUMAN BEINGS EXPOSED TO ARSENIC

Many countries have focused on arsenic health effects and prevention. But, the truth is *at the date there is not an effective treatment for human arsenicosis*. Arsenic contamination of drinking and irrigation water is passed on to the vegetable foods for animals and human beings and so to the trophic levels of food chain. At date, only preventive and palliative treatments have been proposed for arsenate and arsenite toxicity. Among these the following ones can be highlighted:

- Evaluation of the effects of vitamin E and selenium on arsenic-induced skin lesions [123].

- Evaluation of vitamin E and organoselenium compounds for the prevention of cocarcinogenic activity of arsenite with solar UV radiation on mouse skin as human model [124].
- Supplementation of some nutrients would increase arsenic methylation and in turn improving the urine arsenic excretion. From this point of view, it is important to consider folic acid, vitamin B12, and folate-related nutrients such as some thiol-amino acids like methionine and cysteine [125–127]. Intakes of others components of vitamin B complex, that is to say, thiamine, niacin, pantothenic acid, and pyridoxine are also associated with increased creatinine-adjusted urinary total arsenic concentration [128].
- The combination of *Spirulina* and vitamin A was found more effective in the prevention of chronic arsenicosis in rats than using these substances separately [129].

Most studies on nutrition and arsenic toxicity have found that specific nutrients appear to affect arsenic-related health outcomes. Therefore, malnutrition is one of the factors that link to the people of the world more affected by arsenic.

On the other hand, some pharmacological properties of certain phytochemicals [130] appear to be cancer preventives and have protective activity against arsenite toxicity. Some known cases are (1) Hesperidin and lipoic acid, the first, a dithiolic antioxidant present in vegetables like spinach, tomato, and broccoli, but also found in animal tissues, and the second one, a flavanone abundant in citrus fruits, as grapefruit and oranges, having protective activity against arsenite acute toxicity [131]. (2) Turmeric (*Curcuma longa*), a spice that is used worldwide as a culinary seasoning and curcumin, the active ingredient of turmeric, have demonstrated to attenuate the production of a major angiogenic factor, vascular endothelial growth factor, in As(III)-treated human colon cancer cells in culture. Curcumin or diferuloymethane is a yellow spice that is used as an ingredient in *curry*. It is a polyphenolic molecule extracted from the rhizome of *C. longa* used over centuries in Ayurvedic, Chinese, and Hindu traditional medicine. Nowadays, it appear as a promising chemopreventive compound able to reverse, inhibit, or prevent the development of cancer by inhibiting specific molecular signaling pathways involved in arsenic-induced carcinogenesis [132,133]. (3) Recent studies have suggested that asthaxantine exhibit about 100–500-times higher antioxidant activity than α-tocopherol. It has been shown that asthaxantine is much most effective than vitamin E in protecting the mitochondria in rat liver cell against lipid peroxidation and presents nephroprotective effect against inorganic As(III)-induced renal injury in wistar rats [134].

9.16.1 Is Arsenicosis or Arsenicism a Mitochondrial Disease? Arsenic and Mitochondria Dysfunction

The four main focuses of research on the cellular mechanisms of arsenic toxicity are mutations inductions and chromosomal aberrations; altered signal transduction, cell-cycle control, cellular differentiation, and apoptosis; alterations in gene expression; and direct damage through oxidative stress. None of these mechanisms are exclusive, and several studies relate arsenic toxicity with the fact that metabolism is associated with increasing of oxidative stress.

Mitochondrial diseases are the result of either inherited or spontaneous mutations in mtDNA or nDNA that lead to altered functions of the proteins or RNA molecules that normally reside in mitochondria. Problems with mitochondrial function, however, may only affect certain tissues as a result of factors occurring during development and growth that we do not yet understand. Mitochondria are the powerhouses of our cells. They are responsible for generating energy as adenosine triphosphate (ATP) and heat. Mitochondria are the only other subcellular structure besides the nucleus to contain DNA, named mtDNA, which lacks the structural protection of histones, and their repair mechanism is quite susceptible to free radical (ROS) damage [135]. Oxidative stress is the elevation of free radicals (ROS) found in cells that accumulate to higher than normal levels or it can also be defined as a physiological condition that occur when there is a significant imbalance between the production of ROS and antioxidant defenses. When the balance of antioxidants to oxidants or cell redox shifts to higher levels of oxidants, the cell undergoes oxidative stress. Oxidative stress is a chain reaction phenomenon, and it produces cell injury, which may be transient or eventually lead to irreversible injury and cell death can occur basically by two mechanisms, necrosis and apoptosis. This chain reaction phenomenon may be reduced and modulated by a variety of secondary antioxidants through diet and supplementation such as vitamin E, vitamin C, selenium, zinc, coenzyme Q-10, α-lipoic acid, and bioflavonoids (carotenoids and polyphenols).

The mitochondria is a target for environmental toxicants, but at present has been underappreciated from the environmental contamination and health point of view [136]. Mitochondria produce ROS [137] such as superoxide radical (O_2^-), singlet oxygen ($^1O_2$), hydrogen peroxide (H_2O_2), and hydroxyl radical (OH$^-$), which attack the most sensitive biochemical macromolecules in cells and impair their functions. If drought stress is prolonged, ROS production will overwhelm the scavenging homeostasis of the antioxidant defense system composed of both nonenzymatic and enzymatic constituents. The enzymatic antioxidant defense mechanism is represented principally by the enzymes catalase (CAT, Fe dependent), the metalloenzyme superoxide dismutase (SOD, Cu and Zn in prosthetic group), and glutathione peroxidase (GPX, Se(IV) dependent). The biochemical malfunction of mitochondria-ribosome cellular tandem due As, involves to cell energy with expression of stress proteins [138–141]. Oxidants are generated as a result of normal intracellular metabolism in mitochondria and peroxisomes, as well as from a variety of cytosolic enzyme systems. In addition, a number of external agents can trigger ROS production also. Oxidative stress has been implicated in several diseases including cancer, atherosclerosis, malaria, chronic fatigue syndrome, and neurodegenerative diseases such as Parkinson's disease, Alzheimer's disease, and Huntington's disease [33,141]. Some physiological activities of ROS are as follows: (1) regulates adaption to hypoxia, (2) regulates autophagy, (3) regulates immunity, (4) mitochondrial ROS regulates differentiation of stem cells, and (5) regulates aging. Therefore, in addition to ATP generation, mROS seems to have a dual function, that is, to both promote cell damage and promote cell adaptation. So, inhibition of mROS by antioxidants does not have a predictable outcome on cell function since the role of mROS changes under different environmental conditions [142]. Metals play important roles in a variety of biological processes in living systems. Homeostasis of metal species under certain speciation pattern, maintained through tightly regulated mechanisms of uptake, storage, and secretion is critical for life if it is maintained within strict limits [32].

Inorganic arsenic includes arsenite [As(III)] and arsenate [As(V)] and can be either methylated to form MMA or dimethylated as in DMA. The metabolism of inorganic arsenic involves two-electron reduction of arsenate, via GSH, followed by oxidative methylation to form pentavalent organometallic arsenic. Arsenic is toxic to the majority of organ systems, but depends on the speciation forms; As(III) is more toxic than methylated organic arsenic and reacts with thiol groups of proteins [143]. The As(V) possess less toxicity, however uncouples oxidative phosphorylation. In spite of this, the methylation of inorganic arsenic has been considered as a detoxication process. However, the results found in the past decade show that human cells are more sensitive to the cytotoxic effects of MMA(III) than arsenite and that DMA(III) is at least as cytotoxic as arsenite in several human cells. Thus, arsenic methylation process would not be considered only as a detoxification mechanism [32]. Many works confirmed the generation of ROS during arsenic metabolism in cells. Oxidative stress has been linked with the development of arsenic-related diseases, including cancers. Arsenic mediates formation of superoxide anion radical, hydroxyl radical, singlet oxygen, the peroxyl radical (ROO·), nitric oxide (NO·), hydrogen peroxide, dimethylarsinic peroxyl radicals ([$(CH_3)_2AsOO·$]), and also the dimethylarsinic radical [$(CH_3)_2As·$]. The precise mechanisms of ROS generation have not yet been clarified, but might involve the formation of hydroxyl radicals. Some evidences suggest that mitochondria are the primary target. Arsenic triggers rapid morphologic changes in this organelle and leads to inactivation of mitochondrial enzymes and the loss of mitochondrial membrane potential. Besides its recognized capacity to induce oxidative stress, arsenic also interact with cellular targets such as the thiol groups of various proteins [83,143]. In fact, S-adenosylmethionine (SAM) and glutathione (GSH) are required at several stages for the metabolic conversion of both [As(III)] and [As(V)]. The thiol trivalent-binding capacity has been suggested as a trigger for the inactivation of some zinc-finger proteins [144].

9.17 CONCLUSIONS

Environmental pollution is the principal cause of heavy metal contamination in the food chain, but arsenic is a highly toxic element of concern to the human well-being. Arsenic contamination has become a major environmental concern because it not only affects humans adversely but also causes highly toxic effects on the metabolic process of plants such as mitotic abnormalities, leaf chlorosis, growth inhibitions, reduced photosynthesis, DNA replication, and inhibition or activation of enzymatic activities.

Arsenic toxicity is an increasing menace across the world and it is causing significant health damage to people living in developing and third world countries. Most studies on nutrition and arsenic toxicity have found that specific nutrients appear to affect arsenic-related health outcomes. Therefore the malnutrition is one of the factors that link to people of the world more affected by arsenic.

Arsenic is being considered a global hazard, in particular, given its presence in drinking water sources and human foods. In summary, arsenic is a dangerous food chain contaminant and the total arsenic content and speciation are also affected by the cooking methods used during the preparation of meals.

Five concepts, that is to say, mitigation, remediation, remedial action plan, remediation criteria, and prevention they should be very clear to win the war against arsenic, to overcome their harmful effects on the human health and to improve the quality of life of millions of people affected for this silent murderer.

Presently, the world needs to take urgent measures to limit the values of arsenic, termed as *a silent global* poison, in foods and alcoholic drinks, in particular the wines, soft drinks, and fruit juices, to safely allowed limits and mitigate its impact on the quality of healthy life. In the twenty-first century, arsenic continues being a global calamity, one of which few governments speak, but only the scientists concerned about healthy human life doing it. Safe drinking water and well-nourished food, free of arsenic, is essential for the prevention of chronic arsenic toxicity named arsenicosis, a multitarget mitochondrial disease. Many works confirmed the generation of ROS during arsenic metabolism in cells. Oxidative stress has been linked with the development of arsenic related diseases, including cancers.

Although sources of arsenic in Latin America are mainly geogenic, anthropogenic activities such as mining have contributed substantially to the irrigation and drinking water arsenic problems in this part of the world. It is recommended that prior to food processing and consumption, the edible roots and tubers (potatoes, carrots, radishes, and turnips) must be washed, because arsenic accumulates mainly in the root peel. When food processing and cooking practices are formulated, it must also be considered that the total arsenic content and arsenic speciation are affected also by the cooking methods used during the preparation of meals.

Effective legislation, regulation, and identification of the areas of the world where excess levels of arsenic is found in drinking water and food are necessary. Exposure monitoring and possible intervention for reduction in further exposure to arsenic can reduce the arsenic toxicity and must be a significant step toward the prevention. The most important action of mitigation is to educate people exposed to arsenic pollution about options available in geomedical areas to fight arsenic contamination or to educate people who live in arsenic-contaminated areas of the world that it is not enough to mitigate soils or water for human consumption and for irrigation and to carry out ancestral agriculture environmentally sustainable, but rather also involves, at the same time, advancing in the health care and the agricultural or productive reconversion of geographical arsenic-contaminated areas without affecting food production for human consumption.

Therefore, the question is, what should the governments and international organizations do to overcome this serious problem of public health? With the present knowledge, the best strategy would be to recognize the problem and make efforts to address them.

REFERENCES

1. Román DA, Pizarro I, Cámara C, Palacios MA, Gómez MM, Solar C. 2011. An approach to the arsenic status in cardiovascular tissues of patients with coronary heart disease. *Hum Exp Toxicol* **30**: 1150–1164.
2. Román DA, Solar C, Pizarro I. 2013. Niveles de elementos traza en tejidos cardiovasculares de pacientes sometidos a cirugía cardiaca en la región de Antofagasta, Chile. *Rev Chil Cardiol* **32**: 214–220.
3. Solar C, Pizarro I, Román AD. 2012. Presencia de altos niveles de arsénico en tejidos cardiovasculares de pacientes de áreas contaminadas en Chile. *Rev Chil Cardiol* **31**: 41–47.
4. Pizarro I, Román DA, Solar C, Cámara C, Palacios MA, Gómez MM. 2013. Identificación y cuantificación del arsénico unido a proteínas de tejidos cardiovasculares de pacientes sometidos a cirugía de revascularización coronaría. *Rev Chil Cardiol* **32**: 123–129.
5. Pizarro I, Román DA, Solar C, Cámara C, Palacios MA, Gómez MM. 2013. Arsenic species-binding proteins in human cardiovascular and muscle tissues. *J Chil Chem Soc* **58**(4): 2071–2076.
6. Palacios J, Román DA, Cifuentes F. 2012. Exposure to low of arsenic and lead in drinking water from Antofagasta city induces gender differences in glucose homeostasis in rats. *Biol Trace Elem Res* **148**: 224–231.
7. Román DA, Pizarro I, Rivera L, Torres C, Ávila J, Cortes P, Gill M. 2012. Urinary excretion of platinum, arsenic and selenium of cancer patients from the Antofagasta region in Chile treated with platinum-based drugs. *BMC Res Notes* **5**: 207.
8. Selinus O, Alloway BJ, Centeno JA, Finkelman RB, Fuge R, Lindh U, Smedley P. 2005. *Medical Geology Impacts of the Natural Environment on Public Health*. Amsterdam, the Netherlands: Elsevier Academic Press, 811pp.
9. Wallinga D. 2006. *Playing Chicken Avoiding Arsenic in Your Meat*. Minneapolis, MN: Institute for Agriculture and Trade Policy Food and Health Program, pp. 1–33.
10. Datta BK, Mishra A, Singh A, Sar TK, Sarkar S, Bhatachary A, Chakraborty AK, Mandal TK. 2010. Chronic arsenicosis in cattle with special reference to its metabolism in arsenic endemic village of Nadia district West Bengal India. *Sci Total Environ* **409**: 284–288.
11. Mandal PH, Bandyopadhyay S, Kumar R, Datta BK, Maji C, Biswas S, Dash JR, Sar TK, Sarkar S, Manna SK, Chakraborty AK. 2012. Quantitative imaging of arsenic and its species in goat following long term oral exposure. *Food Chem Toxicol* **50**: 1946–1950.
12. Bera AK, Rana T, Bhattachrya D, Das S, Pan D, Das SK. 2012. Chronic arsenicosis in goats with special reference to its exposure, excretion and deposition in an arsenic contaminated zone. *Toxicol Pharmacol* **33**: 372–376.
13. Yao L, Huang L, He Z, Zhou C, Li G, Yang B, Deng X. 2014. Roxarsone and its metabolites in chicken manure significantly enhance the uptake of As species by vegetables. *Chemosphere* **100**: 57–62.
14. Nachman KE, Baron PA, Raber G, Francesconi KA, Navas-Acien A, Love DC. 2013. Roxarsone, inorganic arsenic, and other arsenic species in chicken: A US-Market Basket Sample. *Environ Health Perspect* **121**(7): 818–824.
15. Akinbile ChO, Haque AMM. 2012. Arsenic contamination in irrigation water for rice production in Bangladesh: A review. *Trends Appl Sci Res* **7**(5): 331–339.
16. Koo-Oshima S, Panaullah G, Duxbury JM. 2007. Remediation of arsenic for agriculture sustainability. Food security and health in Bangladesh. FAO Water. Rome, Italy: FAO, pp. 5–28.
17. Duker AA, Carranza EJM, Hale M. 2005. Arsenic geochemistry and health. *Environ Int* **31**: 631–641.

18. Correa JA, Ramírez MA, De la Harpe JP, Román DA, Rivera L. 2000. Copper, copper mining effluents and grazing as potential determinants of algal abundance and diversity in northern Chile. *Environ Monitor Assess* **61**: 265–281.
19. U.S. Geological Survey. 2014. *Mineral Commodity Summaries*.
20. Román DA, Rivera L, Morales T, Ávila J, Cortes P. 2003. Determination of trace elements in environmental and biological samples using improved sample introduction in flame atomic absorption spectrometry (HHPN-AAS; HHPN-FF-AAS). *Intern J Environ Anal Chem* **83**(4): 327–341.
21. Gentes I. 2002. Entre "Propiedad Ambiental" y nueva acción social. Contribuciones al mejoramiento del manejo de los conflictos sobre recursos naturales, 29pp.
22. Barros A, Pereira G. 2013. Informe pericial el impacto del tranque de Talabre en la comunidad Lickanantai de San Francisco de Chiu–Chiu, 24pp.
23. Kabata-Pendias A, Mukherjee AB. 2007. *Trace Elements from Soil to Human*. Berlin, Germany: Springer, 550pp.
24. Seoánez-Calvo M. 2002. *Tratado de la contaminación atmosférica. Problemas, tratamiento y gestión*. Madrid, Spain: Ediciones Mundi-Prensa, 1111pp.
25. Mansouri N, Nouri J. 2004. Development of particulate matter and heavy metal emission factors for Kerman Copper industries. *Iran J Publ Health* **33**(1): 22–26.
26. Al-Khashman OA. 2004. Heavy metal distribution in dust, street dust and soils from the work place in Karak Industrial Estate, Jordan. *Atmos Environ* **38**: 6803–6812.
27. Al-Khashman OA, Shawabkeh RA. 2006. Metals distribution in soils around the cement factory in southern Jordan. *Environ Pollut* **140**: 387–394.
28. Menz FC, Seip HM. 2004. Acid rain in Europe and the United States: An update. *Environ Sci Policy* **7**: 253–265.
29. Bundschuh J, Nath B, Bhattacharya P, Liu C-W, Armienta MA, Moreno MV, Lopez DL, Jean J-S, Cornejo L, Lauer LF, Tenuta A. 2013. Arsenic in the human food chain: The Latin American perspective. *Sci Total Environ* **429**: 92–106.
30. Palacios MA, Pizarro I, León J, Román DA, Gómez M. 2014. Bioavailability, bioaccessibility and speciation of As and others heavy metals in contaminated area of Chile. Personal communication.
31. Meharg A, Norton G, Mestrot A, Deacon C, Feldmann J. 2012. Arsenic speciation in fruit and vegetables grown in the UK. Food Standards Agency Final Report, pp. 1–75.
32. Shekhar HU, Rahman T, Hosen I, Towhidul MM. 2012. Oxidative stress and health. *Adv Biosci Biotechnol* **3**: 997–1019.
33. Jomova K, Valko M. 2011. Advances in metal-induced oxidative stress and human disease. *Toxicology* **283**: 65–87.
34. Zhu Y-G, Williams PN, Meharg AA. 2008. Exposure to inorganic arsenic from rice: A global health issue? *Environ Pollut* **154**: 169–171.
35. Anamika S. 2014. Arsenic—21st Century calamity—A short review. *Res J Recent Sci* **3**: 7–13.
36. Doyle D. 2009. Notoriety to respectability: A short history of arsenic prior to its present day use in haematology. *Brit J Haematol* **145**: 309–314.
37. Zhu Y-G, Zhang Y-N, Sun G-X, Huang Q, Williams PN. 2011. A cultural practice of drinking Realgar wine leading to elevated urinary arsenic and its potential health risk. *Environ Int* **37**: 889–892.
38. Shab A, Catharina C. 2011. Arsenic—Pesticides with an ambivalent character. In: Stoytcheva M (ed.). *Pesticides in the Modern World—Risk and Benefits*. InTech, Chapter 10, pp. 182–196.
39. Moreira CM, Duarte FA, Lebherz J, Pozebon D, Flores EM, Dressler VL. 2011. Arsenic speciation in white wine by LC-ICP-MS. *Food Chem* **126**: 1406–1411.
40. Dessler VL, Moreira CM, Duarte FA, Lebherz J, Pozabon D, Flores MM. 2011. Arsenic speciation in White wine by LC-ICP-MS. *Food Chem* **126**: 1406–1411.

41. Pozebon D, Bentlin FR, Pulgati F, Dressler VL. 2011. Elemental analysis of wines from South America and their classification according to country. *J Braz Chem Soc* **22**(2): 327–336.
42. Bertoldi D, Román T, Larcher R, Santato A, Nicolini G. 2013. Arsenic present in the soil-vine-wine chain in vineyards situated in an old mining area in Trentino, Italy. *Environ Toxicol Chem* **32**(4): 773–779.
43. Lachenmeir D, Przybylski MC, Rehm J. 2012. Comparative risk assessment of carcinogens in alcoholic beverages using the margin of exposure approach. *Int J Cancer* **131**: E995–E1003.
44. Cameán M, Herce-Pagliai C, Moreno I, González G, Reppeto M. 2002. Determination of total arsenic, inorganic and organic arsenic species in wine. *Food Addit Contam* **19**(6): 542–546.
45. US-FDA. 2013. Guidance for Industry Arsenic in Apple Juice—Action Level.
46. Zanoni LZ, Cordoba VL, Melnikov P. 2012. Trace elements in fruit juices. *Biol Trace Elem Res* **146**: 256–261.
47. Ashraf W, Jaffar M, Masud K. 2000. Heavy trace metal and macronutrient levels in various soft drinks and juices. *J Chem Soc Pak* **22**(2): 119–124.
48. Maduabuchi J-M, Adigba EO, Nzegwu CN, Oragwu CI, Okonkwo IP, Orisakwe OE. 2007. Arsenic and chromium in canned and non-canned beverages in Nigeria: A potential public health concern. *Int J Environ Res Public Health* **4**(1): 28–33.
49. Roberge J, Abalos AT, Skinner JM, Kopplin M, Harris RB. 2009. Presence of arsenic in commercial beverages. *Am J Environ Sci* **5**(6): 688–694.
50. Bingöl M, Yentür G, Er B, Öktem AB. 2010. Determination of some heavy metal levels in soft drinks from Turkey using ICP-OES method. *Czech J Food Sci* **28**(3): 213–216.
51. Dehelean A, Magdas DA. 2013. Analysis of mineral an heavy metal content of some commercial fruit juices by inductively coupled plasma mass spectrometry. *Scient World J* **2013**: Article ID 215423, 1–6.
52. Chen T-G, Li Y-X. 2005. Concentration of additive arsenic in Beijing pig feeds and the residues in pig manure resources. *Conserv Recycl* **45**: 356–367.
53. Markris KC, Quazi S, Punamiya P, Sarkar D, Datta R. 2008. Fate of arsenic in swine waste from concentrated animal feeding operation. Technical Reports. *Waste Management* **37**: 1626–1633.
54. U.S. Environmental Protection Agency. 2010. Relative bioavailability of arsenic in soils at 11 hazardous waste sites using an in vivo juvenile swine method. OSWER Directive No. 9200.0-76.
55. Wares MA, Awal MA, Das SK, Alam J. 2013. Environmentally persistent toxicant arsenic affects uterus grossly and histologically. *Bangl J Vet Med* **11**(1): 61–68.
56. Mandal TK, Dash JR, Datta BK, Sarkar S. 2013. Chronic arsenicosis in cattle: Possible mitigation with Zn and Se. *Ecotoxicol Environ Safety* **92**: 119–122.
57. Ashrafihelan J, Amoli JS, Alamdari M, Esfshsni TA, Mozafari M, Nourian AR, Bahari AA. 2013. Arsenic toxicosis in sheep: The first report from Iran. *Interdiscip Toxicol* **6**(2): 93–98.
58. Santra SC, Samal AC, Bhattacharya P, Banerjee S, Biswas A, Majumdar J. 2013. Arsenic in food chain and community health risk: A study in Gangetic West Bengal. *Proc Environ Sci* **18**: 2–13.
59. Genuis SJ, Schwalfenberg G, Siy A-KJ, Rodushkin I. 2012. Toxic element contamination of natural health products and pharmaceutical preparations. *PLoS ONE* **7**: 1–12.
60. Fuh C-B, Lin H-I, Tsai H. 2003. Determination of lead, cadmium, chromium, and arsenic in 13 herbs of tocolysis formulation using atomic absorption spectrometry. *J Food Drug Anal* **11**(1): 39–45.
61. Ernst E. 2002. Toxic heavy metals and undeclared drugs in Asian herbal medicines. *Trends Pharmacol Sci* **23**(3): 136–139.

62. Hanjani NM, Fender AB, Mercurio MG. 2007. Chronic arsenicism from Chinese herbal medicine. *Cutis* **80**: 305–308.
63. Saper RB, Phillips RS, Sehgal A, Khouri N, Davis RB, Paquin J, Thuppil V, Kales SN. 2008. Lead, mercury, and arsenic in US- and Indian-manufactured ayurvedic medicines sold via the Internet. *JAMA* **300**(8): 915–923.
64. Thing A, Chow Y, Tan W. 2013. Microbial and heavy metal contamination in commonly consumed traditional Chinese herbal medicines. *J Trad Chin Med* **33**: 119–124.
65. Liu W-J, Zhao Q-L, Sun G-X, Williams P, Lu X-J, Cai J-Z, Liu X-J. 2013. Arsenic speciation in Chinese herbal medicines and health implication for inorganic arsenic. *Environ Pollut* **172**: 149–154.
66. Kumar G, Kumar Y. 2012. Evidence for safety of ayurvedic herbal, herbo-metallic and Bhasma preparations on neurobehavioral activity and oxidative stress in rats. *Ayu* **33**(4): 569–575.
67. Matthews M, Jack M. 2011. *Spices and Herbs for Home and Market*. Rome, Italy: Rural Infrastructure and Agro-Industries Division, Food and Agriculture Organization of the United Nations.
68. Krejpcio Z, Król E, Sionkowski S. 2007. Evaluation of heavy metals contents in spices and herbs available on the polish market. *Polish J Environ Stud* **16**: 97–100.
69. Denholm JT. 2010. Complementary medicine and heavy metal toxicity in Australia. *Webmed Central Toxicol* **1**(9): WMC00535.
70. Nkansah MA, Opoku AC. 2010. Heavy metal content of some common spices available in markets in the Kumasi metropolis of Ghana. *Am J Scient Indust Res* **1**(2): 158–163.
71. Kandu R, Bhattacharyya K, Pal S. 2012. Arsenic intake and dietary risk assessment of coriander (*Coriandrum sativum* L.) leaves in the Gangetic basin of West Bengal. *J Spice Aromat Crop* **21**(2): 125–129.
72. Adekunle I, Lanre-Iyanda TY. 2012. Assessment of heavy metals and their estimated daily intakes from two commonly consumed foods (Kulikuli and Robo) found in Nigeria. *African J Food Agric Nutr Develop* **12**(3): 6157–6169.
73. Inam F, Deo S, Narkhede N. 2013. Analysis of minerals and heavy metals in some collected from local market. *IOSR J Pharm Biol Sci* **8**: 40–43.
74. Belay K, Tadesse A, Kebede T. 2014. Validation of a method for determining heavy metals in some Ethiopian spices by dry ashing using atomic absorption spectroscopy. *Int J Innov Appl Stud* **5**(4): 327–332.
75. Davarynejad G, Zarep M, Nagay PT. 2013. Identification and quantification of heavy metals concentration in Pistacia. *Not Scient Biol* **5**(4): 438–444.
76. AHPA. 2009. *Heavy Metals: Analysis and Limit in Herbal Dietary Supplements*. The American Herbal Products Association, pp. 1–9.
77. Yentür G, Bingöl M, Er B, Öktem AB. 2010. Determination of some heavy metal levels in Soft drinks from Turkey using ICP-OES method. *Czech J Food Sci* **28**(3): 213–216.
78. Kozul CD, Ely KH, Enelow RI, Hamilton JW. 2009. Low-dose arsenic compromises the immune response to influenza A infection in vivo. *Environ Health Perspect* **117**(9): 1141–1447.
79. Liao C-M, Chio C-P, Cheng Y-H, Hsieh N-H, Chen W-Y, Chen S-C. 2011. Quantitative links between arsenic exposure and influenza A (H1N1) infection-associated lung functions exacerbations risk. *Risk Anal* **31**(8): 1281–1294.
80. Kitchin K, Conolly R. 2010 Arsenic-induced carcinogenesis-oxidative stress as a possible mode of action and future research needs for more biologically based risk assessment. *Chem Res Toxicol* **23**: 327–335.
81. Fraser B. 2012. Cancer cluster in Chile linked to arsenic contamination. *World Report* **379**: 603.

82. Haraguchi H. 2011. Metallomics research related to arsenic, biological chemistry of arsenic, antimony and bismuth. In: Sun H (ed.). *Biological Chemistry of Arsenic, Antimony and Bismuth*. John Wiley & Sons, Chapter 4, pp. 83–112.

83. Pizarro I, Gómez M, Cámara C, Palacios MA, Román DA. 2004. Evaluation of arsenic species-protein binding in cardiovascular tissues by bidimensional chromatography with ICP-MS detection. *J Anal Atom Spectrom* **19**: 292–296.

84. Gergen I, Harmanescu M, Alda LM, Bordean DM, Gogoasa I. 2011. Heavy metals health risk assessment for population via consumption of vegetables grown in old mining area; a case study: Banat County Romania. *Chem Centr J* **5**: 1–10.

85. Brusseau ML, Ramirez-Androtta MD, Beamer P, Maier RM. 2013. Home gardening near a mining site in an arsenic-endemic region of Arizona. Assessing arsenic exposure dose and risk via ingestion of home garden vegetables, soils, and water. *Sci Total Environ* **454–455**: 373–382.

86. Kumar R, Navin S, Das J, Verna C, Kumar M. 2013. Arsenic in the environment effectuates human health: An imperative need to focus. *Int Res J Environ Sci* **2**(11): 101–105.

87. Reddy KR, Danda S, Yukselen-Aksoy Y, Al-Hamdan A. 2010. Sequestration of heavy metals in soils from two polluted industrial sites: Implications for remediation. *Land Contam Reclam* **18**(1): 1–23.

88. Crozier RD. 1992. *Flotation, Theory, Reagents and Ore Testing*. Santiago, Chile: Pergamon Press, p. 343.

89. Chakraborti D, Rahman MM, Paul K, Chowdhury UK, Sengupta MK, Lodh D, Chanda CR, Saha KC, Mukherjee SC. 2002. Arsenic calamity in the Indian subcontinent. What lessons have been learned? *Talanta* **58**: 3–22.

90. Plant JA, Korre A, Reeder S, Smith B, Voulvoulis N. 2005. Chemicals in the environment: Implication for global sustainability. *Appl Earth Sci* (*Trans Inst Min Metall B*) **114**: 65–97.

91. Jiang J-Q, Ashekuzzaman SM, Jiang A, Sharifuzzaman SM, Chowdhury SR. 2013. Arsenic contaminated groundwater and its treatment options in Bangladesh. *Int J Environ Res Public Health* **10**: 18–46.

92. Sharma AK, Tjell JCh, Slot JJ, Holm PE. 2014. Review of arsenic contamination, exposure, through water and food and low cost mitigation options for rural areas. *Appl Geochem* **41**: 11–33.

93. Meagher RB, Heaton CP. 2005. Strategies for the engineered phytoremediation of toxic element pollution: Mercury and arsenic. *J Ind Microbiol Biotechnol* **32**: 502–513.

94. Koenigsberg S, Hazen TC, Peacok AD. 2005. *Environmental Biotechnology: A Bioremediation Perspective*. Published on line in Wiley Interscience. Wiley Periodicals Inc. DOI: 10.1002/REM.20057.

95. Hosamane SN. 2012. Removal of arsenic by phytoremediation—A study of two plant spices. *Int J Scient Eng Technol* **1**(5): 218–224.

96. Mahmood Q, Mirza N, Shah MM, Pervez A, Sultan S. 2014. Plants as useful vectors to reduce environmental toxic arsenic content. *Scient World J* **2014**: Article ID 921581, 1–11.

97. Rajkumar M, Ae N, Vara Prasad MN, Freitas H. 2010. Potential of siderophore-producing bacteria for improving heavy metal phytoextraction. *Trends Biotechnol* **28**(3): 142–149.

98. Ali SS, Vidhale NN. 2013. Bacterial siderophore and their application: A review. *Int J Curr Microbiol Appl Sci* **2**(12): 303–312.

99. Hooda V. 2007. Phytoremediation of toxic metals from soil and waste water. *J Environ Biol* **28**(2): 367–376.

100. Attia-Ismail S. 2008. Role of minerals in halophyte feeding to ruminants. In: Prasad MNV (ed.). *Trace Elements as Contaminants and Nutrients. Consequences in Ecosystems and Human Health*. Hoboken, NJ: John Wiley & Sons, Inc., pp. 701–720, 777pp.

101. Song W-Y, Park J, Mendoza-Cozati DG, Suter-Grotemeyer M, Shim D, Hörtensteiner S, Geisler M et al. 2010. Arsenic tolerance in *Arabidopsis* is mediated by two ABCC-type phytochelatin transporters. *Proc Natl Acad Sci USA* **107**(49): 21187–21192.

102. Rathinasabapathi B, Ma LQ, Srivastava M. 2006. Arsenic hyperaccumulating ferns and their application to phytoremediation of arsenic contaminated sites. *Floricult Ornament Plant Biotechnol* **3**: 304–311.

103. Zhao FJ, Ma JF, Meharg AA, McGrath SP. 2009. Arsenic uptake and metabolism in plants. *New Phytol* **181**: 777–794.

104. Haydon MJ, Cobbett CS. 2007. Transporters of ligands for essential metal ions in plants. *New Phytol* **174**: 499–506.

105. Yadav SK. 2010. Heavy metals toxicity in plants: An overview on the role of glutathione and phytochelatins in heavy metal stress tolerance of plants. *South African J Bot* **76**: 167–179.

106. Odjacova M, Zagorchev L, Seal CE, Kranner I. 2013. A central role for thiols in plant tolerance to abiotic stress. *Int J Mol Sci* **14**: 7405–7432.

107. Ackermann J, Vetterlein D, Kaiser K, Mattusch J, Jahn R. 2010. The bioavailability of arsenic in floodplain soils: A simulation of water saturation. *Eur J Soil Sci* **61**: 84–96.

108. Záray G, Mihucz VG, Tatár E, Virág I, Zang C, Jao Y. 2007. Arsenic removal from rice by washing and cooking with water. *Food Chem* **105**: 1718–1725.

109. Meharg AA, Zhu Y-G, Williams P. 2008. Exposure to inorganic arsenic from rice: A global health issue? *Environ Pollut* **154**(2): 169–171.

110. Zhao FJ, McGrath SP, Meharg AA. 2010. Arsenic as a food chain contaminant: Mechanisms of plant uptake and metabolism and mitigation strategies. *Annu Rev Plant Biol* **61**: 535–539.

111. EFSA panel on contaminants in the food chain, 2009. *EFSA J* **7**(10): 199.

112. Singh N, Kumar D, Sahu AP. 2007. Arsenic in the environment: Effects on human health and possible prevention. *J Environ Biol* **28**(2): 359–365.

113. Flora S, Pachauri V. 2010. Chelation in metal intoxication. *Int J Environ Res Public Health* **7**: 2745–2788.

114. Sears ME. 2013. Chelation: Harnessing and enhancing heavy metal detoxification—A review. *Scient World J* **2013**: 1–13.

115. Fuortes L, Breeher L, Gerr F. 2013. A case report of adult lead toxicity following use of ayurvedic herbal medication. *J Occup Med Toxicol* **8**: 26.

116. Sharma B, Singh S, Siddiqi NJ. 2014. Biomedical implications of heavy metals induced imbalances in redox systems. *Biomed Res Int* **2014**: Article ID 640754, 1–26.

117. Slaveykova VI, Karadjova IB, Tsalev DL. 2007. The biouptake and toxicity of arsenic species on the green microalga *Chlorella salina* in seawater. *Aquat Toxicol* **87**: 264–271.

118. Sayre R, Rajamani S, Torres M, Falcao V, Gray JE, Coury DA, Colepicolo P. 2014. Noninvasive evaluation of heavy metal uptake and storage in microalgae using a fluorescence resonance energy transfer-based heavy metal biosensor. *Plant Physiol* **164**: 1059–1067.

119. Rayman M. 2012. Selenium and health, review. *Lancet* **379**: 1256–1258.

120. Garelick H, Jones H. 2008. Mitigating arsenic pollution: Bridging the gap between knowledge and practice. *Chem Int* **30**: 1–13.

121. Sauve S. 2014. Time to revisit arsenic regulations: Comparing drinking water and rice. *BMC Public Health* **14**: 465.

122. Eyre GE, Spottiswoode W. 1851. An act to regulate the sale of arsenic. *Victoriae Reginae* **13**: 173–175.

123. Ahsan H, Verret WJ, Chen Y, Ahmed A, Islam T, Parvez F, Kibriya MG, Graziano JH. 2005. A randomized, double-blind placebo-controlled trial evaluating the effects of vitamin E and selenium on arsenic-induced skin lesions in Bangladesh. *J Occup Environ Med* **47**: 1026–1035.

124. Uddin AN, Burns FJ, Rossman TG. 2005. Vitamin E and organoselenium prevent the carcinogenic activity of arsenite with solar UVR in mouse skin. *Cardiogenesis* **26**(12): 2179–2186.

125. Gamble MV, Liu X, Ahsan H, Pilsner JR, Ilievski V, Slavkovich V, Parvez F, Chen Y, Levy D, Factor-Litvak P, Graziano JH. 2006. Folate and arsenic metabolism: A double-blind, placebo-controlled folic acid-supplementation trial in Bangladesh. *Am J Clin Nutr* **84**: 1093–1101.

126. Ahsan H, Heck JE, Gamble V, Chen Y, Graziano JH, Slavkovich V, Parvez F, Baron JA, Howe GR. 2007. Consumption of folate-related and metabolism of arsenic in Bangladesh. *Am J Clin Nutr* **85**: 1367–1374.

127. Mitra S, Kundu M, Ghosh P, Das JK, Sau TJ, Banerjee S, States JC, Giri AK. 2010. Precancerous and non-cancer disease endpoints of chronic arsenic exposure: The level of chromosomal damage and XRCC3 T241M polymorphism. *Mutat Res* **706**(1–2): 7–12.

128. Ahsan H, Argos M, Rathouz PJ, Pierce BL, Karla T, Parvez F, Slavkovich V, Ahmed A, Chen Y. 2010. Dietary B vitamin intakes and urinary total arsenic concentration in the health effects of arsenic longitudinal study (HEALS) cohort, Bangladesh. *Eur J Nutr* **49**(8): 473–481.

129. Hossain FMA, Hossain MM, Kabir MG, Fasina FO. 2013. Effectiveness of combined treatment using Spirulina and vitamin A against chronic arsenicosis in rats. *African J Pharm Pharmacol* **7**(20): 1260–1266.

130. Russo M, Spagnuolo C, Tedesco I, Russo GL. 2010. Phytochemicals in cancer prevention and therapy: Truth or dare? *Toxins* **2**: 517–551.

131. Pereira ML, Pires RN, Carvalho F, Carvalho M, Fernandes E, Soares E, Bastos M. 2004. Protective activity of hesperidin and lipoic acid against sodium arsenite acute toxicity in mice. *Toxicol Pathol* **32**: 527–535.

132. Diederich M, Teiten M-H, Eifes S, Dicato M. 2010. Curcumin—The paradigm of a multi-target natural compound with applications in cancer prevention and treatment. *Toxins* **2**: 128–162.

133. Beevers C, Huang S. 2011. Pharmacological and clinical properties of curcumin. *Bot: Targets Ther* **1**: 5–18.

134. Zhang Z, Wang X, Zhao H, Shao Y, Wang P, Wei Y, Zhang W, Jiang J. 2014. Nephroprotective effect of astaxanthin against trivalent inorganic arsenic-induced renal injury in wistar rats. *Nutr Res Pract* **8**(1): 46–53.

135. Pieczenik SR, Neustadt J. 2006. Mitochondrial dysfunction and molecular pathways of disease. *Exp Mol Pathol* **83**: 84–92.

136. Meyer JN, Leung MC, Rooney J, Sendoel A, Hengartner M, Kisby G, Bess A. 2013. Mitochondria as a target of environmental toxicants. *Toxicol Sci* **134**(1): 1–17.

137. Murphy MP. 2009. How mitochondria produce reactive oxygen species? *Biochem J* **417**: 1–13.

138. Hei TK, Partridge MA, Huang SXL, Hernandez-Rosa E, Davidson MM. 2007. Arsenic induced mitochondrial DNA damage and altered mitochondrial oxidative function: Implication for genotoxic mechanisms in mammalian cells. *Cancer Res* **67**(11): 5239–5247.

139. Pourahmad J, Hosseini M-J, Shaki F, Khansari MG. 2013. Toxicity of arsenic(III) on isolated liver mitochondria: A new mechanistic approach. *Iran J Pharmaceut Res* **12**(supplement): 121–138.

140. Ray K, Sinha S, Giri A, Chowdhury R. 2014. Mitochondrial genome variations among arsenic exposed individuals and potential correlation with apoptotic parameters. *Environ Mol Mutagen* **55**: 70–76.

141. Majumdar S, Maiti A, Karmakar S, Das AS, Mukherjee S, Das D, Mitra Ch. 2012. Antiapoptotic efficacy of folic acid and vitamin B12 against arsenic-induced toxicity. *Environ Toxicol* DOI 10.002/tox, 20648, 351–363.

142. Sena L, Chandel NS. 2012. Physiological roles of mitochondrial reactive oxygen species. *Mol Cell* **48**: 158–167.

143. Le XC, Shen S, Li X-F, Cullen WR, Weinfel M. 2013. Arsenic binding to proteins. *Chem Rev* **113**(10): 7769–7792.

144. Henkler F, Brinkmann J, Luch A. 2010. The role of oxidative stress in carcinogenesis induced by metals and xenobiotics. *Cancers* **2**: 376–396.

145. Román DA, Román HA. 2004. *Agreement of Collaboration between University of Antofagasta Technical Attendance*. Antofagasta, Chile: Agricultural and Cattle Service.

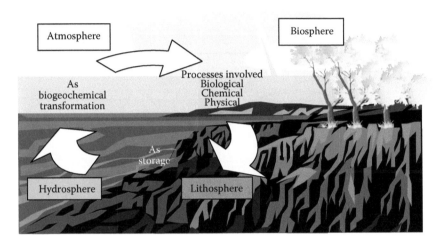

FIGURE 4.1 Simplified geochemical cycle of As in the natural environment.

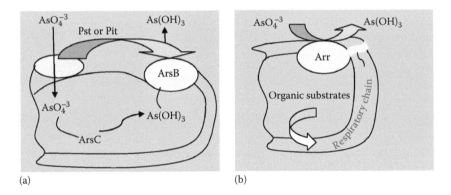

FIGURE 4.2 Cellular functions and locations of bacterial (a) cytoplasmic arsenate reductase and (b) respiratory arsenate reductase. (Modified from Silver, S. and Phung, L.T., *J. Ind. Microbiol. Biotechnol.*, 32(11–12), 2005, 587.)

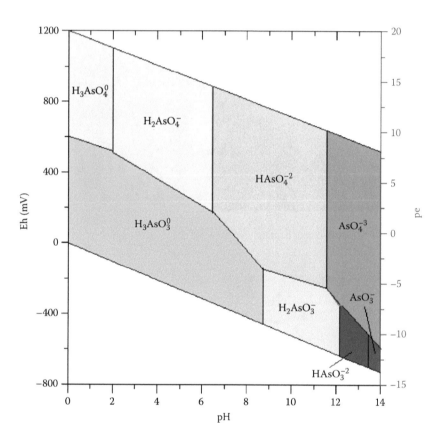

FIGURE 4.4 Eh–pH diagram of aqueous arsenic species in water at 25°C and 1 bar total pressure. (Modified from Smedley, P. and Kinniburgh, D., *Appl. Geochem.*, 17(5), 2002, 517.)

Brake fern (*Pteris vittata*)

FIGURE 7.5 Chinese brake fern (*P. vittata* L.) grown in the greenhouse (left) and in the field. (NV Hue's personal images.)

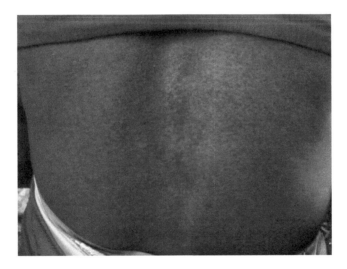

FIGURE 13.1 Middle-aged lady showing raindrop pigmentation and leucomelanosis on the back.

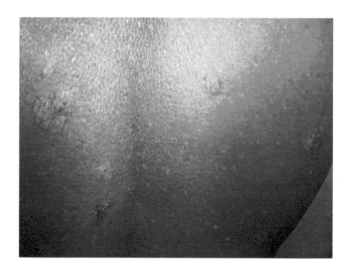

FIGURE 13.2 Elderly gentleman having lesions of Bowen's disease over a background of raindrop pigmentation.

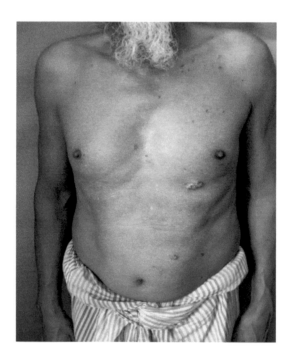

FIGURE 13.3 Elderly gentleman with long-standing features of chronic arsenicosis, eventually developing squamous cell carcinoma on the trunk.

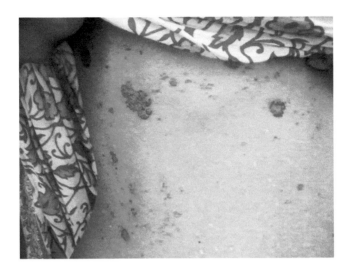

FIGURE 13.4 Young lady with pigmentary changes, Bowen's disease, and basal cell carcinoma on the trunk.

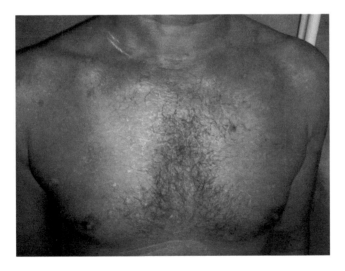

FIGURE 13.5 Development of Bowen's disease on a background of raindrop pigmentation in an elderly male.

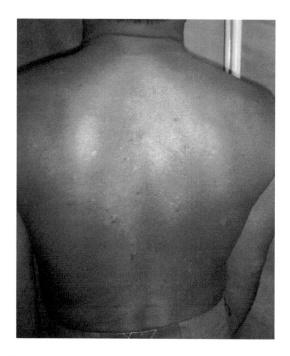

FIGURE 13.6 Younger brother of the patient shown in Figure 13.5; note the lesions of Bowen's disease on the trunk.

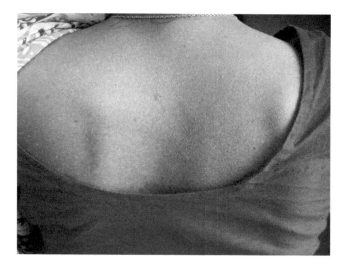

FIGURE 13.7 A young lady with pigmentary changes suggestive of chronic arsenicosis. Note the leucomelanosis pattern.

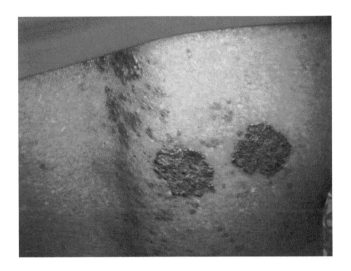

FIGURE 13.8   A middle-aged female developing multicentric Bowen's disease, with a long-standing history of pigmentary changes.

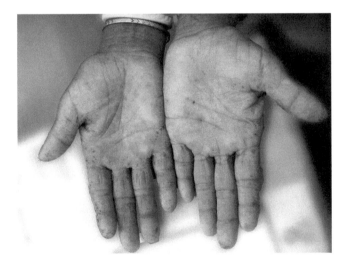

FIGURE 13.9   Note the grit-like skin with minute papules on the palms.

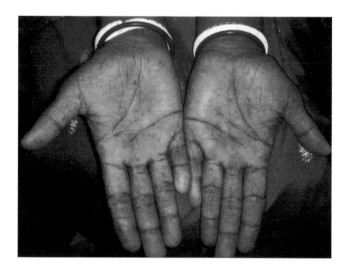

FIGURE 13.10 Note the grit-like skin with minute papules on the palms.

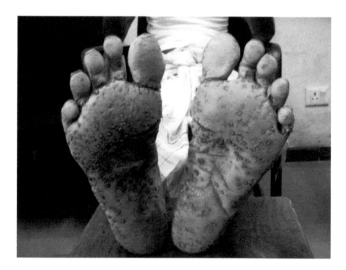

FIGURE 13.11 Characteristic lesions of moderate arsenic keratosis on the soles of a middle-aged gentleman.

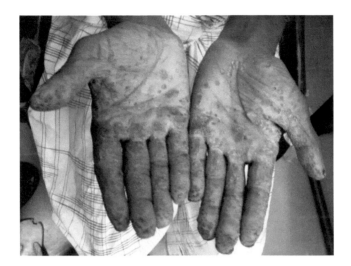

FIGURE 13.12 Palms of the same patient as shown in Figure 13.11; note the punctuate and wart-like lesions.

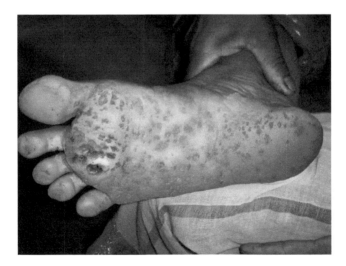

FIGURE 13.13 Moderate to severe arsenic keratosis on the plantar surface of the feet of a middle-aged female. Note the confluence of small papules, which defines severe keratosis.

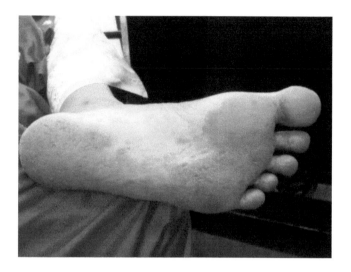

FIGURE 13.14 Severe arsenic keratosis on the sole.

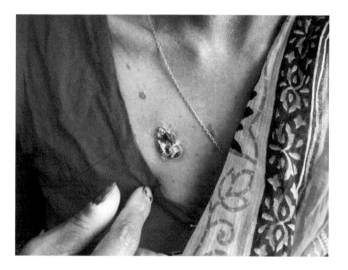

FIGURE 13.15 Development of squamous cell carcinoma in a patient with a history of pigmentary changes for 15 years.

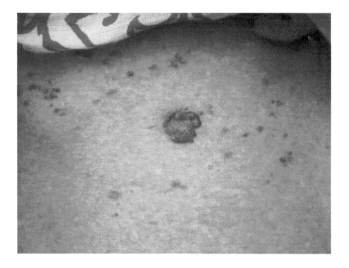

FIGURE 13.16 Note the characteristic rolled-out border of the lesion, which is diagnostic of basal cell carcinoma.

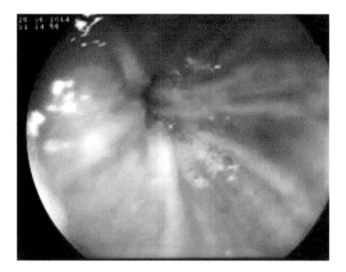

FIGURE 13.18 Esophageal varices due to noncirrhotic portal fibrosis as seen in upper GI endoscopy.

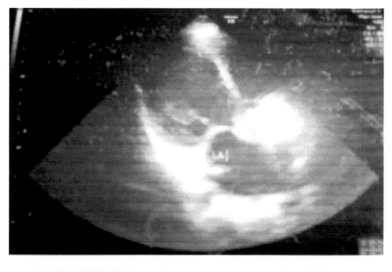

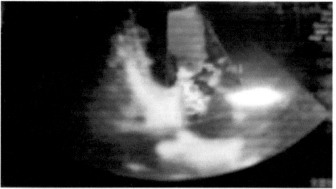

FIGURE 13.19 Echocardiography showing arsenic-induced dilated cardiomyopathy.

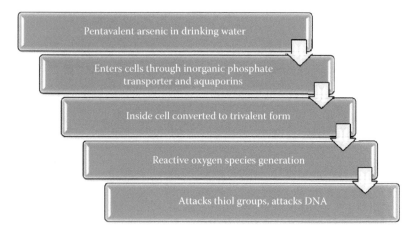

FIGURE 13.23 Possible mechanism of development of malignancies in a long-standing case of arsenicosis.

10 Arsenic in the Soil–Plant System

Phytotoxicity and Phytoremediation

Rebeca Manzano, Eduardo Moreno-Jiménez, and Elvira Esteban

CONTENTS

10.1 INTRODUCTION

There are around 1,400,000 hot spots contaminated with metals in East Europe alone, mainly coming from the usual application of pesticides, fertilizers, and sewage sludge alongside emissions to the atmosphere caused by mining and metal smelting, waste incineration, etc.

The reclamation of trace elements–contaminated soils is an environmental challenge since metals and metalloids are persistent and nondegradable. The risks they pose are associated to their transfer to noncontaminated areas and their inclusion in the trophic chain through plant absorption or direct ingestion. Not all the approaches in the topic of soil remediation have exhibited the same social acceptance. Ex situ chemical decontamination techniques can destroy soil structure, cause a biological inactivation, and are not generally cost-effective, especially when big areas of a low economic value are involved. Phytoremediation appeared as a promising technique able to potentially remediate large areas of moderate contamination with the use of plants. Compared to other methods, this one is cost-effective, environmentally friendly, and is socially accepted. At least three factors should be taken into account for the selection of a phytoremediation technology: the level of contamination, the metal(loid) bioavailability, and the plant's capacity to accumulate trace elements in their aboveground tissues or to stabilize them in their roots. Phytoremediation is suitable for treating a large range of pollutants including hydrocarbons, chlorinated solvents, pesticides, explosives, heavy metals, and radionuclides. For highly contaminated soils, phytoremediation is usually the final step in a process where a previous and more intensive treatment is usually required. In moderate and low levels of contamination, phytoremediation of the soil may be sufficient.

10.2 ARSENIC METABOLISM IN PLANTS

10.2.1 ARSENIC ABSORPTION

Studies reveal that, after arsenate uptake, it is rapidly reduced into arsenite, the latter being the major species of arsenic in plants, with the participation of arsenate reductase and consuming reduced glutathione [1]. The transport of arsenate is mediated through the phosphate transport system family 1 (PHT1) involving a cotransport of $2H^+$ for each molecule of $H_2AsO_4^-$ or $H_2PO_4^-$ [2]. PHT1 is comprised of nine members (detected in *Arabidopsis thaliana*: PHT1;1-PHT1;9) [3]. Those proteins are highly expressed in roots although they have also been detected in aerial parts as flowers and leaves. Arsenite is present as the neutral species (undissociated) in a normal range of water and soil pH. As the pK_a of the arsenous acid is high (9.2, 12.1, and 13.4), arsenite is normally absorbed as an uncharged molecule. Strong evidence has suggested arsenite uptake is mediated through a plant aquaporin subfamily: the nodulin26-like intrinsic proteins (NIPs). Members of the NIP subfamily responsible for the transport of boron and silicon were identified in *A. thaliana* and *Oryza sativa* (AtNIP5;1 and OsNIP2;1 respectively). Isayenkov et al. [4] proved that the loss of expression of the aquaglyceroporin AtNIP7;1 increased tolerance to As(III) and reduced total arsenic in the plants of *A. thaliana*, evidencing that aquaglyceroporin is involved in As(III) absorption. Arsenite and silicic acid have similar pK_a (9.2 and 9.3 respectively) and molecular diameter (4.11 and 4.38 Å respectively). Thus, it was hypothesized that both molecules could be introduced into root cells by the same transport system [5]. It was observed Lsi1 (OsNIP2;1), that encodes a silicon influx transporter located on the plasma membrane on the distal side of the exodermis and the endodermis, mediated the extrusion of arsenite from rice roots, but only in a small proportion, suggesting the existence of other NIPs playing this role [6]. The gen Lsi2 that encodes another silicon transporter in rice plants is located on the proximal side of the exodermis and endodermis cells [7]. It plays a more important role than Lsi1 in the transport of arsenite to the shoot and grain in rice plants, releasing arsenite in the cortex and into the stele, but it has a minor role in arsenite efflux to the external medium also compared to Lsi1 [5,6]. Recently, the role of OsNRAMP1, a member of rice NRAMP (natural resistance-associated macrophage protein) transporter in arsenic accumulation in xylem has been reported, expressed in yeast and *Arabidopsis* [8]. Some *Arabidopsis* NRAMPs (AtNRAMP1, AtNRAMP3, and AtNRAMP4) have been identified as high-affinity iron and manganese transporters but rice NRAMPs have not been identified. The latter work hypothesized OsNRAMP1 may assist arsenic and cadmium xylem loading from root to shoot.

The organic species of arsenic MMAA(V) and DMAA (V) are known to be taken up by the aquaporin channel Lsi1 in rice plants and, in contrast to the role of Lsi1 in arsenite transport, Lsi2 is not involved in the transport of these molecules [9]. It is known that microorganisms can methylate arsenite, and a recent report reveals that plants can absorb the arsenic methylated by microorganisms, but plants are not able to methylate inorganic arsenic [10].

In arsenic-resistant plants, the suppression of the high-affinity phosphate/arsenate uptake system has been observed, allowing the reduction of arsenate influx within plants. Most of the plant species are considered excluders of arsenic, that is, plants with restricted uptake and restricted translocation through their organs. *Pteris vittata* and other ferns from the Pteridacea family are identified as arsenic hyperaccumulators, accumulating an aboveground arsenic concentration of up to 2% with a transfer factor higher than 1. Arsenic uptake in the hyperaccumulator *P. vittata* is also mediated through the phosphate transport system, and applying phosphate in the growth medium inhibits arsenate uptake. When compared with a nonaccumulator fern, *P. vittata* kinetics showed a higher density of transporters for arsenate uptake on the plasma membranes of root cells [11].

10.2.2 Arsenic Accumulation

Most plants appear to have high levels of arsenate reductase activity, which could explain why arsenite is the major species within plant tissues. Arsenite transport from roots to other organs is not very effective, tending to remain in roots, excepting the accumulator and hyperaccumulators plants species like *P. vittata*, which has a high ability to load arsenic in the xylem and translocate it to the fronds.

Once arsenate is reduced to arsenite within the plant cell, it can remain within the root cell. Arsenite forms complexes with thiol groups of glutathione and phytochelatins. The formation of complexes of As–S is believed to contribute to arsenic sequestration in vacuoles [12]. Those are stable complexes within the pH range 1.5 and 7–7.5, being dissociated at higher pHs. It has been reported that plants exposed to arsenic synthesized large amounts of phytochelatins (PCs), and this fact is considered an important mechanism of detoxification in nonaccumulator plants. It was believed that the sequestration of As–S complexes into the vacuole was mediated through the vacuolar transporters observed in the yeast *Schizosaccharomyces pombe* (HTM; heavy metal tolerance 1) but the identification of these transporters in plants has been unsuccessful. However, a recent discovery has identified two ABC transporters required for arsenic tolerance, AtABCC1 and AtABCC2 in *Arabidopsis* plants [13]. In this study, the synthesis of PCs was correlated with the presence of PC transporters. In rice plants, it was observed, using high-resolution secondary ion mass spectrometry in combination with TEM, that arsenic was stored in the vacuoles of the endodermis, pericycle, and xylem parenchyma cells in roots in the form of As–PC complexes [14]. In *P. vittata*, arsenic is rapidly transported from roots to fronds, where it is likely to be sequestered into the vacuoles of the upper and lower epidermal cells and trichomes. The form in which arsenic is transported is arsenite, as it was reported that arsenate reductase is highly expressed in root cells [15] and arsenite is the major species found in the xylem sap regardless of the form of arsenic added to the growth medium, arsenate or arsenite [16]. In contrast to nonaccumulator plants, little As(III)-thiol complexation is observed in *P. vittata*. In a recent study, the gen ACR3 was identified as being similar to a yeast arsenite efflux transporter. This gen (ACR3) encoded for a transporter localized in the vacuolar membrane in gametophytes that showed enhanced As(III) tolerance and accumulation. Authors reported that this transporter is essential for arsenite detoxification in *P. vittata*, contributing to the search of additional arsenic transporters that are related to arsenic tolerance in hyperaccumulators [17].

10.2.3 ARSENIC TRANSPORT TO ABOVEGROUND TISSUES

Arsenic in xylem sap occurs in the forms As(V), As(III), MMAA, and DMAA. However arsenite is the major species of arsenic present in the xylem [12,18].

Arsenate is loaded into the xylem through the phosphate transport system [19]. Further studies have been conducted on arsenite loading. In rice plants, the silicon transporter Lsi2 has been identified for arsenite loading into the xylem [5]. This is a critical point for rice plants, as arsenite translocation to the edible parts of rice plants can be hampered by Si. In turn, the influx of the methylated species MMAA(V) and DMAA(V) into the xylem is not mediated through Lsi2 but by another pathway, as at the cytoplasmic pH (approximately 7.5) both molecules are negatively charged [9].

There has not been evidence using x-ray absorption spectrometry of As–S complexes in the xylem sap of *Brassica juncea*. In fact, complexation with thiols can decrease arsenite mobility from roots to shoots [12]. According to these authors, other complexes could be involved. Synchrotron-based fluorescence-x-ray absorption near-edge spectroscopy (fluorescence-XANES) in wheat and rice roots has strengthened this evidence. Kopittke et al. [20] detected a whole pool of As(III) in the root cells complexed with thiol groups, and uncomplexed As(III) was the major As species transported within the xylem and phloem. According to those authors, the rapid conversion of arsenate to arsenite and its complexation with thiol groups would explain why arsenic movement from the roots to the shoots is highly restricted in nonaccumulator plants.

In rice plants, it has been observed that arsenic uptake and transfer into grains strongly depends on rice cultivar and bioavailability of arsenic in soil [21]. Phytochelatin concentration increases when plants are exposed to higher arsenic concentrations, with a concomitant reduction of arsenic translocation, suggesting that the modification of the level of PCs through overexpression of PC synthase in rice roots would lead to reduced arsenic in rice grains [22].

New evidence was found regarding arsenic and phosphorus transportation in the hyperaccumulator *P. vittata* [23]. This novel work shows that As(V) is the major species from epidermis to cortex and As(III) increases from endodermis to the vascular bundle in rhizoids, suggesting As(V) conversion to As(III) takes place before the endodermis. It was also found that phosphorous deficiency promoted arsenic movement from the epidermis to cortex but it was less obvious from the endodermis to the vascular bundle. Authors reported an arsenic and phosphorous cotransport before the endodermis, that is not different in nonaccumulator plants, but after As(V) reduction to As(III) at the endodermis, that process no longer acts. This suggests the more efficient translocation of As(III) to fronds in *P. vitatta* could be mediated through a passive process mainly depending on aquaporins.

Arsenic transport into the phloem has been little investigated due to technical difficulties in the sampling of phloem. Recent studies show that GSH and PCs for As(III), cadmium, zinc, and mercury, and nicotianamine for iron, copper, manganese and zinc are the main ligands found in phloem sap [24]. According to some authors, the presence of PCs in the phloem was unlikely, since PCs have been considered molecules that mediate metal transport from the cytosol to the vacuole [25]. However, analyses in inductively coupled plasma–mass spectrometry (ICP-MS) revealed the presence of PCs in the phloem sap of *Brassica napus* and *Ricinus communis* [25]. Complexes of phytochelatins with trace elements are likely to be unstable in the alkaline phloem sap [12]. In agreement with that, the techniques high-resolution ICP-MS and accurate mass electrospray mass spectrometry coupled to high-performance liquid chromatography revealed high concentrations of GSH and GSSH and some oxidized phytochelatins in phloem sap, but As(III)-thiol complexes were not identified [25]. A current study with barley revealed the silicon transporter HvLsi6 in the outer parenchyma cells surrounding the phloem regions and the xylem parenchyma cells are probably involved in xylem unloading in the leaves [26]. According to other authors, considering the similarity with silicon transporter OsLsi2, HvLsi6 might be implicated in the xylem unloading of As(III) in graminaceous plants [27].

10.2.4 ACTIVE EFFLUX

Arsenite and arsenate efflux has been reported from root cell to the external medium. Efflux occurs mainly in the form of arsenite. Arsenite efflux has been observed in several plant species, such as *A. thaliana*, *Holcus lanatus*, wheat, barley, and maize, but little arsenic efflux has been reported for the As hyperaccumulator *P. vittata* [1,16]. It seems that arsenite efflux by roots is very rapid after arsenate uptake and diminishes once arsenic supply is withheld. In *Escherichia coli*, arsenite efflux has been identified as a detoxification mechanism [7]. Intracellular As(III) is exuded out of fungal cells under As(V) exposure after its reduction to As(III) [28]. Further, it was observed by synchrotron radiation spectroscopic analysis that As(III) is exuded outside the rhizosphere from the extraradical hyphae of *Rhizophagus intraradices* [29]. In the aquatic fern *Azolla*, arsenic accumulation and efflux differ between two strains, *A. caroliniana* and *A. filiculoides* [30]. In contrast with the great number of studies that reported the major form of As extrusion is arsenite, it was observed that arsenate efflux was ninefold higher than arsenite extrusion in tomato plants, probably because arsenite is complexed with thiol groups, making arsenate extrusion a passive process [18].

In rice plants, arsenic efflux has been intensively studied. Lsi1 is the protein transporter that mediates As(III) efflux [6]. In the same way as the complexation of As(III) with PCs reduces the root-to-shoot translocation, this mechanism also contributes to a lower extrusion of As(III) to the external medium of *Arabidopsis* plants [12].

It has been questioned whether arsenic efflux can be considered a detoxification mechanism. Studies made with resistant and nonresistant ecotypes of *H. lanatus* revealed arsenic accumulation was lower in the resistant plants but there were no differences in As(III) efflux within the phenotypes, suggesting arsenite efflux is not enhanced in the resistant ecotype [31].

10.3   ARSENIC TOXICITY IN PLANTS

Arsenic does not present a biological function, although it has been observed that low arsenic doses could induce growth. This effect on growth has been attributed to several factors: an adequate phosphate nutrition that leads to a downregulation of the Pi/arsenate transport system; the displacement of phosphate by arsenate on soils, or high levels of cytoplasmic Pi, which reduces Pi/Asi competition in the biochemical processes, and hence lower toxicity. Arsenic toxicity for plants depends on its chemical speciation and availability in soil.

In soils, arsenite is more toxic than arsenate. The high pk_a of arsenite (9.22) allows it to have a high solubility and mobility. In plants, arsenite easily reacts with sulfhydryl groups of enzymes and proteins disrupting plant metabolism. Arsenate can interfere with the ATP molecule replacing phosphate and generating ADP-As that is unstable, leading to the disruption of energy flows in cells and to cell death. However, as arsenate is rapidly reduced inside the plant cell, it is more feasible that arsenite causes the major detrimental effects on plant cells. Further, inorganic compounds of arsenic are considered more toxic than the organic ones. However, this assertion should not be taken as a general statement, as the contrary has also been reported. Hydroponics experiments have allowed to study the diverse responses of plants to different combinations of organic and inorganic compounds of arsenic. Studies on *Spartina* species reported DMAA(V) was the most toxic species followed by MMAA(V) and the inorganic species As(III) and As(V), despite the uptake of DMAA(V) was the lowest [32].

10.3.1   Visual Toxic Symptoms

Plants that are not resistant to arsenic can suffer disorders from root inhibition to death. Frequent visual symptoms of arsenic are poor seed germination, loss of plant weight, weak root development (reduced elongation, reduced root branching, color changes), wilting and stunting of the aerial parts, foliar chlorosis and necrosis, curling leaves, and loss of water [33,34]. The beneficial role of phosphorous on mitigating toxic symptoms on plants exposed to arsenic is well known. When plants were grown within solutions containing both elements, rice plants did not shown any toxicity symptoms due to the downregulation of the high affinity Pi/As transport system [34].

10.3.2 Nutritional Disorders

Phosphorous is the nutrient most largely affected by arsenic. Other nutritional disorders triggered by arsenic are scarcely reported in the literature. When assessing plant response to arsenic, their nutritional status is not frequently a target point of study although a compromised nutritional level can limit plants' phytoremediation potential. Some authors reported relations between arsenic and nutrients, such as manganese, iron, zinc, copper, and magnesium [35]. Among them, the As–Fe relation could be one of the most important after phosphorous. Shaibur et al. [36] showed high levels of arsenic induced iron deficiency in leaves and hence chlorosis together with other nutritional disorders. An increased concentration of micronutrients in plants growing with high arsenic concentrations can be normally found because of a concentration effect, as plant growth generally declines with increasing arsenic concentrations. This effect can undergo negative consequences if micronutrient levels are too high and surpass the toxic level.

10.3.3 Reactive Oxygen Species

Arsenic in plant can cause the production of reactive oxygen species (ROS). ROS, as the superoxide radical ($O_2^{\cdot-}$), the hydroxyl radical ($^{\cdot}OH$), and the hydrogen peroxide (H_2O_2), comes from the reduction of free oxygenated radicals that accept electrons from other molecules. Those molecules are responsible for the oxidative damage at the cellular level to proteins, amino acids, purine nucleotides, nucleic acids, and the lipid peroxidation of cellular membranes. It can be induced under external biotic and abiotic changes in the environment such as light and temperature and also by nutritional disorders. Under arsenic exposure, ROS can be produced through the conversion of arsenate to arsenite, using oxygen as the final acceptor of electrons [37]. Damaged tissues present an unusual concentration of malondialdehyde (MDA). This latter is typically used to assess oxidative stress, which is an indicator of lipid peroxidation [38]. In response to oxidative stress, plants increase the level of antioxidant enzymes in order to repair the damage caused by ROS. Among them, catalase, superoxide dismutase (SOD), peroxidase, glutathione peroxidase, and glutathione reductase (GR) concentrations can be modified when plants are exposed to contaminants. Glutathione is a nonenzymatic antioxidant and plays an important role in plant protection against oxidative stress. The ascorbate–glutathione cycle uses reduced glutathione (GSH) as an electron donor to regenerate ascorbate from its oxidized form, dehydroascorbate; H_2O_2 is accumulated in the cell; and APX reduces it to H_2O [39]. The oxidized glutathione GSSH acts as a substrate for GR that catalyzes the regeneration of Cys thiol groups in order to maintain a highly reduced glutathione pool necessary for active protein functions [40]. Glutathione-S-transferases (GST) and glutathione peroxidases (GPX) respond to ROS. The antioxidant role of GST is related to the dismutation of hydroxyl radicals ($\cdot$OH) and hydrogen peroxide (H_2O_2) [41]. Furthermore, phytochelatins are synthetized using glutathione as a substrate and it is well known that As(III) forms complexes with GSH and PCs previously to their sequestration into the cell vacuoles. Liu et al. [42] stated these enzyme activities need to increase in synchrony in order to scavenge the free radicals; otherwise the oxidative damage will not decrease.

10.3.4 PHOTOSYNTHESIS INHIBITION

The decrease in chlorophyll content by arsenic has been demonstrated. It has been confirmed that arsenic decreases leaf conductance at the stomata and mesophyll levels. A recent study has shown silicon helps to preserve the photosynthetic pathway in plants growing with arsenic [43].

10.4 PHYTOREMEDIATION

10.4.1 PRINCIPAL TECHNIQUES OF PHYTOREMEDIATION AND NATURAL ATTENUATION

10.4.1.1 Phytoextraction

Phytoextraction is the use of plant species able to accumulate significant amounts of pollutants in their aboveground tissues. The objective is to reduce efficiently trace element concentration in soils. The discovery of plant species that accumulate high amounts of essential and nonessential elements in their harvestable parts successfully directed phytoextraction to its higher development. Those plants are called hyperaccumulators and a particular pollutant can be found in a concentration higher than 1% of its dry weight. *P. vittata* was the first arsenic hyperaccumulator described. To date, hyperaccumulator plants that have been described extract a specific metal, and to a great majority only nickel, while the plant species that are able to extract copper, lead, zinc, and arsenic are less in number. Plant species need to have the following characteristics in order to consider the phytoextraction of trace elements as a profitable technique: fast growing, high biomass, multiple tolerance to metal(loid)s, absorption and translocation capacity to aboveground tissues, and easy harvest. A limiting factor is the resupply of trace elements from the solid phase of the soil to the soil solution. The activity of the metal(oid) in the soil solution and in the vicinity of the rhizosphere must be maintained for a quick resupply so that plant absorption could be high. This resupply, or the response time of a soil to the depletion of an analyte in the soil solution, is dependent on the metal speciation. In addition, the low biomass yield is one of the major limitations of this technique, influencing the total extraction time. These factors condition the selection of this technique among other quicker practices.

One of the great advances in this field has been achieved in the field of genetic engineering, aiming to reduce the physiological barriers that threaten metal(loid) translocation from roots to shoots. The goal is to obtain cultivars more resistant to toxicity, with higher biomass and greater ability to extract contaminants, so that extraction time could be reduced and this technique made more attractive from a commercial perspective. This research is primarily based on the introduction and overexpression of genes of hyperaccumulators to nonaccumulators, these genes being related to metabolism, absorption, and transport of contaminants. The arguments for accepting the genetic manipulation as a valuable practice among the scientific community are concerned to the fact that cultivars subjected to experimentation are not edible, do not reach the reproductive age, and the flowers are female and cannot produce pollen. Even though the implementation of genetically modified crops is possible from a scientific point of view, present legislation does not allow its application. Further, its necessity is questioned; gene transfer does not always result in an increased extraction capacity but understanding the physiological processes of hyperaccumulation would help to increase trace element concentrations within harvestable parts.

10.4.1.2 Phytostabilization

Phytostabilization consists in the use of plants with the aim of reducing the risk of trace elements mobilization and transfer of contaminants to other environmental compartments. Trace elements accumulation in plant roots or chemical precipitation are some of the techniques included in phytostabilization. Plants growing in contaminated soils help to fix the soil, diminishing the dispersion of trace elements by air and/or reducing the direct contact between pollutants and biota. Plant species must be multiple tolerant to metal(loid)s, have a high biomass, and exclude the contaminant.

The rhizosphere is changing continuously, modified by plant–soil–microorganism interactions. Compounds excreted by plant roots (amino acids, carboxylic acids, sugars, flavonoids, phenols) may form stable complexes with metal cations, control rhizosphere acidification, and modify contaminant availability in the soil solution. Phytoextraction has been progressively displaced by phytostabilization although phytostabilization requires longer times. It can be found in the literature that the majority of works in phytoremediation with nonaccumulator species consideres phytostabilization as the main technique. Phytostabilization uses amendments to reduce trace elements solubility and mobility, as well as to improve soil fertility for plant establishment.

10.4.1.3 Phytovolatilization

In this technique, contaminants are absorbed by plant roots, transported to leaves and volatilized into the atmosphere through the stomata. It has been demonstrated for mercury, selenium, and arsenic in the plant–microorganism system [44].

Arsenic volatilization can occur also naturally, although there is little information and data available to date are still being discussed. Arsenic biomethylation is carried out by microorganisms, but in a previous step arsenate is reduced to arsenite. Some studies have shown that in the absence of roots in the soil, the concentration of volatile arsenical compounds was up to 1% of total soil arsenic [45]. The application of compost to the soil did not increase arsenic volatilization as significantly as compared to the control [46].

10.4.1.4   Rhizofiltration

Rhizofiltration removes contaminants from water using plants. Trace elements absorption or adsorption requires a high root surface area in order to consider this technology profitable. A recent study on five *Salix* species in hydroponics showed their potential to accumulate arsenic, absorbing 50% of the total arsenic in solution [47]. In a recently reported extended work, 18 aquatic plant species growing in fresh waters from natural environment were analyzed for arsenic concentration [48]. The authors reported that the plant species belonging to the Callitrichaceae family had the ability to concentrate arsenic several orders of magnitude higher than the water (up to 2346 mg/kg in waters ranging 0.15–40.2 µg/L), showing a high potential to phytofiltrate arsenic.

10.4.2   PHYTOSTABILIZATION BASED ON AMENDMENT APPLICATION

Practices like soil removal by excavation, landfill, or soil replacement have progressively become less attractive because of high cost and/or disruptive nature. The progress of other more profitable and environmentally sustainable remediation techniques has risen a great interest. The processes that can take place adding soil amendments are based on pollutants adsorption or absorption, complexation or (co)precipitation. Given that techniques such as phytoextraction are not viable when contamination is too high, other more realistic techniques like amendment application into soils in combination with plants, that is, assisted phytostabilization, has gained attention. The choice of the ameliorants doses should be made cautiously and after conducting preliminary tests on minor scales than the field.

The number of scientific publications focused on assisted stabilization using amendments has exponentially increased, showing that this technique has greater approval due to its efficiency, the possibility of giving an added value to industrial by-products, its nondestructive nature, and its capacity to create favorable conditions for soil biological activity and to recover the physicochemical properties of soil.

The choice of the appropriate amendment depends on several factors, including soil physicochemical properties, type and level of contamination, soil organic matter content, soil lime requirements, and the future use of the target soil. Special attention should be paid to multicontaminated sites, such as those containing arsenic and metals, or inorganic and organic pollutants. It is particularly important when metals and metalloids (like arsenic) are present within the same mineral structure (parent material). Waste sludge from water depuration has been intensively used to improve soil physicochemical properties and to supply sufficient nutrients for crops. These materials are considered to be a slow release nutrient source and furthermore, they are characterized by a high content of organic matter. However, these materials contain heavy metals and when added to the soil, they contribute to the increase in the amount of metals present in that soil. The most common metals found in sewage sludge are cadmium, chromium, and nickel. In the literature, a large number of publications can be found showing the benefits of providing organic matter through these wastes but doses should be set in order to not exceed the heavy metal limit established by law. Sludges coming from municipal solid wastes are also a source of organic matter and can be found in combination with water treatment residues. Livestock production has also generated a large volume of solid and liquid wastes, like manure and poultry. More advanced processes of wastewater decontamination and changes in the diet for animal production have led to reduced level of contaminants in the wastes generated in these industries. After being processed, all organic wastes undergo subsequent analysis of their physicochemical characteristics in order to satisfy the legal standards concerning environmental parameters. The European Union regulates the use of sewage sludge in agriculture to prevent harmful effects on soil, vegetation, animals, and humans. In particular, it sets maximum values of concentrations of heavy metals and bans the spreading of sewage sludge when the concentration of certain substances in the soil exceeds these values. In the European Union, the Council Directive 86/278/ EEC [49] establishes limits for metal content in sewage sludge applied to agricultural soils. In United States, the EPA 503 rules describe the standards for the use or disposal of sewage sludge. Regarding arsenic mobilization, soluble organic compounds or soluble phosphorus contained in these materials can displace arsenic from soil exchange sites, triggering its mobility [50]. Dissolved organic carbon (DOC) is the most mobile fraction of organic ligands, and Buckingham et al. [51] described it as the fraction of organic matter that passes through a filter of 0.45 μm pore size. It is characterized by rapid mobilization that requires additional supply of organic matter in the long term to maintain appropriate levels of organic carbon, which may also imply a rapid comobilization with contaminants. It has been reported that arsenic mobilization by DOC is mainly attributed to competition between anions for retention sites in the soil, like the arsenic sorbed on to iron oxide surfaces [46].

Arsenic presents a high affinity for manganese, aluminum, and specially iron oxide surface [52]. Oxide surfaces can adsorb arsenate, forming iron arsenate (III) amorphous compounds ($FeAsO_4 \cdot H_2O$) [53]. Goethite, hematite, amorphous iron hydroxides, or elemental iron such as iron rolling mill scale are other iron sources that can efficiently retain arsenic [54,55]. The addition of iron salts such as iron sulfate (II) acidifies the growth medium, releasing protons with the formation of iron (III) hydroxide. Thus, the combination of iron sources with alkaline materials allows to mitigate the pH drop triggered by the release of protons after the application of ferrous sulfate. This indicates that the single addition of oxy-hydroxides of iron in a certain dose is not recommended since decline of pH below a certain value induces the increase of the toxic elements solubility, like aluminum, preventing plant growth and deteriorating the soil microbial activity. Hartley and Lepp [56] observed the addition of iron sulfate (II) or (III) with an alkaline amendment contributed to increase arsenic concentration into the fraction bound to iron and aluminum oxides, reducing its mobility. Recycled de-inking paper sludge has been used in recent decades to enhance agricultural yields and as an amendment in degraded soils. Paper sludge has a high content in calcium oxide, high carbon and low nitrogen, so it should be accompanied with other nutrient input. It has been utilized in combination with iron sulfate to reduce arsenic in pore water [57]. Sugar beet sludge results from sugar manufacturing, which is an alkaline material and has up to 70%–80% of calcium carbonate content. This material has also been used to retain metals in soils [58]. Compost from the solid olive-mill waste has gained attention, especially in the Mediterranean regions, where olive oil extraction industry is largely developed. It has a high lignocellulose content and is alkaline [59]. The compost from olive mill can reduce the metals' mobility due to its liming effect and it also improves soil fertility with the supply of carbon, nitrogen, and phosphorous [59]. Increased arsenic extractability has been observed, mainly explained as a consequence of the increase in pH and phosphorous [60] or to an increase in DOC after adding this compost into soils [61]. Also, a negligible effect on arsenic extractability with the olive mill compost has been reported [59]. The mushroom industries produce by-products like spent mushroom substrate (the unutilized substrate and the mushroom mycelium left after harvest) and are in continuous expansion, with their residue volume increasing year after year. This by-product contains high levels of humic substances that can adsorb analytes from the soil solution and has a high affinity for metal cations. Arsenic and metals retention effects have been reported [62,63].

Recently the use of *biochar* has received special attention because of its large amount of organic carbon content. Although the main application of biochar has originally been described as carbon sequestration, it has revealed a great potential to significantly modify the solubility, mobility, transport and spatial distribution of metals. Beesley and Marmiroli [64] defined biochar as a low-density material obtained by burning biomass at low temperature under low oxygen concentrations. During the biochar carbonization its surface area can be increased several thousand times. Because of this feature, biochar application can improve soil physicochemical properties, favoring the adsorption of nutrients, especially calcium, phosphorous, zinc, and manganese [65] and increasing soil porosity and water retention capacity. Regarding the biochar–arsenic relationship, contradictory effects have been reported. Hartley et al. [66] observed that the input of biochar to multicontaminated soils increased the soluble fraction of iron and carbon in the soil solution, which contributed to higher arsenic concentration in the soil soluble phase. Brennan et al. [67] observed differences depending on the type of biochar, but there was also an increase of extractable arsenic, partially explained by the arsenic presence in the material. Also, Kloss et al. [68] found significant increase of extractable arsenic in soil leachate. Generally, the elucidation for this behavior is related to the fact that biochar causes a liming effect triggering soil functional groups deprotonation and charging soil particles negatively. Gregory et al. [69] observed a temporal variation of water extractable arsenic with biochar, not finding significant differences at the end of the experiment compared to the control. It is advisable to study biochar retention capacity over time, as exchange soil positions may be easily occupied by other soil components, such as organic matter. In addition, as with all amendments, but more especially because of the low density of biochar, attention should be paid to not confound the observed effects with the dilution effect of the contaminated soil amended with biochar. New advances in the research of biochar have been made. A recent study on biochar impregnated with iron salt showed a high ability to retain water-soluble arsenic controlled by a chemisorption mechanism [70].

REFERENCES

1. Zhao, F.J., J.F. Ma, A.A. Meharg, and S.P. McGrath. Arsenic uptake and metabolism in plants. *New Phytologist* 181 (2009): 777–794.

2. Ullrich-Eberius, C.I, A. Sanz, and J. Novacky. "Evaluation of Arsenate- and Vanadate-Associated Changes of Electrical Membrane Potential and Phosphate Transport in Lemna Gibba-G1." *Journal of Experimental Botany* 40 (1989): 119–128.

3. Nussaume, L., S. Kanno, H. Javot, E. Marin, N. Pochon, A. Ayadi et al. Phosphate import in plants: Focus on the PHT1 transporters. *Frontiers in Plant Science* 2 (2011): 1–12.

4. Isayenkov, S.V. and F.J.M. Maathuis. The *Arabidopsis thaliana* Aquaglyceroporin AtNIP7;1 Is a pathway for arsenite uptake. *FEBS Letters* 582 (2008): 1625–1628.

5. Ma, J.F., N. Yamaji, N. Mitani, X-Y. Xu, Y-H. Su, S.P. McGrath et al. Transporters of arsenite in rice and their role in arsenic accumulation in rice grain. *Proceedings of the National Academy of Sciences of the United States of America* 105 (2008): 9931–9935.

6. Zhao, F.-J., Y. Ago, N. Mitani, R.-Y. Li, Y.-H. Su, N. Yamaji et al. The role of the rice aquaporin Lsi1 in arsenite efflux from roots. *The New Phytologist* 186 (2010): 392–399.

7. Ma, J.F., N. Yamaji, N. Mitani, K. Tamai, S. Konishi, T. Fujiwara et al. An efflux transporter of silicon in rice. *Nature* 448 (2007): 209–212.

8. Tiwari, M., D. Sharma, S. Dwivedi, M. Singh, R. Deo Tripathi, and P.K. Trivedi. Expression in *Arabidopsis* and cellular localization reveal involvement of rice NRAMP, OsNRAMP1, in arsenic transport and tolerance. *Plant, Cell and Environment* 37 (2014): 140–152.

9. Li, R.-Y., Y. Ago, W.-J. Liu, N. Mitani, J. Feldmann, S.P. McGrath et al. The rice aquaporin Lsi1 mediates uptake of methylated arsenic species. *Plant Physiology* 150 (2009): 2071–2080.

10. Zangi, R. and M. Filella. Transport routes of metalloids into and out of the cell: A review of the current knowledge. *Chemico–Biological Interactions* 197 (2012): 47–57.

11. Caille, N., F.J. Zhao, and S.P. McGrath. Comparison of root absorption, translocation and tolerance of arsenic in the hyperaccumulator *Pteris vittata* and the nonhyperaccumulator *Pteris tremula*. *New Phytologist* 165 (2005): 755–761.

12. Liu, W.-J., B. Alan Wood, A. Raab, S.P. McGrath, F.-J. Zhao, and J. Feldmann. Complexation of arsenite with phytochelatins reduces arsenite efflux and translocation from roots to shoots in *Arabidopsis*. *Plant Physiology* 152 (2010): 2211–2221.

13. Song, W.-Y., J. Park, D.G. Mendoza-Cózatl, M. Suter-Grotemeyer, D. Shim, S. Hörtensteiner et al. Arsenic tolerance in *Arabidopsis* is mediated by two ABCC-type phytochelatin transporters. *Proceedings of the National Academy of Sciences of the United States of America* 107 (2010): 21187–21192.

14. Moore, K.L., M. Schröder, Z. Wu, B.G.H. Martin, C.R. Hawes, S.P. McGrath et al. High-resolution secondary ion mass spectrometry reveals the contrasting subcellular distribution of arsenic and silicon in rice roots. *Plant Physiology* 156 (2011): 913–924.

15. Duan, G.L., Y.G. Zhu, Y.P. Tong, C. Cai, and R. Kneer. Characterization of arsenate reductase in the extract of roots and fronds of Chinese brake fern, an arsenic hyperaccumulator. *Plant Physiology* 138 (2005):461–469.

16. Su, Y.H., S.P. McGrath, Y.G. Zhu, and F.J. Zhao. Highly efficient xylem transport of arsenite in the arsenic hyperaccumulator *Pteris vittata*. *New Phytologist* 180 (2008): 434–441.

17. Indriolo, E., G.N. Na, D. Ellis, D.E. Salt, and J.A. Banks. A vacuolar arsenite transporter necessary for arsenic tolerance in the arsenic hyperaccumulating fern *Pteris vittata* is missing in flowering plants. *The Plant Cell* 22 (2010): 2045–2057.

18. Xu, X.Y., S.P. McGrath, and F.J. Zhao. Rapid reduction of arsenate in the medium mediated by plant roots. *New Phytologist* 176 (2007): 590–599.

19. Catarecha, P., M.D. Segura, J.M. Franco-Zorrilla, B. García-Ponce, M. Lanza, R. Solano et al. A mutant of the *Arabidopsis* phosphate transporter PHT1;1 displays enhanced arsenic accumulation. *The Plant Cell* 19 (2007): 1123–1133.

20. Kopittke, P.M, M.D. de Jonge, P. Wang, B.A. McKenna, E. Lombi, D.J. Paterson et al. Laterally resolved speciation of arsenic in roots of wheat and rice using fluorescence-XANES imaging. *New Phytologist* 201 (2014): 1251–1262.

21. Batista, B.L., M. Nigar, A. Mestrot, B. Alves Rocha, F. Barbosa Júnior, A.H. Price et al. Identification and quantification of phytochelatins in roots of rice to long-term exposure: Evidence of individual role on arsenic accumulation and translocation. *Journal of Experimental Botany* 65 (2014): 1467–1479.

22. Duan, G.-L., Y. Hu, W.-J. Liu, R. Kneer, F.-J. Zhao, and Y.-G. Zhu. Evidence for a role of phytochelatins in regulating arsenic accumulation in rice grain. *Environmental and Experimental Botany* 71 (2011): 416–421.

23. Lei, M., X.-M. Wan, Z.-C. Huang, T.-B. Chen, X.-W. Li, and Y.-R. Liu. First evidence on different transportation modes of arsenic and phosphorus in arsenic hyperaccumulator *Pteris vittata*. *Environmental Pollution* 161 (2012): 1–7.

24. Mendoza-Cózatl, D.G., T.O. Jobe, F. Hauser, and J.I. Schroeder. Long-distance transport, vacuolar sequestration, tolerance, and transcriptional responses induced by cadmium and arsenic. *Current Opinion in Plant Biology* 14 (2011): 554–562.

25. Ye, W.-L., B. Alan Wood, J.L. Stroud, P. John Andralojc, A. Raab, S.P. McGrath et al. Arsenic speciation in phloem and xylem exudates of castor bean. *Plant Physiology* 154 (2010): 1505–1513.

26. Yamaji, N., Y. Chiba, N. Mitani-Ueno, and J.F. Ma. Functional characterization of a silicon transporter gene implicated in silicon distribution in barley. *Plant Physiology* 160 (2012): 1491–1497.

27. Kutrowska, A. and M. Szelag. Low-molecular weight organic acids and peptides involved in the long-distance transport of trace metals. *Acta Physiologiae Plantarum* 36 (2014): 1957–1968.

28. Zeng, X., S. Su, Q. Feng, X. Wang, Y. Zhang, L. Zhang et al. Arsenic speciation transformation and arsenite influx and efflux across the cell membrane of fungi investigated using HPLC–HG–AFS and in-situ XANES. *Chemosphere* 119 (2015): 1163–1168.

29. González-Chávez, M.D.C.A., B. Miller, I.E. Maldonado-Mendoza, K. Scheckel, and R. Carrillo-González. Localization and speciation of arsenic in *Glomus Intraradices* by synchrotron radiation spectroscopic analysis. *Fungal Biology* 118 (2014): 444–452.

30. Zhang, X., A.-J. Lin, F.-J. Zhao, G.-Z. Xu, G.-L. Duan, and Y.-G. Zhu. Arsenic accumulation by the aquatic fern azolla: Comparison of arsenate uptake, speciation and efflux by *A. Caroliniana* and *A. Filiculoides*. *Environmental Pollution* 156 (2008): 1149–1155.

31. Logoteta, B., X.Y. Xu, M.R Macnair, S.P. McGrath, and F.J. Zhao. Arsenite efflux is not enhanced in the arsenate-tolerant phenotype of *Holcus lanatus*. *New Phytologist* 183 (2009): 340–348.

32. Carbonell-Barrachina, A.A., M.A. Aarabi, R.D. Delaune, R.P. Gambrell, and W.H. Patrick Jr. The influence of arsenic chemical form and concentration on *Spartina patens* and *Spartina alterniflora* growth and tissue arsenic concentration. *Plant & Soil* (1998): 33–43.

33. Singh, N. and L.Q. Ma. Arsenic speciation, and arsenic and phosphate distribution in arsenic hyperaccumulator *Pteris vittata* L. and non-hyperaccumulator *Pteris ensiformis* L. *Environmental Pollution* 141 (2006): 238–246.

34. Wang, L. and G. Duan. Effect of external and internal phosphate status on arsenic toxicity and accumulation in rice seedlings. *Journal of Environmental Sciences* 21 (2009): 346–351.

35. Mascher, R., B. Lippmann, S. Holzinger, and H. Bergman, H. Arsenate toxicity: Effects on oxidative stress response molecules and enzymes in red clover plants. *Plant Science* 163 (2002):961–969.

36. Shaibur, M.R., N. Kitajima, R. Sugawara, T. Kondo, S. Alam, S.M. Imamul Huq et al. Critical toxicity level of arsenic and elemental composition of arsenic-induced chlorosis in hydroponic sorghum. *Water, Air, and Soil Pollution* 191 (2008): 279–92.

37. Tamaki, S. and W.T. Frankenberger Jr. Environmental biochemistry of arsenic. *Reviews of Environmental Contamination and Toxicology* 124 (1992): 79–110.

38. Wang, L.-H., X.-Y. Meng, B. Guo, and G.-L. Duan. Reduction of arsenic oxidative toxicity by phosphate is not related to arsenate reductase activity in wheat plants. *Journal of Plant Nutrition* 30 (2007): 2105–2117.

39. Shri, M., S. Kumar, D. Chakrabarty, P.K. Trivedi, S. Mallick, P. Misra et al. Effect of arsenic on growth, oxidative stress, and antioxidant system in rice seedlings. *Ecotoxicology and Environmental Safety* 72 (2009): 1102–1110.

40. Foyer, C.H., G. Noctor. Oxidant and antioxidant signalling in plants: A re-evaluation of the concept of oxidative stress in a physiological context. *Plant, Cell and Environment* 28 (2005): 1056–1071.

41. Kim, K.T., S.J. Klaine, J. Cho, S.-H. Kim, and S.D. Kim. Oxidative stress responses of daphnia magna exposed to TiO_2 nanoparticles according to size fraction. *The Science of the Total Environment* 408 (2010): 2268–2272.

42. Liu, X., S. Zhang, X.-Q. Shan, and P. Christie. Combined toxicity of cadmium and arsenate to wheat seedlings and plant uptake and antioxidative enzyme responses to cadmium and arsenate co-contamination. *Ecotoxicology and Environmental Safety* 68 (2007): 305–313.

43. Sanglard, L.M.V.P., S.C.V. Martins, K.C. Detmann, P.E.M. Silva, A.O. Lavinsky, and M.M. Silva. Silicon nutrition alleviates the negative impacts of arsenic on the photosynthetic apparatus of rice leaves: An analysis of the key limitations of photosynthesis. *Physiologia Plantarum* 152 (2014): 355–366.

44. Brooks, R.R. Phytoremediation by volatilisation. In Brooks, R.R. ed. *Plants That Hyperaccumulate Heavy Metals: Their Role in Phytoremediation, Microbiology, Archaeology, Mineral Exploration and Phytomining*, pp. 289–312. Wallingford, UK, CAB International, 1998.

45. Prohaska, T., M. Pfeffer, M. Tulipan, G. Stingeder, A. Mentler, and W.W. Wenzel. Speciation of arsenic of liquid and gaseous emissions from soil in a microcosmos experiment by liquid and gas chromatography with inductively coupled plasma mass spectrometer (ICP-MS) detection. *Fresenius' Journal of Analytical Chemistry* 364 (1999): 467–470.

46. Moreno-Jiménez, E., R. Clemente, A. Mestrot, and A.A. Meharg. Arsenic and selenium mobilisation from organic matter treated mine spoil with and without inorganic fertilisation. *Environmental Pollution* 173 (2013): 238–244.

47. Chen, G., X. Zou, Y. Zhou, J. Zhang, and G. Owens. A short-term study to evaluate the uptake and accumulation of arsenic in Asian willow (*Salix* Sp.) from arsenic-contaminated water. *Environmental Science and Pollution Research International* 21 (2014): 3275–3284.

48. Favas, P.J.C., J. Pratas, and M.N.V. Prasad. Accumulation of arsenic by aquatic plants in large-scale field conditions: Opportunities for phytoremediation and bioindication. *Science of the Total Environment* 433 (2012): 390–397.

49. Council Directive 86/278/EEC of 12 June 1986 on the protection of the environment, and in particular of the soil, when sewage sludge is used in agriculture. *Official Journal of the European Communities* 181 (1986): 6–12.

50. Wang, S. and C.N. Mulligan. Effect of natural organic matter on arsenic release from soils and sediments into groundwater. *Environmental Geochemistry and Health* 28 (2006): 197–214.

51. Buckingham, S., E. Tipping, and J. Hamilton-Taylor. Dissolved organic carbon in soil solutions: A comparison of collection methods. *Soil Use and Management* 24 (2008): 29–36.

52. Smedley, P. and D. G. Kinniburgh. A review of the source, behaviour and distribution of arsenic in natural waters. *Applied Geochemistry* 17 (2002): 517–568.

53. Carlson, L., J.M. Bigham, U. Schwertmann, A. Kyek, and F. Wagner. Scavenging of as from acid mine drainage by schwertmannite and ferrihydrite: A comparison with synthetic analogues. *Environment Science and Technology* 36 (2002): 1712–1719.

54. Artiola, J.F., D. Zabcik, and S.H. Johnson. In situ treatment of arsenic contaminated soil from a hazardous industrial site: Laboratory studies. *Waste Management* 10 (1990): 73–78.

55. Vangronsveld, J., R. Herzig, N. Weyens, J. Boulet, K. Adriaensen, A. Ruttens et al. Phytoremediation of contaminated soils and groundwater: Lessons from the field. *Environmental Science and Pollution Research* 16 (2009): 765–794.

56. Hartley, W. and N.W. Lepp. Effect of in situ soil amendments on arsenic uptake in successive harvests of ryegrass (*Lolium perenne* Cv Elka) grown in amended as-polluted soils. *Environmental Pollution* 156 (2008): 1030–1040.

57. Manzano, R., J.M. Peñalosa, and E. Esteban. Amendment application in a multicontaminated mine soil: Effects on trace element mobility. *Water, Air, & Soil Pollution* 225 (2014): 1–10.

58. Alvarenga, P., P. Palma, A.P. Gonçalves, N. Baião, R. M. Fernandes, A. de Varennes et al. Assessment of chemical, biochemical and ecotoxicological aspects in a mine soil amended with sludge of either urban or industrial origin. *Chemosphere* 72 (2008): 1774–1781.

59. Alburquerque, J.A., C. de la Fuente, and M.P. Bernal. Improvement of soil quality after 'Alperujo' compost application to two contaminated soils characterised by differing heavy metal solubility. *Journal of Environmental Management* 92 (2011): 733–741.

60. Pardo, T., R. Clemente, and M. Pilar Bernal. Effects of compost, pig slurry and lime on trace element solubility and toxicity in two soils differently affected by mining activities. *Chemosphere* 84 (2011): 642–650.

61. Beesley, L., O.S. Inneh, G.J. Norton, E. Moreno-Jimenez, T. Pardo, R. Clemente, and J.J.C. Dawson. Assessing the influence of compost and biochar amendments on the mobility and toxicity of metals and arsenic in a naturally contaminated mine soil. *Environmental Pollution* 186 (2014): 195–202.

62. Chen G.Q., G.M. Zeng, X. Tu, G.H., Huang , Y.N. Cheng. A novel biosorbent: characterization of the spent mushroom compost and its application for removal of heavy metals. *Journal of Environmental Sciences* 17 (2005): 756–760.

63. Koo, N., S.-H. Lee, J.-G. Kim, Arsenic mobility in the amended mine tailings and its impact on soil enzyme activity. *Environmental Geochemistry and Health* 34 (2012): 337–348.

64. Beesley, L. and M. Marmiroli. The immobilisation and retention of soluble arsenic, cadmium and zinc by biochar. *Environmental Pollution* 159 (2011): 474–480.

65. Laird, D., P. Fleming, B. Wang, R. Horton, and D. Karlen. Biochar impact on nutrient leaching from a midwestern agricultural soil. *Geoderma* 158 (2010): 436–442.

66. Hartley, W., N.M. Dickinson, P. Riby, and N.W. Lepp. Arsenic mobility in brownfield soils amended with green waste compost or biochar and planted with *Miscanthus*. *Environmental Pollution* 157 (2009): 2654–2662.

67. Brennan, A., E.M. Jiménez, M. Puschenreiter, J.A. Alburquerque, and C. Switzer. Effects of biochar amendment on root traits and contaminant availability of maize plants in a copper and arsenic impacted soil. *Plant and Soil* 379 (2014): 351–360.

68. Kloss, S., F. Zehetner, E. Oburger, J. Buecker, B. Kitzler, W.W. Wenzel et al. Trace element concentrations in leachates and mustard plant tissue (*Sinapis alba* L.) after biochar application to temperate soils. *Science of the Total Environment* 481 (2014): 498–508.

69. Gregory, S.J., C.W.N. Anderson, M. Camps Arbestain, and M.T. McManus. Response of plant and soil microbes to biochar amendment of an arsenic-contaminated soil. *Agriculture, Ecosystems & Environment* 191 (2014): 133–141.

70. Hu, X., Z. Ding, A.R. Zimmerman, S. Wang, and B. Gao. Batch and column sorption of arsenic onto iron-impregnated biochar synthesized through hydrolysis. *Water Research* 68 (2015): 206–216.

11 Mitigation of Arsenic Toxicity by Plant Products

Tetsuro Agusa

CONTENTS

11.1 INTRODUCTION

Arsenic (As) is widely distributed in the environment. On the other hand, it is well known that As and its compounds are carcinogens. However, the specific mechanism of induction of carcinogenesis by As is still unknown due to its complexity and the scarcity of model experimental animals. Probably, oxidative stress caused by inorganic As (IAs) as well as some organoarsenic compounds (e.g., dimethylarsinic acid [DMA(V)] and dimethylarsinous acid [DMA(III)]) is one of the triggers that can induce cancers [1–3]. Indeed, DNA damage by oxidative stress was observed in humans exposed to IA through the consumption of contaminated groundwater [4–6]. Arsenite (As(III)) and arsenate (As(V)) are highly toxic compounds, but the

235

induction mechanisms of toxicity are different among species; As(III) readily binds to sulfhydryl group of various enzymes inhibiting the activities, whereas As(V) substitutes phosphate due to the similarity of their structures, leading to disruption of metabolic reactions related to phosphorylation [7]. Symptoms of acute intoxication in humans by IA include severe gastrointestinal disorders, hepatic and renal failure, and cardiovascular disturbances, whereas chronic exposure causes skin pigmentation, hyperkeratosis, and cancers in the lung, bladder, liver, kidney, and skin [8,9]. According to the Agency for Toxic Substances and Disease Registry [10], As has been ranked at the top of the substance priority list due to its severe toxicity on human health [11].

Arsenic pollution in groundwater, with concentrations above the WHO guidelines value (10 µg/L) for drinking [11], has been widely reported in the world [12]. The main source of As is naturally derived from aquifer sediment. In the alluvial/lacustrine aquifers of many inland basins and deltaic systems in the South and Southeast Asia, high concentrations of As have been detected [13]. Particularly, a significant risk to about 36 million people in the Bengal Delta consisting of Bangladesh and the state of West Bengal, India, by As exposure through the consumption of groundwater is of great concern [14]. In Latin America, it is estimated that the number of As-exposed people is about 14 million [15]. Human epidemiological studies conducted in those As-affected areas have reported significant increase cases of skin lesions, cancers, Blackfoot disease, and vascular disease by As exposure [14–20]. Furthermore, chronic As exposure has caused excess spontaneous abortion, stillbirth, and preterm birth in pregnant women [21,22].

Therefore, the establishment of a functional system to reduce or prevent further As exposure in human is an urgent task in the As-contaminated groundwater areas, especially in the developing countries. In addition, effective drug discovery for patients with arsenicosis is required. Particularly, those should be simple and available at low price in the developing countries.

Removal of As by filtration of raw groundwater or development of alternative water resources for drinking has been introduced in some places with As contamination. These methods are effective [23–26], but it may be difficult to cover the areas of As contamination entirely by these methods.

As another approach, As elimination from the body of people exposed daily to As through the consumption of groundwater is considered. For this objective, chelating agents might be useful. It is known that As has been used as poison for many years and chemical warfare agents like dichlorovinyl arsine (Lewisite) were used during the World War II. Therefore, there have been continuing efforts to find antidotes for these chemicals. The studies on treatment of As toxicity are restricted mainly to some sulfhydryl chelating agents such as meso 2,3-dimercaptosuccinic acid (DMSA), 2,3-dimercaptopropane-1-sulfonate (DMPS), or 2,3-dimercaprol (British anti-Lewisite [BAL]) [27,28] administered either individually or in combination with antioxidants such as vitamin C, vitamin E [29,30], or N-acetyl cysteine [31], and some micronutrients like zinc and selenium [32]. Most of the conventional metal chelating agents and antioxidants have been reported to possess toxic side effects or disadvantages [33,34].

Therefore, natural materials like plants with antioxidant properties to potentially reduce As toxicity as well as with no side effects are of great concern [35,36]. In this chapter, previous studies regarding natural therapeutic drugs against As toxicity are reviewed. Here, for comparison, detailed results of those studies are summarized in Table 11.1.

11.2 TEA

Tea (*Camellia sinesis*), which is one of the most commonly consumed beverages in the world, contains high concentrations of polyphenols. Tea polyphenols have antioxidative property, leading to potential human health benefits [37]. Green and black teas may protect against several types of cancers and cardiovascular diseases [38].

As(III) or As(V) as well as tea or polyphenols ((−)-epigallocatechingallate and theaflavin) were cotreated in Chinese hamster male lung fibroblast cells (CH V-99) and then the cytotoxicity was assessed [39] and found that tea and polyphenols decreased cytotoxicity by As compounds. Later, the same research group confirmed that increased chromosomal aberrations and catalase (CAT) and superoxide dismutase (SOD) activities by As(III) were also reduced by teas, especially Darjeeling tea [40]. Furthermore, they have investigated the effect of tea on micronuclei formation by As(III), As(V), and DMA(III) [41]. Although statistical significances of the results were not clearly shown, the authors have mentioned that teas and the polyphenols reduced the frequency of micronuclei increased by As compounds, while DNA repair activity was also enhanced [41]. In similar assays, teas and polyphenols inhibited DNA damage caused by As species, but statistical results were not indicated [42].

It was found in an *in vitro* study using lymphocyte from healthy human blood that teas and their polyphenols afforded reduction of As-induced DNA damage and oxidative stress [43]. In addition, reduced protein expression level of DNA repair enzyme, poly (ADP-ribose) polymerase (PARP) by As treatment was significantly rescued by teas and polyphenols [43].

There are several *in vivo* studies using mouse, rat, and rabbit on the protective role of tea against As toxicity. Mukherjee et al. [44] tested effects of black tea in reducing the chromosomal damage caused by As(III) in mice. As(III) exposure induced chromosomal aberrations in bone marrow cells, while black tea treatment reduced the aberrations [44]. Similar results were observed in Swiss albino mice concomitantly treated with As(III) and black tea [45]. Sinha et al. [46] investigated the role of teas against the oxidative damage caused by As on DNA, protein, and lipid in Swiss albino mice. Increased DNA damage, reactive oxygen species (ROS), lipid peroxidation, and protein carbonyl by As were effectively reduced with tea treatments. On the other hand, As-induced depletion of antioxidants were protected by teas, suggesting that tea-induced antioxidants provide further protection against oxidative stress caused by As [46].

Oxidative stress in blood of New Zealand rabbits treated with arsenic trioxide for 14 days were significantly removed by posttreatment of green and black tea [47]. In addition, remediation capacity of green tea was higher than that of black tea, resulting from slightly higher content of polyphenols in green tea [47].

TABLE 11.1
Results of Studies on Mitigation Effect of As Toxicity by Natural Plants

Plant Material	References	Cell/Animal	Experiment	Measurement	Results
Tea	[39]	Chinese hamster male lung fibroblast cells (CH V-79)	[Arsenite, 500 μM, for 24 h] + [GT, 0–200 μg/mL, for 24 h] or [DT, 0–200 μg/mL, for 24 h] or [AT, 0–200 μg/mL, for 24 h] or [EGCG, 0–50 μM, for 24 h] or [TF; 0–100 μM, for 24 h]	Cytotoxicity	Induced cytotoxicity was suppressed by tea and polyphenols (Significant results are not shown)
			[Arsenate, 500 μM, for 24 h] + [GT, 0–200 μg/mL, for 24 h] or [DT, 0–200 μg/mL, for 24 h] or [AT, 0–200 μg/mL, for 24 h] or [EGCG, 0–50 μM, for 24 h] or [TF; 0–100 μM, for 24 h]	Cytotoxicity	Induced cytotoxicity was suppressed by tea and polyphenols (Significant results are not shown)
	[40]	Chinese hamster lung fibroblast cells (CH V79)	[Arsenite, 500 μM, for 1 h] + [GT, 0–200 μg/mL, for 1 h] or [DT, 0–200 μg/mL, for 1 h] or [AT, 0–200 μg/mL, for 1 h]	Chromosomal aberration	Induced chromosomal aberration was suppressed by tea, especially DT (Significant results are not shown)
			[[GT, 0–200 μg/mL, for 1 h] or [DT, 0–200 μg/mL, for 1 h] or [AT, 0–200 μg/mL, for 1 h]] → [arsenite, 500 μM, for 1 h]	Chromosomal aberration	Induced chromosomal aberration was suppressed by tea, especially DT (Significant results are not shown)
			[Arsenite, 500 μM, for 1 h] → [control, for 2 h] or [GT, 200 μg/mL, for 2 h] or [DT, 200 μg/mL, for 2 h] or [AT, 200 μg/mL, for 2 h]	% of DNA repair activity	Repair activity was enhanced by tea (Significant results are not shown)
			[Arsenite, 500 μM, for 1 h] → [control, for 4 h] or [GT, 200 μg/mL, for 4 h] or [DT, 200 μg/mL, for 4 h] or [AT, 200 μg/mL, for 4 h]	% of DNA repair activity	Repair activity was enhanced by tea (Significant results are not shown)

(Continued)

TABLE 11.1 (Continued)
Results of Studies on Mitigation Effect of As Toxicity by Natural Plants

Plant Material	References	Cell/Animal	Experiment	Measurement	Results
			[Control, for 1 h]+[GT, 200 µg/mL, for 1 h] or [DT, 200 µg/mL, for 1 h] or [AT, 200 µg/mL, for 1 h]	CAT	Reduced CAT was raised by tea
			[Arsenite, 500 µM, for 1 h]+[GT, 200 µg/mL, for 1 h] or [DT, 200 µg/mL, for 1 h] or [AT, 200 µg/mL, for 1 h]	SOD	Reduced SOD was raised by tea
	[41]	Chinese hamster lung fibroblast cells (CH V79)	[Arsenite, 50–500 µM, for 1 h] or [arsenate, 50–500 µM, for 1 h] or [dimethylarsinic acid, 50–500 µM, for 1 h]+[GT, 0–200 µg/mL, for 1 h] or [DT, 0–200 µg/mL, for 1 h] or [AT, 0–200 µg/mL, for 1 h] or [EGCG, 0–100 µM, for 1 h] or [TF, 0–200 µM, for 1 h]	MN frequency	Induced MN frequency was suppressed by tea and polyphenols
			[GT, 0–200 µg/mL, for 1 h] or [DT, 0–200 µg/mL, for 1 h] or [AT, 0–200 µg/mL, for 1 h] or [EGCG, 0–100 µM, for 1 h] or [TF, 0–200 µM, for 1 h] → [arsenite, 500 µM, for 1 h] or [arsenate, 500 µM, for 1 h] or [dimethylarsinic acid, 500 µM, for 1 h]	MN frequency	Induced MN frequency was suppressed by tea and polyphenols
			[Arsenite, 500 µM, for 1 h] or [arsenate, 500 µM, for 1 h] or [dimethylarsinic acid, 500 µM, for 1 h] → [GT, 0–200 µg/mL, for 2 h] or [DT, UK µg/mL, for 2 h] or [AT, UK µg/mL, for 2 h] or [EGCG, UK µM, for 2 h] or [TF, UK µM, for 2 h]	% of DNA repair activity	Repair activity was enhanced by tea and polyphenols

(Continued)

TABLE 11.1 (Continued)

Results of Studies on Mitigation Effect of As Toxicity by Natural Plants

Plant Material	References	Cell/Animal	Experiment	Measurement	Results
			[Arsenite, 500 µM, for 1 h] or [Arsenate, 500 µM, for 1 h] or [dimethylarsinic acid, 500 µM, for 1 h] → [GT, 0–200 µg/mL, for 4 h] or [DT, UK µg/mL, for 4 h] or [AT, UK µg/mL, for 4 h] or [EGCG, UK µM, for 4 h] or [TF, UK µM, for 4 h]	% of DNA repair activity	Repair activity was enhanced by tea and polyphenols
	[42]	Chinese hamster lung fibroblast cells (CH V79)	[Arsenite, 100–1000 µM, for 1 h] or [Arsenate, 100–1000 µM, for 1 h] or [dimethylarsinic acid, 100–1000 µM, for 1 h] + [GT, 0–200 µg/mL, for 1 h] or [DT, 0–200 µg/mL, for 1 h] or [AT, 0–200 µg/mL, for 1 h] or [EGCG, 0–100 µM, for 1 h] or [TF, 0–200 µg/mL, for 1 h]	Comet tail moment	Induced chromosomal aberration was suppressed by teas and polyphenols (Significant results are not shown)
			[GT, 0–200 µg/mL, for 1 h] or [DT, 0–200 µg/mL, for 1 h] or [AT, 0–200 µg/mL, for 1 h] or [EGCG, 0–100 µM, for 1 h] or [TF, 0–200 µg/mL, for 1 h] → [arsenite, 1000 µM, for 1 h] or [arsenate, 1000 µM, for 1 h] or [dimethylarsinic acid, 1000 µM, for 1 h]	Comet tail moment	Induced chromosomal aberration was suppressed by teas and polyphenols (Significant results are not shown)
			[Arsenite, 1000 µM, for 1 h] or [arsenate, 1000 µM, for 1 h] or [dimethylarsinic acid, 1000 µM, for 1 h] → [GT, 200 µg/mL, for 2 h] or [DT, 200 µg/mL, for 2 h] or [AT, 200 µg/mL, for 2 h] or [EGCG, 100 µM, for 2 h] or [TF, 200 µg/mL, for 2 h]	% of DNA repair activity	Repair activity was enhanced by teas and polyphenols (Significant results are not shown)

(Continued)

TABLE 11.1 (Continued)

Results of Studies on Mitigation Effect of As Toxicity by Natural Plants

Plant Material	References	Cell/Animal	Experiment	Measurement	Results
	[43]	Human lymphocytes	[Arsenite, 1000 μM, for 1 h] or [arsenate, 1000 μM, for 1 h] or [dimethylarsinic acid, 1000 μM, for 1 h] → [GT, 200 μg/mL, for 4 h] or [DT, 200 μg/mL, for 4 h] or [AT, 200 μg/mL, for 4 h] or [EGCG, 100 μM, for 4 h] or [TF, 200 μg/mL, for 4 h]	[3H]T in corporation	Repair activity was enhanced by teas and polyphenols (Significant results are not shown)
			[Arsenite, 1000 μM, for 1 h] + [EGCG, 0–100 μM, for 1 h] or [TF, 0–200 μg/mL, for 1 h]	Comet tail	Induced comet tail was dose-dependently suppressed by polyphenol
			[EGCG, 0–15 μM, for 24 h] or [TF, 0–15 μg/mL, for 24 h] → [arsenite, 1000 μM, for 1 h]	Comet tail	Induced comet tail was dose-dependently suppressed by polyphenol
			[Arsenite, 1000 μM, for 1 h] + [GT, 0–200 μg/mL, for 1 h] or [DT, 0–200 μg/mL, for 1 h] or [AT, 0–200 μg/mL, for 1 h]	Comet tail	Induced comet tail was dose-dependently suppressed by polyphenol
			[GT, 0–200 μg/mL, for 24 h] or [DT, 0–200 μg/mL, for 24 h] or [AT, 0–200 μg/mL, for 24 h] → [arsenite, 1000 μM, for 1 h]	Comet tail	Induced comet tail was dose-dependently suppressed by polyphenol
			G1; control	ROS	G2>G1; G2>G3; G2>G4; G2>G5; G2>G6; G2>G7
			G2; arsenite, 1000 μM, for 1 h	Lipid peroxidation	G2>G1; G2>G3; G2>G4; G2>G5; G2>G6; G2>G7
			G3; G2+[EGCG, 100 μM, for 1 h] G4; G2+[TF, 200 μg/mL, for 1 h]		

(Continued)

TABLE 11.1 (*Continued*)
Results of Studies on Mitigation Effect of As Toxicity by Natural Plants

Plant Material	References	Cell/Animal	Experiment	Measurement	Results
			G5; G2 + [GT, 200 µg/mL, for 1 h] G6; G2 + [DT, 200 µg/mL, for 1 h] G7; G2 + [AT, 200 µg/mL, for 1 h] G1-1; control	CAT	G1-1>G1-2; G2-1>G1-1; G3-1>G1-1; G4-1>G1-1; G5-1>G1-1; G6-1>G1-1; G2-1>G2-2; G3-1>G3-2; G4-1>G4-2; G5-1>G5-2; G6-1>G6-2
			G1-2; arsenite, 1000 µM, for 1 h	SOD	G1-1>G1-2; G3-1>G1-1; G2-1>G2-2; G3-1>G3-2; G4-1>G4-2; G5-1>G5-2; G6-1>G6-2
			G2-1; EGCG, 100 µM, for 1 h	GPx	G1-1>G1-2; G2-1>G1-1; G2-1>G2-2; G3-1>G3-2; G4-1>G4-2; G5-1>G5-2; G6-1>G6-2
			G2-2; G1-2 + [EGCG, 100 µM, for 1 h] G3-1; TF, 200 µg/mL, for 1 h G3-2; G1-2 + [TF, 200 µg/mL, for 1 h] G4-1; GT, 200 µg/mL, for 1 h G4-2; G1-2 + [GT, 200 µg/mL, for 1 h] G5-1; DT, 200 µg/mL, for 1 h		

(Continued)

TABLE 11.1 (*Continued*)
Results of Studies on Mitigation Effect of As Toxicity by Natural Plants

Plant Material	References	Cell/Animal	Experiment	Measurement	Results
			G5-2; G1-2 + [DT, 200 μg/mL, for 1 h]		
			G6-1; AT, 200 μg/mL, for 1 h		
			G6-2; G1-2 + [AT, 200 μg/mL, for 1 h]		
			G1; arsenite, 1000 μM, for 1 h	% of DNA repair activity	G2>G1; G3>G1; G4>G1; G5>G1; G6>G1
			G2; G1 → [EGCG, 100 μM, for 2 h]	[H^3] thymidine incorporation	G2>G1; G3>G1; G4>G1; G5>G1; G6>G1
			G3; G1 → [TF, 200 μg/mL, for 2 h]		
			G4; G1 → [GT, 200 μg/mL, for 2 h]		
			G5; G1 → [DT, 200 μg/mL, for 2 h]		
			G6; G1 → [AT, 200 μg/mL, for 2 h]		
			G1; arsenite, 1000 μM, for 1 h	% of DNA repair activity	G2>G1; G3>G1; G4>G1; G5>G1; G6>G1
			G2; G1 → [EGCG, 100 μM, for 4 h]	[H^3] thymidine incorporation	G2>G1; G3>G1; G4>G1; G5>G1; G6>G1
			G3; G1 → [TF, 200 μg/mL, for 4 h]		
			G4; G1 → [GT, 200 μg/mL, for 4 h]		
			G5; G1 → [DT, 200 μg/mL, for 4 h]		
			G6; G1 → [AT, 200 μg/mL, for 4 h]		
			L1; control	PARP protein	L3>L2; L4>L2
			L2; arsenite, 1000 μM, for 1 h	DNA polymerase β	Not significant

(*Continued*)

TABLE 11.1 (*Continued*)
Results of Studies on Mitigation Effect of As Toxicity by Natural Plants

Plant Material	References	Cell/Animal	Experiment	Measurement	Results
			L3; L2 → [EGCG, 100 µM, for 2 h] or [TF, 200 µg/mL, for 2 h] or [GT, 200 µg/mL, for 2 h] or [DT, 200 µg/mL, for 2 h] or [AT, 200 µg/mL, for 2 h] L4; L2 → [EGCG, 100 µM, for 4 h] or [TF, 200 µg/mL, for 4 h] or [GT, 200 µg/mL, for 4 h] or [DT, 200 µg/mL, for 4 h] or [AT, 200 µg/mL, for 4 h]		
	[44]	Mouse (unknown strain)	G1; distilled water, for 7 days G2; black tea, 1.756 mg/0.16 mL in distilled water, twice daily, for 7 days G3; sodium arsenite, 2.5 mg/kg/day in distilled water, o.p., for 24 h G4; [black tea, 1.756 mg/0.16 mL in distilled water, twice daily, for 6 days] → G3	Chromosomal aberrations in bone marrow cells	G3>G4>G1, G2
	[52]	Swiss albino mouse (both sexes)	G1; control, distilled water, for 7 days G2; ferrous sulfate, 152 mg/kg, p.o., for 7 days G3; sodium arsenite, 2.5 mg/kg bw, p.o., for 24 h G4; tea, 10 mL/kg bw/day, p.o., for 7 days G5; G2+G4 G6; [G2+G4] → G3	Chromosomal aberrations in bone marrow	G3>G6>G4>G3>G1; G3>G6>G4, G5>G1; G3>G6>G5, G2

TABLE 11.1 (Continued)

Results of Studies on Mitigation Effect of As Toxicity by Natural Plants

Plant Material	References	Cell/Animal	Experiment	Measurement	Results
	[45]	Swiss albino mouse (Male)	G1-4; distilled water, for 4 weeks	Chromosomal aberrations in bone marrow	G3-4, G3-6>G4-4>G2-5, G3-5>G2-6, G4-6>G1-4>G1-5, G1-6
			G1-5; distilled water, for 5 weeks		G3-4, G3-6>G4-4>G2-5, G3-5, G4-5>G1-4>G1-5, G1-6
			G1-6; distilled water, for 6 weeks		G3-4, G3-6>G4-4>G3-5>G2-4> G1-4>G1-5, G1-6
			G2-4; black tea, 1.756 mg/0.16 mL/day, o.p., for 4 weeks		G3-4, G3-6>G2-4, G2-6, G4-5, G4-6>G1-4>G1-5, G1-6
			G2-5; black tea, 1.756 mg/0.16 mL/day, o.p., for 5 weeks		G3-4, G3-6>G2-4, G2-5, G4-5>G1-4>G1-5, G1-6
			G2-6; black tea, 1.756 mg/0.16 mL/day, o.p., for 6 weeks		
			G3-4; sodium arsenite, 2.5 mg/kg bw, o.p., the 7th day of 4 weeks		
			G3-5; sodium arsenite, 2.5 mg/kg bw, o.p., the 7th day of 5 weeks		
			G3-6; sodium arsenite, 2.5 mg/kg bw, o.p., the 7th day of 6 weeks		
			G4-4; G2-4 + G3-4		
			G4-5; G2-5 + G3-5		
			G4-6; G2-6 + G3-6		

(Continued)

TABLE 11.1 (*Continued*)
Results of Studies on Mitigation Effect of As Toxicity by Natural Plants

Plant Material	References	Cell/Animal	Experiment	Measurement	Results
	[47]	New Zealand rabbit	G1-1; arsenic trioxide, 3 mg/kg/day, o.p., for 14 days	GSH in blood	G1-2>G1-1; G2-2>G2-1; G3-2>G3-1
			G1-2; G1-1 → [black tea, 100 μL/day in 3.75 g/100 mL, o.p., for 14 days]	TBARS in serum	G1-1>G1-2; G2-2>G2-1; G4-1>G4-2
			G2-1; arsenic trioxide, 3 mg/kg/day, o.p., for 14 days	NOx in serum	G1-1>G1-2; G2-2>G2-1; G3-1>G3-2; G4-1>G4-2
			G2-2; G2-1 → [green tea, 100 μL/day in 3.75 g/100 mL, o.p., for 14 days]		
			G3-1; arsenic trioxide, 3 mg/kg/day, o.p., for 14 days		
			G3-2; G3-1 → [placebo, o.p., for 14 days]		
			G4-1; arsenic trioxide, 3 mg/kg/day, o.p., for 14 days		
			G4-2; G4-1 + [black tea, 100 μL/day in 3.75 g/100 mL, o.p., for 14 days]		
	[50]	Sprague Dawley rat (female)	G1; control	GSH in liver	G1>G2; G3>G2; G4>G2; G5>G2; G6>G2; G4>G7; G5>G7; G6>G7; G4>G8; G5>G8; G6>G8
			G2; sodium arsenite, 100 ppm, drinking water, for 4 weeks	LPO in liver	G2>G1; G2>G3; G2>G4; G2>G5; G2>G6; G7>G4; G7>G5; G7>G6; G8>G4; G8>G5; G8>G6

(Continued)

TABLE 11.1 (Continued)
Results of Studies on Mitigation Effect of As Toxicity by Natural Plants

Plant Material	References	Cell/Animal	Experiment	Measurement	Results
			G3; G2 + [vitamin C, 150 mg/kg/day, o.p., for 4 weeks]	SOD in liver	G1>G2; G5>G2; G6>G2; G4>G7; G5>G7; G6>G7; G4>G8; G5>G8; G6>G8
			G4; G2 + [green tea crude, 30 mg/kg/day, o.p., for 4 weeks]	GPx in liver	G1>G2; G3>G2; G4>G2; G5>G2; G6>G2; G4>G7; G5>G7; G6>G7; G4>G8; G5>G8; G6>G8
			G5; G2 + [green tea crude, 100 mg/kg/day, o.p., for 4 weeks]	Nitrite in liver	G1>G2; G4>G2; G5>G2; G6> G2; G4>G7; G5>G7; G6>G7; G4>G8; G5>G8; G6>G8
			G6; G2 + [green tea crude, 150 mg/kg/day, o.p., for 4 weeks]	As in liver	G2>G1; G2>G3; G2>G4; G2>G5; G2>G6; G2>G7; G7>G5; G7>G6; G8>G4; G8>G5; G8>G6
			G7; G2 + [detannified green tea, 100 mg/kg/day, o.p., for 4 weeks]	Histopathological finding in liver	Treatment of green tea crude enhanced the protection against As toxicity in comparison with the treatment of detannified green tea
			G8; G2 + [detannified green tea, 150 mg/kg/day, o.p., for 4 weeks]	GSH in kidney	G1>G2; G3>G2; G4>G2; G5>G2; G6>G2; G4>G7; G5>G7; G6>G7; G4>G8; G5>G8; G6>G8

(Continued)

TABLE 11.1 (*Continued*)
Results of Studies on Mitigation Effect of As Toxicity by Natural Plants

Plant Material	References	Cell/Animal	Experiment	Measurement	Results
				LPO in kidney	G2>G1; G2>G3; G2>G4; G2>G5; G2>G6; G2>G7; G2>G8; G4>G7; G7>G6; G8>G6
				SOD in kidney	G1>G2; G3>G2; G4>G2; G5>G2; G6>G2; G7>G2; G8>G2; G6>G7
				GPx in kidney	G1>G2; G3>G2; G4>G2; G5>G2; G6>G2; G4>G7; G5>G7; G6>G7; G4>G8; G5>G8; G6>G8
				Nitrite in kidney	G1>G2; G5>G2; G6>G2; G5>G7; G6>G7; G6>G8
				As in kidney	G2>G1; G2>G4; G2>G5; G2>G6; G7>G4; G7>G5; G7>G6; G8>G4; G8>G5; G8>G6
				Histopathological finding in kidney	Treatment of green tea crude enhanced the protection against As toxicity in comparison with the treatment of detannified green tea

(Continued)

TABLE 11.1 (*Continued*)
Results of Studies on Mitigation Effect of As Toxicity by Natural Plants

Plant Material	References	Cell/Animal	Experiment	Measurement	Results
	[46]	Swiss albino mouse (Male)	G1; normal tap water, for 22 days	Comet tail moment on the 7th day	G2>G 1; G3>G1; G4>G1
			G2; arsenite, 50 μg/L, drinking water, for 22 days	Comet tail moment on the 15th day	G2>G 1; G3>G1; G4>G1
			G3; arsenite, 250 μg/L, drinking water, for 22 days	Comet tail moment on the 22nd day	G2>G 1; G3>G1; G4>G1
			G4; arsenite, 500 μg/L, drinking water, for 22 days	Comet tail moment on the 7th day	G2>G3; G2>G4; G2>G5
			G1; normal tap water, for 22 days	Comet tail moment on the 15th day	G2>G3; G2>G4; G2>G5
			G2; arsenite, 500 μg/L, drinking water, for 22 days	Comet tail moment on the 22nd day	G2>G3; G2>G4; G2>G5
			G3; G2 + [green tea, 2.5%, drinking water, for 22 days]	ROS in total liver homogenate on the 7th day	G2>G4
			G4; G2 + [Darjeeling tea, 2.5%, drinking water, for 22 days]	ROS in nuclear fraction on the 7th day	Not significant
			G5; G2 + [Assam tea, 2.5%, drinking water, for 22 days]	ROS in mitochondrial fraction on the 7th day	Not significant
				ROS in cytosolic and microsomal fraction on the 7th day	Not significant

(Continued)

TABLE 11.1 (*Continued*)
Results of Studies on Mitigation Effect of As Toxicity by Natural Plants

Plant Material	References	Cell/Animal	Experiment	Measurement	Results
				ROS in total liver homogenate on the 15th day	G2>G4
				ROS in nuclear fraction on the 15th day	Not significant
				ROS in mitochondrial fraction on the 15th day	G2>G3; G2>G4; G2>G5
				ROS in cytosolic and microsomal fraction on the 15th day	Not significant
				ROS in total liver homogenate on the 22nd day	G2>G3; G2>G4
				ROS in nuclear fraction on the 22nd day	Not significant
				ROS in mitochondrial fraction on the 22nd day	G2>G3; G2>G4; G2>G5
				ROS in cytosolic and microsomal fraction on the 22nd day	Not significant
				Lipid peroxidation in liver on the 7th day	G2>G3; G2>G4; G2>G5
				Lipid peroxidation in liver on the 15th day	G2>G3; G2>G4; G2>G5

(Continued)

TABLE 11.1 (Continued)
Results of Studies on Mitigation Effect of As Toxicity by Natural Plants

Plant Material	References	Cell/Animal	Experiment	Measurement	Results
				Lipid peroxidation in liver on the 22nd day	G2>G3; G2>G4; G2>G5
				Protein carbonyl in liver on the 7th day	G2>G3; G2>G4; G2>G5
				Protein carbonyl in liver on the 15th day	G2>G3; G2>G4; G2>G5
				Protein carbonyl in liver on the 22nd day	G2>G3; G2>G4; G2>G5
				CAT in liver on the 7th day	G3>G2; G4>G2; G5>G2
				CAT in liver on the 15th day	G3>G2; G4>G2; G5>G2
				CAT in liver on the 22nd day	G3>G2; G4>G2; G5>G2
				SOD in liver on the 7th day	G3>G2; G4>G2; G5>G2
				SOD in liver on the 15th day	G3>G2; G4>G2; G5>G2
				SOD in liver on the 22nd day	G3>G2; G4>G2; G5>G2
				GPx in liver on the 7th day	Not significant
				GPx in liver on the 15th day	Not significant
				GPx in liver on 22nd day	G3>G2; G4>G2
				GR in liver on the 7th day	Not significant
				GR in liver on the 15th day	G4>G2; G5>G2
				GR in liver on the 22nd day	Not significant
				GST in liver on the 7th day	Not significant

(Continued)

TABLE 11.1 (*Continued*)
Results of Studies on Mitigation Effect of As Toxicity by Natural Plants

Plant Material	References	Cell/Animal	Experiment	Measurement	Results
			G1; normal tap water, for 22 days	GST in liver on the 15th day	G3>G2; G4>G2; G5>G2
				GST in liver on the 22nd day	G3>G2
				GSH in liver on the 7th day	Not significant
				GSH in liver on the 15th day	G3>G2; G4>G2; G5>G2
				GSH in liver on 22nd day	G3>G2; G4>G2; G5>G2
			G2; [G1 for 11 days] → [arsenite, 500 µg/L, drinking water, for 11 days]	Comet tail moment on the 7th day	Not significant
			G3A; [green tea, 2.5%, drinking water, for 11 days] → [green tea, 2.5%, drinking water, for 11 days] + [arsenite, 500 µg/L, drinking water, for 11 days]	Comet tail moment on the 15th day	G2>G3A; G2>G3B; G2>G4A; G2>G4B; G2>G5A
			G3B; [green tea, 2.5%, drinking water, for 11 days] → [arsenite, 500 µg/L, drinking water, for 11 days]	Comet tail moment on the 22nd day	G2>G3A; G2>G3B; G2>G4A; G2>G4B; G2>G5A; G2>G5B
			G4A; [Darjeeling tea, 2.5%, drinking water, for 11 days] → [Darjeeling tea, 2.5%, drinking water, for 11 days]+[arsenite, 500 µg/L, drinking water, for 11 days]		

(Continued)

TABLE 11.1 (*Continued*)

Results of Studies on Mitigation Effect of As Toxicity by Natural Plants

Plant Material	References	Cell/Animal	Experiment	Measurement	Results
			G4B; [Darjeeling tea, 2.5%, drinking water, for 11 days] → [arsenite, 500 µg/L, drinking water, for 11 days]		
			G5A; [Assam tea, 2.5%, drinking water, for 11 days] → [Assam tea, 2.5%, drinking water, for 11 days] + [arsenite, 500 µg/L, drinking water, for 11 days]		
			G5B; [Assam tea, 2.5%, drinking water, for 11 days] → [arsenite, 500 µg/L, drinking water, for 11 days]		
			G1; normal tap water, for 22 days	ROS in total liver homogenate on the 7th day	Not significant
			G2; [G1 for 11 days] → [arsenite, 500 µg/L, drinking water, for 11 days]	ROS in nuclear fraction on the 7th day	Not significant
			G3A; [green tea, 2.5%, drinking water, for 11 days] → [green tea, 2.5%, drinking water, for 11 days] + [arsenite, 500 µg/L, drinking water, for 11 days]	ROS in mitochondrial fraction on the 7th day	Not significant
			G4A; [Darjeeling tea, 2.5%, drinking water, for 11 days] → [Darjeeling tea, 2.5%, drinking water, for 11 days] + [arsenite, 500 µg/L, drinking water, for 11 days]	ROS in cytosolic and microsomal fraction on the 7th day	Not significant

(*Continued*)

TABLE 11.1 (*Continued*)
Results of Studies on Mitigation Effect of As Toxicity by Natural Plants

Plant Material	References	Cell/Animal	Experiment	Measurement	Results
			G5A; [Assam tea, 2.5%, drinking water, for 11 days] → [Assam tea, 2.5%, drinking water, for 11 days] + [arsenite, 500 μg/L, drinking water, for 11 days]	ROS in total liver homogenate on the 15th day	Not significant
				ROS in nuclear fraction on the 15th day	Not significant
				ROS in mitochondrial fraction on the 15th day	G2>G3A; G2>G4A; G2>G5A
				ROS in cytosolic and microsomal fraction on the 15th day	Not significant
				ROS in total liver homogenate on the 22nd day	G2>G3A; G2>G4A; G2>G5A
				ROS in nuclear fraction on the 22nd day	Not significant
				ROS in mitochondrial fraction on the 22nd day	G2>G3A; G2>G4A; G2>G5A
				ROS in cytosolic and microsomal fraction on the 22nd day	Not significant
				Lipid peroxidation in liver on the 7th day	Not significant

(*Continued*)

TABLE 11.1 (Continued)

Results of Studies on Mitigation Effect of As Toxicity by Natural Plants

Plant Material	References	Cell/Animal	Experiment	Measurement	Results
				Lipid peroxidation in liver on the 15th day	G2>G5A
				Lipid peroxidation in liver on the 22nd day	G2>G5A
				Protein carbonyl in liver on the 7th day	Not significant
				Protein carbonyl in liver on the 15th day	G2>G3A
				Protein carbonyl in liver on the 22nd day	G2>G3A; G2>G4A
				CAT in liver on the 7th day	Not significant
				CAT in liver on the 15th day	G5A>G2
				CAT in liver on the 22nd day	G3A>G2; G4A>G2; G5A>G2
				SOD in liver on the 7th day	Not significant
				SOD in liver on the 15th day	G3A>G2; G4A>G2; G5A>G2
				SOD in liver on the 22nd day	G3A>G2; G4A>G2
				GPx in liver on the 7th day	Not significant
				GPx in liver on the 15th day	Not significant
				GPx in liver on the 22nd day	G4A>G2
				GR in liver on the 7th day	Not significant
				GR in liver on the 15th day	Not significant

(Continued)

TABLE 11.1 (*Continued*)
Results of Studies on Mitigation Effect of As Toxicity by Natural Plants

Plant Material	References	Cell/Animal	Experiment	Measurement	Results
	[49]	Swiss albino mouse (male)	G1; control (normal tap water), for 22 days	GR in liver on the 22nd day	G3A>G2; G4A>G2; G5A>G2
			G2; Arsenite, 500 µg/L, drinking water, for 22 days	GST in liver on the 7th day	Not significant
			G3; G2+[GT, 2.5%, drinking water, for 22 days]	GST in liver on the 15th day	Not significant
			G4; G2+[DT, 2.5%, drinking water, for 22 days]	GST in liver on the 22nd day	G3A>G2; G4A>G2; G5A>G2
			G5; G2+[AT, 2.5%, drinking water, for 22 days]	GSH in liver on the 7th day	Not significant
			G1; normal tap water, for 22 days	GSH in liver on the 15th day	Not significant
				GSH in liver on the 22nd day	G3A>G2
			G2; [arsenite, 500 µg/L, drinking water, for 11 days] → [normal tap water, for 11 days]	8-OHdG in liver on the 9th day	G2>G3; G2>G4; G2>G5
			G3; G2 → [GT, 2.5%, drinking water, for 11 days]	8-OHdG in liver on the 15th day	G2>G3; G2>G4; G2>G5
				8-OHdG in liver on the 23rd day	G2>G3; G2>G4; G2>G5
				Comet tail in blood on the 9th day	Not significant
				Comet tail in blood on the 15th day	Not significant
				Comet tail in blood on the 23rd day	G2>G3; G2>G4; G2>G5

(*Continued*)

TABLE 11.1 (Continued)
Results of Studies on Mitigation Effect of As Toxicity by Natural Plants

Plant Material	References	Cell/Animal	Experiment	Measurement	Results
			G4; G2 → [DT, 2.5%, drinking water, for 11 days]	DNA repair in blood on the 23rd day	G3>G2; G4>G2; G5>G2
			G5; G2 → [AT, 2.5%, drinking water, for 11 days]	DNA β pol protein in plasma	G1>G2; G3>G2; G4>G2; G5>G2
				XRCC1 protein in plasma	G1>G2; G3>G2; G4>G2; G5>G2
				PARP1 protein in plasma	G1>G2; G3>G2; G4>G2; G5>G2
				DNA ligase III protein in plasma	G1>G2; G3>G2; G4>G2; G5>G2
				OGG1 protein in plasma	G2>G1; G2>G3; G2>G4; G2>G5
				DNA PKcs protein in plasma	G1>G2; G3>G2; G4>G2; G5>G2
				XRCC4 protein in plasma	G1>G2; G3>G2; G4>G2; G5>G2
				DNA ligase IV protein in plasma	G1>G2; G3>G2; G4>G2; G5>G2
				Topo IIβ protein in plasma	G1>G2; G3>G2; G4>G2; G5>G2
				DNA β pol protein in blood	G1>G2; G3>G2; G5>G2
				XRCC1 protein in blood	G1>G2; G3>G2; G4>G2; G5>G2

(Continued)

TABLE 11.1 (*Continued*)
Results of Studies on Mitigation Effect of As Toxicity by Natural Plants

Plant Material	References	Cell/Animal	Experiment	Measurement	Results
				PARP1 protein in blood	G1>G2; G3>G2; G4>G2; G5>G2
				DNA ligase III protein in blood	G1>G2; G3>G2; G4>G2; G5>G2
				OGG1 protein in blood	G2>G1; G2>G3; G2>G4; G2>G5
				DNA PKcs protein in blood	G1>G2; G3>G2; G4>G2; G5>G2
				XRCC4 protein in blood	G1>G2; G3>G2; G4>G2; G5>G2
				DNA ligase IV protein in blood	G1>G2; G3>G2; G4>G2; G5>G2
				Topo IIβ protein in blood	G1>G2; G3>G2; G4>G2; G5>G2
	[48]	Wister rat (male)	G1: [control, for 6 weeks] → [0.9% (w/v) NaCl, i.p., for 3 weeks] G2: [control, for 6 weeks] → [NaAsO₂, 5.55 mg/ kg bw/day, i.p., for 3 weeks] G3: [green tea, 66 g/L in drinking water, for 6 weeks] → [green tea, 66 g/L in drinking water, for 3 weeks] + [0.9% (w/v) NaCl, i.p., for 3 weeks]	Body weight	G1>G3; G4>G3
				Absolute liver weight	G3>G1; G3>G4
				Relative liver weight	G3>G1; G3>G4

(Continued)

TABLE 11.1 (*Continued*)
Results of Studies on Mitigation Effect of As Toxicity by Natural Plants

Plant Material	References	Cell/Animal	Experiment	Measurement	Results
			G4: [green tea, 66 g/L in drinking water, for 6 weeks] → [green tea, 66 g/L in drinking water, for 3 weeks] + [NaAsO₂, 5.55 mg/kg bw/day, i.p., for 3 weeks]	Absolute kidney weight	Not significant
				Relative kidney weight	G3>G1; G3>G4
				Absolute testis weight	Not significant
				Relative testis weight	Not significant
				RBC in blood	G1>G3; G4>G3
				Hb in blood	G1>G3; G4>G3
				Ht in blood	G1>G3; G4>G3
				MCV in blood	Not significant
				MCHC in blood	Not significant
				MCH in blood	Not significant
				Total protein in serum	G1>G3; G3>G4
				Albumin in serum	G1>G3; G4>G3
				Total bilirubin in serum	G1>G3; G3>G4
				Urea in serum	G1>G3; G3>G4
				Creatinine in serum	G1>G3; G3>G4
				Cholesterol in serum	G1>G3; G3>G4
				Glucose in serum	G1>G3; G3>G4
				AST in serum	G3>G1; G3>G4
				ALT in serum	G3>G1; G3>G4
				ALP in serum	G3>G1; G3>G4

(Continued)

TABLE 11.1 (Continued)
Results of Studies on Mitigation Effect of As Toxicity by Natural Plants

Plant Material	References	Cell/Animal	Experiment	Measurement	Results
	[54]	As-exposed population in West Bengal, India	G1; drink tea G2; do not drink tea	TBRAS in liver TBRAS in kidney TBRAS in testis Micronuclei in oral mucosal cell	G3>G1; G3>G4 G3>G1; G3>G4 G3>G1; G3>G4 G2>G1 (but statistical significance is not shown)
Curcumin	[56]	Human lymphocyte (Male)	AsIII, 0, 100, 250, 500, 1000 μM AsIII, 1000 μM and curcumin 0, 10, 25, 50 μM, for 1 h	DNA damage DNA damage ROS production Lipid peroxidation CAT SOD GPx	As(III) dose dependant increase Curcumin dose dependent decrease Curcumin dose dependent decrease Curcumin dose dependent decrease Curcumin dose dependent increase Curcumin dose dependent increase Curcumin dose dependent increase

(Continued)

TABLE 11.1 (Continued)
Results of Studies on Mitigation Effect of As Toxicity by Natural Plants

Plant Material	References	Cell/Animal	Experiment	Measurement	Results
			[Curcumin 0, 5, 10, 15 µM, for 24 h] → [AsIII, 1000 µM, for 1 h]	DNA damage	Curcumin dose dependent decrease
			[AsIII, 1000 µM, for 1 h] → [curcumin 0, 5, 10, 15 µM, for 2 or 4 h]	Repair activity	Curcumin dose dependent increase
				DNA repair synthesis	Curcumin dose dependent increase
				PARP protein	Curcumin dose dependent increase
				DNA β polymerase	Insignificant effect of curcumin
	[57]	Human peripheral blood lymphocytes	G1; control	Structural aberrations in chromosome	G2>G1; G2>G5; G3>G1; G3>G6; G4>G1; G4>G7; G5>G1, G6>G1; G7>G1; G9>G1
			G2; As$_2$O$_3$, 1.4 µM, for 24 h	Numerical aberrations in chromosome	G2>G1; G2>G5; G3>G1; G3> G6; G4>G1; G4>G7; G5>G1, G6>G1; G7>G1; G9>G1
			G3; NaF, 34 µM, for 24 h	Comet length	G2>G1; G2>G5; G3>G1; G3>G6; G4>G1; G4>G7; G5>G1, G6>G1; G7>G1; G9>G1
			G4; G2+G3	%DNA in comet head	G2>G1; G2>G5; G3>G1; G3>G6; G4>G1; G4>G7; G5>G1, G6>G1; G7>G1; G9>G1

(Continued)

TABLE 11.1 (*Continued*)
Results of Studies on Mitigation Effect of As Toxicity by Natural Plants

Plant Material	References	Cell/Animal	Experiment	Measurement	Results
			G5; G2+G8	%DNA in comet tail	G1>G2; G1>G3; G1>G4; G1>G5; G1>G6; G1>G7; G1>G9; G5>G2; G6>G3; G7>G4
			G6; G3+G8		
			G7; G4+G8		
			G8; curcumin, 7.7 μM, for 24 h		
			G9; Ethyl methane sulfonate, 1.93 mM, for 24 h		
	[58]	Wister rat (male)	G1; control	AST in plasma	G3>G1, G4>G2
			G2; curcumin, 15 mg/kg, p.o., daily, for 30 days	ALT in plasma	G3>G4>G1>G2
			G3; arsenite (NaAsO₂), 5 mg/kg, p.o., daily, for 30 days	AIP in plasma	G3>G1, G4>G2
			G4; G2+G3	AcP in plasma	G3>G1, G2, G4
				AChE in plasma	G3>G1, G4>G2
				AST in liver	G2>G1, G4>G3
				ALT in liver	G1, G2>G4>G3
				AcP in liver	G2>G1, G4>G3
				AChE in brain	G1>G2, G3; G4>G3
				Total protein in plasma	G1, G2, G4>G3
				Albumin in plasma	G1, G2, G4>G3
				Glucose in plasma	G3>G4>G1>G2
				Urea in plasma	G3>G4>G1>G2

(Continued)

TABLE 11.1 (Continued)

Results of Studies on Mitigation Effect of As Toxicity by Natural Plants

Plant Material	References	Cell/Animal	Experiment	Measurement	Results
	[62]	Wister rat (female)	G1; arsenite ($NaAsO_2$), 20 mg/kg, p.o., daily, for 28 days G2; curcumin, 100 mg/kg, p.o., daily, for 28 days G3; G1 + G2 G4; distilled water, p.o., daily, for 28 days	Creatinine in plasma	G3>G4>G1>G2
				Bilirubin in plasma	G3>G1, G2, G4
				TL in plasma	G3>G1, G4>G2
				Cholesterol in plasma	G3>G1, G4>G2
				TG in plasma	G3>G4>G1>G2
				HDL–C in plasma	G2>G3, G4
				LDL–C in plasma	G3, G4>G1>G2
				Body weight	G4>G1, G3
				Brain weight	Not significant
				Locomotor activity	
				Total distance traveled	G4>G3>G1
				Resting time	G4>G1, G3
				Stereotypic time	Not significant
				Time moving	G4>G3>G1
				Rearing	G4>G1
				Rota-rod performance	G4>G3>G1
				Forelimb grip strength	G4>G3>G1
				^{3}H-spiperone binding to striatal membrane	G3, G4>G1
				Kd of ^{3}H-spiperone binding to striatal membrane	G4>G3>G1

(Continued)

TABLE 11.1 (*Continued*)
Results of Studies on Mitigation Effect of As Toxicity by Natural Plants

Plant Material	References	Cell/Animal	Experiment	Measurement	Results
				Bmax of ^{3}H-spiperone binding to striatal membrane	Not significant
				MDA in frontal cortex	G1>G3, G4
				MDA in corpus striatum	G1>G3, G4
				MDA in hippocampus	G1, G3>G4
				Protein carbonyl in frontal cortex	G1>G3, G4
				Protein carbonyl in corpus striatum	G1>G3>G4
				Protein carbonyl in hippocampus	G1, G3>G4
				GSH in frontal cortex	G3, G4>G1
				GSH in corpus striatum	G4>G3>G1
				GSH in hippocampus	G3, G4>G1
				SOD in frontal cortex	G3, G4>G1
				SOD in corpus striatum	G3, G4>G1
				SOD in hippocampus	Not significant
				CAT in frontal cortex	G4>G1, G3
				CAT in corpus striatum	G3, G4>G1
				CAT in hippocampus	Not significant
				GPx in frontal cortex	G4>G1
				GPx in corpus striatum	G4>G1

(*Continued*)

TABLE 11.1 (*Continued*)

Results of Studies on Mitigation Effect of As Toxicity by Natural Plants

Plant Material	References	Cell/Animal	Experiment	Measurement	Results
				GPx in hippocampus	G3, G4>G1
				As level in frontal cortex	G1>G3>G4
				As level in corpus striatum	G1>G3>G4
				As level in hippocampus	G1>G3>G4
				Tyrosine hydroxylase expression in striatal sections	G4>G1
	[59]	Sprague Dawley rat (Male)	G1; control	GST in plasma	G2>G1>G3; G2>G4
			G2; curcumin, 15 mg/kg, p.o., daily, for 30 days	SOD in plasma	G2>G1>G3; G2>G4
			G3; arsenite (NaAsO$_2$), 5 mg/kg, p.o., daily, for 30 days	CAT in plasma	G2>G1, G4>G3
			G4; G2 + G3	GST in liver	G2>G1>G3; G2>G4
				SOD in liver	G2>G1; G4>G3
				CAT in liver	G2>G1; G4>G3
				GST in testes	G2>G1>G3; G2>G4
				SOD in testes	G2>G1>G3; G4>G3
				CAT in testes	G2>G1, G4>G3
				GST in brain	G2>G3; G1, G4>G3
				SOD in brain	G2>G3; G1, G4>G3
				CAT in brain	G1, G2>G3
				GST in kidney	G2>G1, G4>G3
				SOD in kidney	G2>G1, G4>G3

(*Continued*)

TABLE 11.1 (Continued)
Results of Studies on Mitigation Effect of As Toxicity by Natural Plants

Plant Material	References	Cell/Animal	Experiment	Measurement	Results
				CAT in kidney	G2>G4>G3; G1>G3
				GST in lung	G2>G3, G4; G1>G3
				SOD in lung	G2>G1>G4>G3
				CAT in lung	G2>G1>G3; G4>G3
				TBARS in plasma	G3>G1, G4>G2
				SH in plasma	G2>G4>G3; G1>G3
				TBARS in liver	G3>G4>G1>G2
				SH in liver	G2>G3, G4; G1>G3
				Protein in liver	G2>G3, G4; G1>G3
				TBARS in testes	G3>G4>G1>G2
				SH in testes	G2>G2, G4>G3
				Protein in testes	G2>G3, G4
				TBARS in brain	G3>G4>G1>G2
				SH in brain	G2>G1>G3; G2>G4
				Protein in brain	Not significant
				TBARS in kidney	G3>G4>G1>G2
				SH in kidney	G2>G1>G4>G3
				Protein in kidney	G1, G2, G4>G3
				TBARS in lung	G3>G1, G4>G2
				SH in lung	G2>G4>G3; G1>G3
				Protein in lung	Not significant
				Body weight	G1, G2>G3, G4
				Relative weight of kidney	Not significant

(Continued)

TABLE 11.1 (*Continued*)
Results of Studies on Mitigation Effect of As Toxicity by Natural Plants

Plant Material	References	Cell/Animal	Experiment	Measurement	Results
	[63]	Wister rat (female)	G1; arsenite (NaAsO₂), 20 mg/kg, p.o., daily, for 28 days	Relative weight of liver	G3>G1, G2, G4
				Relative weight of lung	Not significant
				Relative weight of brain	Not significant
				Relative weight of testes	Not significant
			G2; curcumin, 100 mg/kg, p.o., daily, for 28 days	Dopamine in corpus striatum	G3, G4>G1
				Norepinephrine in corpus striatum	G4>G1, G3
			G3; G1 + G2	Epinephrine in corpus striatum	Not significant
			G4; distilled water, p.o., daily, for 28 days	Serotonin in corpus striatum	G3, G4>G1
				3,4-Dihydroxyphenylacetic acid in corpus striatum	G3, G4>G1
				Homovanillic acid in corpus striatum	G3, G4>G1
				Dopamine in frontal cortex	G4>G1, G3
				Norepinephrine in frontal cortex	Not significant
				Epinephrine in frontal cortex	G3, G4>G1
				Serotonin in frontal cortex	G3, G4>G1
				3,4-Dihydroxyphenylacetic acid in frontal cortex	G3, G4>G1

(Continued)

TABLE 11.1 (*Continued*)

Results of Studies on Mitigation Effect of As Toxicity by Natural Plants

Plant Material	References	Cell/Animal	Experiment	Measurement	Results
	[64]	Wister rat (female)	G1; arsenite (NaAsO$_2$), 20 mg/kg, p.o., daily, for 28 days	Homovanillic acid in frontal cortex	G3, G4>G1
				Dopamine in hippocampus	G4>G1, G3
				Norepinephrine in hippocampus	G4, G3>G1
			G2; curcumin, 100 mg/kg, p.o., daily, for 28 days	Epinephrine in hippocampus	G4>G1, G3
				Serotonin in hippocampus	G3, G4>G1
				3,4-Dihydroxyphenylacetic acid in hippocampus	Not significant
			G3; G1+G2	Homovanillic acid in hippocampus	G4>G1, G3
				NO in corpus striatum	G1>G3, G4
				NO in frontal cortex	G1>G3, G4
				NO in hippocampus	G1>G3, G4
			G4; distilled water, p.o., daily, for 28 days	Difference in transfer latency time between acquisition and retention	Significant in G2-4; No significant in G1
				AChE activity in hippocampus	G4>G3>G1
				AChE activity in frontal cortex	G4, G3>G1
				^{3}H-QNB receptor binding in hippocampus	G4>G3>G1

(*Continued*)

TABLE 11.1 (*Continued*)

Results of Studies on Mitigation Effect of As Toxicity by Natural Plants

Plant Material	References	Cell/Animal	Experiment	Measurement	Results
		Kunming mouse (female)	G1; control	^{3}H-QNB receptor binding in frontal cortex	G4, G3>G1
				ChAT protein in hippocampus	G4>G3>G1
				ChAT immunoreactivity in denate gyrus area of hippocampus	G4>G3>G1
				Nissl staining in denate gyrus area of hippocampus	G4, G3>G1
	[60]		G2; arsenite (NaAsO$_2$), 10 mg/L, drinking, 6 weeks	ALT in serum	G2>G1; G3>G1; G4>G1; G2>G5; G3>G6; G4>G7
			G3; arsenite (NaAsO$_2$), 50 mg/L, drinking, 6 weeks	AST in serum	G2>G1; G3>G1; G4>G1; G2>G5; G3>G6
			G4; arsenite (NaAsO$_2$), 100 mg/L, drinking, 6 weeks	MDA in liver	G2>G1; G3>G1; G4>G1; G3>G6; G4>G7
			G5; G2 + [curcumin, 200 mg/kg, gavage, twice 1 week for 6 weeks]	GSH in liver	G1>G3; G1>G4; G2>G5; G6>G3; G7>G4
			G6; G3 + [curcumin, 200 mg/kg, gavage, twice 1 week for 6 weeks]	GSH in liver	G1>G3; G1>G4; G2>G5; G6>G3; G7>G4
			G7; G4 + [curcumin, 200 mg/kg, gavage, twice 1 week for 6 weeks]	GSH-Px in liver	G1>G3; G1>G4
				IAs in urine	G5>G2; G3>G6; G4>G7

(Continued)

TABLE 11.1 (Continued)
Results of Studies on Mitigation Effect of As Toxicity by Natural Plants

Plant Material	References	Cell/Animal	Experiment	Measurement	Results
				MMA in urine	G2>G5; G3>G6; G7>G4
				DMA in urine	G2>G5; G6>G3; G7>G4
				T-As in urine	G2>G5; G7>G4
				IAs% in urine	G3>G6; G4>G7
				MMA% in urine	G3>G6; G7>G4
				DMA% in urine	G6>G3; G7>G4
				PMR in urine	G6>G3; G7>G4
				SMR in urine	G6>G3; G7>G4
				Nrf2 protein distribution in liver	Clear enrichment of Nrf2 protein both in the nuclear and cytoplasmic fractions in curcumin treatment
				Nrf2 protein expression in liver	Clear induction of Nrf2 protein in curcumin treatment
				NQO1 protein expression in liver	Clear induction of NQO1 protein in curcumin treatment
				HO-1 protein expression in liver	Clear induction of HO-1 protein in curcumin treatment
	[69]	Wister rat (male)	G1; normal (no treatment)	WBC in blood	G1, G3, G4>G4, G5, G6>G2
			G2; arsenite (NaAsO$_2$), 2 mg/kg, p.o., for 4 weeks	RBC in blood	Not significant
			G3; curcumin, 15 mg/kg, p.o., for 4 weeks	Hb in blood	Not significant
			G4; nanocurcumin, 15 mg/kg, p.o., for 4 weeks	HCT in blood	Not significant

(Continued)

TABLE 11.1 (Continued)
Results of Studies on Mitigation Effect of As Toxicity by Natural Plants

Plant Material	References	Cell/Animal	Experiment	Measurement	Results
			G5; G2+G3	MCV in blood	Not significant
			G6; G2 + [nanocurcumin, 1.5 mg/kg (low dose), p.o., for 4 weeks]	MCH in blood	Not significant
			G7; G2 + [nanocurcumin, 15 mg/kg (high dose), p.o., for 4 weeks]	MCHC in blood	G2>G1, G3-7
				PLT in blood	G1, G3, G4, G6>G4, G5>G2
				ROS in blood	G2, G5>G1, G3, G4, G6, G7
				GSH in blood	G1, G3, G4>G6, G7>G2, G5
				ALAD in blood	G1, G3, G4>G5-7>G2
				ROS in liver	G2, G5>G1, G3, G4, G6, G7
				GSH in liver	G1, G3, G4>G5-7>G2
				GSSG in liver	G2, G5>G6, G7>G1, G3, G4
				TBARS in liver	G2>G5>G1, G3, G4, G6, G7
				SOD in liver	G1, G3, G5>G6, G7>G2, G5
				CAT in liver	G1, G3, G4>G6, G7>G2, G5
				ROS in brain	G2>G5-7>G1, G3, G4
				GSH in brain	G1, G3, G4>G2, G5-7
				GSSG in brain	G2, G5-7>G1, G3, G4
				TBARS in brain	G2>G1, G3-7
				SOD in brain	G1, G3, G4>G5-7>G2
				CAT in brain	G1, G3, G4>G5-7>G2
				NE in brain	G1, G3, G4, G6, G7>G2, G5
				DA in brain	G1, G3, G4, G6, G7>G2, G5

(Continued)

TABLE 11.1 (Continued)

Results of Studies on Mitigation Effect of As Toxicity by Natural Plants

Plant Material	References	Cell/Animal	Experiment	Measurement	Results
				5-HT in brain	G2, G5, G6>G1, G3, G4, G7
				TH expression in brain	Treatment of ECNPs and As recovered the decreased expression by As
				TPH expression in brain	Treatment of ECNPs and As recovered the decreased expression by As
				As in blood	G2, G5>G6, G7>G1, G3, G4
				As in brain	G2, G5, G6>G7>G1, G3, G4
				As in kidney	G2, G5, G6>G7>G1, G3, G4
				As in liver	G2>G5>G6, G7>G1, G3, G4
	[70]	Wistar rat (male)	G1; control	Spleen weight	Not significant
			G2; arsenite (NaAsO$_2$), 25 ppm, drinking, daily for 42 days	Relative spleen weight	G2, G3>G1
			G3; G2 + [empty nanoparticles, p.o., during the last 14 days (29th to 42nd day of As exposure)]	DTH response	G1>G2, G3; G5>G2
			G4; G2 + [curcumin, 100 mg/kg bw, p.o., during the last 14 days (29th to 42nd day of As exposure)]	T cell stimulation index	G1, G5>G2, G3
			G5; G2 + [curcumin in nanoparticle, 100 mg/kg bw, p.o., during the last 14 days (29th to 42nd day of As exposure)]	KLH antibody titre	G1>G5>G3>G2; G1>G4>G2
				Nitrite production	G1>G2, G3

(Continued)

TABLE 11.1 (*Continued*)

Results of Studies on Mitigation Effect of As Toxicity by Natural Plants

Plant Material	References	Cell/Animal	Experiment	Measurement	Results
	[71]	As-exposed population in West Bengal, India	G1; non-As-exposed population (control)	Double-stranded DNA in lymphocyte after 1 month	Not significant
			G2; As-exposed population, curcumin (in capsule), 500 mg, p.o., twice daily for 3 months	Double-stranded DNA in lymphocyte after 2 months	G2>G3
			G3; As-exposed population, placebo, p.o., twice daily for 3 months	Double-stranded DNA in lymphocyte after 3 months	G2>G3
				Comet tail moment in lymphocyte after 1 month	G3>G2
				Comet tail moment in lymphocyte after 2 months	G3>G2
				Comet tail moment in lymphocyte after 3 months	G3>G2
				ROS in lymphocyte after 1 month	G3>G2
				ROS in lymphocyte after 2 months	G3>G2
				ROS in lymphocyte after 3 months	G3>G2
				MDA in lymphocyte after 1 month	G3>G2

(*Continued*)

TABLE 11.1 (Continued)
Results of Studies on Mitigation Effect of As Toxicity by Natural Plants

Plant Material	References	Cell/Animal	Experiment	Measurement	Results
				MDA in lymphocyte after 2 months	G3 > G2
				MDA in lymphocyte after 3 months	G3 > G2
				Protein carbonyl formation in lymphocyte after 1 month	Not significant
				Protein carbonyl formation in lymphocyte after 2 months	Not significant
				Protein carbonyl formation in lymphocyte after 3 months	G3 > G2
				CAT in plasma after 1 month	G2 > G3
				CAT in plasma after 2 months	G2 > G3
				CAT in plasma after 3 months	G2 > G3
				SOD in plasma after 1 month	G2 > G3
				SOD in plasma after 2 months	G2 > G3

(Continued)

TABLE 11.1 (*Continued*)

Results of Studies on Mitigation Effect of As Toxicity by Natural Plants

Plant Material	References	Cell/Animal	Experiment	Measurement	Results
				SOD in plasma after 3 months	G2>G3
				GSH in plasma after 1 month	G2>G3
				GSH in plasma after 2 months	G2>G3
				GSH in plasma after 3 months	G2>G3
				GR in plasma after 1 month	G2>G3
				GR in plasma after 2 months	G2>G3
				GR in plasma after 3 months	G2>G3
				GST in plasma after 1 month	G2>G3
				GST in plasma after 2 months	G2>G3
				GST in plasma after 3 months	G2>G3
				NADPH in plasma after 1 month	Not significant
				NADPH in plasma after 2 months	Not significant
				NADPH in plasma after 3 months	G2>G3

(Continued)

TABLE 11.1 (*Continued*)
Results of Studies on Mitigation Effect of As Toxicity by Natural Plants

Plant Material	References	Cell/Animal	Experiment	Measurement	Results
	[72]	As-exposed population in West Bengal, India	G1: non-As-exposed population (control)	ROS in lymphocytes	G2>G3; G2>G4; G2>G5
			G2: As-exposed population	8-OhdG in blood	G2>G3; G2>G4; G2>G5
			G3: As-exposed population, curcumin (in capsule), 500 mg, p.o., twice daily for 1 month	DNA β pol protein in leukocytes	G1>G2; G5>G2
			G4: As-exposed population, curcumin (in capsule), 500 mg, p.o., twice daily for 2 months	PARP protein in leukocytes	G1>G2; G5>G2
			G5: As-exposed population, curcumin (in capsule), 500 mg, p.o., twice daily for 3 months	XRCC1 protein in leukocytes	G1>G2; G5>G2
				Ligase III protein in leukocytes	G1>G2; G5>G2
				DNA PKcs protein in leukocytes	G1>G2; G5>G2
				XRCC4 protein in leukocytes	G1>G2; G5>G2
				Topo II β protein in leukocytes	G1>G2; G5>G2
				Ligase IV protein in leukocytes	G1>G2; G5>G2
				OGG1 protein in leukocytes	G2>G1; G2>G5
				DNA β pol mRNA in blood	G1>G2; G5>G2

(*Continued*)

TABLE 11.1 (*Continued*)
Results of Studies on Mitigation Effect of As Toxicity by Natural Plants

Plant Material	References	Cell/Animal	Experiment	Measurement	Results
				PARP mRNA in blood	G1 > G2; G5 > G2
				XRCC1 mRNA in blood	G1 > G2; G5 > G2
				Ligase III mRNA in blood	G1 > G2; G5 > G2
				DNA PKcs mRNA in blood	G1 > G2; G5 > G2
				XRCC4 mRNA in blood	G1 > G2; G5 > G2
				Topo II β mRNA in blood	G1 > G2; G5 > G2
				Ligase IV mRNA in blood	G1 > G2; G5 > G2
				OGG1 mRNA in blood	G2 > G1; G2 > G5
				% of DNA repair in lymphocytes (2 h)	G3 > G2; G4 > G2; G5 > G2
				% of DNA repair in lymphocytes (4 h)	G3 > G2; G4 > G2; G5 > G2
Garlic	[74]	Swiss albino mouse (unknown sex)	G1; [distilled water, for 30 days] + [mitomycin C, 1.5 mg/kg bw, i.p., on the 30th day]	Chromosomal aberration in bone marrow	G4 > G5
			G2; saline, 1 mL/100 g bw, S.C., on the 7th, 14th, 21st, and 30th day	Damaged cells in bone marrow	G4 > G5
			G3; garlic extract, 100 mg/kg bw/day, p.o., for 30 days		
			G4; $NaAsO_2$, 0.1 mg/kg bw, S.C., on the 7th, 14th, 21st, and 30th day		
			G5; G3 + G4		

(*Continued*)

TABLE 11.1 (Continued)
Results of Studies on Mitigation Effect of As Toxicity by Natural Plants

Plant Material	References	Cell/Animal	Experiment	Measurement	Results
	[76]	Swiss albino mouse (male)	G1A; NaAsO$_2$, 2.5 mg/kg bw/day, o.p., for 24 h	Chromosomal aberration in bone marrow	G1A>G2, G3A>G4; G1B>G3B>G2>G4; G1C>G2>G4; G3C>G4
			G1B; NaAsO$_2$, 0.833 mg/kg bw/day, o.p., for 24 h	Damaged cells in bone marrow	G1A>G2, G3A>G4; G1B>G2, G3B>G4; G1C>G2, G3C>G4
			G1C; NaAsO$_2$, 0.5 mg/kg bw/day, o.p., for 24 h		
			G2; garlic extract, 100 mg/kg bw/day, o.p., for 24 h		
			G3A; G1A+G2		
			G3B; G1B+G2		
			G3C; G1C+G2		
			G4; deionized water for 24 h		
			G5; G4+[mitomycin, 1.5 mg/kg bw, i.p., for 24 h]		
	[75]	Swiss albino mouse (male)	G1; distilled water, for 24 h	Chromosomal aberration in bone marrow	G3>G2, G4>G1
			G2; garlic extract, 100 mg/kg bw/day, o.p., for 24 h	Damaged cells in bone marrow	G3>G2, G4>G1
			G3; NaAsO$_2$, 2.5 mg/kg bw/day, o.p., for 24 h		
			G4; G2+G4		

(Continued)

TABLE 11.1 (*Continued*)
Results of Studies on Mitigation Effect of As Toxicity by Natural Plants

Plant Material	References	Cell/Animal	Experiment	Measurement	Results
	[78]	Swiss albino mouse (both sexes); F0 mice were treated with G1–G4; F1 were obtained these F0 mice	G1: distilled water (negative control), for 30 days	Chromosomal aberration in bone marrow of F1 males	G2, G4>G1, G3
			G2; NaAsO$_2$, 0.1 mg/kg bw, i.p., on days 7, 14, 21, 30 in 30 days	Damaged cell in the bone marrow of F1 males	G2, G4>G1, G3
			G3; garlic extract, 100 mg/kg bw/day, p.o., for 30 days	Chromosomal aberration in bone marrow of F1 females	G2, G4>G1, G3
			G4; G2 + G3	Damaged cell in bone marrow of F1 females	G2, G4>G1, G3
				Chromosomal aberration in bone marrow between F0 and F1 of G1–4 (both sexes)	F0>F1
				Damaged cell in bone marrow between F0 and F1 of G1–4 (both sexes)	F0>F1
	[77]	Swiss albino mouse (both sexes)	G1; mitomycin C (positive control), 1.5 mg/kg bw, for 30 days	Chromosomal aberration in bone marrow (both sexes)	G3>G5>G4>G2

(Continued)

TABLE 11.1 (*Continued*)
Results of Studies on Mitigation Effect of As Toxicity by Natural Plants

Plant Material	References	Cell/Animal	Experiment	Measurement	Results
			G2; distilled water (negative control), for 30 days	Damaged cell in bone marrow (both sexes)	G3>G5>G4>G2
			G3; NaAsO$_2$, 0.1 mg/kg bw, i.p., on days 7, 14, 21, and 30 in 30 days		
			G4; garlic extract, 100 mg/kg bw/day, p.o., for 30 days		
			G5; G3+G4		
			G1; mitomycin C (positive control), 1.5 mg/kg bw, for 60 days	Chromosomal aberration in bone marrow (both sexes)	G3>G4, G5>G2
			G2; distilled water (negative control), for 60 days	Damaged cell in bone marrow (both sexes)	G3>G4, G5>G2
			G3; NaAsO$_2$, 0.1 mg/kg bw, s.c., at intervals of 7 days in 60 days		
			G4; garlic extract, 100 mg/kg bw/day, p.o., for 60 days		
			G5; G3+G4		
	[79]	Swiss albino mouse (both sexes)	G1; mitomycin C (positive control), 1.5 mg/kg bw, for 30 days	Chromosomal aberration in bone marrow of males	G5>G8>G4, G7>G2, G3, G6, G9
			G2; distilled water (negative control), for 30 days	Damaged cell in bone marrow of males	G5>G8>G4, G7>G3, G6, G2; G5>G8>G4, G7>G3, G6, G9; G5>G8>G4, G7>G9>G2

(Continued)

TABLE 11.1 (Continued)
Results of Studies on Mitigation Effect of As Toxicity by Natural Plants

Plant Material	References	Cell/Animal	Experiment	Measurement	Results
			G3; mustard oil, 0.643 mg/kg bw, p.o., for 30 days	Chromosomal aberration in bone marrow of females	G5>G8>G4, G7>G6, G9>G2; G5>G8>G4, G7>G9>G2, G3; G5>G8>G4, G7>G3, G6
			G4; garlic extract, 100 mg/kg bw/day, p.o., for 30 days	Damaged cell in bone marrow of females	G5>G4, G7>G3, G6, G9>G8; G5>G4, G7>G2, G3, G6>G8
			G5; NaAsO₂, 0.1 mg/kg bw, s.c., on days 7, 14, 21, and 30 in 30 days		
			G6; G3+G4		
			G7; G3+G5		
			G8; G4+G5		
			G9; G3+G4+G5		
	[80]	Malignant melanoma (A375)	[NaAsO₂, 0–40 μM, for 24 h] + [garlic extract, 2 mg/mL, for 24 h]	Cytotoxicity	Dose-dependently induced cytotoxicity was suppressed by garlic extract
			[NaAsO₂, 10 μM, for 24 h] + [garlic extract, 0 or 2 mg/mL, for 24 h]	ROS	Induced ROS was suppressed by garlic extract
			G1; control	p53 mRNA	Induced p53 was suppressed by garlic extract
			G2; [NaAsO₂, 5 μM, for 24 h]	hsp90 mRNA	Induced hsp90 was suppressed by garlic extract
			G3; [NaAsO₂, 10 μM, for 24 h]		
			G4; G2 + [garlic extract, 2 mg/mL, for 24 h]		

(Continued)

TABLE 11.1 (*Continued*)
Results of Studies on Mitigation Effect of As Toxicity by Natural Plants

Plant Material	References	Cell/Animal	Experiment	Measurement	Results
		Nontumorigenic human keratinocyte (HaCaT)	G5; G3 + [garlic extract, 2 mg/mL, for 24 h] [NaAsO$_2$, 0–40 µM, for 24 h] + [garlic extract, 2 mg/mL, for 24 h]	Cytotoxicity	Dose-dependently induced cytotoxicity was suppressed by garlic extract
			[NaAsO$_2$, 10 µM, for 24 h] + [garlic extract, 0 or 2 mg/mL, for 24 h]	ROS	Induced ROS was suppressed by garlic extract
		Dermal fibroblast	[NaAsO$_2$, 0–40 µM, for 24 h] + [garlic extract, 2 mg/mL, for 24 h]	Cytotoxicity	Dose-dependently induced cytotoxicity was suppressed by garlic extract
			[NaAsO$_2$, 10 µM, for 24 h] + [garlic extract, 0 or 2 mg/mL, for 24 h]	ROS	Induced ROS was suppressed by garlic extract
		Sprague Dawley rat (Female)	G1; distilled water (control), for 5 days	WBC in blood	G2>G1; G2>G3
			G2; NaAsO$_2$ in deionized water, 5 mg/kg bw/day, for 5 days	RBC in blood	G2>G1; G2>G3
			G3; G2 + [garlic extract, 20 mg/kg bw/day, for 5 days]	PLT in blood	Not significant
				MCV in blood	Not significant
				MCH in blood	Not significant
				MCHC in blood	Not significant
				HGB in blood	G2>G1; G2>G3
				As in blood	G2>G1; G2>G3

(*Continued*)

TABLE 11.1 (*Continued*)
Results of Studies on Mitigation Effect of As Toxicity by Natural Plants

Plant Material	References	Cell/Animal	Experiment	Measurement	Results
				As in hair	G2>G1; G2>G3
				As in urine	G2>G1; G3>G2
				As in ovary	G2>G1; G2>G3
				As in liver	G2>G1; G2>G3
				As in kidney	G2>G1; G2>G3
				Catalase in ovary	G1>G2; G3>G2
				Catalase in kidney	G1>G2; G3>G2
				Catalase in liver	G1>G2; G3>G2
				SOD in ovary	G1>G2; G3>G2
				SOD in kidney	G1>G2; G3>G2
				SOD in liver	G1>G2; G3>G2
				MDA in ovary	G2>G1; G2>G3
				MDA in kidney	G2>G1; G2>G3
				MDA in liver	G2>G1; G2>G3
				MPO in peripheral blood mononuclear cells	G2>G1; G2>G3
				GSH in blood	G1>G2; G3>G2
				GSH in ovary	G1>G2; G3>G2
				GSH in liver	Not significant
				GSH in kidney	G1>G2
				Thiol group in ovary	Not significant
				Thiol group in liver	G1>G2; G3>G2
				Thiol group in kidney	G1>G2; G3>G2

(Continued)

TABLE 11.1 (*Continued*)
Results of Studies on Mitigation Effect of As Toxicity by Natural Plants

Plant Material	References	Cell/Animal	Experiment	Measurement	Results
			[NaAsO$_2$, 0.05 M, for overnight] + [garlic extract, 0, 1, 2, 5, and 10 mg/mL, for overnight]	As in supernatant	As was dose-dependently decreased with garlic extract
				As in precipitation	As was dose-dependently increased with garlic extract
				AsV in supernatant	AsV was dose-dependently increased with garlic extract
				AsV in precipitation	AsV was dose-dependently increased with garlic extract
	[81]	Swiss mouse (male)	G1: no treatment	ROS in liver	G3>G4>G1, G2, G5
			G2: garlic extract, 500 mg/kg, oral, daily	Mitochondrial membrane potential in liver	G1, G2, G5>G3>G4
			G3: arsenite (NaAsO$_2$), 2.5 mg/kg, i.p., daily	ATP in liver	G1, G2, G5>G3>G4
			G4: G3 + [garlic extract, 250 mg/kg, oral, daily]	bax/bcl2 mRNA expression in liver	G3>G4>G1, G2, G5
			G5: G3 + [garlic extract, 500 mg/kg, oral, daily]	Caspase 3 in liver	G3>G4>G1, G2, G5
				G6PDH in liver	G1, G2, G5>G3, G4
				SOD in liver	G2>G1>G5>G3, G4
				Catalase in liver	G2>G1>G3, G4, G5
				GPx in liver	G2>G1>G5>G4>G3
				TBARS in liver	G1, G2, G5>G3, G4
				AST in serum	G1, G2, G5>G3, G4
				ALT in serum	G1, G2, G4, G5>G3

(*Continued*)

TABLE 11.1 (Continued)
Results of Studies on Mitigation Effect of As Toxicity by Natural Plants

Plant Material	References	Cell/Animal	Experiment	Measurement	Results
	[82]	A375 (malignant melanoma cell line)	Arsenite ($NaAsO_2$), 10 μM, 24 h	ALAD in liver	G1, G2>G5>G4>G3
				GSH/GSSG in liver	G1, G2, G5>G4>G3
				As in liver	G3, G4>G5>G1, G2
				As in urine	G5>G3, G4
				Zn in liver	Not significant
				Cu in liver	Not significant
				Cytotoxicity	Decreased with 0–60 μM arginine
			Arsenite ($NaAsO_2$), 10 μM + arginine, 10–60 μM, 24 h		
			G1; control	ROS	G2>G3
			G2: arsenite ($NaAsO_2$), 10 μM, 24 h	SOD	G3>G2
			G3: G2 + G4	CAT	G3>G2
			G4: arginine, 60 μM, 24 h	GSH	G3>G2
				MDA	G2>G3
				Comet tail length	G2>G3
				PARP protein expression	G3>G2
Mushroom lectin	[87]	Hepatocytes of Wister rat (male)	G1; control, for 24 h	Adherent ability	Number of As-induced deformed cells was decreased by mushroom lectin
			G2: sodium arsenite, 5 μM, for 24 h	Morphological alteration	G2>G1; G2>G3; G2>G4; G2>G5; G3>G1; G4>G1

(Continued)

TABLE 11.1 (*Continued*)

Results of Studies on Mitigation Effect of As Toxicity by Natural Plants

Plant Material	References	Cell/Animal	Experiment	Measurement	Results
			G3; G2 + [mushroom lectin, 10 µg/mL, for 24 h]	Cell proliferation	G4>G2; G5>G2
			G4; G2 + [mushroom lectin, 20 µg/mL, for 24 h]	Phagocytic activity	G1>G2; G1>G3; G1>G4; G1>G5; G3>G2; G4>G2; G5>G2
			G5; G2 + [mushroom lectin, 50 µg/mL, for 24 h]	NO production	G2>G1; G2>G3; G2>G4; G2>G5; G3>G1; G4>G1; G5>G1
				DNA fragmentation	G2>G1; G2>G3; G2>G4; G2>G5; G3>G1; G4>G1; G5>G1
				Caspase-3 activity	G2>G1; G2>G3; G2>G4; G2>G5; G3>G1; G4>G1; G5>G1
	[86]	Hepatocytes of Wister rat (male)	G1; control, for 72 h	Cell proliferation at 24 h	G1>G2; G3>G2; G4>G2; G5>G2; G6>G2
			G2; sodium arsenite, 5 µM, for 72 h	Cell proliferation at 48 h	G1>G2; G3>G2; G4>G2; G5>G2; G6>G2
			G3; ascorbic acid, 10 µg/mL, for 72 h	Cell proliferation at 72 h	G1>G2; G3>G2; G4>G2; G5>G2; G6>G2
			G4; mushroom lectin, 50 µg/mL, for 72 h	Phagocytic activity at 24 h	G1>G2; G1>G5; G1>G6; G3>G2; G4>G2; G5>G2; G6>G2

(*Continued*)

TABLE 11.1 (*Continued*)

Results of Studies on Mitigation Effect of As Toxicity by Natural Plants

Plant Material	References	Cell/Animal	Experiment	Measurement	Results
			G5; G2 + G3	Phagocytic activity at 48 h	G1 > G2; G1 > G5; G1 > G6; G3 > G2; G4 > G2; G5 > G2; G6 > G2
			G6; G2 + G4	Phagocytic activity at 72 h	G1 > G2; G1 > G5; G3 > G2; G4 > G2; G5 > G2; G6 > G2
				NO production at 24 h	G2 > G1; G2 > G3; G2 > G4; G2 > G5; G2 > G6
				NO production at 48 h	G2 > G1; G2 > G3; G2 > G4; G2 > G5; G2 > G6
				NO production at 72 h	G2 > G1; G2 > G3; G2 > G4; G2 > G5; G2 > G6
				SOD at 24 h	G1 > G2; G1 > G5; G1 > G6; G3 > G2; G4 > G2; G5 > G2; G6 > G2
				SOD at 48 h	G1 > G2; G3 > G2; G4 > G2; G5 > G2; G6 > G2
				SOD at 72 h	G1 > G2; G3 > G2; G4 > G2; G5 > G2; G6 > G2
				SOD2 mRNA expression at 72 h	G1 > G2; G1 > G6; G5 > G2; G6 > G2
	[88]	Wister albino rat (male)	G1; control, for 12 weeks	Body weight at 0 week	Not significant
			G2; sodium arsenite, 20 ppm in drinking water, for 12 weeks	Body weight at 2 weeks	Not significant

(Continued)

TABLE 11.1 (*Continued*)
Results of Studies on Mitigation Effect of As Toxicity by Natural Plants

Plant Material	References	Cell/Animal	Experiment	Measurement	Results
			G3; G2+ [L-ascorbate, 25 mg/kg bw, oral, for 12 weeks]	Body weight at 4 weeks	Not significant
			G4; G2+ [mushroom lectin, 150 mg/kg bw, oral, for 12 weeks]	Body weight at 6 weeks	Not significant
				Body weight at 8 weeks	Not significant
				Body weight at 10 weeks	G1>G2
				Body weight at 12 weeks	G1>G2; G1>G3; G1>G4
				Relative kidney weight at 2 weeks	G2>G1; G2>G3; G2>G4; G3>G1; G4>G1
				Relative kidney weight at 4 weeks	G2>G1; G2>G3; G2>G4
				Relative kidney weight at 6 weeks	G2>G1; G2>G3; G2>G4
				Relative kidney weight at 8 weeks	G2>G1; G2>G3; G2>G4
				Relative kidney weight at 10 weeks	G2>G1; G2>G3; G2>G4
				Relative kidney weight at 12 weeks	G2>G1; G2>G3; G2>G4; G3>G1
				As concentration in kidney at 2 weeks	G2>G1; G2>G3; G3>G1; G4>G1
				As concentration in kidney at 4 weeks	G2>G1; G3>G1; G4>G1

(*Continued*)

TABLE 11.1 (*Continued*)
Results of Studies on Mitigation Effect of As Toxicity by Natural Plants

Plant Material	References	Cell/Animal	Experiment	Measurement	Results
				As concentration in kidney at 6 weeks	G2>G1; G2>G3; G2>G4; G3>G1; G4>G1
				As concentration in kidney at 8 weeks	G2>G1; G2>G3; G2>G4; G3>G1; G4>G1
				As concentration in kidney at 10 weeks	G2>G1; G3>G1; G4>G1
				As concentration in kidney at 12 weeks	G2>G1; G2>G3; G2>G4; G3>G1; G4>G1
				SOD in kidney at 2 weeks	G1>G4; G2>G1; G2>G3; G2>G4
				SOD in kidney at 4 weeks	G1>G3; G1>G4; G2>G1; G2>G3; G2>G4
				SOD in kidney at 6 weeks	G1>G3; G1>G4; G2>G1; G2>G3; G2>G4
				SOD in kidney at 8 weeks	G2>G1; G2>G3; G2>G4
				SOD in kidney at 10 weeks	G1>G2; G3>G2; G4>G2
				SOD in kidney at 12 weeks	G1>G2; G3>G2; G4>G2
				CAT in kidney at 2 weeks	G2>G1; G2>G4
				CAT in kidney at 4 weeks	G2>G1; G2>G4
				CAT in kidney at 6 weeks	G2>G1; G2>G3; G2>G4
				CAT in kidney at 8 weeks	G2>G1; G2>G3; G2>G4
				CAT in kidney at 10 weeks	G2>G1; G2>G3; G2>G4

(*Continued*)

TABLE 11.1 (*Continued*)
Results of Studies on Mitigation Effect of As Toxicity by Natural Plants

Plant Material	References	Cell/Animal	Experiment	Measurement	Results
				CAT in kidney at 12 weeks	G1>G3; G2>G1; G2>G3; G2>G4
				MDA in kidney at 2 weeks	G2>G1; G2>G3; G2>G4; G3>G1; G4>G1
				MDA in kidney at 4 weeks	G2>G1; G2>G3; G2>G4; G3>G1; G4>G1
				MDA in kidney at 6 weeks	G2>G1; G2>G3; G2>G4; G3>G1; G4>G1
				MDA in kidney at 8 weeks	G2>G1; G2>G3; G2>G4; G3>G1; G3>G4
				MDA in kidney at 10 weeks	G2>G1; G2>G3; G2>G4; G3>G1
				MDA in kidney at 12 weeks	G2>G1; G2>G3; G2>G4
				Protein carbonyl in kidney at 2 weeks	G2>G1; G3>G1; G4>G1
				Protein carbonyl in kidney at 4 weeks	G2>G1; G2>G3; G2>G4; G3>G1; G4>G1
				Protein carbonyl in kidney at 6 weeks	G2>G1; G2>G3; G2>G4; G3>G1; G4>G1
				Protein carbonyl in kidney at 8 weeks	G2>G1; G2>G3; G2>G4; G3>G1; G4>G1
				Protein carbonyl in kidney at 10 weeks	G2>G1; G2>G3; G2>G4; G3>G1; G4>G1

(*Continued*)

TABLE 11.1 (*Continued*)
Results of Studies on Mitigation Effect of As Toxicity by Natural Plants

Plant Material	References	Cell/Animal	Experiment	Measurement	Results
				Protein carbonyl in kidney at 12 weeks	G2>G1; G2>G3; G2>G4; G3>G1; G4>G1
				Nitric oxide in kidney at 2 weeks	G2>G1; G3>G1; G4>G1
				Nitric oxide in kidney at 4 weeks	G2>G1; G2>G3; G2>G4; G3>G1; G4>G1
				Nitric oxide in kidney at 6 weeks	G2>G1; G2>G3; G2>G4; G3>G1; G4>G1
				Nitric oxide in kidney at 8 weeks	G2>G1; G2>G3
				Nitric oxide in kidney at 10 weeks	G2>G1; G2>G3
				Nitric oxide in kidney at 12 weeks	G2>G1; G2>G3
				SOD2 mRNA in kidney at 4 weeks	G1>G2; G1>G3; G1>G4
				SOD2 mRNA in kidney at 8 weeks	G1>G2; G1>G3; G1>G4; G3>G2; G4>G2
				SOD2 mRNA in kidney at 12 weeks	G1>G2; G3>G1; G3>G2; G4>G1; G4>G2
	[89]	Wistar albino rat (male)	G1; control, for 12 weeks	Relative liver weight at 2 weeks	Not significant

(Continued)

TABLE 11.1 (Continued)
Results of Studies on Mitigation Effect of As Toxicity by Natural Plants

Plant Material	References	Cell/Animal	Experiment	Measurement	Results
			G2; sodium arsenite, 20 ppm in drinking water, for 12 weeks	Relative liver weight at 4 weeks	G2>G1; G2>G4
			G3; G2 + [L-ascorbate, 25 mg/kg bw, oral, for 12 weeks]	Relative liver weight at 6 weeks	G2>G1; G2>G3; G2>G4; G3>G4
			G4; G2 + [mushroom lectin, 150 mg/kg bw, oral, for 12 weeks]	Relative liver weight at 8 weeks	G2>G1; G2>G3; G2>G4; G3>G1; G3>G4
				Relative liver weight at 10 weeks	G2>G1; G2>G3; G2>G4; G3>G1; G3>G4
				Relative liver weight at 12 weeks	G2>G1; G2>G3; G2>G4; G3>G1; G3>G4
				As concentration in liver at 2 weeks	G2>G1; G3>G1; G4>G1
				As concentration in liver at 4 weeks	G2>G1; G2>G3; G2>G4; G3>G1; G4>G1
				As concentration in liver at 6 weeks	G2>G1; G2>G3; G3>G1; G4>G1
				As concentration in liver at 8 weeks	G2>G1; G2>G3; G2>G4; G3>G1; G4>G1
				As concentration in liver at 10 weeks	G2>G1; G2>G3; G2>G4; G3>G1; G4>G1
				As concentration in liver at 12 weeks	G2>G1; G2>G3; G2>G4; G3>G1; G4>G1
				SOD in liver at 2 weeks	Not significant

(Continued)

TABLE 11.1 (Continued)
Results of Studies on Mitigation Effect of As Toxicity by Natural Plants

Plant Material	References	Cell/Animal	Experiment	Measurement	Results
				SOD in liver at 4 weeks	G1>G4; G2>G1; G2>G3; G2>G4; G3>G4
				SOD in liver at 6 weeks	G3>G4
				SOD in liver at 8 weeks	Not significant
				SOD in liver at 10 weeks	G1>G2; G3>G2; G4>G2; G4>G3
				SOD in liver at 12 weeks	G1>G2; G3>G2; G4>G2; G4>G3
				CAT in liver at 2 weeks	G2>G1; G2>G3; G2>G4
				CAT in liver at 4 weeks	G2>G1; G2>G3; G2>G4; G3>G1; G4>G1; G4>G3
				CAT in liver at 6 weeks	G4>G1; G4>G3
				CAT in liver at 8 weeks	G3>G1; G3>G2; G4>G1; G4>G2
				CAT in liver at 10 weeks	G1>G2; G3>G2; G4>G1; G4>G2
				CAT in liver at 12 weeks	G1>G2; G3>G1; G3>G2; G4>G1; G4>G2
				MDA in liver at 2 weeks	G2>G1; G2>G3; G2>G4
				MDA in liver at 4 weeks	G2>G1; G2>G3; G2>G4
				MDA in liver at 6 weeks	G1>G3; G1>G4; G2>G1; G2>G3; G2>G4

(Continued)

TABLE 11.1 (*Continued*)
Results of Studies on Mitigation Effect of As Toxicity by Natural Plants

Plant Material	References	Cell/Animal	Experiment	Measurement	Results
				MDA in liver at 8 weeks	G1>G3; G1>G4; G2>G1; G2>G3; G2>G4
				MDA in liver at 10 weeks	G1>G3; G1>G4; G2>G1; G2>G3; G2>G4; G3>G4
				MDA in liver at 12 weeks	G1>G4; G2>G1; G2>G3; G2>G4; G3>G4
				Protein carbonyl in liver at 2 weeks	G2>G1; G2>G3; G3>G1; G4>G1
				Protein carbonyl in liver at 4 weeks	G2>G1; G2>G3; G2>G4; G3>G1; G4>G1
				Protein carbonyl in liver at 6 weeks	G2>G1; G2>G3; G2>G4; G3>G1; G4>G1
				Protein carbonyl in liver at 8 weeks	G2>G1; G2>G3; G2>G4; G3>G1; G4>G1
				Protein carbonyl in liver at 10 weeks	G2>G1; G2>G3; G2>G4; G3>G1; G3>G4; G4>G1
				Protein carbonyl in liver at 12 weeks	G2>G1; G2>G3; G2>G4; G3>G1; G3>G4; G4>G1
				Nitric oxide in liver at 2 weeks	G2>G1; G3>G1; G4>G1
				Nitric oxide in liver at 4 weeks	G2>G1; G2>G3; G2>G4; G3>G1; G4>G1

(Continued)

TABLE 11.1 (*Continued*)
Results of Studies on Mitigation Effect of As Toxicity by Natural Plants

Plant Material	References	Cell/Animal	Experiment	Measurement	Results
				Nitric oxide in liver at 6 weeks	G2>G1; G2>G3; G2>G4; G3>G1; G4>G1
				Nitric oxide in liver at 8 weeks	G2>G1; G2>G3; G2>G4
				Nitric oxide in liver at 10 weeks	G2>G1; G2>G3; G2>G4
				Nitric oxide in liver at 12 weeks	G2>G1; G2>G3; G2>G4
				SOD2 mRNA in liver at 4 weeks	G1>G2; G3>G2; G4>G1; G4>G2; G4>G3
				SOD2 mRNA in liver at 8 weeks	G1>G2; G3>G2; G4>G2; G4>G3
				SOD2 mRNA in liver at 12 weeks	G1>G2; G2>G3; G2>G4; G4>G3
	[90]	Wister albino rat (male)	G1; control, for 12 weeks	Cell adhesion in kidney at 2 weeks	G1>G2
			G2; sodium arsenite, 20 ppm in drinking water, for 12 weeks	Cell adhesion in kidney at 4 weeks	G1>G2; G1>G3; G1>G4; G3>G2
			G3; G2 + [L-ascorbate, 25 mg/kg bw, oral, for 12 weeks]	Cell adhesion in kidney at 6 weeks	G1>G2; G3>G2; G4>G2
			G4; G2 + [mushroom lectin, 150 mg/kg bw, oral, for 12 weeks]	Cell adhesion in kidney at 8 weeks	G1>G2; G3>G2; G4>G2; G4>G3

(*Continued*)

TABLE 11.1 (Continued)
Results of Studies on Mitigation Effect of As Toxicity by Natural Plants

Plant Material	References	Cell/Animal	Experiment	Measurement	Results
				Cell adhesion in kidney at 10 weeks	G1>G2; G3>G2; G4>G1; G4>G2; G4>G3
				Cell adhesion in kidney at 12 weeks	G1>G2; G3>G2; G4>G2
				Cell polarity in kidney at 2 weeks	G2>G1; G2>G4; G3>G1; G4>G1
				Cell polarity in kidney at 4 weeks	G2>G1; G2>G3; G2>G4; G3>G1
				Cell polarity in kidney at 6 weeks	G2>G1; G2>G3; G2>G4; G3>G1
				Cell polarity in kidney at 8 weeks	G2>G1; G2>G3; G2>G4; G3>G1
				Cell polarity in kidney at 10 weeks	G2>G1; G2>G3; G2>G4
				Cell polarity in kidney at 12 weeks	G2>G1; G2>G3; G2>G4
				Cell proliferation index in kidney at 2 weeks	G1>G2; G1>G3
				Cell proliferation index in kidney at 4 weeks	G1>G2; G1>G3; G1>G4; G3>G2; G4>G2
				Cell proliferation index in kidney at 6 weeks	G1>G2; G1>G3; G1>G4; G3>G2; G4>G2

(Continued)

TABLE 11.1 (*Continued*)
Results of Studies on Mitigation Effect of As Toxicity by Natural Plants

Plant Material	References	Cell/Animal	Experiment	Measurement	Results
				Cell proliferation index in kidney at 8 weeks	G1>G2; G1>G3; G1>G4; G3>G2; G4>G2
				Cell proliferation index in kidney at 10 weeks	G1>G2; G1>G3; G1>G4; G3>G2; G4>G2
				Cell proliferation index in kidney at 12 weeks	G1>G2; G1>G3; G1>G4; G3>G2; G4>G2
				Cell apoptosis in kidney at 4 weeks	G2>G1; G2>G3; G3>G1; G4>G1
				Cell apoptosis in kidney at 8 weeks	G2>G1; G2>G3; G2>G4; G3>G1; G4>G1
				Cell apoptosis in kidney at 12 weeks	G2>G1; G2>G3; G2>G4; G3>G1; G4>G1
				Cell caspase-3 activity in kidney at 4 weeks	G2>G1; G2>G3; G2>G4
				Cell caspase-3 activity in kidney at 8 weeks	G2>G1; G2>G3; G2>G4
				Cell caspase-3 activity in kidney at 12 weeks	G2>G1; G2>G3; G2>G4
				Cell DNA fragmentation in kidney at 2 weeks	G2>G1; G2>G3; G2>G4; G3>G1; G4>G1
				Cell DNA fragmentation in kidney at 4 weeks	G2>G1; G2>G3; G2>G4; G3>G1; G4>G1

(*Continued*)

TABLE 11.1 (*Continued*)
Results of Studies on Mitigation Effect of As Toxicity by Natural Plants

Plant Material	References	Cell/Animal	Experiment	Measurement	Results
				Cell DNA fragmentation in kidney at 6 weeks	G2>G1; G2>G3; G2>G4; G3>G1; G4>G1
				Cell DNA fragmentation in kidney at 8 weeks	G2>G1; G2>G3; G2>G4; G3>G1; G4>G1
				Cell DNA fragmentation in kidney at 10 weeks	G2>G1; G2>G3; G2>G4; G3>G1; G4>G1
				Cell DNA fragmentation in kidney at 12 weeks	G2>G1; G2>G3; G2>G4; G3>G1; G4>G1
Horseradish tree	[97]	Chicken liver homogenate	Control	Protein	Significant amelioration of PF, MO, or TA
			Arsenic trioxide, 3 μg/mL, 30 min	TBARS	Significant amelioration of PF, MO, or TA
			Plant extract (PF, TA, or MO), 100 μg/mL, 30 min [Arsenic trioxide, 3 μg/mL, 30 min] + [Plant extract (PF, TA, or MO), 100 μg/mL, 30 min]		
	[94]	Wister rat (male)	G1; control (no treatment)	ALAD in blood	G1>G2; G3>G2
			G2; [sodium arsenite, 100 ppm, drinking water, for 4 months] → [no treatment for 5 days]	GSH in blood	G1>G2
			G3; [sodium arsenite, 100 ppm, drinking water, for 4 months] → [seed powder of *M. oleifera*, 500 mg/kg/day, p.o., for 5 days]	ROS in blood	G2>G1
				WBC in blood	G2>G1; G2>G3

(Continued)

TABLE 11.1 (Continued)
Results of Studies on Mitigation Effect of As Toxicity by Natural Plants

Plant Material	References	Cell/Animal	Experiment	Measurement	Results
				RBC in blood	G1 > G2
				Hemoglobin in blood	G1 > G2
				Hematocrit in blood	G1 > G2
				MCV in blood	Not significant
				MCH in blood	Not significant
				MCHC in blood	Not significant
				ALAD in liver	G1 > G2
				ALAS in liver	G2 > G1; G2 > G3
				TBARS in liver	G2 > G1; G2 > G3
				SOD in liver	Not significant
				Catalase in liver	G1 > G2; G3 > G2
				GSH in liver	Not significant
				GSSG in liver	Not significant
				ROS in liver	G2 > G1; G2 > G3
				TBARS in kidney	G2 > G1; G2 > G3
				SOD in kidney	Not significant
				Catalase in kidney	G1 > G2; G3 > G2
				GSH in kidney	G1 > G2; G3 > G2
				GSSG in kidney	Not significant
				SOD in brain	G1 > G2
				GPx in brain	G1 > G2
				GSH in brain	Not significant
				GSSG in brain	G2 > G1

(*Continued*)

TABLE 11.1 (*Continued*)
Results of Studies on Mitigation Effect of As Toxicity by Natural Plants

Plant Material	References	Cell/Animal	Experiment	Measurement	Results
				Metallothionein in liver	G2>G1
				Metallothionein in kidney	G2>G1; G2>G3
				As in blood	G2>G1; G2>G3
				As in liver	G2>G1; G2>G3
				As in kidney	G2>G1; G2>G3
				Zn in blood	Not significant
				Zn in liver	Not significant
				Zn in kidney	Not significant
				Zn in brain	Not significant
				Mn in blood	Not significant
				Mn in liver	G3>G2
				Mn in kidney	Not significant
				Mn in brain	Not significant
	[91]	Swiss mouse (male)	G1; no treatment	ROS in blood	G3>G1, G2, G4, G5
			G2; *M. oleifera* seed powder, 500 mg/kg, o.p., daily, for 6 weeks	ROS in liver	G3>G1, G2, G4, G5
			G3; sodium arsenite, 3.5 mg/kg, i.p., daily, for 6 weeks	ROS in kidney	G3, G4>G1, G2, G5
			G4; G3+[*M. oleifera* seed powder, 250 mg/kg, o.p., daily, for 6 weeks]	ROS in brain	G3>G1, G2, G4, G5
			G5; G3+G2	ALAD in blood	G1, G2, G5>G3, G4
				ALAD in liver	G1, G2, G5>G3, G4

(Continued)

TABLE 11.1 (*Continued*)
Results of Studies on Mitigation Effect of As Toxicity by Natural Plants

Plant Material	References	Cell/Animal	Experiment	Measurement	Results
				ALA in urine	G3>G1, G2, G4, G5
				GSH in blood	G1, G2>G3-5
				Catalase in blood	G1, G2, G5>G3, G4
				RBC in blood	Not significant
				WBC in blood	Not significant
				Hb in blood	Not significant
				HCT in blood	Not significant
				PLT in blood	G3, G4>G1, G2, G5
				SOD in liver	G1, G2, G4, G5>G3
				Catalase in liver	G1, G2, G5>G3, G4
				GPx in liver	G1, G2, G5>G3, G4
				ALAS in liver	Not significant
				TBARS in liver	G3>G1, G2, G4, G5
				ACP in liver	G3>G1, G2, G4, G5
				ALP in liver	G1, G2, G5>G3, G4
				ALT in liver	Not significant
				AST in liver	G1, G2>G3-5
				Catalase in kidney	G1, G2, G5>G3, G4
				SOD in kidney	Not significant
				TBARS in kidney	G3, G4>G1, G2, G5
				GSH in kidney	Not significant
				GSSG in kidney	Not significant
				SOD in brain	G4>G1, G2>G5>G3

(Continued)

TABLE 11.1 (*Continued*)
Results of Studies on Mitigation Effect of As Toxicity by Natural Plants

Plant Material	References	Cell/Animal	Experiment	Measurement	Results
				TBARS in brain	G3, G4>G1, G2, G5
				GSH in brain	Not significant
				GSSG in brain	Not significant
				Metallothionein	G3>G1, G2, G4, G5
				As in blood	G3>G4>G1, G2, G5
				As in liver	G3>G4>G1, G2, G5
				As in kidney	G3>G4>G1, G2, G5
				As in brain	G3, G4>G5>G1, G2
				Zn in blood	Not significant
				Zn in liver	Not significant
				Zn in kidney	Not significant
				Zn in brain	Not significant
				Fe in blood	Not significant
				Fe in liver	Not significant
				Fe in kidney	Not significant
				Fe in brain	Not significant
				Cu in blood	Not significant
				Cu in liver	Not significant
				Cu in kidney	Not significant
				Cu in brain	Not significant

(*Continued*)

TABLE 11.1 (*Continued*)

Results of Studies on Mitigation Effect of As Toxicity by Natural Plants

Plant Material	References	Cell/Animal	Experiment	Measurement	Results
	[96]	Swiss mouse (male)	G1; normal drinking water, for 6 months	Body weight gain	G1>G2A-D
			G2; sodium arsenite, 100 ppm in drinking water, for 6 months	Food intake	No significant
			G2A; G2 → [saline, for 10 days]	Water intake	No significant
			G2B; G2 → [MiADMSA, 50 mg/kg/day, i.p., for 10 days]	GSH in blood	No significant
			G2C; G2 → [powder solution of *M. oleifera* in saline, 500 mg/kg/day, p.o., for 10 days]	WBC in blood	G2A-D>G1
			G2D; G2 → G2B+G2C	RBC in blood	G1, G2B, G2D>G2A, G2C
				Hb in blood	G1, G2B, G2D>G2A, G2C
				HCT in blood	G1, G2B, G2D>G2A, G2C
				MCV in blood	No significant
				MCH in blood	G1, G2D>G2A-C
				MCHC in blood	No significant
				PLT in blood	No significant
				ROS in blood	G2A>G1, G2B-D
				ROS in liver	G2A>G1, G2B-D
				ROS in kidney	G2A-C>G1, G2D
				ROD in brain	G2A>G2B, G2C>G1, G2D
				ALAD in blood	G1, G2C, G2D>G2B>G2A
				ALAD in liver	G2D>G1, G2B, G2C>G2A
				SOD in liver	G1, G2B>G2A, G2C, G2D

(Continued)

TABLE 11.1 (*Continued*)
Results of Studies on Mitigation Effect of As Toxicity by Natural Plants

Plant Material	References	Cell/Animal	Experiment	Measurement	Results
				Catalase in liver	No significant
				GPx in liver	G1, G2B-D>G2A
				TBARS in liver	G2A-C>G1, G2D
				GSH in liver	G1>G2B>G2A, G2C, G2D
				GSSG in liver	G2A, G2C>G2B, G2D>G1
				ACP in liver	No significant
				ALP in liver	No significant
				AST in liver	G1, G2B-D>G2A
				ALT in liver	G1, G2B-D>G2A
				SOD in kidney	No significant
				Catalase in kidney	G1, G2D>G2A-C
				GPx in kidney	G1, G2D>G2A-C
				TBARS in kidney	G2A-D>G1
				GSH in kidney	No significant
				GSSG in kidney	No significant
				GSH:GSSG in kidney	G1, G2B-D>G2A
				SOD in brain	G1, G2B-D>G2A
				GPx in brain	No significant
				TBARS in brain	G2A>G1, G2B-D
				GSH in brain	No significant
				GSSG in brain	No significant
				GSH:GSSG in brain	No significant
				MT in liver	G2A-C>G1, G2D

(Continued)

TABLE 11.1 (*Continued*)
Results of Studies on Mitigation Effect of As Toxicity by Natural Plants

Plant Material	References	Cell/Animal	Experiment	Measurement	Results
				DNA fragmentation in liver	G2A, G2B>G1, G2C, G2D; (statistical significance is not shown)
				Histopathological observation in liver	Disturbed structure by G2A was recovered by G2B-D, especially G2D was most effective
				As in blood	G2A>G2B, G2C>G1, G2D
				As in liver	G2A>G2B-D>G1
				As in kidney	G2A, G2C>G2B, G2D>G1
				As in brain	G2A, G2C>G2B, G2D>G1
				Cu in blood	G1, G2A, G2C, G2D>G2B
				Cu in liver	G1, G2A, G2C, G2D>G2B
				Cu in kidney	G1, G2C, G2D>G2A, G2B
				Cu in brain	G1, G2A, G2C, G2D>G2B
				Zn in blood	G2A>G1, G2B-D
				Zn in liver	No significant
				Zn in kidney	G1, G2C, G2D>G2A, G2B
				Zn in brain	G1, G2A, G2C, G2D>G2B
				Fe in blood	G1>G2C>G2A, G2B, G2D
				Fe in liver	G2B, G2D>G1, G2A, G2C
				Fe in kidney	G2C>G1, G2A, G2B, G2D
				Fe in brain	G2C>G1, G2A, G2B>G2D

(*Continued*)

(Continued)

TABLE 11.1 (*Continued*)
Results of Studies on Mitigation Effect of As Toxicity by Natural Plants

Plant Material	References	Cell/Animal	Experiment	Measurement	Results
	[95]	Wister rat (female)	G1; distilled water, 0.6 mL/100 g bw/day, administration in drinking water, for 24 days	Body weight	Not significant
			G2; [sodium arsenite, 0.4 ppm/100 g bw/day, administration in drinking water, for 24 days] + [distilled water, 0.1 mL/100 g bw/day, administration in drinking water, for 24 days]	Hepatosomatic index	G2, G3>G1
			G3; [sodium arsenite, 0.4 ppm/100 g bw/day, administration in drinking water, for 24 days] + [extract of *M. oleifera*, 500 mg/0.1 mL/100 g bw/day, administration in drinking water, for 24 days]	ALT in plasma	G2>G3>G1
				AST in plasma	G2>G3>G1
				ALP in plasma	G1, G2>G3
				Total protein in plasma	G3>G1>G2
				CHL in plasma	G2>G1, G3
				TG in plasma	G2>G3>G1
				HDL in plasma	Not significant
				LDL in plasma	G2>G3>G1
				MDA in liver	G2>G3>G1
				CD in liver	G2>G1, G3
				SOD in liver	G1, G3>G2
				Catalase in liver	G1, G3>G2

TABLE 11.1 (Continued)
Results of Studies on Mitigation Effect of As Toxicity by Natural Plants

Plant Material	References	Cell/Animal	Experiment	Measurement	Results
Tossa jute	[101]	Wister rat (male)	G1: distilled water, 2.0 mL/kg, p.o., daily for 15 days	Histology in liver	Disarrangement by G2 was partially protected by G3
				DNA fragment in liver	DNA fragmentation by G2 was partially protected by G3
			G2: arsenite (NaAsO$_2$), 10 mg/kg, p.o., daily for 15 days	DNA fragmentation in liver	G2>G1; G2>G3, G4, G5
			G3: [AECO, 50 mg/kg, p.o., daily for 15 days] → [arsenite (NaAsO$_2$), 10 mg/kg, p.o., daily for 10 days]	TBARS in liver	G2>G1; G2>G3, G4, G5
			G4: [AECO, 100 mg/kg, p.o., daily for 15 days] → [arsenite (NaAsO$_2$), 10 mg/kg, p.o., daily for 10 days]	SOD in liver	G1>G2; G3, G4, G5>G2
			G5: [querectin, 10 mg/kg, p.o., daily for 15 days] → [arsenite (NaAsO$_2$), 10 mg/kg, p.o., daily for 10 days]	CAT in liver	G1>G2; G4, G5>G2
				GST in liver	G1>G2; G3, G4, G5>G2
				GPx in liver	G1>G2; G3, G4, G5>G2
				GR in liver	G1>G2; G3, G4, G5>G2
				GSH in liver	G1>G2; G3, G4, G5>G2
				GSSG in liver	G2>G1; G2>G3, G4, G5

(Continued)

TABLE 11.1 (*Continued*)

Results of Studies on Mitigation Effect of As Toxicity by Natural Plants

Plant Material	References	Cell/Animal	Experiment	Measurement	Results
				Structure in liver	Disorganization by arsenite was reduced by the prior treatment of AECO
				DNA fragmentation in kidney	G2>G1; G2>G3, G4, G5
				TBARS in kidney	G2>G1; G2>G4, G5
				SOD in kidney	G1>G2; G3, G4, G5>G2
				CAT in kidney	G1>G2; G3, G4, G5>G2
				GST in kidney	G1>G2; G4, G5>G2
				GPx in kidney	G1>G2; G4, G5>G2
				GR in kidney	G1>G2; G3, G4, G5>G2
				GSH in kidney	G1>G2; G3, G4, G5>G2
				GSSG in kidney	G2>G1; G2>G3, G4, G5
				Structure in kidney	Disorganization by arsenite was reduced by the prior treatment of AECO
	[102]	Swiss albino mouse (male)	G1: distilled water, 2.0 mL/kg, p.o., daily for 15 days	Total cholesterol in serum	G2>G1; G2>G3, G4, G5
		Wister rat (male)	G2: arsenite (NaAsO₂), 10 mg/kg, p.o., daily for 10 days	HDL cholesterol in serum	G2>G1; G2>G3, G4, G5
			G3: [AECO, 50 mg/kg, p.o., daily for 15 days] → [arsenite (NaAsO₂), 10 mg/kg, p.o., daily for 10 days]	DNA fragmentation in myocardial tissues	G2>G1; G2>G3, G4, G5

(*Continued*)

TABLE 11.1 (Continued)
Results of Studies on Mitigation Effect of As Toxicity by Natural Plants

Plant Material	References	Cell/Animal	Experiment	Measurement	Results
			G4: [AECO, 100 mg/kg, p.o., daily for 15 days] → [arsenite (NaAsO$_2$), 10 mg/kg, p.o., daily for 10 days]	MDA in myocardial tissues	G2>G1; G2>G3, G4, G5
			G5: [querectin, 10 mg/kg, p.o., daily for 15 days] → [arsenite (NaAsO$_2$), 10 mg/kg, p.o., daily for 10 days]	Protein carbonylation in myocardial tissues	G2>G1; G2>G3, G4, G5
				SOD in myocardial tissues	G1>G2; G3, G4, G5>G2
				GST in myocardial tissues	G1>G2; G3, G4, G5>G2
				CAT in myocardial tissues	G1>G2; G3, G4, G5>G2
				GPx in myocardial tissues	G1>G2; G3, G4, G5>G2
				GR in myocardial tissues	G1>G2; G3, G4, G5>G2
				GSH in myocardial tissues	G1>G2; G3, G4, G5>G2
				GSSG in myocardial tissues	G2>G1; G2>G3, G4, G5
				Structure in myocardial tissues	Disorganization by arsenite was reduced by the prior treatment of AECO
	[103]	Albino Wistar rat (male)	G1: distilled water, 2.0 mL/kg, p.o., daily for 15 days	TBARS in brain	G2>G1; G2>G4, G5
			G2: arsenite (NaAsO$_2$), 10 mg/kg, p.o., daily for 10 days	SOD in brain	G1>G2; G3, G4, G5>G2
			G3: [AECO, 50 mg/kg, p.o., daily for 15 days] → [arsenite (NaAsO$_2$), 10 mg/kg, p.o., daily for 10 days]	CAT in brain	G1>G2; G3, G4, G5>G2

(Continued)

TABLE 11.1 (Continued)
Results of Studies on Mitigation Effect of As Toxicity by Natural Plants

Plant Material	References	Cell/Animal	Experiment	Measurement	Results
			G4: [AECO, 100 mg/kg, p.o., daily for 15 days] → [arsenite (NaAsO₂), 10 mg/kg, p.o., daily for 10 days]	GST in brain	G1>G2; G3, G4, G5>G2
			G5: [quercetin, 10 mg/kg, p.o., daily for 15 days] → [arsenite (NaAsO₂), 10 mg/kg, p.o., daily for 10 days]	GPx in brain	G1>G2; G3, G4, G5>G2
				GR in brain	G1>G2; G3, G4, G5>G2
				GSH in brain	G1>G2; G3, G4, G5>G2
				GSSG in brain	G2>G1; G2>G3, G4, G5
				Structure in brain	Disorganization by arsenite was reduced by the prior treatment of AECO
Sea-buckthorn	[106]	Swiss albino mouse (male)	G1: no treatment, for 3 months	ALAD in blood	G1>G2c>G2a, G2b, G2d
			G2: sodium arsenite, 25 ppm in drinking water, for 3 months	GSH in blood	G1>G2b, G2d>G2a, G2c
			G2a; G2 → [saline, for 10 days]	WBC in blood	G2b, G2c>G1, G2d>G2a
			G2b; G2 → [HF-WRC, 500 mg/kg/day, p.o., for 10 days]	RBC in blood	Not significant
			G2c; G2 → [HF-WRT, 500 mg/kg/day, p.o., for 10 days]	Hemoglobin in blood	Not significant
			G2d; G2 → [HF-EtOH, 500 mg/kg/day, p.o., for 10 days]	Hematocrit in blood	G1, G2b-d>G2a

(Continued)

TABLE 11.1 (*Continued*)
Results of Studies on Mitigation Effect of As Toxicity by Natural Plants

Plant Material	References	Cell/Animal	Experiment	Measurement	Results
				TBARS in liver	G2a, G2b>G1, G2c, G2d
				GSH/GSSG in liver	G1, G2c, G2d>G2a, G2b
				SOD in liver	G2b, G2c>G1>G2a, G2d
				ALT in liver	G1>G2a-d
				AST in liver	G1>G2a-d
				ALP in liver	G1, G2b>G2a, G2c, G2d
				TBRAS in kidney	G2a-d>G1
				GSH/GSSG in kidney	Not significant
				TBRAS in brain	G2a>G2b>G1, G2c, G2d
				GSH/GSSG in brain	G2d>G1, G2b, G2c>G2a
				As in blood	G2a-d>G1
				As in liver	G2a-d>G1
				As in kidney	G2a-d>G1
				As in brain	G2d>G2a-c>G1
	[107]	Swiss albino mouse (male)	G1; normal animals	ALAD in blood	G1, G4, G5>G2, G3
			G2; sodium arsenite, 2.5 mg/kg/day, i.p., for 3 weeks	GSH in blood	Not significant
			G3; G2 + [HF-WRC, 500 mg/kg/day, p.o., for 3 weeks]	ROS in blood	G2>G1, G3-5
			G4; G2 + [HF-WRT, 500 mg/kg/day, p.o., for 3 weeks]	WBC in blood	G2, G3>G1, G4, G5

(*Continued*)

TABLE 11.1 (*Continued*)
Results of Studies on Mitigation Effect of As Toxicity by Natural Plants

Plant Material	References	Cell/Animal	Experiment	Measurement	Results
			G3; G2 + [HF-EtOH, 500 mg/kg/day, p.o., for 3 weeks]	RBC in blood	Not significant
				Hemoglobin in blood	G1, G4, G5>G2, G3
				Hematocrit in blood	G1, G4, G5>G2, G3
				MCV in blood	Not significant
				MCH in blood	Not significant
				MCHC in blood	G1, G3, G4>G2, G5
				SOD in liver	G5>G1, G3, G4>G2
				Catalase in liver	G1, G4, G5>G2, G3
				GPx in liver	G1, G3-5>G2
				GST in liver	G1>G2-5
				TBARS in liver	G2>G1, G3-5
				GSH in liver	G1, G3-5>G2
				GSSG in liver	G3>G1, G2, G4, G5
				TBARS in kidney	G2>G1, G3-5
				GSH in kidney	Not significant
				GSSG in kidney	G2-5>G1
				TBARS in brain	G2>G1, G3-5
				GSH in brain	Not significant
				GSSG in brain	Not significant
				Metallothionein in liver	G2-5>G1
				As in blood	G2-5>G1
				As in liver	G2-5>G1

(Continued)

TABLE 11.1 (*Continued*)

Results of Studies on Mitigation Effect of As Toxicity by Natural Plants

Plant Material	References	Cell/Animal	Experiment	Measurement	Results
				As in kidney	G2-5>G1
				As in brain	G2-5>G1
				Zn in blood	G4>G1-3, G5
				Zn in liver	G2-5>G1
				Zn in kidney	Not significant
				Zn in brain	G4, G5>G1-3
				Cu in blood	G4>G1-3, G5
				Cu in liver	G2, G5>G1, G3, G4
				Cu in kidney	G5>G14
				Cu in brain	Not significant
Aloe	[35]	Wister rat (male)	G1; normal	Body weight gain	Not significant
			G2; *Aloe vera*, 1% in drinking water (w/v), for 3 weeks	Liver weight	Not significant
			G3; *Aloe vera*, 2% in drinking water (w/v), for 3 weeks	Kidney weight	Not significant
			G4; *Aloe vera*, 5% in drinking water (w/v), for 3 weeks	Food intake	Not significant
			G5; sodium arsenite, 0.2 mg/kg, i.p., daily, for 3 weeks	Water intake	Not significant
			G6; G2+G5	ALAD in blood	G1>G5; G7>G5; G8>G5
			G7; G3+G5	ZPP in blood	Not significant
			G8; G4+G5	GSH in blood	G1>G5

(Continued)

TABLE 11.1 (Continued)
Results of Studies on Mitigation Effect of As Toxicity by Natural Plants

Plant Material	References	Cell/Animal	Experiment	Measurement	Results
				As in blood	G5>G1
				GSH in liver	Not significant
				GSSG in liver	Not significant
				TBRAS in liver	G5>G1; G5>G6; G5>G7; G5>G8
				ALP in liver	G1>G5; G8>G5
				ACP in liver	Not significant
				AST in liver	G1>G5; G8>G5
				ALT in liver	G1>G5; G7>G5; G8>G5
				Catalase in liver	G1>G5; G8>G5
				SOD in liver	G1>G5; G8>G5
				As in liver	G5>G1
				GSH in kidney	G1>G5
				GSSG in kidney	Not significant
				SOD in kidney	G1>G5
				As in kidney	G5>G1
Argentinian medicinal plants	[108]	African green monkey kidney cells (Vero cell)	G1; 0.2% DMSO, for 2 h	Aqueous hydroperoxides	G2>G3-1; G2>G5-1; G2>G5-4; G2>G6-1; G2>G6-4; G2>G7-2; G2>G7-3; G2>G7-4; G4-1>G2; G4-2>G2; G4-3>G2; G6-3>G2; G7-1>G2

(Continued)

TABLE 11.1 (Continued)

Results of Studies on Mitigation Effect of As Toxicity by Natural Plants

Plant Material	References	Cell/Animal	Experiment	Measurement	Results
			G2; G1 + [NaAsO$_2$, 200 µM, for 2 h]	Lipid hydroperoxides	G2>G3-1; G2>G3-2; G2> G3-3; G2>G4-1; G2>G4-3; G2>G5-1; G2>G5-2; G2> G5-3; G2>G5-4; G2>G6-1; G2> G6-2; G2>G6-4; G2> G7-1; G2>G7-2; G2>G7-3; G2>G7-4; G3-4>G2; G6-3>G2
			G3-1; G2 + [petroleum ether extract of E. buniifolium, 200 µg/mL, for 2 h]		
			G3-2; G2 + [dichloromethane extract of E. buniifolium, 200 µg/mL, for 2 h]		
			G3-3; G2 + [methanol extract of E. buniifolium, 200 µg/mL, for 2 h]		
			G3-4; G2 + [water extract of E. buniifolium, 200 µg/mL, for 2 h]		
			G4-1; G2 + [petroleum ether extract of H. alienus, 200 µg/mL, for 2 h]		
			G4-2; G2 + [dichloromethane extract of H. alienus, 200 µg/mL, for 2 h]		
			G4-3; G2 + [methanol extract of H. alienus, 200 µg/mL, for 2 h]		

(Continued)

TABLE 11.1 (Continued)
Results of Studies on Mitigation Effect of As Toxicity by Natural Plants

Plant Material	References	Cell/Animal	Experiment	Measurement	Results
			G4-4; G2+ [water extract of *H. alienus*, 200 µg/mL, for 2 h]		
			G5-1; G2+ [petroleum ether extract of *L. grisebachii*, 200 µg/mL, for 2 h]		
			G5-2; G2+ [dichloromethane extract of *L. grisebachii*, 200 µg/mL, for 2 h]		
			G5-3; G2+ [methanol extract of *L. grisebachii*, 200 µg/mL, for 2 h]		
			G5-4; G2+ [water extract of *L. grisebachii*, 200 µg/mL, for 2 h]		
			G6-1; G2+ [petroleum ether extract of *M. pentlandiana*, 200 µg/mL, for 2 h]		
			G6-2; G2+ [dichloromethane extract of *M. pentlandiana*, 200 µg/mL, for 2 h]		
			G6-3; G2+ [methanol extract of *M. pentlandiana*, 200 µg/mL, for 2 h]		
			G6-4; G2+ [water extract of *M. pentlandiana*, 200 µg/mL, for 2 h]		
			G7-1; G2+ [petroleum ether extract of *S. commersoniana*, 200 µg/mL, for 2 h]		
			G7-2; G2+ [dichloromethane extract of *S. commersoniana*, 200 µg/mL, for 2 h]		

(Continued)

TABLE 11.1 (*Continued*)

Results of Studies on Mitigation Effect of As Toxicity by Natural Plants

Plant Material	References	Cell/Animal	Experiment	Measurement	Results
			G7-3: G2 + [methanol extract of *S. commersoniana*, 200 µg/mL, for 2 h]		
			G7-4: G2 + [water extract of *S. commersoniana*, 200 µg/mL, for 2 h]		
Flaxseed oil	[110]	Wister rat (male)	G1: [normal diet for 14 days] → [distilled water, i.p., for 4 days]	Creatinine in serum	G2>G1; G2>G3; G3>G1
			G2: [normal diet for 14 days] → [sodium arsenite, 20 mg/kg bw/day in distilled water, i.p., for 4 days]	BUN in serum	G2>G1; G2>G3; G1>G4
			G3: [flaxseed oil (15%) diet for 14 days] → [distilled water, i.p., for 4 days]	Cholesterol in serum	G2>G1
			G4: [flaxseed oil (15%) diet for 14 days] → [sodium arsenite, 20 mg/kg bw/day in distilled water, i.p., for 4 days]	Phospholipid in serum	G1>G4; G2>G1; G2>G3; G3>G1
				Phosphate in serum	G1>G2; G3>G2; G4>G1
				Glucose in serum	G1>G2; G3>G2
				Urine flow rate	G2>G1; G2>G3; G3>G1
				Creatinine clearance	G1>G2; G1>G3; G3>G2
				Phosphate in urine	G2>G1; G2>G3
				Protein in urine	G1>G4; G2>G1; G2>G3; G3>G1
				Glucose in urine	G2>G1; G2>G3; G3>G1
				ALP in renal cortex	G1>G2

(*Continued*)

TABLE 11.1 (*Continued*)
Results of Studies on Mitigation Effect of As Toxicity by Natural Plants

Plant Material	References	Cell/Animal	Experiment	Measurement	Results
				GGTase in renal cortex	G1 > G2
				LAP in renal cortex	G1 > G2; G3 > G2; G4 > G1
				ACPase in renal cortex	G2 > G1; G2 > G3
				ALP in renal medulla	Not significant
				GGTase in renal medulla	G1 > G2; G3 > G2
				LAP in renal medulla	G1 > G2; G3 > G1; G3 > G3; G4 > G1
				ACPase in renal medulla	Not significant
				ALP in cortical brush border membrane vesicles of kidney	G1 > G2; G1 > G3; G4 > G1
				GGTase in cortical brush border membrane vesicles of kidney	G1 > G2; G1 > G3; G3 > G2; G4 > G1
				LAP in cortical brush border membrane vesicles of kidney	G1 > G2; G3 > G2; G4 > G1
				LDH in renal cortex	G1 > G4; G2 > G1; G2 > G3
				MDH in renal cortex	G1 > G2; G1 > G3; G3 > G2
				HK in renal cortex	Not significant
				G6Pase in renal cortex	G1 > G2; G3 > G2
				FBPase in renal cortex	G1 > G2; G1 > G3; G3 > G2
				ME in renal cortex	G2 > G1; G2 > G3

(*Continued*)

TABLE 11.1 (*Continued*)
Results of Studies on Mitigation Effect of As Toxicity by Natural Plants

Plant Material	References	Cell/Animal	Experiment	Measurement	Results
				G6PDH in renal cortex	G1>G2; G3>G2
				LDH in renal medulla	G2>G1; G2>G3; G3>G1
				MDH in renal medulla	G1>G2
				HK in renal medulla	Not significant
				G6Pase in renal medulla	G1>G2
				FBPase in renal medulla	G1>G2
				ME in renal medulla	G2>G1
				G6PDH in renal medulla	G1>G2
				Lipid peroxidation in renal cortex	G2>G1; G2>G3
				Total SH in renal cortex	G1>G2
				SOD in renal cortex	G1>G2; G1>G3; G3>G2
				Catalase in renal cortex	G1>G2; G3>G2; G4>G1
				GSH-Px in renal cortex	G1>G2; G3>G2; G4>G1
				Lipid peroxidation in renal medulla	G1>G4; G2>G1; G2>G3
				Total SH in renal medulla	G1>G2
				SOD in renal medulla	G1>G2; G1>G3; G3>G2
				Catalase in renal medulla	G1>G2; G4>G1
				GSH-Px in renal medulla	G1>G2; G1>G3; G1>G4; G3>G2

(Continued)

TABLE 11.1 (*Continued*)

Results of Studies on Mitigation Effect of As Toxicity by Natural Plants

Plant Material	References	Cell/Animal	Experiment	Measurement	Results
Indian gooseberry	[114]	Swiss albino mouse	G1; distilled water, for 24 h	Histopathology of kidney	Arsenic-induced damage to renal corpuscles and tubules were protected by flaxseed oil
			G2; sodium arsenite, 2.5 mg/kg bw/day, p.o., for 24 h	Chromosomal aberration in femur	G2>G4a>G1, G3a; G2>G4b>G1, G3b
			G3a: extract of *E. officinalis*, 685 mg/kg bw/day, for 7 days	Chromosomal breaks	G2>G4a>G1, G3a; G2>G4b>G1, G3b
			G4a; G3 → G2	Damaged cells	G2>G4a>G1, G3a; G2>G4b>G1, G3b
			G3b; extract of *E. officinalis*, 685 mg/kg bw/day, for 14 days		
			G4b; G3b → G2		
Indian pennywort	[115]	Wister rat (male)	G1; normal	Body weight gain	Not significant
			G2: *C. asiatica*, 100 mg/kg/day, o.p., for 4 weeks	Food intake	Not significant
			G3: *C. asiatica*, 200 mg/kg/day, o.p., for 4 weeks	Water intake	Not significant
			G4: *C. asiatica*, 300 mg/kg/day, o.p., for 4 weeks	WBC in blood	Not significant
			G5; sodium arsenite, 20 ppm in drinking water, for 4 weeks	RBC in blood	Not significant
			G6; G2+G5	HGB in blood	Not significant
			G7; G2+G6	HCT in blood	Not significant
			G8; G2+G7	MCV in blood	Not significant

(*Continued*)

TABLE 11.1 (Continued)

Results of Studies on Mitigation Effect of As Toxicity by Natural Plants

Plant Material	References	Cell/Animal	Experiment	Measurement	Results
				MCH in blood	Not significant
				MCHC in blood	Not significant
				PLT in blood	Not significant
				ALAD in blood	G1>G5; G6>G5; G7>G5; G8>G5
				ZPP in blood	G5>G1
				GSH in blood	G1>G5
				As in blood	G5>G1; G5>G8
				As in liver	G5>G1; G5>G8
				As in kidney	G5>G1
				As in brain	G5>G1; G5>G7; G5>G8
				GSH in liver	G1>G5; G5>G7; G5>G8
				GSSG in liver	G5>G1; G5>G8
				TBARS in liver	G5>G1; G5>G8
				GSH in kidney	G1>G5
				GSSG in kidney	G5>G1; G5>G6; G5>G7; G5>G8
				TBRAS in kidney	G4>G1; G5>G6
				GSH in brain	Not significant
				GSSG in brain	Not significant
				TBRAS in brain	G5>G1; G5>G7; G5>G8
				SOD in brain	G5>G1; G6>G5; G7>G5
				Catalase in brain	G1>G5; G6>G5

(Continued)

TABLE 11.1 (*Continued*)
Results of Studies on Mitigation Effect of As Toxicity by Natural Plants

Plant Material	References	Cell/Animal	Experiment	Measurement	Results
				Cu in blood	Not significant
				Cu in liver	Not significant
				Cu in kidney	G5>G1; G6>G5; G7>G5; G8>G5
				Zn in blood	G7>G5; G8>G5
				Zn in liver	Not significant
				Zn in kidney	Not significant
				Fe in blood	Not significant
				Fe in liver	Not significant
				Fe in kidney	Not significant
Jaggery	[116]	Swiss albino mouse (male)	G1; control, for 180 days	Initial body weight	G2>G1, G3>G1
			G2; sodium arsenite, 0.05 ppm in distilled water, gavage, for 180 days	Final body weight	G1>G2, G1>G3, G1>G4, G1>G5, G4>G2, G5>G3
			G3; sodium arsenite, 5 ppm in distilled water, gavage, for 180 days	TAS in serum	G1>G2, G1>G3, G4>G2, G5>G3
			G4; G2+[jaggery, 250 mg/kg in distilled water, gavage, for 180 days]	GPx in blood	G1>G2, G1>G3, G4>G2, G5>G3
			G5; G3+[jaggery, 250 mg/kg in distilled water, gavage, for 180 days]	GR in serum	G1>G2, G1>G3, G4>G2, G5>G3
				TNF-α in serum	G2>G1, G3>G1, G2>G4, G3>G5

(*Continued*)

TABLE 11.1 (*Continued*)
Results of Studies on Mitigation Effect of As Toxicity by Natural Plants

Plant Material	References	Cell/Animal	Experiment	Measurement	Results
				IL-1β in serum	G2>G1, G3>G1, G2>G4, G3>G5
				IL-6 in serum	G2>G1, G3>G1, G2>G4, G3>G5
				DNA tail moment in bone marrow cells	G2>G1, G3>G1, G2>G4, G3>G5
				Histopathology in lung	Arsenic-induced necrosis and degenerative changes in bronchiolar epithelium with emphysema and thickening of alveolar septa were antagonized by jaggery
Mango	[118]	Proximal tubule epithelial cell line from normal adult human kidney (HK-2)	[MSBE, 0, 50, and 100 μg/mL, for 2 h] → [arsenite, 0–100 μM, for 24 h] [MG, 0 and 100 μM, for 2 h] → [arsenite, 0–100 μM, for 24 h] [CTCH, 0 and 100 μM, for 2 h] → [arsenite, 0–100 μM, for 24 h]	Cell viability	Reduced cell viability by arsenite was mitigated by MSBE, GA, CTCH, MG, MG+Fe

(*Continued*)

TABLE 11.1 (Continued)
Results of Studies on Mitigation Effect of As Toxicity by Natural Plants

Plant Material	References	Cell/Animal	Experiment	Measurement	Results
			[GA, 0 and 100 μM, for 2 h] → [arsenite, 0–100 μM, for 24 h]		
			[QCT, 0 and 25 μM, for 2 h] → [arsenite, 0–100 μM, for 24 h]		
			[MG+Fe, 0 and 100 μM, for 2 h] → [arsenite, 0–100 μM, for 24 h]		
			G1; control	Cell morphology	Apoptosis by arsenite was reduced by MSBE treatment
			G2; arsenite, 100 μM, for 6 h		
			G3; G2+[MSBE, 100 μg/mL, for 6 h]		
			G4; [arsenite, 100 μM, for 24 h]+[MSBE, 100 μg/mL, for 24 h]		
			[MSBE, 0 and 100 μg/mL, for 3 h] → [arsenite, 0–200 μM, for 2 h]	Cell survival	Cell survival was elevated by all treatments compared with arsenite
			[MG, 0 and 100 μM, for 3 h] → [arsenite, 0–200 μM, for 2 h]		
			[MG+Fe, 0 and 100 μM, for 3 h] → [arsenite, 0–200 μM, for 2 h]		
			[CTCH, 0 and 100 μM, for 3 h] → [arsenite, 0–200 μM, for 2 h]		
			[GA, 0 and 100 μM, for 3 h] → [arsenite, 0–200 μM, for 2 h]		

(Continued)

TABLE 11.1 (*Continued*)
Results of Studies on Mitigation Effect of As Toxicity by Natural Plants

Plant Material	References	Cell/Animal	Experiment	Measurement	Results
			G1; control	ROS	G1>G6; G1>G7; G1>G8; G1>G9; G1>G10; G1>G11; G1>G12; G1>G13; G2>G1
			G2; H_2O_2, 10 μM, for 6 h		
			G3; arsenite, 20 μM, for 6 h		
			G4; arsenite, 60 μM, for 6 h		
			G5; arsenite, 100 μM, for 6 h		
			G6; MG, 100 μM, for 6 h		
			G7; G6+G3		
			G8; G6+G4		
			G9; G6+G5		
			G10; MSBE, 100 μg/mL, for 6 h		
			G11; G10+G3		
			G12; G10+G4		
			G13; G10+G5		
			[MSBE, 0, 50, and 100 μg/mL, for 72 h] → [arsenite, 0–100 μM, for 24 h]	Cell viability	Reduced cell viability by arsenite was mitigated by MSBE, GA, CTCH, MG, MG+Fe
			[MG, 0 and 100 μM, for 72 h] → [arsenite, 0–100 μM, for 24 h]		
			[CTCH, 0 and 100 μM, for 72 h] → [arsenite, 0–100 μM, for 24 h]		

(Continued)

TABLE 11.1 (*Continued*)
Results of Studies on Mitigation Effect of As Toxicity by Natural Plants

Plant Material	References	Cell/Animal	Experiment	Measurement	Results
			[GA, 0 and 100 μM, for 72 h] → [arsenite, 0–100 μM, for 24 h]		
			[QCT, 0 and 25 μM, for 72 h] → [arsenite, 0–100 μM, for 24 h]		
			[MG + Fe, 0 and 100 μM, for 72 h] → [arsenite, 0–100 μM, for 24 h]		
			[Control, for 72 h] → [arsenite, 0–100 μM, for 24 h]	Cell viability	Pretreatments of arsenite and ALB resulted in more resistance and sensitivity, respectively, to cytotoxicity by arsenite.
			[ALB, 15 mg/mL, for 72 h] → [arsenite, 0–100 μM, for 24 h]		
			[Arsenite, 2 μM, for 72 h] → [arsenite, 0–100 μM, for 24 h]		
			Control, for 72 h	P-gp protein expression	Increased with arsenite treatment
			ALB, 15 mg/mL, for 72 h	ABCB1 mRNA expression	Increased with arsenite treatment
			Arsenite, 2 μM, for 72 h	Cell viability	G1 > G2; G1 > G3; G1 > G4; G3 > G2
			G1; control		

(*Continued*)

TABLE 11.1 (*Continued*)
Results of Studies on Mitigation Effect of As Toxicity by Natural Plants

Plant Material	References	Cell/Animal	Experiment	Measurement	Results
			G2; arsenite, 60 μM, for 24 h	P-gp protein expression	Increased with MG treatment
			G3; [MG, 10 μM, for 2 months] → G2	ABCB1 mRNA expression	No discussion
			G4; [MSBE, 100 mg/mL, for 2 months] → G2		
			Control		
			MG, 10 μM, for 2 months		
			MSBE, 100 mg/mL, for 2 months		
Marine algae	[121]	Wister rat (both sex)	G1; control	Growth rate of body weight at 1 week	G2>G6; G2>G7; G2>G8
			G2; arsenic trioxide, 5 mg/kg bw/day, p.o., for 6 weeks	Growth rate of body weight at 2 weeks	Not significant
			G3; G2+[DMPS, 5 mg/kg bw/day, i.p., for 3 continual days a week in 6 weeks]	Growth rate of body weight at 3 weeks	Not significant
			G4; G2+[*L. japonica*, 1000 mg/kg bw, p.o., for 6 continual days a week in 6 weeks]	Growth rate of body weight at 4 weeks	G9>G2
			G5; G2+[*L. japonica*, 500 mg/kg bw, p.o., for 6 continual days a week in 6 weeks]	Growth rate of body weight at 5 weeks	G1>G2
			G6; G2+[*L. japonica*, 100 mg/kg bw, p.o., for 6 continual days a week in 6 weeks]	Growth rate of body weight at 6 weeks	G1>G2
			G7; G2+[*P. haitanensis*, 1000 mg/kg bw, p.o., for 6 continual days a week in 6 weeks]	TP in serum	G2>G1; G2>G3; G2>G4; G2>G5; G2>G6; G2>G7; G2>G8; G2>G9
			G8; G2+[*P. haitanensis*, 500 mg/kg bw, p.o., for 6 continual days a week in 6 weeks]	ALT in serum	G2>G1; G2>G3; G2>G5; G2>G6; G2>G7; G2>G8

(Continued)

TABLE 11.1 (Continued)

Results of Studies on Mitigation Effect of As Toxicity by Natural Plants

Plant Material	References	Cell/Animal	Experiment	Measurement	Results
			G9: G2 + [P. haitanensis, 100 mg/kg bw, p.o., for 6 continual days a week in 6 weeks]	ALP in serum	G2>G1; G2>G3; G2>G4; G2>G5; G2>G6; G2>G8; G2>G9
				BUN in serum	G2>G1; G2>G3; G2>G5; G2>G6; G2>G7; G2>G8
				Cr in serum	G2>G1; G2>G3; G2>G7; G2>G8
				TC in serum	G1>G2; G3>G2; G4>G2; G5>G2; G6>G2; G7>G2; G8>G2
				TG in serum	G1>G2; G3>G2; G5>G2; G8>G2; G9>G2
				HDL–C in serum	G2>G1; G2>G3; G2>G5; G2>G8
				LDL–C in serum	G1>G2; G5>G2; G6>G2; G7>G2; G8>G2; G9>G2
				SOD in serum	G1>G2; G3>G2; G4>G2; G5>G2; G7>G2
				GPx in serum	G7>G2; G8>G2
				MDA in serum	G2>G1; G2>G3; G2>G4; G2>G5; G2>G6
				SH in serum	G1>G2; G3>G2; G4>G2; G5>G2; G6>G2; G7>G2; G8>G2; G9>G2

Influences of green tea on the concentration of lipid peroxidation products and on certain hematological and biochemical parameters in Wister rats exposed to As were investigated by Messarah et al. [48]. Increased lipid peroxidation in liver, kidney, and testis, and renal and hepatic injury markers in serum by As exposure were decreased by green tea.

Sinha and Roy [49] investigated the protective effects of tea against oxidative damage by As exposure at mRNA and protein levels. Treatments of green, Darjeeling, and Assam teas could suppress an oxidative DNA damage marker, 8-hydroxy-2′-deoxyguanosine (8-OHdG) induced by As in liver of Swiss albino mice. Similarly, As-induced OGG1 protein in plasma and mRNA in blood were decreased by tea treatments. Furthermore, at both protein and genetic levels, several DNA-repair enzymes were inhibited by As, while the expressions increased by tea treatments.

Chandronitha et al. [50] focused on tannins in green tea because of their capacity as metal chelator, protein precipitating agents, and antioxidants [51]. Their research group compared the effects of antiarsenic toxicity between crude green tea and detannified green tea. Green tea extract had higher remission in oxidative stress and lower accumulation of As and frequency of tissue abnormality in liver and kidney than detannified green tea, suggesting that tannin in green tea is beneficial to reduce As toxicity.

Specific scenario study on the interaction of supplemental iron (Fe) on anti-As toxicity (chromosome damage) of black tea by using mice was conducted by Poddar [52]. Because in 1988, United Nations International Children's Emergency Fund (UNICEF), WHO, and the International Nutritional Anemia Consultative Group (INACG) published guidelines for Fe supplementation in children aged 6–24 months in developing countries [53]. Their results showed that black tea significantly reduced chromosomal breakage in bone marrow induced by As, while ferrous sulfate did not modify the protective effect of black tea [52].

A recent study targeting the As-exposed population in West Bengal, India, investigated relationship between micronuclei in oral mucosal cells by As exposure and tea consumption habit [54]. Frequency of micronuclei in As-exposed group showed significantly higher than that in the reference group. Authors mentioned that individuals with tea drinking habit had lower percentage of micronuclei compared with non–tea drinkers, but the result of statistical significant analysis was not shown.

11.3 CURCUMIN/TURMERIC

Curcumin (diferuloymethane) is obtained from *Curcuma longa*, which is called turmeric. This yellow colored powder is commonly used as spice, food preservative, and coloring material in the world, especially, in Asian countries. Having polyphenolic structure and β-diketone functional group, it is known that curcumin has antioxidant property [55]. Therefore, several researchers have investigated the detoxification role of curcumin against As-related toxicity.

Mukherjee et al. [56] tested modulation of As toxicity by curcumin using human lymphocytes. DNA damage induced by As could be efficiently reduced by curcumin. Interestingly, pretreatment with curcumin prior to As exposure was more effective than the restoration. Curcumin reduced ROS and lipid per-oxidation, and increased phase II detoxification enzymes as well as DNA repair activity and expression of poly (ADP ribose) polymerase, a repair enzyme in As-exposed lymphocytes. Protective effect by curcumin against genotoxicity of As and fluoride (F) was evaluated by Tiwari and Rao [57]. They coexposed As, F, and curcumin in human blood lymphocytes and found increased chromosomal aberrations and DNA damage by As and F were not completely, but significantly suppressed by curcumin.

Similar approach has been performed in *in vivo* studies using mouse or rat. Yousef et al. [58] orally treated Wistar rats with As(III), curcumin, and both As(III) and curcumin, respectively, and measured biochemical parameters in liver, brain, and blood. Curcumin reduced As-induced transaminases, phosphatases, glucose, urea, creatinine, bilirubin, total lipid (TL), cholesterol, and triglyceride, while induced As-reduced liver transaminases and phosphatases, plasma and brain acetylcholin-esterase, and the levels of total protein (TP) and albumin (Alb) [58]. As-induced oxidative stress and lipid peroxidation in plasma, liver, testes, brain, kidney, and lung of Sprague Dawley rat were suppressed by curcumin [59]. To understand the detoxification mechanism of curcumin to As toxicity, Gao et al. [60] investigated the interaction between oxidative stress by As and induction of nuclear factor (erythroid-2 related) factor 2 (Nrf2) by curcumin. Nfr2 and its signal pathway are known to have a critical role in antioxidative stress [61]. According to the *in vivo* study by Gao et al. [60], curcumin treatment upregulated protein expression levels of hepatic Nrf2 as well as Nrf2 downstream genes, NADP(H) quinine oxidoreduc-tase 1 (NQO1) and heme oxygenase-1 (HO-1) in As-treated Kumming mice. On the other hand, As-induced hepatic damage and oxidative stress were suppressed by curcumin, indicating that Nrf2 activation by curcumin may be associated with defense effect against As toxicity. Interestingly, they found for the first time that curcumin treatment significantly increased methylation capacity of As and excre-tion to urine. Although the mechanism is still unknown, this enhancement of As methylation and its related As excretion by curcumin may be associated with the reduction of As toxicity.

For protective effect of curcumin against neurotoxicity of As, Yadav et al. [62–64] carried out experiments using Wister rats. In their experiments, curcumin showed significant mitigation effects on the behavior, binding of striatal dopamine recep-tors, and tyrosine hydroxylase expression reduced by As exposure [62]. In addition, As accumulation and oxidative stress in the brain of As-treated rats were decreased by curcumin [62]. Brain biogenic amines and their metabolites were prohibited and nitric oxide level was induced by As, but curcumin remedied these effects [63]. Furthermore, As-induced impaired learning and cholinergic dysfunction in brain was confirmed to be suppressed by curcumin [64].

However, one of the limitations for curcumin usage is its low bioavailability. Curcumin is poorly absorbed from the intestine and rapidly eliminated through the bile [65,66]. Recently, novel chemotherapeutic method using biodegradable polymer nanoparticles have been developed [67]. In this system, by increasing bioavailability, solubility, and retention time, detoxification function of curcumin became more effective [68]. Therefore, usage of nanocapsuled curcumin may be a useful way to increase the bioavailability and reduce the amount of its dose to prevent As toxicity. Yadav et al. [69] synthesized nanoparticles (less than 50 nm in diameter) including curcumin and compared the difference in therapeutic effect of As toxicity between curcumin and encapsulated curcumin nanoparticles treatment in As-administrated Wister rats. In blood, liver, and brain of rats, both free curcumin and encapsulated one effectively recovered adverse effects of oxidative stress, lipoperoxidation, biogenic, and amines by As exposure. In addition, As concentrations in blood, brain, and liver of rats treated with As significantly decreased with treatments of curcumin and nanocurcumin. Interestingly, the amount of curcumin nanoparticles needed for reduction of As toxicity was lower than free curcumin. Effectiveness of free curcumin and nanocurcumin (130 nm) on immune dysfunction by As in Wister rats was investigated by Sankar et al. [70]. They found that nanocurcumin more effectively removed toxic effects on cellular and humoral immune responses by As compared with free curcumin. From both studies results by Yadav et al. [69] and Sankar et al. [70], it is indicated that nanocurcumin can be more effective compared with free curcumin on enhanced antioxidant and chelating potency at lower dose.

Although the number is limited, there is some field trial experiments conducted in As-exposed population of West Bengal, India [71,72]. According to Biswas et al. [71], As exposure induced severe DNA damage with increased ROS and lipid peroxidation levels in blood. However, intake of curcumin for three months reduced the DNA damage, retarded ROS generation and lipid peroxidation, and raised antioxidant activity in the blood of population from As-contaminated areas in West Bengal, India. Later, the role of curcumin in reducing 8-hydroxy-2′-deoxyguanosine (8-OHdG) formation and enhancing DNA repair capacity in As-exposed population was investigated by Roy et al. [72]. Arsenic exposure increased 8-OHdG concentration and mRNA and protein expression levels of OGG1, which encodes the enzyme responsible for the excision of 8-oxoguanine, a mutagenic base byproduct which occurs as a result of exposure to ROS, while these parameters were significantly suppressed by curcumin. In addition, curcumin mitigated suppressed DNA repair genes and proteins by As. The authors concluded that curcumin intervention may be a useful modality for the prevention of As-induced cancers.

These results by *in vitro*, *in vivo*, and human trial studies shows that curcumin can be an available tool to significantly reduce the risk of As toxicity in rural population from As-contaminated groundwater areas in Asia. Furthermore, for the remediation of damages induced by As, nanocurcumin is more effective than free curcumin.

11.4   GARLIC

Garlic (*Allium sativum*) is widely used as a valuable spice in a variety of foods and a popular remedy for various ailments and physiological disorders. Sulfur-bearing compounds in the garlic have antioxidative property [73].

Das et al. [74] cotreated garlic extract and As to Swiss albino mice for 4 weeks and measured the chromosome aberration in the bone marrow. Arsenic treatment induced abnormal chromosome, while extract of garlic reduced the abnormality. This protective function against As toxicity may be due to the organosulfur compounds in garlic [74]. Similar results were obtained in previous short time exposure study conducted by Das et al. [75]. The same research group has further investigated the effect of garlic extract on cytotoxicity by three doses of As(III) (2.5, 0.833, and 0.5 mg/kg bw corresponding to 1/10, 1/30, and 1/50 of 50% lethal dose [LD_{50}], respectively) in the same mouse strains for 24 h and found that garlic extract could suppress As-induced chromosomal damage [76]. Mitigation effect of garlic to clastogenic toxicity of chronic As exposure (30 and 60 days) in mice was assessed by Choudhury et al. [77]. In this case, chromosomal aberrations and cell damage in bone marrow decreased when garlic extract was cotreated with As(III) [77]. Roy Choudhury et al. [78] investigated whether the protective effects found in mice treated with As and garlic extract were observed in the next generation. Chromosome damage by As treatment was observed in F1 generation, but it was significantly lower than the parents (F0). Garlic extract was not effective on clastogenic effect of As in the next generation, probably indicating that the components of the extract are not able to cross the transplacental barrier.

Diet is a complex mixture and cooperation or interaction of individual components in the diet is a confounding but interesting possibility. Choudhury et al. [79] assessed the inhibitory effect of the two dietary supplements, garlic and mustard oil on clastogenicity of As in mice. The mixture of garlic and mustard oil enhanced the protective effect against As-induced clastogenicity than single treatments.

Chowdhury et al. [80] evaluated the therapeutic efficacy of garlic extract on As toxicity and accumulation through *in vitro* and *in vivo* approaches. *In vitro* studies using human malignant melanoma cells (A375), human keratinocyte cells (HaCaT), and human dermal fibroblast cells cotreated with As(III) and garlic extract revealed that As-induced cytotoxicity and ROS generation and As-repressed p53 and hsp90 mRNA expressions were recovered by garlic extract. In *in vivo* experiment, garlic extract reduced oxidative stress, ROS, and myeloperoxidase activity as well as accumulation of As. Interestingly, a mixture of garlic extract and As(III) solution incubated overnight at room temperature produced a precipitate containing As. Concentration of As in the precipitate dose-dependently increased with garlic extract. Furthermore, in the supernatant and precipitate, As(III) was converted to As(V) by treatment of garlic extract. It is expected that arsenic sulfide (As_2S_3 and As_2S_5) are formed in the precipitate.

More comprehensive study on the protective effects of garlic extracts against As toxicity and accumulation in mice was conducted by Flora et al. [81]. In the liver, As treatment induced ROS generation causing apoptosis through mitochondria-mediated pathway, oxidative stress, and injuries, while garlic extracts cured such effects. In addition, garlic extract could enhance the excretion of As into urine and reduce the accumulation of As in liver without the disturbance of essential elements like Zn and Cu.

The latest *in vitro* study revealed that arginine was one of the significant components in aqueous garlic extract and formed a precipitate with As(III) [82]. Arginine could mitigate As toxicity such as cell death, ROS generation, oxidative stress, and DNA damage. Furthermore, protein expression level of a DNA repair enzyme, poly-ADP ribose polymerase, suppressed by As(III) was increased by arginine treatment.

These results indicates that intake of garlic is beneficial for preventing As poisoning. However, in several studies, garlic extract alone also increased chromosomal aberrations compared with negative control, but there was not enough explanation [74–77,79].

11.5 MUSHROOM LECTIN

Mushrooms are functional food for nutrition and beneficial medicines [83]. In India, oyster mushroom (*Pleuritus florida*) is used as traditional indigenous drug. Methanol extract of *P. florida* has antioxidant and antitumor activity [84]. Lectin in mushroom has significant immune-enhancing effects and health promoting activity [85].

A research group of Asit Kumar Berȧ has evaluated whether mushroom lectin is available for detoxification of As-mediated toxicity by *in vitro* [86,87] and *in vivo* [88–90] studies. Results using hepatocytes of rat showed that mushroom lectin mitigated As-induced apoptosis [86,87]. Antiapoptotic properties of the lection were also observed in renal cells of rat treated with As for 12 weeks [90]. Similarly, significant therapeutic effects of *P. florida* lectin on oxidative stress in liver [89] and kidney [88] of As-treated rats were confirmed.

11.6 HORSERADISH TREE

Horseradish tree (*Moringa oleifera*) originated from India is widely distributed in tropic countries in Africa, Asia, and South America [91]. It is cultivated for use as a vegetable (leaves, green pods, flowers, and roasted seeds), spice (mainly roots), for cooking and cosmetic oil (seeds), and as a medicinal plant (all plant organs) [92]. The extracts of leaf, fruit, and seed have protective activity against oxidative DNA damage [93].

Gupta et al. [94] revealed that the protective effect of *M. oleifera* against As toxicity for the first time. They treated seed powder of *M. oleifera* in post As-administrated Wister rats and found that oxidative stress in blood, liver, kidney, and brain induced by As was suppressed. Furthermore, *M. oleifera* could significantly remove As burden in blood and organs although the capacity was moderate. Later, the same research group further investigated the effect of concomitant administration of As and *M. oleifera* in Swiss mice [91]. Treatment of *M. oleifera* reduced oxidative stress in blood, liver, kidney, and brain and hepatic damage as well as body burden of As. Similarly, arsenic-induced hepatic function, oxidative stress, and DNA fragmentation in Wister rats were significantly protected by *M. oleifera* [95].

To understand the combination effect of chelating agents and herbal extracts on As toxicity, Mishra et al. [96] coadministrated a thiol chelator, monoisoammyl DMSA (MiADMSA) and seed powder of *M. oleifera* in Swiss mice treated with As(III) for 6 months. This combination treatment showed better protective effect against toxicities related to ROS, ALAD, oxidative stress, enzymes of antioxidant defense, MT, and hepatic injures, and disturbance of metal accumulation caused by As exposure than separate MiADMSA and *M. oleifera* treatment [96].

One *in vitro* study on mitigation effect of horseradish tree against As conducted by Verma et al. [97] showed that coexposure of arsenic trioxide and extract of *M. oleifera* in chicken liver homogenate ameliorated lipid peroxidation, although only As treatment showed no significant variation.

From these results, it can be concluded that *M. oleifera* has two beneficial characteristics: (1) to restore As-induced oxidative stress and (2) to prevent accumulation of As in the target organs/tissues.

11.7 TOSSA JUTE

Tossa jute, *Corchorus olitorius* Linn. is one of the major agricultural products of West Bengal, India and Bangladesh, which are severe As-contaminated groundwater areas in the world. It is known that the leaves contain antioxidants namely carotenoids, flavonoids, and vitamin C [98–100].

Research group of Das et al. conducted a set of studies on the efficiency of *C. olitorius* against As-induced toxicity [101–103]. Prior to As(III) exposure, aqueous extract of *C. olitorius* was daily administrated to rats for 15 days. This pretreatment by *C. olitorius* prevented As-induced oxidative stress in the liver, kidney, brain, and heart of rats [101–103]. Measurement of As concentration in the heart of rats showed that *C. olitorius* could reduce the As content [103]. Such protective effect may be due to the presence of phytophenols and flavonoids.

11.8 SEABUCKTHORN

Seabuckthorn (*Hippophae rhamnoides* L.) is deciduous shrub that is distributed from northwestern Europe through Central Asia to the Altai Mountain, Western Northern China, and the northern Himalayas [104]. An extract of *H. rhamnoides* L. has the function of scavenging superoxide radicals and preventing lipid peroxidation, probably due to existence of polyphenols in the extract [105].

Given the potential effects of antioxidative stress by *H. rhamnoides*, Gupta and Flora [106] investigated the therapeutic efficacy of this plant against As toxicity in blood, liver, kidney, and brain of mice. In liver and brain, pretreatment by extracts of *H. rhamnoides* recovered oxidative stress disturbed by As(III). On the contrary, significant depression of As was not observed in blood, liver, kidney, and brain in mice pretreated with the extracts; moreover, ethanolic extract enhanced As accumulation in brain. Similar results were obtained when As(III) and extracts of *H. rhamnoides* were concomitantly treated in mice [107]. Extracts of *H. rhamnoides* were effective for the reduction of As-induced oxidative damage but not for the accumulation of As in the blood and organs. Therefore, it can be concluded that *H. rhamnoides* has anti-As-induced oxidative stress, but does not have chelating capacity for As.

11.9 ALOE VERA

Aloe vera (*Aloe barbadensis*) is widely used as a traditional medicine. Gupta and Flora [35] investigated the oxidative stress and As accumulation in Wister rats concomitantly exposed to As and Aloe vera. Results of analyses of oxidative stress showed significant protective effect by Aloe vera at higher dose in liver, but not in kidney. Interestingly, unlike other natural plant materials discussed in this review, Aloe vera had no effect on As concentrations in blood, liver, and kidney of rats, indicating that Aloe vera has protective value against As oxidative damage, but these effects are independent of As depletion from body.

11.10   ARGENTINIAN MEDICAL PLANTS

By using African green monkey kidney cells (Vero cells) treated by As(III), Soria et al. [108] investigated protective antioxidant activity of five traditional medical plants in Argentina, *Eupatorium buniifolium*, *Lantana grisebachii*, *Mandevilla pentlandiana*, *Sebastiania commersoniana*, and *Heterothalamus alienus*. The cells were treated with extracts of the plants by petroleum ether (PE), dichloromethane (DCM), methanol (OL), and water (W) and the resulting aqueous and lipid hydroperoxides were analyzed. As(III)-induced hydroperoxides formation in both aqueous and lipid fraction was significantly suppressed by extracts of PE of *E. buniifolium*, PE and W of *L. grisebachii*, PE and W of *M. pentlandiana*, and DCM, OL, and W of *S. commersoniana*. Only increased lipid hydroperoxide was prevented from PE, DCM, and OL extracts of *H. alienus*. Finally, the authors concluded that W extract of *L. grisebachii* may be a suitable and useful preparation because this is used as a tea in Argentina. However, although some extracts significantly increased aqueous or lipid hydroperoxides compared with single As(III) treatment, there is no discussion about them.

11.11   FLAXSEED OIL

Flaxseed oil is used as/in food in Asia, Europe, and Africa. Seed oil of flax (*Linum usitatissimum*) contains high concentrations of ω-3 polyunsaturated fatty acids (PUFAs), especially α-linolenic acid and lignan. These components have beneficial effects on human health because of its antioxidant property [109]. To understand the protective effect of flaxseed oil on renal As toxicity comprehensively, Rizwan et al. [110] pretreated flaxseed oil in rats for 14 days and then exposed them to As(III) for 4 days. Severe nephrotoxicity and profound damage in plasma membranes of the renal proximal tubules as well as the disruption of enzymes including oxidative carbohydrate metabolism, gluconeogenesis, and brush border membrane were induced by As(III). On the other hand, pretreatment of flaxseed oil significantly attenuated the As-induced toxic effects. Probably, enriched PUFAs and lignans in flaxseed oil enhanced resistance to free radical attack generated by As(III) [110]. However, only flaxseed oil treatment significantly influences several characteristics of kidney, but there was no discussion about these results in the study.

11.12   INDIAN GOOSEBERRY

A fruit extract of Indian gooseberry, *Emblica officinalis* Gaertn, is known to be vitamin C rich and has the capacity to reduce cytotoxicity of Zn, Pb, Al, and Ni

[111–113]. The extract has been used as a traditional medicine in India. Biswas et al. [114] found this extract reduce the extent of cell damage by As exposure *in vivo*. In their experiments, the increased chromosomal aberrations and damaged cells by As(III) exposure were suppressed by short (7 days)- and long (14 days)-time treatments of the crude extract of the fruit of *E. officinalis* Gaertn in mice [114].

11.13 INDIAN PENNYWORT

Centella asiatica is a small herbaceous plant that occurs in marshy places throughout the country up to an altitude of 200 m. It is used as a medicinal herb in traditional medicine. Effects of the crude extract of *C. asiatica* on As-induced oxidative stress and biochemical changes, and removal of As were investigated by Gupta and Flora [115]. *C. asiatica* significantly recovered oxidative stress and ALAD, and marginal chelation effect were observed in blood, liver, and brain at the high dose treatment. However, as a significant side effect of As and *C. asiatica* cotreatment, it was found that the Cu concentration in kidney increased.

11.14 JAGGERY

Jaggery is a natural sweetener made from sugar cane (*Saccharum officinarum*) and contains polyphenols, vitamin C, carotene, and other biologically active components. According to a study by Singh et al. [116], cotreatment of jaggery antagonized oxidative stress and DNA damage as well as immune toxicity by As exposure for 180 days in mice. Although the mechanism of prevention is still unknown, biologically active compounds in jaggery may be playing an important role in the inhibition of As toxicity.

11.15 MANGO

Polyphenols in mango, *Mangifera indica* L. have antioxidants and anti-inflammatory, immunomodulatory, chemopreventive, and anticancer functions [117]. Garrido et al. [118] investigated whether the extract of the stem bark of mango as well as polyphenols including mangiferin, catechin, gallic acid, and quercetin can mitigate cytotoxicity by As in proximal tubule cell line (HK-2). The crude extract effectively alleviated As-induced cytotoxicity, followed by gallic acid, catechin, and mangiferin. Induction of P-glycoprotein (P-gp), a potential transporter of As, was observed by mango extract and mangiferin treatments, suggesting that the interaction between mango polyphenols and P-gp is associated with As resistance. As(III) (up to 100 μM) did not induce ROS in this cell line, but extract of mango and mangiferin significantly reduced ROS compared with control [118].

11.16 MARINE ALGAE

Marine algae are commonly ingested by Chinese, Japanese, Korean, and some other Southeast Asian people. Antioxidative effect of *Laminaria japonica* and *Porphyrya haitanesis*, which are economically important edible algae, has been reported in several studies [119,120]. According to a recent study by Jin et al. [119], ingestion of *L. japonica* and *P. haitanesis* powder could attenuate liver and kidney malfunctions in rats treated with arsenic trioxide. Arsenic-induced disruption of lipid profile and oxidative stress in serum were protected by both algae treatments. In addition, as no side effects were detected in rats treated with both algae, the authors concluded that these algae can be used as a safe and effective regimen for treating As poisoning [121]. However, the dose-dependent mitigation effects of As toxicity by both algae were not clear in this study.

11.17 CONCLUSIONS

In this review, mitigation effects of As-induced toxicity by natural products were summarized. To resolve the severe situation of As poisoning through the consumption of As-polluted groundwater, many studies have been conducted by research groups in India. Through these studies, it is suggested that many plant materials, which have been commonly or traditionally consumed in the local areas, are useful for mitigating As poisoning. Among those studies, tea may be a useful tool due to its commercial availability in the wide areas of the affected regions. Nanoparticle of curcumin, which was recently developed, can be beneficial considering its mitigation efficiency, but the high production cost is still a barrier in the developing countries.

One of the critical factors is that the mechanism of mitigation effects by natural plants remains still unknown. Therefore, more molecular biology studies focusing on the protection mechanism are needed. In addition to this, human case studies on mitigation effects of As toxicity by pre-/aftertreatment of these effective natural plants are still lacking. More clinical studies, in which the efficiency of mitigation together with the confirmed absence of side effects, are required for using plant materials for treatment by people in the As-contaminated areas.

ACKNOWLEDGMENTS

I am grateful to Professor A. Subramanian in the CMES, Ehime University, Japan, for critical reading of the manuscript. This study was supported by Grants-in-Aid for Scientific Research (S) (No. 26220103) from the Japan Society for the Promotion of Science (JSPS).

ABBREVIATION LIST

AChE	Acetylcholinesterase
ACP	Acid phosphatase
ACPase	Acid phosphatase
AECO	Aqueous extract of *Corchorus olitonus* leaves
ALAD	Aminolevulinic acid dehydratase
ALB	Albumin
ALP	Alkaline phosphatase
ALT	Alanine aminotransferase
AST	Aspartate aminotransferase
AT	Assam tea

CAT	Catalase
CD	Conjugated dienes
ChAT	Choline acetyltransferase
CHL	Cholesterol
Cr	Creatinine
CTCH	Catechin
DA	Dopamine
DMA	Dimethylarsinic acid
DMPS	2,3-Dimercaptopropane-1-sulfonate
DT	Darjeeling tea
DTH	Delayed-type hypersensitivity
EGCG	(–)-Epigallocatechingallate
FBP	Fructose-1,6-bisphosphatase
G6Pase	Glucose-6-phosphatase
G6PDH	Glucose-6-phosphate dehydrogenase
GA	Gallic acid
GGTase	γ-Glutamyltransferase
GPx	Glutathione peroxidase
GR	Glutathione reductase
GSH	Reduced glutathione
GSH-Px	Glutathione peroxidase
GT	Green tea
5-HT	5-Hydroxy-tryptamine
Hb	Hemoglobin
HCT	Hematocrit
HDL-C	High density lipoprotein cholesterol
HF-EtOH	Concentration of ethanol extract of *H. rhamnoides* L. fruit by using a vacuum pressure
HF-WRC	Concentration of water extract of *H. rhamnoides* L. fruit by using a reflux method
HF-WRT	Concentration of water extract of *H. rhamnoides* L. fruit by using a rotary evaporator
HGB	Hemoglobin
HK	Hexokinase
HO-1	Heme oxygenase-1
IAs	Inorganic As
KLH	Keyhole-limpet hemocyanin
LAP	Leucine aminopeptidase
LDH	Lactate dehydrogenase
LDL-C	Low density lipoprotein cholesterol
LPO	Lipid peroxidation
MCH	Mean cell hemoglobin
MCHC	Mean cell hemoglobin concentration
MCV	Mean cell volume
MDH	Malate dehydrogenase

ME	Malic enzyme
MG	Mangiferin
MMA	Monomethylarsonic acid
MN	Micronuclei
MO	*Moringa oleifera*
MPO	Myeloperoxidase
MSBE	*M. indica* stem bark extract
NE	Norepinephrine
NQO1	NADP(H) quinine oxidoreductase 1
Nrf2	Nuclear factor (erythroid-2 related) factor 2
PARP	Poly (ADP-ribose) polymerase
PF	*P. fraternus*
P-gp	P-glycoprotein
PLT	Platelet count
PMR	(MMA + DMA)/T-As
QCT	Querectin
QNB	Quinuclidinyl benzilate
RBC	Red blood cells
SMR	DMA/(MMA + DMA)
SOD	Superoxide dismutase
TA	*Terminalia arjuna*
TAS	Total antioxidant status
T-As	Total As
TBARS	Thiobarbituric acid reactive substances
TC	Total cholesterol
TF	Theaflavin
TG	Triglyceride, Triglycerides
TH	Tyrosine hydroxylase
TL	Total lipid
TP	Total protein
TPH	Tryptophan hydroxylase
WBC	White blood cells
ZPP	Zinc protoporphyrin

REFERENCES

1. Kitchin, K.T., 2001. Recent advances in arsenic carcinogenesis: Modes of action, animal model systems, and methylated arsenic metabolites. *Toxicology and Applied Pharmacology* 172, 249–261.
2. Kitchin, K.T., Ahmad, S., 2003. Oxidative stress as a possible mode of action for arsenic carcinogenesis. *Toxicology Letters* 137, 3–13.
3. Hei, T.K., Filipic, M., 2004. Role of oxidative damage in the genotoxicity of arsenic. *Free Radical Biology and Medicine* 37, 574–581.
4. Feng, Z., Xia, Y., Tian, D., Wu, K., Schmitt, M., Kwok, R.K., Mumford, J.L., 2001. DNA damage in buccal epithelial cells from individuals chronically exposed to arsenic via drinking water in Inner Mongolia, China. *Anticancer Research* 21, 51–58.

5. Basu, A., Som, A., Ghoshal, S., Mondal, L., Chaubey, R.C., Bhilwade, H.N., Rahman, M.M., Giri, A.K., 2005. Assessment of DNA damage in peripheral blood lymphocytes of individuals susceptible to arsenic induced toxicity in West Bengal, India. *Toxicology Letters* 159, 100–112.

6. Kubota, R., Kunito, T., Agusa, T., Fujihara, J., Monirith, I., Iwata, H., Subramanian, A., Tana, T.S., Tanabe, S., 2006. Urinary 8-hydroxy-2'-deoxyguanosine in inhabitants chronically exposed to arsenic in groundwater in Cambodia. *Journal of Environmental Monitoring* 8, 293–299.

7. Cox, P.A., 1995. *The Elements on Earth: Inorganic Chemistry in the Environment*. Oxford University Press, Oxford, U.K., p. 287.

8. Gorby, M.S., 1994. Arsenic in human medicine. In: Nriagu, J.O. (ed.), *Arsenic in the Environment, Part II: Human Health and Ecosystem Effects*. Wiley, New York, pp. 1–16.

9. WHO, 2001. *Environmental Health Criteria 224: Arsenic and Arsenic Compounds*, 2nd edn. World Health Organization, Geneva, Switzerland.

10. Agency for Toxic Substances and Disease Registry (ATSDR), 2013. Top 20 hazardous substances: ATSDR/EPA priority list for 2013. http://www.atsdr.cdc.gov/spl/. Accessed on March 31, 2015.

11. WHO, 2004. *Guidelines for Drinking Water Quality*, 3rd edn. World Health Organization, Geneva, Switzerland.

12. Mandal, B.K., Suzuki, K.T., 2002. Arsenic round the world: A review. *Talanta* 58, 201–235.

13. Winkel, L., Berg, M., Amini, M., Hug, S.J., Johnson, C.A., 2008. Predicting groundwater arsenic contamination in Southeast Asia from surface parameters. *Nature Geoscience* 1, 536–542.

14. Nordstrom, D.K., 2002. Public health. Worldwide occurrences of arsenic in ground water. *Science* 296, 2143–2145.

15. Bundschuh, J., Litter, M.I., Parvez, F., Román-Ross, G., Nicolli, H.B., Jean, J.S., Liu, C.W. et al., 2012. One century of arsenic exposure in Latin America: A review of history and occurrence from 14 countries. *Science of the Total Environment* 429, 2–35.

16. Chakraborti, D., Mukherjee, S.C., Pati, S., Sengupta, M.K., Rahman, M.M., Chowdhury, U.K., Lodh, D., Chanda, C.R., Chakraborti, A.K., Basu, G.K., 2003. Arsenic groundwater contamination in Middle Ganga Plain, Bihar, India: A future danger? *Environmental Health Perspectives* 111, 1194–1201.

17. Chowdhury, U.K., Biswas, B.K., Chowdhury, T.R., Samanta, G., Mandal, B.K., Basu, G.C. et al., 2000. Groundwater arsenic contamination in Bangladesh and West Bengal, India. *Environmental Health Perspectives* 108, 393–397.

18. Rahman, M.M., Chowdhury, U.K., Mukherjee, S.C., Mondal, B.K., Paul, K., Lodh, D., Biswas, B.K. et al., 2001. Chronic arsenic toxicity in Bangladesh and West Bengal, India—A review and commentary. *Journal of Toxicology—Clinical Toxicology* 39, 683–700.

19. Tseng, W.P., 1977. Effects and dose response relationships of skin cancer and blackfoot disease with arsenic. *Environmental Health Perspectives* 19, 109–119.

20. Wang, C.H., Hsiao, C.K., Chen, C.L., Hsu, L.I., Chiou, H.Y., Chen, S.Y., Hsueh, Y.M., Wu, M.M., Chen, C.J., 2007. A review of the epidemiologic literature on the role of environmental arsenic exposure and cardiovascular diseases. *Toxicology and Applied Pharmacology* 222, 315–326.

21. Ahmad, S.A., Salim Ullah Sayed, M.H., Barua, S., Haque Khan, M., Faruquee, M.H., Jalil, A., Abdul Hadi, S., Kabir Talukder, H., 2001. Arsenic in drinking water and pregnancy outcomes. *Environmental Health Perspectives* 109, 629–631.

22. Milton, A.H., Smith, W., Rahman, B., Hasan, Z., Kulsum, U., Dear, K., Rakibuddin, M., Ali, A., 2005. Chronic arsenic exposure and adverse pregnancy outcomes in Bangladesh. *Epidemiology* 16, 82–86.

23. Berg, M., Luzi, S., Trang, P.T.K., Viet, P.H., Giger, W., Stüben, D., 2006. Arsenic removal from groundwater by household sand filters: Comparative field study, model calculations, and health benefits. *Environmental Science and Technology* 40, 5567–5573.

24. Agusa, T., Iwata, H., Fujihara, J., Kunito, T., Takeshita, H., Minh, T.B., Trang, P.T., Viet, P.H., Tanabe, S., 2009. Genetic polymorphisms in AS3MT and arsenic metabolism in residents of the Red River Delta, Vietnam. *Toxicology and Applied Pharmacology* 236, 131–141.

25. Agusa, T., Kunito, T., Tue, N.M., Lan, V.T.M., Fujihara, J., Takeshita, H., Minh, T.B. et al., 2012. Individual variations in arsenic metabolism in Vietnamese: The association with arsenic exposure and GSTP1 genetic polymorphism. *Metallomics* 4, 91–100.

26. Agusa, T., Trang, P.T.K., Lan, V.M., Anh, D.H., Tanabe, S., Viet, P.H., Berg, M., 2014. Human exposure to arsenic from drinking water in Vietnam. *Science of the Total Environment* 488–489, 562–569.

27. Aposhian, H.V., Aposhian, M.M., 1990. Meso-2,3-dimercaptosuccinic acid: Chemical, pharmacological and toxicological properties of an orally effective metal chelating agent. *Annual Review of Pharmacology and Toxicology* 30, 279–306.

28. Flora, S.J.S., Bhadauria, S., Pant, S.C., Dhaked, R.K., 2005. Arsenic induced blood and brain oxidative stress and its response to some thiol chelators in rats. *Life Sciences* 77, 2324–2337.

29. Kannan, G.M., Flora, S.J.S., 2004. Chronic arsenic poisoning in the rat: Treatment with combined administration of succimers and an antioxidant. *Ecotoxicology and Environmental Safety* 58, 37–43.

30. Ramanathan, K., Balakumar, B.S., Panneerselvam, C., 2002. Effects of ascorbic acid and α-tocopherol on arsenic-induced oxidative stress. *Human and Experimental Toxicology* 21, 675–680.

31. Flora, S.J.S., 1999. Arsenic-induced oxidative stress and its reversibility following combined administration of N-acetylcysteine and meso 2,3-dimercaptosuccinic acid in rats. *Clinical and Experimental Pharmacology and Physiology* 26, 865–869.

32. Modi, M., Pathak, U., Kalia, K., Flora, S.J.S., 2005. Arsenic antagonism studies with monoisoamyl DMSA and zinc in male mice. *Environmental Toxicology and Pharmacology* 19, 131–138.

33. Mehta, A., Flora, S.J.S., 2001. Possible role of metal redistribution, hepatotoxicity and oxidative stress in chelating agents induced hepatic and renal metallothionein in rats. *Food and Chemical Toxicology* 39, 1029–1038.

34. Nocentini, S., Guggiari, M., Rouillard, D., Surgis, S., 2001. Exacerbating effect of vitamin E supplementation on DNA damage induced in cultured human normal fibroblasts by UVA radiation. *Photochemistry and Photobiology* 73, 370–377.

35. Gupta, R., Flora, S.J.S., 2005. Protective value of Aloe vera against some toxic effects of arsenic in rats. *Phytotherapy Research* 19, 23–28.

36. Koleva, I.I., Van Beek, T.A., Linssen, J.P.H., De Groot, A., Evstatieva, L.N., 2002. Screening of plant extracts for antioxidant activity: A comparative study on three testing methods. *Phytochemical Analysis* 13, 8–17.

37. Mukhtar, H., Ahmad, N., 2000. Tea polyphenols: Prevention of cancer and optimizing health. *American Journal of Clinical Nutrition* 71, 1698S–1704S.

38. Khan, N., Mukhtar, H., 2007. Tea polyphenols for health promotion. *Life Sciences* 81, 519–533.

39. Sinha, D., Roy, M., Dey, S., Siddiqi, M., Bhattacharya, R.K., 2003. Modulation of arsenic induced cytotoxicity by tea. *Asian Pacific Journal of Cancer Prevention* 4, 223–237.

40. Sinha, D., Bhattacharya, R.K., Siddiqi, M., Roy, M., 2005. Amelioration of sodium arsenite-induced clastogenicity by tea extracts in Chinese hamster V79 cells. *Journal of Environmental Pathology, Toxicology and Oncology* 24, 129–139.

41. Sinha, D., Roy, M., Siddiqi, M., Bhattacharya, R.K., 2005. Arsenic-induced micronuclei formation in mammalian cells and its counteraction by tea. *Journal of Environmental Pathology, Toxicology and Oncology* 24, 45–56.

42. Sinha, D., Roy, M., Siddiqi, M., Bhattacharya, R.K., 2005. Modulation of arsenic induced DNA damage by tea as assessed by single cell gel electrophoresis. *International Journal of Cancer Prevention* 2, 143–154.

43. Sinha, D., Dey, S., Bhattacharya, R.K., Roy, M., 2007. In vitro mitigation of arsenic toxicity by tea polyphenols in human lymphocytes. *Journal of Environmental Pathology, Toxicology and Oncology* 26, 207–220.

44. Mukherjee, P., Poddar, S., Talukder, G., Sharma, A., 1999. Protection by black tea extract against chromosome damage induced by two heavy metals in mice. *Pharmaceutical Biology* 37, 243–247.

45. Patra, M., Halder, A., Bhowmik, N., De, M., 2005. Use of black tea in modulating clastogenic effects of arsenic in mice in vivo. *Journal of Environmental Pathology, Toxicology and Oncology* 24, 201–210.

46. Sinha, D., Roy, S., Roy, M., 2010. Antioxidant potential of tea reduces arsenite induced oxidative stress in Swiss albino mice. *Food and Chemical Toxicology* 48, 1032–1039.

47. Raihan, S.Z., Chowdhury, A.K.A., Rabbani, G.H., Marni, F., Ali, M.S., Nahar, L., Sarker, S.D., 2009. Effect of aqueous extracts of black and green teas in arsenic-induced toxicity in rabbits. *Phytotherapy Research* 23, 1603–1608.

48. Messarah, M., Saoudi, M., Boumendjel, A., Kadeche, L., Boulakoud, M.S., Feki, A.E., 2013. Green tea extract alleviates arsenic-induced biochemical toxicity and lipid peroxidation in rats. *Toxicology and Industrial Health* 29, 349–359.

49. Sinha, D., Roy, M., 2011. Antagonistic role of tea against sodium arsenite-induced oxidative DNA damage and inhibition of DNA repair in Swiss albino mice. *Journal of Environmental Pathology, Toxicology and Oncology* 30, 311–322.

50. Chandronitha, C., Ananthi, S., Ramakrishnan, G., Lakshmisundaram, R., Gayathri, V., Vasanthi, H.R., 2010. Protective role of tannin-rich fraction of *Camellia sinensis* in tissue arsenic burden in Sprague Dawley rats. *Human and Experimental Toxicology* 29, 705–719.

51. Nam, S., Smith, D.M., Dou, Q.P., 2001. Tannic acid potently inhibits tumor cell proteasome activity, increases p27 and bax expression, and induces G1 arrest and apoptosis. *Cancer Epidemiology Biomarkers and Prevention* 10, 1083–1088.

52. Poddar, S., 2004. Dietary intervention with iron and black tea infusion in reducing cytotoxicity of arsenic. *Indian Journal of Experimental Biology* 42, 900–903.

53. Brabin, B., 1999. Iron pots for cooking: Wishful thinking or traditional common sense? *Lancet* 353, 690–691.

54. Chakraborty, T., De, M., 2013. Study of the effect of tea in an arsenic exposed population using micronuclei as a biomarker. *International Journal of Human Genetics* 13, 47–51.

55. Kapoor, S., Priyadarsini, K.I., 2001. Protection of radiation-induced protein damage by curcumin. *Biophysical Chemistry* 92, 119–126.

56. Mukherjee, S., Roy, M., Dey, S., Bhattacharya, R.K., 2007. A mechanistic approach for modulation of arsenic toxicity in human lymphocytes by curcumin, an active constituent of medicinal herb *Curcuma longa* Linn. *Journal of Clinical Biochemistry and Nutrition* 41, 32–42.

57. Tiwari, H., Rao, M.V., 2010. Curcumin supplementation protects from genotoxic effects of arsenic and fluoride. *Food and Chemical Toxicology* 48, 1234–1238.

58. Yousef, M.I., El-Demerdash, F.M., Radwan, F.M.E., 2008. Sodium arsenite induced biochemical perturbations in rats: Ameliorating effect of curcumin. *Food and Chemical Toxicology* 46, 3506–3511.

59. El-Demerdash, F.M., Yousef, M.I., Radwan, F.M.E., 2009. Ameliorating effect of curcumin on sodium arsenite-induced oxidative damage and lipid peroxidation in different rat organs. *Food and Chemical Toxicology* 47, 249–254.
60. Gao, S., Duan, X., Wang, X., Dong, D., Liu, D., Li, X., Sun, G., Li, B., 2013. Curcumin attenuates arsenic-induced hepatic injuries and oxidative stress in experimental mice through activation of Nrf2 pathway, promotion of arsenic methylation and urinary excretion. *Food and Chemical Toxicology* 59, 739–747.
61. He, X., Chen, M.G., Lin, G.X., Ma, Q., 2006. Arsenic induces NAD(P)H-quinone oxidoreductase I by disrupting the Nrf2·Keap1·Cul3 complex and recruiting Nrf2·Maf to the antioxidant response element enhancer. *Journal of Biological Chemistry* 281, 23620–23631.
62. Yadav, R.S., Sankhwar, M.L., Shukla, R.K., Chandra, R., Pant, A.B., Islam, F., Khanna, V.K., 2009. Attenuation of arsenic neurotoxicity by curcumin in rats. *Toxicology and Applied Pharmacology* 240, 367–376.
63. Yadav, R.S., Shukla, R.K., Sankhwar, M.L., Patel, D.K., Ansari, R.W., Pant, A.B., Islam, F., Khanna, V.K., 2010. Neuroprotective effect of curcumin in arsenic-induced neurotoxicity in rats. *Neurotoxicology* 31, 533–539.
64. Yadav, R.S., Chandravanshi, L.P., Shukla, R.K., Sankhwar, M.L., Ansari, R.W., Shukla, P.K., Pant, A.B., Khanna, V.K., 2011. Neuroprotective efficacy of curcumin in arsenic induced cholinergic dysfunctions in rats. *Neurotoxicology* 32, 760–768.
65. Wahlstrom, B., Blennow, G., 1978. A study on the fate of curcumin in the rat. *Acta Pharmacologica et Toxicologica* 43, 86–92.
66. Maiti, K., Mukherjee, K., Gantait, A., Saha, B.P., Mukherjee, P.K., 2007. Curcumin–phospholipid complex: Preparation, therapeutic evaluation and pharmacokinetic study in rats. *International Journal of Pharmaceutics* 330, 155–163.
67. Yallapu, M.M., Jaggi, M., Chauhan, S.C., 2010. β-Cyclodextrin-curcumin self-assembly enhances curcumin delivery in prostate cancer cells. *Colloids and Surfaces B: Biointerfaces* 79, 113–125.
68. Shaikh, J., Ankola, D.D., Beniwal, V., Singh, D., Kumar, M.N.V.R., 2009. Nanoparticle encapsulation improves oral bioavailability of curcumin by at least 9-fold when compared to curcumin administered with piperine as absorption enhancer. *European Journal of Pharmaceutical Sciences* 37, 223–230.
69. Yadav, A., Lomash, V., Samim, M., Flora, S.J.S., 2012. Curcumin encapsulated in chitosan nanoparticles: A novel strategy for the treatment of arsenic toxicity. *Chemico-Biological Interactions* 199, 49–61.
70. Sankar, P., Telang, A.G., Suresh, S., Kesavan, M., Kannan, K., Kalaivanan, R., Nath Sarkar, S., 2013. Immunomodulatory effects of nanocurcumin in arsenic-exposed rats. *International Immunopharmacology* 17, 65–70.
71. Biswas, J., Sinha, D., Mukherjee, S., Roy, S., Siddiqi, M., Roy, M., 2010. Curcumin protects DNA damage in a chronically arsenic-exposed population of West Bengal. *Human and Experimental Toxicology* 29, 513–524.
72. Roy, M., Sinha, D., Mukherjee, S., Biswas, J., 2011. Curcumin prevents DNA damage and enhances the repair potential in a chronically arsenic-exposed human population in West Bengal, India. *European Journal of Cancer Prevention* 20, 123–131.
73. Chung, L.Y., 2006. The antioxidant properties of garlic compounds: Alyl cysteine, alliin, allicin, and allyl disulfide. *Journal of Medicinal Food* 9, 205–213.
74. Das, T., Choudhury, A.R., Sharma, A., Talukder, G., 1993. Modification of cytotoxic effects of inorganic arsenic by a crude extract of *Allium sativum* L. in mice. *International Journal of Pharmacognosy* 31, 316–320.
75. Das, T., Roychoudhury, A., Sharma, A., Talukder, G., 1993. Modification of clastogenicity of three known clastogens by garlic extract in mice in vivo. *Environmental and Molecular Mutagenesis* 21, 383–388.

76. Roychoudhury, A., Das, T., Sharma, A., Talukder, G., 1993. Use of crude extract of garlic (*Allium sativum* L.) in reducing cytotoxic effects of arsenic in mouse bone marrow. *Phytotherapy Research* 7, 163–166.

77. Choudhury, A.R., Das, T., Sharma, A., Talukder, G., 1997. Inhibition of clastogenic effects of arsenic through continued oral administration of garlic extract in mice in vivo. *Mutation Research—Genetic Toxicology and Environmental Mutagenesis* 392, 237–242.

78. Roy Choudhury, A., Das, T., Sharma, A., Talukder, G., 1996. Dietary garlic extract in modifying clastogenic effects of inorganic arsenic in mice: Two-generation studies. *Mutation Research—Environmental Mutagenesis and Related Subjects* 359, 165–170.

79. Choudhury, A.R., Das, T., Sharma, A., 1997. Mustard oil and garlic extract as inhibitors of sodium arsenite-induced chromosomal breaks in vivo. *Cancer Letters* 121, 45–52.

80. Chowdhury, R., Dutta, A., Chaudhuri, S.R., Sharma, N., Giri, A.K., Chaudhuri, K., 2008. In vitro and in vivo reduction of sodium arsenite induced toxicity by aqueous garlic extract. *Food and Chemical Toxicology* 46, 740–751.

81. Flora, S.J.S., Mehta, A., Gupta, R., 2009. Prevention of arsenic-induced hepatic apoptosis by concomitant administration of garlic extracts in mice. *Chemico-Biological Interactions* 177, 227–233.

82. Das, B., Mandal, S., Chaudhuri, K., 2014. Role of arginine, a component of aqueous garlic extract, in remediation of sodium arsenite induced toxicity in A375 cells. *Toxicology Research* 3, 191–196.

83. Wasser, S.P., Weis, A.L. 1999. Medicinal properties of substances occurring in higher basidiomycetes mushrooms: Current perspectives. *International Journal of Medicinal Mushrooms* 1, 31–62.

84. Jose, N., Janardhanan, K.K., 2000. Antioxidant and antitumour activity of *Pleurotus florida*. *Current Science* 79, 941–943.

85. Borchers, A.T., Keen, C.L., Gershwin, M.E., 2004. Fatalities following allergen immunotherapy. *Clinical Reviews in Allergy and Immunology* 27, 147–158.

86. Bera, A.K., Rana, T., Bhattacharya, D., Das, S., Pan, D., Das, S.K., 2011. Sodium arsenite-induced alteration in hepatocyte function of rat with special emphasis on superoxide dismutase expression pathway and its prevention by mushroom lectin. *Basic and Clinical Pharmacology and Toxicology* 109, 240–244.

87. Rana, T., Bera, A.K., Das, S., Bhattacharya, D., Pan, D., Bandyopadhyay, S., De, S., Subrata Kumar, D., 2011. Mushroom lectin protects arsenic induced apoptosis in hepatocytes of rodents. *Human and Experimental Toxicology* 30, 307–317.

88. Bera, A.K., Rana, T., Das, S., Bhattacharya, D., Pan, D., Bandyopadhyay, S., Das, S.K., 2011. Mitigation of arsenic-mediated renal oxidative stress in rat by *Pleurotus florida* lectin. *Human and Experimental Toxicology* 30, 940–951.

89. Rana, T., Bera, A.K., Das, S., Bhattacharya, D., Pan, D., Bandyopadhyay, S., Mondal, D.K., Samanta, S., Bandyopadhyay, S., Das, S.K., 2012. *Pleurotus florida* lectin normalizes duration dependent hepatic oxidative stress responses caused by arsenic in rat. *Experimental and Toxicologic Pathology* 64, 665–671.

90. Rana, T., Bera, A.K., Bhattacharya, D., Das, S., Pan, D., Das, S.K., 2012. Evidence of antiapoptotic properties of *Pleurotus florida* lectin against chronic arsenic toxicity in renal cells of rats. *Journal of Environmental Pathology, Toxicology and Oncology* 31, 39–48.

91. Gupta, R., Dubey, D.K., Kannan, G.M., Flora, S.J.S., 2007. Concomitant administration of *Moringa oleifera* seed powder in the remediation of arsenic-induced oxidative stress in mouse. *Cell Biology International* 31, 44–56.

92. Oliveira, J.T.A., Silveira, S.B., Vasconcelos, I.M., Cavada, B.S., Moreira, R.A., 1999. Compositional and nutritional attributes of seeds from the multiple purpose tree *Moringa oleifera* Lamarck. *Journal of the Science of Food and Agriculture* 79, 815–820.

93. Singh, B.N., Singh, B.R., Singh, R.L., Prakash, D., Dhakarey, R., Upadhyay, G., Singh, H.B., 2009. Oxidative DNA damage protective activity, antioxidant and anti-quorum sensing potentials of *Moringa oleifera*. *Food and Chemical Toxicology* 47, 1109–1116.

94. Gupta, R., Kannan, G.M., Sharma, M., Flora, S.J.S., 2005. Therapeutic effects of Moringa oleifera on arsenic-induced toxicity in rats. *Environmental Toxicology and Pharmacology* 20, 456–464.

95. Chattopadhyay, S., Maiti, S., Maji, G., Deb, B., Pan, B., Ghosh, D., 2011. Protective role of *Moringa oleifera* (Sajina) seed on arsenic-induced hepatocellular degeneration in female albino rats. *Biological Trace Element Research* 142, 200–212.

96. Mishra, D., Gupta, R., Pant, S.C., Kushwah, P., Satish, H.T., Flora, S.J.S., 2009. Co-administration of monoisoamyl dimercaptosuccinic acid and *Moringa oleifera* seed powder protects arsenic-induced oxidative stress and metal distribution in mice. *Toxicology Mechanisms and Methods* 19, 169–182.

97. Verma, A.R., Vijayakumar, M., Mathela, C.S., Rao, C.V., 2009. In vitro and in vivo anti-oxidant properties of different fractions of *Moringa oleifera* leaves. *Food and Chemical Toxicology* 47, 2196–2201.

98. Azuma, K., Nakayama, M., Koshioka, M., Ippoushi, K., Yamaguchi, Y., Kohata, K., Yamauchi, Y., Ito, H., Higashio, H., 1999. Phenolic antioxidants from the leaves of *Corchorus olitorius* L. *Journal of Agricultural and Food Chemistry* 47, 3963–3966.

99. Khan, M.S.Y., Bano, S., Javed, K., Asad Mueed, M., 2006. A comprehensive review on the chemistry and pharmacology of Corchorus species—A source of cardiac glycosides, triterpenoids, ionones, flavonoids, coumarins, steroids and some other compounds. *Journal of Scientific and Industrial Research* 65, 283–298.

100. Zeid, A.H.S.A., 2002. Stress metabolites from *Corchorus olitorius* L. leaves in response to certain stress agents. *Food Chemistry* 76, 187–195.

101. Das, A.K., Bag, S., Sahu, R., Dua, T.K., Sinha, M.K., Gangopadhyay, M., Zaman, K., Dewanjee, S., 2010. Protective effect of *Corchorus olitorius* leaves on sodium arsenite-induced toxicity in experimental rats. *Food and Chemical Toxicology* 48, 326–335.

102. Das, A.K., Dewanjee, S., Sahu, R., Dua, T.K., Gangopadhyay, M., Sinha, M.K., 2010. Protective effect of *Corchorus olitorius* leaves against arsenic-induced oxidative stress in rat brain. *Environmental Toxicology and Pharmacology* 29, 64–69.

103. Das, A.K., Sahu, R., Dua, T.K., Bag, S., Gangopadhyay, M., Sinha, M.K., Dewanjee, S., 2010. Arsenic-induced myocardial injury: Protective role of *Corchorus olitorius* leaves. *Food and Chemical Toxicology* 48, 1210–1217.

104. Rousi, A., 1971. The genus Hippophae L.: A taxonomic study. *Annales Botanici Fennici* 8, 177–227.

105. Costantino, L., Rastelli, G., Rossi, T., Bertoldi, M., Albasini, A., 1994. Composition, superoxide radicals scavenging and antilipoperoxidant activity of some edible fruits. *Fitoterapia* 65, 44–47.

106. Gupta, R., Flora, S.J.S., 2005. Therapeutic value of *Hippophae rhamnoides* L. against subchronic arsenic toxicity in mice. *Journal of Medicinal Food* 8, 353–361.

107. Gupta, R., Flora, S.J.S., 2006. Protective effects of fruit extracts of *Hippophae rhamnoides* L. against arsenic toxicity in Swiss albino mice. *Human and Experimental Toxicology* 25, 285–295.

108. Soria, E.A., Goleniowski, M.E., Cantero, J.J., Bongiovanni, G.A., 2008. Antioxidant activity of different extracts of Argentinian medicinal plants against arsenic-induced toxicity in renal cells. *Human and Experimental Toxicology* 27, 341–346.

109. Kitts, D.D., Yuan, Y.V., Wijewickreme, A.N., Thompson, L.U., 1999. Antioxidant activity of the flaxseed lignan secoisolariciresinol diglycoside and its mammalian lignan metabolites enterodiol and enterolactone. *Molecular and Cellular Biochemistry* 202, 91–100.

110. Rizwan, S., Naqshbandi, A., Farooqui, Z., Khan, A.A., Khan, F., 2014. Protective effect of dietary flaxseed oil on arsenic-induced nephrotoxicity and oxidative damage in rat kidney. *Food and Chemical Toxicology* 68, 99–107.

111. Dhir, H., Kumar Roy, A., Sharma, A., Talukder, G., 1990. Protection afforded by aqueous extracts of Phyllanthus species against cytotoxicity induced by lead and aluminium salts. *Phytotherapy Research* 4, 172–176.

112. Dhir, H., Roy, A.K., Sharma, A., Talukder, G., 1990. Modification of clastogenicity of lead and aluminium in mouse bone marrow cells by dietary ingestion of *Phyllanthus emblica* fruit extract. *Mutation Research/Genetic Toxicology* 241, 305–312.

113. Agarwal, K., Dhir, H., Sharma, A., Talukder, G., 1992. The efficacy of two species of Phyllanthus in counteracting nickel clastogenicity. *Fitoterapia* 63, 49–54.

114. Biswas, S., Talukder, G., Sharma, A., 1999. Protection against cytotoxic effects of arsenic by dietary supplementation with crude extract of *Emblica officinalis* fruit. *Phytotherapy Research* 13, 513–516.

115. Gupta, R., Flora, S.J.S., 2006. Effect of *Centella asiatica* on arsenic induced oxidative stress and metal distribution in rats. *Journal of Applied Toxicology* 26, 213–222.

116. Singh, N., Kumar, D., Lal, K., Raisuddin, S., Sahu, A.P., 2010. Adverse health effects due to arsenic exposure: Modification by dietary supplementation of jaggery in mice. *Toxicology and Applied Pharmacology* 242, 247–255.

117. Martin, M., He, Q., 2009. Mango bioactive compounds and related nutraceutical properties—A review. *Food Reviews International* 25, 346–370.

118. Garrido, G., Romiti, N., Tramonti, G., de la Fuente, F., Chieli, E., 2012. Polyphenols of *Mangifera indica* modulate arsenite-induced cytotoxicity in a human proximal tubule cell line. *Brazilian Journal of Pharmacognosy* 22, 325–334.

119. Jin, D.Q., Li, G., Kim, J.S., Yong, C.S., Kim, J.A., Huh, K., 2004. Preventive effects of *Laminaria japonica* aqueous extract on the oxidative stress and xanthine oxidase activity in streptozotocin-induced diabetic rat liver. *Biological and Pharmaceutical Bulletin* 27, 1037–1040.

120. Zhao, G.L., Liu, C.C., Xie, J., Li, Y.S., Li, J.L., 2010. Antioxidant effects of the Soxhlet extraction product from *Porphyra haitanensis* and its different solvent-soluble fractions. *Food Science* 31, 186–191.

121. Jiang, Y., Wang, L., Yao, L., Liu, Z., Gao, H., 2013. Protective effect of edible marine algae, laminaria japonica and *Porphyra haitanensis*, on subchronic toxicity in rats induced by inorganic arsenic. *Biological Trace Element Research* 154, 379–386.

Section III

Treatment

12 Treatment of Arsenic Poisoning

Diagnosis with Biomarkers

Michael F. Hughes

CONTENTS

12.1 INTRODUCTION

Metalloid arsenic is a well-known poisonous substance affecting multiple organs such as the skin and lungs, the cardiovascular and nervous systems, and others [1]. The manifold toxic effects of arsenic occur from both acute and chronic exposure to it. Arsenic has a long and interesting history as an acute toxic substance [2]. From its use during the Middle Ages to discreetly assassinate European royalty, arsenic earned the expression *King of Poisons* [3]. To assist in the prosecution of individuals accused of using arsenic as a homicidal agent, James Marsh developed a chemical procedure in 1832 to detect arsenic in various substances [2,4]. The development of this method by Marsh could be considered a defining moment in biomarkers of exposure for arsenic. This is because a fairly simple analytical method could be used to crudely detect and quantify arsenic in biological samples. Today, with the advancements in analytical technology, multiple arsenicals can be identified in one biological sample at the ppb level or lower.

Chronic exposure to arsenic is primarily a result of it being an environmental contaminant [5,6]. Arsenic is found naturally in soil, water, air, and food in various structural forms, oxidation states, and concentrations [7]. Its presence in drinking water has raised many concerns, particularly in developing countries, regarding consumption of arsenic-contaminated water and the increasing incidence of cancer in several organs such as skin, bladder, and kidney [1,8]. Noncancerous effects such as cardio- and peripheral vascular diseases, diabetes, neurodevelopmental effects, and others are attributed to chronic exposure to arsenic in drinking water [1,6].

There are needs in the medical, scientific, and public health communities for methods to detect human exposure to arsenic to aid in diagnosis of potential adverse health effects from this metalloid. One way to diagnose arsenic poisoning from acute and chronic exposure is the use of biological markers (i.e., biomarkers).

12.1.1 BIOMARKERS

As humans are exposed to a vast array of chemicals, in many cases unintentionally, adverse health effects may develop from exposure to one or more of them. Clinicians, public health officials, and regulatory scientists want to determine how an adverse health effect developed following a chemical exposure to treat affected individuals and prevent further occurrences. Biomarkers are one of many tools that clinicians and other scientists can use to understand exposure to and effect of chemicals. A biomarker can be defined as a qualitative or quantitative measurement that reveals an interaction between a biological system and an impending risk, which may be chemical, physical, or biological [9]. It is generally recognized that biomarkers are classified as biomarkers of exposure, effect, or susceptibility. These biomarkers can be used for clinical diagnosis, for monitoring exposures to populations, in epidemiology studies, and in human health risk assessments.

Biomarkers must be carefully selected and validated before they are used to assure they are specific for the intended exposure or effect and sensitive enough to detect the desired response. There are several factors that must be considered that can impact the interface between the individual exposed and the agent of concern. These factors include source of the agent (e.g., soil), properties of the agent, exposure route (e.g., inhalation), agent specifics (e.g., concentration), individual specifics (e.g., age), and response (e.g., acute) [9]. Practical aspects about the selection of a biomarker include how it will be measured (e.g., quantifying analyte in blood), stability of the biomarker, the connection between exposure and biomarker, quality assurance considerations (plan, control of samples, reproducibility of the method, etc.), and other factors [9]. Finally, ethical and social considerations of the biomarker need to be considered. Collection of samples can be invasive (e.g., blood sampling), which may deter an individual from participating and the norms of some cultures may not allow for the collection of some biological samples. Also, it is necessary to consider how the results will be communicated to the individual and population.

12.1.2 METABOLISM

An understanding of the metabolism of arsenic is important so that biomarkers of exposure for this metalloid can be developed and used in clinical diagnosis, monitoring exposures to populations, epidemiology studies, and human health risk assessments. Arsenic is found in the environment in inorganic and organic forms and several oxidation states. The oxidation states of arsenic include $-$III, 0, III, and V. Elemental arsenic is in the 0 oxidation state. For environmental arsenic exposures, the III and V oxidation states are the most important. Examples of inorganic arsenicals (iAs) in these oxidation states are arsenite (iAs^{III}) and arsenate (iAs^V), respectively. Examples of organic arsenicals include monomethylarsonous acid ($MMAs^{III}$) and dimethylarsinic acid ($DMAs^V$). Arsine is an example of an arsenical in the $-$III oxidation state and exposure to it is primarily occupational.

Inorganic arsenic is metabolized by the enzyme arsenic (+3 oxidation state) methyltransferase (As3MT) [10]. Although its primary biological function is not known, species ranging from sea squirts to humans possess a gene that encodes for variants of As3MT [10]. As3MT is thought to metabolize inorganic arsenic by oxidative methylation [10], although an alternative proposal involves reductive methylation [11]. In vitro studies indicate that arsenite is the substrate for As3MT. Thus for arsenate to be methylated, it must be first reduced to arsenite. In mammals, there does not appear to be a specific arsenate reductase. However, arsenate can be reduced to arsenite both nonenzymatically by thiols [12,13] and enzymatically by polynucleotide phosphorylase (PNP)–catalyzed arsenolysis of ADP [14]. The proposed pathway for oxidative methylation entails sequential reduction of the oxidized methylated arsenical, and the process repeats itself (Figure 12.1). Arsenite is oxidatively methylated to $MMAs^V$. $MMAs^V$ is reduced to $MMAs^{III}$, which is then oxidatively methylated to $DMAs^V$. In humans, $DMAs^V$ is the primary metabolite of inorganic arsenic excreted

Oxidative methylation pathway for arsenic metabolism

$$iAs^V \xrightarrow{\text{GSH or PNP?}} iAs^{III} \xrightarrow{\text{As3MT + CH}_3} MMAs^V \xrightarrow{\text{As3MT + 2e}^-} MMAs^{III} \xrightarrow{\text{As3MT + CH}_3}$$

$$DMAs^V \xrightarrow{\text{As3MT + 2e}^-} DMAs^{III} \xrightarrow{\text{As3MT + CH}_3} TMAs^V O$$

FIGURE 12.1 Pathway of oxidative methylation of arsenic by arsenic (+3 oxidation) state methyltransferase.

in urine [15]. Administered DMAs^V can be further metabolized to trimethylarsine oxide (TMAO^V) [16] via the intermediate DMAs^{III} [10]. Most populations exposed to inorganic arsenic excrete in urine 10%–30% inorganic arsenic, 20%–30% mono-methyl arsenic, and 60%–70% dimethyl arsenic [17]. However, the metabolism of arsenic appears to be influenced by genetic polymorphisms in the metabolizing enzymes [17], which may render an individual susceptible to the adverse health effects of arsenic [18]. The methylation of arsenic facilitates excretion of arsenic, which aids in detoxification of it [19]. The oxidized methyl arsenicals are also less acutely toxic than trivalent arsenicals [20]. However, the reduction of arsenic to more potent trivalent arsenical species indicates the oxidative methylation pathway is one of detoxification and activation. Thiolated arsenicals have also been detected in bio-logical samples, suggesting alternative pathways of arsenic metabolism [21,22].

12.2 BIOMARKERS OF EXPOSURE

Biomarkers of exposure can provide information that a recent or past exposure to a xenobiotic has occurred. This type of biomarker does not reveal information on a mechanism of action or predict a potential adverse outcome. Regarding arsenic, biomarkers of exposure are the best developed of the three biomarker classifications. This is primarily due to advancements in analytical chemistry in the ability to speci-ate and detect arsenic at sub-ppb levels. Biomarkers of exposure for arsenic include urine, blood, hair, and nails and are of tremendous use to confirm an exposure to arsenic had occurred and for the diagnosis of arsenic poisoning. The use of this biomarker type may aid in the treatment of individuals adversely affected by arsenic exposure.

12.2.1 URINE

The analysis of urine for arsenic is the most utilized method of the four biological samples that are commonly collected and analyzed as biomarkers of arsenic expo-sure. The reasons for this are urine is an easy biological sample to collect, urinary excretion is the main route of elimination of absorbed arsenic, a positive result pro-vides an indication of internal exposure to arsenic, and unless handled carelessly, it is unlikely to be contaminated with external sources of arsenic as can occur with hair and nails. Most importantly, urinary arsenic concentration correlates strongly with exposure to arsenic, particularly with arsenic in drinking water. Once collected, the urine can be processed by a variety of methods to detect and quantify arsenic in terms of total and speciated arsenic. Background levels of arsenic in urine range from 5 to 50 μg/L [1]. Approximately, 45%–85% of a dose of iAs is excreted by humans in urine 1–3 days following oral exposure [23].

12.2.1.1 Method

Good laboratory practices require the use of protocols and standard procedures for the collection of urine and its analysis for arsenic [24]. This would include details on sample collection and storage, methods of sample preparation and analysis, use of standard reference materials, and quality assurance and control. The timing of the urine collection needs to be considered. Basically, urine can be obtained as a spot collection, such as the first morning void or over a longer period, such as a 24-h collection. The spot collection is easier for a study subject to collect as opposed to a 24-h collection. With a longer collection period, the subject would need to have the collection vessel nearby throughout the day, which may be socially stigmatizing. The subject may also forget to collect a sample when urinating at some point within the 24-h collection period. Calderon et al. [25] reported that urinary arsenic levels remain stable throughout the day and over a 5-day collection period, which supports spot collection of urine. The urine should be kept cold after collection. This will aid in preventing oxidation of trivalent arsenicals if the analysis will include their speciation. Keeping urine samples cold is easier for spot than 24-h collections. Other issues to consider are if the urinary arsenic concentration will be adjusted to the specific gravity of the urine or urinary creatinine, gathering information about dietary history, chain of custody for quality assurance and control purposes, and standard reference materials for the analysis. Urinary arsenic concentration can be normalized to urinary creatinine or specific gravity of the urine, particularly for spot urine samples, because the volume of urine excreted can vary throughout the day [26]. The volume of urine is affected by kidney function (e.g., glomerular filtration, tubular section, and reabsorption), diet and ingestion of liquids, perspiration, and other factors [26]. Depending on these factors, a substance in urine collected as a spot sample may be diluted or concentrated, which would under- or overestimate the 24-h result. However, urinary creatinine appears to be highly correlated to urinary arsenic, suggesting that this adjustment should not be done [27,28]. Assessing the diet of the individual provides information on sources of arsenic other than drinking water. This information would be critical if reporting total arsenic, because some seafood contains organic arsenicals such as arsenobetaine (Figure 12.2), which

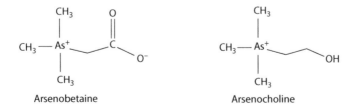

FIGURE 12.2 Structure of seafood arsenicals that can confound total arsenic exposure.

is relatively nontoxic compared to inorganic arsenic [29]. If total urinary arsenic is assessed, and an individual had consumed seafood containing arsenobetaine 2–3 days before urine collection, and dietary information is not determined before urinary arsenic analysis, the exposure to toxic forms of arsenic may be overestimated. Arsenobetaine and other seafood-derived arsenicals can be detected in speciated analysis, so the dietary history may not be as critical in this case. However, seaweed contains arsenosugars, which are metabolized to dimethyl arsenic and excreted in urine. Urinary dimethyl arsenic levels in humans who consume seaweed increase [30,31]. As dimethyl arsenic is the main metabolite of arsenic excreted in urine by humans, consumption of seaweed by urine donors can complicate the arsenic exposure estimate.

Arsenic in urine can be measured in terms of total or speciated arsenic. Total arsenic is the sum of all arsenic species in the urine sample analyzed, regardless of its source. This includes both inorganic (i.e., arsenate and arsenite) and organic species of arsenic. In the latter group, this would comprise the metabolites of inorganic arsenic, mono- and dimethylated arsenic and if seafood is consumed, potentially other organic arsenicals such as arsenobetaine or arsenocholine (Figure 12.2). These seafood-derived arsenicals such as arsenobetaine are absorbed, are generally resistant to metabolism, and excreted intact in urine [31]. However, arsenosugars, which can be found in seaweed, are metabolized to $DMAs^V$ and excreted in urine [30]. Speciated arsenic indicates that arsenicals are separated using, for example, high pressure liquid chromatography and then quantitated. The arsenic species can also be summed as a check for total arsenic. For example, some studies speciate iAs^{III}, iAs^V, $MMAs^V$, and $DMAs^V$. The concentrations of these species can then be summed, but it does not necessarily give an indication of total arsenic exposure. Francesconi and Kuehnelt [32] have reviewed the analytical methods available to determine arsenic species in biological samples.

12.2.1.1.1   Total Arsenic in Urine

Total arsenic in urine can be measured by methods including neutron activation [33–35], inductively coupled mass spectrometry (ICP-MS) [25,30,36], atomic absorption spectrometry (AAS) [22,37–39], atomic fluorescence spectrometry (AFS) [40,41], and graphite furnace atomic absorption spectrometry [42]. Neutron activation can require minimal preparation of the sample. The volume collected should be measured and an aliquot or the whole sample can be analyzed. What is needed for the analysis is a neutron source to irradiate the sample and a germanium detector for spectral analysis of the irradiated sample. The availability of a neutron source would hamper this type of analysis. In some cases, hydrides of the arsenicals are formed before irradiation [34]. For some biological samples, solvent extraction may be used to improve detection of the irradiated arsenic [43]. For the spectrophotometric methods, the urine sample would need to be processed so that organic arsenicals are degraded to iAs, which is the species detected. One method to degrade or decompose organic arsenic is adding concentrated acid to urine and heating the solution to high temperatures. This process can be somewhat time consuming, but, with microwave technology, can be done in a reasonable amount of time.

12.2.1.1.2 Speciated Arsenic in Urine

Arsenicals detected in urine collected from arsenic-exposed individuals using speciated analysis include in various amounts iAs^V, iAs^{III}, MMAs^V, MMAs^{III}, DMAs^V, DMAs^{III}, arsenobetaine, and others [44–46]. In speciated urinary analysis, the arsenicals can be separated on a chromatography column using ion exchange or high performance liquid chromatography (HPLC) and detected by coupling with an ICP-MS [22,36,44,45,47], AAS [48], or AFS [41,49,50]. HPLC-ICP-MS is the most dominant method to speciate and quantitate arsenicals [32], but it can be the most expensive method too. Speciated arsenic can be determined using pH selective hydride generation cold-trap atomic absorption spectrometry (HG-CT-AAS) (Figure 12.3) [46,51]. In this method sodium borohydride is used to reduce the arsenicals to arsines, which are volatile and cold-trapped on an absorbent column. The column is heated and the arsines are speciated based on boiling point and detected in the spectrometer. Depending on the pH of the reaction buffer, different arsenic species can be separated. At pH 6, trivalent arsenicals and trimethylarsine oxide form arsines when sodium borohydride is added [51]. At pH 1, pentavalent arsenicals are reduced to trivalent arsenicals and these species form arsines following addition of sodium borohydride. The concentration of pentavalent arsenicals in the original sample is determined by the difference between the arsenic concentrations of the two reactions. One drawback to the methods using hydride generation is that it is difficult to generate hydrides of arsenobetaine and other seafood arsenicals. Using a nebulizer between the end of the column and an ICP-MS will allow detection of the seafood arsenicals [52]. Schmeisser et al. [53] reported that a hydride generation method with HPLC-ICP-MS did detect arsenosugars from marine organisms. The efficiency of the detection of the arsenosugars was dependent on the hydride generation system used.

Lindberg et al. [54] evaluated the three most commonly used analytical methods to speciate arsenic in urine. These methods were HPLC-HG-ICP-MS, HPLC-HG-AFS, and HG-AAS. This evaluation was done because of the high cost to operate the ICP-MS. The arsenicals speciated were iAs^{III}, iAs^V, MMAs^V, and DMAs^V.

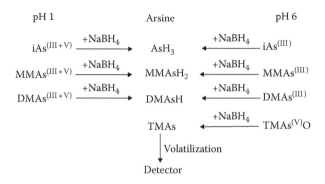

FIGURE 12.3   Arsenicals volatilized by pH selective hydride generation.

They reported low variation (i.e., high correlation, $r^2 \geq 0.9$) between two separate laboratories in the use of ICP-MS and AFS. (AAS was not compared between laboratories.) The within-laboratory comparisons for all three methods were even more agreeable than the between-laboratory comparisons ($r^2 \geq 0.95$). The authors suggested using AFS if the arsenic concentration in urine exceeds 10 µg/L because it is less expensive to use than the ICP-MS.

12.2.1.2 Correlation of Urinary Arsenic to Exposure

Urinary arsenic reflects internal exposure to this metalloid and is correlated to arsenic exposure (Table 12.1). Ahmed et al. [37] investigated the relationship between groundwater arsenic concentration and total urinary excretion of arsenic in a Pakistani population. Spot urine samples were collected from individuals ≥15 years of age. Dietary history was determined and those who had consumed seafood within 3 days of urine collection were excluded from the study. Total arsenic was determined in the water and urine using HG-AAS. The median arsenic concentration in groundwater was 2.1 µg/L; the arsenic concentrations in water ranged from 0.1 to 350 µg/L. Approximately 32% of the water samples were above the WHO guideline value of 10 µg/L [55]. Urinary arsenic concentration ranged from 0.1 to 848 µg/L. A significant positive correlation was found between arsenic in groundwater and arsenic in urine (Spearman $r^2 = 0.314$, $p < 0.01$) and arsenic in groundwater and daily intake of arsenic (Spearman $r^2 = 0.285$, $p < 0.01$).

Agusa et al. [56] investigated the exposure of arsenic in groundwater to a population in four districts of northern Vietnam. Arsenic concentrations in raw groundwater ranged from <1 to 632 µg/L. Three of the districts treated filtered the arsenic-contaminated water in sand, which lowered the arsenic concentration. For the population studied, the water consumed had concentrations ranging from <1 to 309 µg/L. Spot urine samples were collected. Arsenicals in urine were speciated using HPLC-ICP-MS. These arsenicals included iAs^{III}, iAs^V, $MMAs^V$, $DMAs^V$,

TABLE 12.1

Representative Studies Describing Exposure–Biomarker Associations for Total Arsenic in Urine

Source of Exposure	Arsenic Exposure Concentration (µg/L)	Association, N	Reference
Groundwater	0.1–350	$r = 0.52$, $p < 0.01$, $N = 465$	Ahmed et al. [37]
Groundwater—sand-filtered	<1–309	$r^2 = 0.072$, $p = 0.002$, $N = 183$ ($\sum As = iAs + MMA + DMA$)	Agusa et al. [56]
Groundwater	<0.02–140	$r^2 = 0.568$,[a] $p < 0.0001$, $N = 110$ ($\sum As = iAs + MMA + DMA$)	Normandin et al. [36]
Groundwater, surface water, and bottled water	0.09–1.39	Spearman $r^2 = 0.1827$, $p < 0.01$, $N = 343$	Rivera-Núñez et al. [47]

[a] Based on estimated daily water intake.

and arsenobetaine. The arsenical detected in urine at the highest concentration was $DMAs^V$ followed by arsenobetaine. There was no difference in arsenobetaine levels excreted in urine between high and low arsenic-exposed groups. A significant positive correlation was reported between the arsenic concentration in sand-filtered groundwater and the sum of iAs^{III}, iAs^V, $MMAs^V$, and $DMAs^V$ excreted in urine ($r^2 = 0.072$, p = 0.002).

A rural Canadian population exposed to arsenic in groundwater was studied by Normandin et al. [36]. A 2-day dietary history was recorded and 12-h urine samples were collected. Arsenic concentrations in the groundwater ranged from <0.02 to 140 µg/L. Urine samples were speciated for arsenicals using HPLC-ICP-MS. The arsenicals included iAs^{III}, iAs^V, $MMAs^{III}$, $MMAs^V$, $DMAs^{III}$, $DMAs^V$, arsenobetaine, and arsenocholine. Total arsenic excreted in urine of the study participants ranged from 2.02 to 137 µg/L. There was a significant association between estimated daily water intake of arsenic and the sum of iAs^{III}, iAs^V, $MMAs^V$, and $DMAs^V$ excreted in urine ($R^2 = 0.568$, p < 0.0001). Daily water intake of arsenic was not associated with the excretion of arsenobetaine.

Rivera-Núñez et al. [47] studied a population in the state of Michigan in the United States exposed to arsenic in drinking water with concentrations ≤50 µg/L. A dietary history and other personal information (e.g., smoking) was assessed; <10% of the individuals included in the analysis did report that they had consumed seafood. Drinking water sources included surface, bottled, and groundwater. Total arsenic in drinking water was analyzed by ICP-MS and determined to be 0.74 µg/L (geometric mean). Spot urine samples were collected from the study population and urinary arsenic was speciated using HPLC-ICP-MS. The urinary sum of arsenic (sum of iAs^{III}, iAs^V, $MMAs^V$, and $DMAs^V$) was 5.02 µg/L. The $\log_{10}$-transformed drinking water arsenic concentration was significantly related to the sum of urinary arsenic concentration (Spearman $r^2 = 0.1827$, p < 0.01). When excluding arsenic drinking water concentrations <1 µg/L, the correlation (r^2) increased to 0.1959.

Not all studies of arsenic exposure via drinking water and urinary arsenic are positively correlated. Adair et al. [40] examined a population in the state of Nevada in the United States. Water samples were analyzed by ICP-MS, or if too turbid, by graphite furnace-AAS. The total arsenic in groundwater ranged from <3–2100 µg/L with a mean concentration of 89 µg/L. A dietary history of the study population was taken. The range of total arsenic in urine ranged from 11.2 to 749 ppb. Unlike many other studies, arsenic in drinking water was not correlated with urinary total arsenic, either in the group that did not consume seafood or combining that data with individuals that did consume seafood. The authors speculate that the reason for the lack of a positive correlation between arsenic exposure and urinary arsenic may have been due to the age of the population studied, which was >45 years old. There may have been some underlying health issues that could have affected the relationship between arsenic exposure and urinary arsenic.

12.2.1.3   Advantages of Urinary Arsenic as a Biomarker of Exposure

The advantages of urinary arsenic as a biomarker of exposure are that urine is an easy sample to collect, it is a clear indicator of internal arsenic exposure and in many studies urinary arsenic correlates with arsenic exposure. The reference range for urinary arsenic is well established (5–50 µg/L) [1], so a judgment whether or not an exposure to arsenic occurred can be done fairly reliably. The collection of urine is not an invasive procedure so more people should be willing to collect their urine and have it analyzed for arsenic. With speciated arsenic analysis, it can be determined if the arsenic exposure was from the consumption of seafoods that contain organic arsenicals that are relatively nontoxic relative to inorganic arsenic [29].

12.2.1.4 Disadvantage of Urinary Arsenic as a Biomarker of Exposure

The disadvantages of urinary arsenic as a biomarker of exposure are that it only gives information on recent exposures, within 2–3 days of exposure. Arsenic is excreted in urine relatively quickly following oral exposure. It has a half-life of 1–3 days following oral exposure in humans [23]. Trying to determine if an exposure to arsenic had occurred beyond 3 days by this method may not provide useful information. Another disadvantage is that organic arsenicals in seafood can add to total arsenic measured in urine. These seafood arsenicals are relatively nontoxic, yet if detected in urine, without the knowledge of their consumption, may overestimate the risk to inorganic arsenic. Finally, adjusting for urinary creatinine can lead to overestimation of urinary arsenic concentration.

12.2.2 BLOOD

The analysis of blood is used as a biomarker of exposure to arsenic, although it is probably the least employed of the four biological samples (urine, blood, hair, and nails) that are analyzed for this purpose. The main reasons for this are that arsenic is cleared fairly rapidly from blood [57] and collecting blood is an invasive procedure. The reference range for background arsenic in blood is 0.3–2 µg/L [1]. Blood may be used in forensic analysis for arsenic poisoning, but blood needs to be collected for analysis within a few days of the exposure because it is cleared rapidly from this tissue. To be of any value as a biomarker of exposure, blood must be collected during the time of arsenic exposure or soon after it occurred.

12.2.2.1 Method

Proper protocols are needed in the collection, storage, processing, and analysis of arsenic in blood [24]. Depending on the method, whole blood, serum, plasma, red blood cells, or hemolyzed red blood cells can be analyzed for arsenic [44,45,58]. Like urine, total and speciated arsenic can be analyzed in blood. If the samples are not processed and analyzed immediately after collection, the blood would need to be kept cold on ice or even frozen. This may limit the ability to analyze different blood components. For whole blood analysis, the red blood cells would need to be hydrolyzed, with, for example, a detergent.

12.2.2.1.1 Total Arsenic in Blood

Methods to measure total arsenic in blood or its components include nuclear neutron activation [41,59,60], AAS [39,61], AFS [40], and ICP-MS [42]. Blood and its components can be digested with acid and then dry ashed [39], and red blood cells may be solubilized with detergents (e.g., Triton X-100) [42]. Because arsenobetaine, a seafood arsenical, can be detected in blood [44], it is critical that the study participants refrain from consuming seafood 2–3 days before the blood is collected and analyzed for total arsenic. If not, exposure to more toxic forms of arsenic may be overestimated.

12.2.2.1.2 Speciated Arsenic in Blood

Arsenicals detected in blood in various amounts include iAs^{III}, $MMAs^V$, $DMAs^V$, and arsenobetaine [44]. Methods for speciated arsenic in blood include HPLC-ICP-MS [44,45] and HPLC-HG-AFS [41]. The methods that use hydride generation use a gas:liquid separator for the arsines generated to volatilize from the reaction buffer. For complex mixtures such as blood, there may be excessive foaming in the separator that can impact the gas exchange [45].

12.2.2.2 Correlation of Blood Arsenic to Exposure

Mandal et al. [44] studied a population in West Bengal, India, who had been exposed to arsenic in drinking water and displayed skin lesions from this exposure. At the time of the blood collection, this population had no longer been drinking water with elevated arsenic content. However, their food continued to be prepared using this same water. The mean total blood arsenic in this population was 26.3 µg/L. Blood from a control group had 2.25 µg/L arsenic. The study population also had higher urinary arsenic levels than the control population, suggesting that the food-borne arsenic contributed to the exposure. The arsenic was not related to seafood as these individuals refrained from consuming this type of food 3 days before blood sampling.

The study of Hall et al. [42] is part of a large investigation in Bangladesh (Health Effects of Arsenic Longitudinal study, HEALS) that is examining the health effects from arsenic drinking water exposure. The average arsenic level in water was 103.1 µg/L. Hall et al. [42] examined the relationship between blood arsenic and the risk of skin lesions, which is a hallmark effect of arsenic exposure. Blood arsenic (average 10.8 µg/L) was correlated with drinking water arsenic (Table 12.2) and with creatinine-adjusted urinary arsenic (Spearman $r=0.85$) for both all study participants and for arsenic drinking water levels in individuals exposed to <50 µg As/L (Spearman $r=0.65$).

Katiyar and Singh [61] investigated a population in India that consumed water with levels of arsenic ranging from 0.6 to 135 µg/L (Table 12.2). The average arsenic

---

TABLE 12.2

Representative Studies Describing Exposure–Biomarker Associations for Total Arsenic in Blood

Source of Exposure	Exposure Concentration (µg/L)	Association, N	Reference
Groundwater	0.1–564	r (Spearman) = 0.76, N = 724	Hall et al. [42]
Groundwater	<50	r (Spearman) = 0.59, N = 341	Hall et al. [42]
Groundwater	0.6–135	r = 0.68, p < 0.05, N = 65	Katiyar and Singh [61]

concentration in blood, measured by AAS, was 0.226 µg/dL. The blood arsenic ranged from 0.16–1.5 µg/dL. These blood levels were significantly higher than an unexposed control group. There was a significant positive correlation ($r = 0.68$, $p < 0.05$) between blood arsenic and drinking water arsenic in the exposed population. It should be noted that these investigators measured total arsenic, but did not mention if the study participants refrained from consuming seafood arsenic before the blood was collected.

Adair et al. [40] analyzed total arsenic in blood of a United States population exposed to 3–2100 µg As/L in groundwater. Total arsenic in blood was measured using HG-AFS. Only 20% of the blood samples were above the method detection limited, ranging from 10–30 µg/L. Because of the low detection rate, the investigators were unable to examine correlation between arsenic exposure and total arsenic in blood. This study shows the difficulty in using blood as a biomarker of arsenic exposure.

12.2.2.3 Advantage of Blood Arsenic as a Biomarker of Exposure

The primary advantage of blood as a biomarker of arsenic exposure is that it represents internal dose of arsenic. It may be used for high dose acute exposures that occurred recently [58]. More recent studies of chronic drinking water exposures have utilized blood arsenic as a biomarker of exposure [42,61].

12.2.2.4 Disadvantage of Blood Arsenic as a Biomarker of Exposure

Collecting blood is an invasive procedure, such that subjects may not want to have their blood withdrawn because of cultural issues or fear of medical procedures. The clearance of arsenic from blood is rapid, with the majority apparently cleared with a half-life of 1 h [1], so collection would have to be done soon after the exposure. However, if exposure is continuous, steady-state levels of arsenic in blood may be achieved and would give an indication of exposure [1]. Measurement of total arsenic in blood can be influenced by dietary arsenic (e.g., arsenobetaine), such that a dietary history 2–3 days before the analysis needs to be known or the risk to arsenic exposure may be overestimated. Blood can also be a difficult matrix to work with regard to sample preparation [58].

12.2.3 Hair

Hair is used as a biomarker of exposure to arsenic, primarily in biomonitoring and epidemiology studies, but also in forensic analysis. Biological factors (e.g., type and location of hair on the body), demographic factors (e.g., age, sex), and environmental factors (e.g., duration and time of exposure) can affect the levels of arsenic and other environmental chemicals in hair [62]. The background levels of arsenic in hair is <1 µg/g [63]. The main form of arsenic bound to hair is inorganic, but some methylated species (mono- and dimethylated arsenic) have been detected at low levels [64]. The use of hair as a biomarker of arsenic exposure is more useful for chronic than a recent acute arsenic exposure. The protein keratin is an important component of hair and it is believed that trivalent arsenic binds to the sulfhydryl groups of keratin [1]. Determination of arsenic in hair has been used for forensic analysis in cases of high dose acute poisoning [63]. Analyzing hair may also be suitable for a past exposure to arsenic. Chemicals in hair reflect what was in blood at the time the hair began to grow [65]. Chemicals in blood can be incorporated into hair as it grows because the hair root is vascularized. How arsenic gets incorporated into hair from blood is not known.

The detection and measurement of arsenic in hair as a biomarker of exposure has a role in the debate regarding the death of Napoleon, the eighteenth century French leader. Forshufvud et al. [66] reported that hair from Napoleon had elevated levels of arsenic. However, Lewin et al. [67] analyzed another sample of Napoleon's hair several years later and reported the arsenic hair levels were not elevated over background. One issue is that during the period of Napoleon's death, arsenic was used in many commercial products. In fact, Jones and Ledingham [68] have proposed that the wallpaper in the home of Napoleon at the time of his death, which contained arsenic, was the means of his alleged exposure to arsenic. Hindmarsh [63] suggests that arsenic was added to Napoleon's hair as a preservative after his death. The possibility exists that the arsenic in Napoleon's hair was from external arsenic exposure.

12.2.3.1 Method

Proper procedures and methods for the collection and analysis of hair for arsenic should be prepared and followed [69]. The method requires first removing hair from an individual. Typically, it is hair on the nape of the neck or on the scalp. The hair should be collected from several different spots, because hair grows at different rates [70]. Approximately 500 mg of hair, cut 3 cm in length, is needed for analysis. Hair can be stored conveniently in, for example, paper envelopes. Washing methods are highly recommended as a first step after collecting the hair samples. Hair may first be washed with deionized water, then an organic solvent such as ethanol or acetone, followed by another wash with deionized water [71]. Others have used 1% Triton-X along with water instead of using an organic solvent [72]. Depending on the method of arsenic analysis to be used, the hair may need to be processed by acid digestion, dry ashing, or other methods.

12.2.3.1.1 Total Arsenic in Hair

With dissolved hair, total arsenic can be measured by ICP-MS [38,71,73,74] or AAS [39]. Total arsenic in charred powdered hair can be measured by radioisotope x-ray fluorescence (XRF) [75]. Some methods do not require dissolving hair to analyze arsenic and include neutron activation analysis [36,76], XRF [49], and x-ray absorbance near edge spectrometry (XANES) [49]. These methods do however require washing the hair before analysis. Neutron activation analysis will only detect total arsenic, whereas XANES will give an indication of arsenic species.

12.2.3.1.2 Speciated Arsenic in Hair

Speciated arsenic can be assessed by HPLC-ICP-MS [64] and HG-AAS [72,77]. However, it is less common that speciated arsenic is analyzed in hair than in urine. Unlike blood and urine, arsenobetaine and other seafood-derived arsenicals do not accumulate in hair [78].

12.2.3.2 Correlation of Hair Arsenic to Exposure

Arsenic in hair is correlated to the levels of arsenic in drinking water (Table 12.3). Phan et al. [73] studied populations in three provinces of Cambodia for arsenic exposure that consumed groundwater. Arsenic in the water and hair was measured by ICP-MS. The highest level of arsenic in the groundwater of each province was 846 μg/L, followed by 22 and 1 μg/L. The range of arsenic in hair in all of the samples was 0.07–57.21 μg/g. A significant positive correlation ($p < 0.001$) was reported between arsenic in groundwater and total arsenic in hair (Spearman $r = 0.75$) as well as with the average daily dose of arsenic and the total arsenic in hair (Spearman $r = 0.74$). In the highly exposed arsenic group, 78% of the hair samples had arsenic levels exceeding the background level of 1 μg/g. Less than 1.3% of the hair samples exceeded 1 μg/g arsenic in the lower-exposed populations.

Normandin et al. [36] studied the arsenic hair levels in a rural population in Canada exposed to arsenic in the groundwater. The levels of arsenic in the water were determined using ICP-MS and in hair by NAA. Arsenic levels in the water ranged from <0.02 to 140 μg/L. Stratifying the exposure groups into <1 μg/L, >1 to ≤10 μg/L, and >10 μg/L, significant differences between the groups were observed with increased arsenic in hair with increasing concentration of arsenic in water. There was also a significant linear relationship ($p < 0.001$, $R^2 = 0.403$) between estimated drinking water intakes of arsenic and the levels of arsenic in hair.

Cui et al. [49] examined arsenic levels in groundwater, food, and biological samples including hair in a Chinese population. Arsenic levels in water were measured by AAS and in hair using XRF and XANES. The average arsenic concentration in groundwater was 174 μg/L and ranged from nondetectable levels to 1160 μg/L. Cui et al. [49] reported a positive correlation between water arsenic levels and concentration of arsenic in hair (Spearman $r = 0.344$, $p < 0.01$).

12.2.3.3 Advantage of Hair Arsenic as a Biomarker of Exposure

Using hair as a biomarker of exposure can be advantageous because the collection of hair is not an invasive procedure. Thus hair samples can be collected fairly easily, and storage of the samples is straightforward because they are light and do not take up a lot of space. Analyzing hair for arsenic is best for chronic or past exposures to

---

TABLE 12.3

Representative Studies Describing Exposure–Biomarker Associations for Total Arsenic in Hair

Source of Exposure	Exposure Concentration (μg/L)	Association, N	Reference
Groundwater	1.28–846.1	Spearman $r = 0.75$, $p < 0.0001$, N = 525	Phan et al. [73]
Groundwater	<0.02–140	$r^2 = 0.403$, $p < 0.0001$, N = 110	Normandin et al. [36]
Groundwater	Nondetectable–1160	Spearman $r = 0.344$, $p < 0.01$, N = 131	Cui et al. [49]

arsenic. If an exposure to arsenic ceases, it can be determined when the exposure concluded because hair can be cut into smaller pieces and analyzed segmentally. Hair grows about 1 cm/month, although the rate of growth can vary based on age and sex [70]. However, the time of exposure can be estimated, depending on which segment of hair arsenic was detected. Hair is biochemically inactive [70]. Thus, arsenic in blood that is incorporated into hair will not be metabolized. Arsenic derived from seafood arsenic does not accumulate in hair [78], so this form of arsenic should not complicate assessments.

The analysis of hair for chemicals can also be used as an adjunct to other biological samples and to confirm to a population that all means are being carried out to assess an exposure to an environmental chemical [79].

12.2.3.4 Disadvantage of Hair Arsenic as a Biomarker of Exposure

The use of hair as a biomarker of arsenic exposure can be somewhat limiting [63]. While the analysis of hair for arsenic can lead to a positive result, the question is whether or not the arsenic detected is a result of internal or external exposure. Using hair arsenic levels without consideration of exogenous binding may overestimate the arsenic exposure. While washing procedures are available, whether or not these procedures remove all of the arsenic in the hair from external exposure is not clear [63]. For example, Harrington et al. [80] studied a population in the state of Alaska in the United States that lived in an area with high levels of arsenic in drinking water. Some of the population had been drinking bottled water with much lower levels of arsenic. However, they bathed in the high arsenic water and were found to have higher hair arsenic levels than expected. Other disadvantages of hair as a biomarker of arsenic exposure are that there is no standard analytical method that is used across laboratories to measure arsenic in hair, there is no standard washing procedure, and the reference range of arsenic in hair is not well defined.

12.2.4 Nails

Arsenic binding to nails can be used as a biomarker of exposure [81]. This includes for acute high dose exposures in forensic analysis to investigate poisonings [82], which is done at the base of the nail, or for chronic exposure to arsenic. The background levels of arsenic in nails is <1 µg/g [23]. The growth rate of toenails (0.03–0.05 mm/day) is slower than the rate for fingernails (0.1 mm/day) [81] and both are slower than the rate for hair (2–5 mm/day) [62]. Thus, arsenic in nails reflects exposures over a longer period than in hair. Nails contain keratin, and it is believed that trivalent inorganic arsenic binds to the sulfhydryl groups of cysteine within keratin. However, Mandal et al. [64] reported iAs^{III}, iAs^{V}, $MMAs^{V}$, $DMAs^{V}$, and $DMAs^{III}$ in water extracts of fingernails from a study population in West Bengal, India, with high arsenic drinking water levels. The time window of exposure reflected by the nail clipping is dependent on the length of the clipping and the rate of nail growth. There are several factors that affect nail growth rate and include length of the finger, sex (faster in males), age (peak growth rate, 10–14 years old), health status, nutrition, and others [81].

12.2.4.1   Method

As with the other biological samples, a protocol should be prepared and followed in the nail collection process and its analysis for arsenic [81]. There are several ways nails can be sampled. Clippings should be from all toenails or fingernails or a nail from the same finger (e.g., thumb) or toe (e.g., big toe). The same-nail sample collected should remain the same from subject to subject due in part to different growth rates of the nails.

The nails should then be washed to remove arsenic bound to the nails from external sources. To minimize the effect of exogenous contamination, it has been suggested to use toenails instead of fingernails [81,83]. Several methods are available, but there is no one established method for the washing of nails that are followed across laboratories. Examples include water only, water and organic solvent such as acetone, or organic solvent alone.

12.2.4.1.1   Total Arsenic in Nails

Neutron activation analysis [40,84–87] is one of several methods to measure total arsenic in nails. This method generally does not require sample digestion, but the nails should be washed before the analysis to remove exogenously bound arsenic. Other methods to measure total arsenic in nails include HG-AFS [40], HG-AAS [77,88], ICP-MS [89–91], graphite furnace AAS [92–94], and others. This latter group of methods requires dissolution of the nails, by, for example, acid digestion and using a source of heat such as a hot plate or microwave.

12.2.4.1.2   Speciated Arsenic in Nails

Mandal et al. [44,64] and Button et al. [95] have used HPLC-ICP-MS to speciate arsenic in nails. For the analysis, the nails were washed and then extracted with water under heated conditions. Species detected in these extracts of arsenic-exposed individuals include predominantly iAs^{III}, with lower amounts of iAs^V, MMAs^V, DMAs^{III}, and DMAs^V [44,65,95].

12.2.4.2   Correlation of Nail Arsenic to Exposure

Arsenic levels in nails are correlated to exposure to this metalloid (Table 12.4). However, for drinking water exposure at low levels, the relationship appears to be nonlinear. Karagas et al. [86] reported that in a U.S. study in the state of New Hampshire, in subjects exposed to arsenic drinking water concentrations ≥ 1 µg/L, the correlation ($r = 0.65$) was significant ($p < 0.001$) with arsenic nail concentration. However, at arsenic concentrations <1 µg/L, the relationship was not significant ($p = 0.31$, $r = 0.08$).

In a study in the state of Nevada in the United States, Adair et al. [40] measured total arsenic toenail concentrations of a population exposed to arsenic in groundwater (9–2100 µg/L for total study population; 9 to approximately 700 µg/L for individuals included in nail analysis). Both NAA and HG-AFS were used to measure arsenic in the nails. They reported similar positive correlations of arsenic in toenails between HG-AFS ($r^2 = 0.4994$, $p < 0.0001$) and NAA ($r^2 = 0.5222$, $p < 0.0001$) with respect to arsenic in groundwater.

TABLE 12.4

Representative Studies Describing Exposure–Biomarker Associations for Total Arsenic in Nails

Source of Exposure	Exposure Concentration (µg/L)	Association, N	Reference
Groundwater	1.28–846.1	Spearman r=0.72, p<0.0001, N=398 (fingernail) Spearman r=0.61, p<0.0001, N=279 (toenail)	Phan et al. [73]
Groundwater	0.002–66.6	r=0.46, p<0.001, N=208 (toenail, overall data) r=0.082, p=0.31, N=166 (As<1 µg/L) r=0.65, p<0.001, N=42 (As ≥ 1 µg/L)	Karagas et al. [86]
Groundwater	88–612	r^2=0.68, p=0.029, N=47 (fingernail)	Mandal et al. [64]
Groundwater	<0.02–140	r^2=0.464, p<0.0001, N=110 (big toenail) r^2=0.441, p<0.0001, N=110 (other toenails)	Normandin et al. [36]
Groundwater	9–700	r^2=0.4994, p<0.0001, N=40 (no fish consumption; method—AFS) r^2=0.5222, p<0.0001, N=49 (no fish consumption; method—NAA)	Adair et al. [40]
Drinking water (includes public, well, and bottled water)	<0.048–99.3	r^2=0.32, p<0.05, N=430	Slotnick et al. [91]

Phan et al. [73] studied populations in three provinces of Cambodia for arsenic exposure that consumed groundwater. The highest level of arsenic in the water was 846 µg/L, followed by 22 and 1 µg/L. Toenail and fingernails were collected from volunteers in the three provinces. The nails were washed and acid-digested. Total arsenic was measured by ICP-MS. Significant positive correlations were reported for arsenic water concentration and arsenic in fingernails (Spearman r=0.72, p<0.0001) and arsenic in toenails (Spearman r=0.61, p<0.0001).

A population in India that consumed arsenic-contaminated groundwater for 3–10 years was examined for arsenic in fingernails by Mandal et al. [64]. The arsenic concentration in groundwater ranged from 88–612 µg/L. The study participants refrained from eating seafood 3 days before sample collection. The nails were washed and incubated in hot water and the extracts were analyzed by HPLC-ICP-MS. Arsenicals analyzed were iAs^{III}, iAs^{V}, MMAs^{III}, MMAs^{V}, DMAs^{III}, and DMAs^{V}. A positive correlation was reported relating arsenic concentration in groundwater and sum of the arsenicals in the fingernails (r^2=0.68, p<0.029).

Slotnik et al. [91] examined toenail arsenic concentrations and their relationship to inorganic arsenic intake from drinking water in a population in the state of Michigan in the United States. The arsenic drinking water levels ranged from below the detection limit (0.048 μg/L) to 100 μg/L. Toenails were collected, washed, and digested in acid. Total arsenic was measured by ICP-MS. All toenail samples were above the detection limit for arsenic; the concentration of total arsenic in the toenails ranged from 0.003 to 1.3 μg/g. The concentration of arsenic drinking water collected at the residences of the study population correlated significantly ($p < 0.05$, $r^2 = 0.32$) with concentration of toenail arsenic. The correlation increased ($r^2 = 0.48$) between arsenic intake (consumption of tap water and beverages prepared from tap water) and toenail arsenic concentration when the drinking water concentration was >1 μg/L.

12.2.4.3   Advantage of Nail Arsenic as a Biomarker of Exposure

The advantages of measuring arsenic in nails as a biomarker of exposure are that collecting nails is a noninvasive procedure and more subjects may be willing to participate in a study; nails are easily collected and stored; due to the slow growth of nails, the detection of arsenic reflects exposure over an extended time period from all routes of exposure; and there are no concerns about dietary exposures to marine animals and shellfish that contain organic arsenicals such as arsenobetaine, because these arsenicals have not been detected in nails. Since nails are not metabolically active, the bound arsenic is not metabolized and reflects what was in blood at the time when arsenic was bound to the nail [91].

12.2.4.4   Disadvantage of Nails Arsenic as a Biomarker of Exposure

A disadvantage to using nails as a biomarker of exposure to arsenic is adsorption to it from external exposures. This type of binding would overestimate arsenic exposure if the nails are not properly cleaned. This is why some investigators prefer to collect and analyze toenails instead of fingernails, because outdoor exposures are generally less for toenails than fingernails [81,83]. Careful cleaning of the nails before analysis is important in decreasing adsorbed arsenic [74]. Another disadvantage is that there is no standard analytical method that can be used across laboratories to determine arsenic concentration in nails. Finally, the reference range for arsenic in nails is not well defined.

12.3   BIOMARKERS OF EFFECT

A biomarker of effect can be defined as a quantifiable change (e.g., biochemical, physiological) within an organism that can be associated with a potential adverse health outcome [9]. This type of biomarker would not necessarily be used in the clinic to diagnose arsenic poisoning, but is more appropriate for quantitative dose–response assessments. Examples of a biomarker effect of arsenic include urinary and blood porphyrins [96] and those resulting from the genotoxicity of arsenic. Arsenic is not mutagenic but is genotoxic [97]. Genotoxic effects of arsenic include micronuclei in lymphocytes and mucosal and urothelial cells [98], chromosomal aberrations [99,100], and sister chromatid exchanges [101,102].

One of the proposed mechanisms of action of arsenic is oxidative stress [5,103,104]. Biomarkers from arsenic-induced oxidative stress have been identified [104]. Some of these markers include reactive oxidants and lipid peroxidative end products in plasma, urinary guanine oxidative products (e.g., 8′-hydroxy-2′-deoxyguanosine), elevation of reactive oxygen species detected by fluorescent or electron spin resonance, and effects on reactive oxidant species defense mechanisms [104]. Arsenic-induced oxidative stress may be involved in the development of cancer; cardiovascular diseases such as myocardial injury, hypertension, and atherosclerosis; skin lesions; and other organ-related diseases [105–108]. How these aforementioned biomarkers are linked to arsenic-induced effects is still not completely clear.

The hallmark effect of chronic arsenic poisoning is changes to the skin displayed by hyperpigmentation and palmoplantar hyperkeratosis [109,110]. Some scientists consider the arsenic-specific skin lesions a biomarker of exposure following long-term (several years) exposure to arsenic in drinking water [111]. These lesions appear to be associated with the risk of skin cancer [112].

While exposure to arsenic can result in an adverse outcome, the mechanism(s) of action for any of the adverse health effects is not completely understood and may include multiple mechanisms of action. The lack of this understanding has limited the development and implementation of biomarkers of effect for the diagnosis of arsenic poisoning.

12.4   BIOMARKERS OF SUSCEPTIBILITY

A biomarker of susceptibility is a marker of an innate or attained ability of an organism to react to the challenge of exposure to a definitive exogenous substance [9]. Biomarkers of susceptibility for arsenic have only more recently been studied and are less likely to be used for the diagnosis of arsenic poisoning, but do provide an indication of the risk an individual may have to the poisonous effects of arsenic.

The biomarkers of susceptibility for arsenic revolve around genetic polymorphisms in the metabolism of arsenic, DNA repair enzymes, and enzymes related to oxidative stress [111,113,114]. A recent review by Antonelli et al. [113] examined the literature on polymorphisms in As3MT, glutathione-S-transferase omega, and PNP. As discussed, in Section 12.2, As3MT is the enzyme that metabolizes arsenic. Three polymorphisms of As3MT (rs3740390, rs11191439 and rs11101453) were significantly associated with in vivo changes in methylation of arsenic, particularly with % urinary monomethyl arsenic. Polymorphisms in glutathione-S-transferase omega and PNP did not impact arsenic methylation.

12.5   FUTURE DIRECTIONS

Technology has been advancing rapidly in the past 20 years, particularly in the area of *omic* technologies such as genomics, metabolomics, and proteomics. Several of these *omic* technologies have been used recently to investigate biomarkers of exposure, effect, and susceptibility for arsenic [115]. For example, Hegedus et al. [116] analyzed the urinary proteome of study participants that had low (<100 µg total urinary As/L) or high (≥100 µg total urinary As/L) exposure to arsenic.

The investigators found two polypeptides that were significantly decreased in the urine of males of the high exposure group. These peptides were identified as being part of human beta-defensin-1. The gene for this protein may have a role in suppression of urological cancers. Arsenic is a known human bladder carcinogen and interfering with tumor suppression may be a mechanism of action for this metalloid.

As *omic* technologies become more sensitive, reliable, and less expensive, their use should benefit research efforts in identifying molecular and biological events that occur upstream to the adverse health effects that can result from arsenic exposure. This information will aid in the development of definitive exposure levels of arsenic in water and air such that the adverse health effects such as carcinogenicity of this metalloid should no longer be of consequence to the world population.

12.6   CONCLUSION

Biomarkers of exposure are the most useful of the three biomarker classifications (exposure, effect, and susceptibility) for aiding clinicians and scientists in diagnosing adverse health outcomes from exposure to arsenic. For exposure to arsenic that occurred in the previous 3 days, analysis of urine is the most advantageous because it is easy to handle, there are many reliable methods to analyze urine for arsenic, and a positive response is indicative of an exposure to arsenic.

Biomarkers of effect are best to study the mechanism of action for arsenic and can aid in future guidance levels of arsenic in drinking water, food, etc., that would be likely not to result in an adverse effect in humans that are exposed to arsenic.

Biomarkers of susceptibility, which are the most recently developed type of biomarker for arsenic, should be beneficial in prediagnosing the susceptibility of an individual to the toxic effects of arsenic. These predisposed individuals could be made aware of their susceptibility and the potential risk to developing disease from arsenic exposure. They can then alter their exposure to arsenic with the hope of not developing a disease from it.

DISCLAIMER

This article has been reviewed in accordance with the policy of the National Health and Environmental Effects Research Laboratory, US Environmental Protection Agency, and approved for publication. Approval does not signify that the contents necessarily reflect the views and policies of the Agency, nor does mention of trade names or commercial products constitute endorsement or recommendation for use.

REFERENCES

1. NRC. *Arsenic in Drinking Water.* (Washington, DC: National Research Council, National Academy Press, 1999.)
2. Cullen, W.R. *Is Arsenic an Aphrodisiac? The Sociochemistry of an Element.* (Cambridge, U.K.: The Royal Society of Chemistry, 2008.)
3. Smith, R. Arsenic: A murderous history. (2012). Available at http://www.dartmouth.edu/~toxmetal/arsenic/history.html, Accessed on February 14, 2014.

4. Marsh, J. Account of a method of separating small quantities of arsenic from substances with which it may be mixed. *Edinburgh New Philosophical Journal* 21 (1836): 229–36. http://www.archive.org/stream/edinburghnewphil21edin#page/228/mode/2up, Accessed on March 1, 2014.

5. Hughes, M.F., B.D. Beck, Y. Chen et al. Arsenic exposure and toxicology: A historical perspective. *Toxicological Sciences* 123 (2) (2011): 305–332.

6. Kapaj, S., H. Peterson, K. Liber et al. Human health effects from chronic arsenic poisoning—A review. *Journal of Environmental Science and Health Part A* 41 (10) (2006): 2399–2428.

7. WHO. *Arsenic and Arsenic Compounds.* (Geneva, Switzerland: World Health Organization, Environmental Health Criteria Monograph 224, International Programme on Chemical Safety, 2001.)

8. Cohen, S.M., L.L. Arnold, B.D. Beck et al. Evaluation of the carcinogenicity of inorganic arsenic. *Critical Reviews in Toxicology* 43 (9) (2006): 711–752.

9. WHO. *Biomarkers and Risk Assessment: Concepts and Principles.* (Geneva, Switzerland: World Health Organization, Environmental Health Criteria Monograph 155, 1993.)

10. Thomas, D.J., J. Li, S.B. Waters et al. Arsenic (+3 oxidation state) methyltransferase and the methylation of arsenicals. *Experimental Biology and Medicine (Maywood)* 232 (1) (2007): 3–13.

11. Hayakawa, T., Y. Kobayashi, X. Cui et al. A new metabolic pathway of arsenite: Arsenic-glutathione complexes are substrate for human arsenic methyltransferase Cyt19. *Archives of Toxicology* 79 (4) (2005): 183–191.

12. Delmondedieu, M., M.M. Basti, J.D. Otvos et al. Reduction and binding of arsenate and dimethylarsinate by glutathione: A magnetic resonance study. *Chemico-Biological Interactions* 90 (2) (1994): 139–155.

13. Scott, N., K.M. Hatlelid, N.E. MacKenzie et al. Reactions of arsenic(III) and arsenic(V) species with glutathione. *Chemical Research in Toxicology* 6 (1) (1993): 102–106.

14. Nemeti, B., M.E. Regonesi, P. Tortora et al. The mechanism of polynucleotide phosphorylase-catalyzed arsenolysis of ADP. *Biochimie* 93 (3) (2011): 624–627.

15. Vahter, M. Mechanisms of arsenic biotransformation. *Toxicology* 181–182 (2002): 211–217.

16. Marafante, E., M. Vahter, H. Norin et al. Biotransformation of dimethylarsinic acid in mouse, hamster and man. *Journal of Applied Toxicology* 7 (2) (1987): 111–117.

17. Vahter, M. Genetic polymorphisms in the biotransformation of inorganic arsenic and its role in toxicity. *Toxicology Letters* 112–113 (2000): 209–217.

18. Valenzuela, O.L., Z. Drobná, E. Hernández-Catellanos et al. Association of *AS3MT* polymorphism and the risk of premalignant arsenic skin lesions. *Toxicology and Applied Pharmacology* 239 (2) (2009): 200–207.

19. Gebel, T.W. Arsenic methylation is a process of detoxification through accelerated excretion. *International Journal of Hygiene and Environmental Health* 205 (6) (2002): 505–508.

20. Styblo, M., L.M. Del Razo, L. Vega et al. Comparative toxicity of trivalent and pentavalent inorganic and methylated arsenicals in rat and human cells. *Archives of Toxicology* 74 (6) (2000): 289–299.

21. Naranmandura, H., K. Rehman, X.C. Le et al. Formation of methylated oxyarsenicals and thioarsenicals in wild-type and arsenic (+3 oxidation state) methyltransferase knockout mice exposed to arsenate. *Analytical and Bioanalytical Chemistry* 405 (6) (2013): 1885–1891.

22. Raml, R., A. Rumpler, W. Goessler et al. Thio-dimethylarsinate is a common metabolite in urine samples from arsenic-exposed women in Bangladesh. *Toxicology and Applied Pharmacology* 222 (3) (2007): 374–380.

23. ATSDR. *Toxicological Profile for Arsenic*. (Atlanta, GA: Agency for Toxic Substances and Disease Registry, Public Health Service, U.S. Department of Health and Human Services, 2007.)

24. Cornelis, R., B. Heinzow, R.F.M. Herber et al. Sample collection guidelines for trace elements in blood and urine. *Pure and Applied Chemistry* 67 (8–9) (1995): 1575–1608.

25. Calderon, R.L., E. Hudgens, X.C. Le et al. Excretion of arsenic in urine as a function of exposure to arsenic in drinking water. *Environmental Health Perspectives* 107 (8) (1999): 663–667.

26. Carrieri, M., A. Trevisan, G.B. Bartolucci. Adjustment of concentrate-dilution of spot urine samples: Correlation between specific gravity and creatinine. *International Archives of Occupational and Environmental Health* 74 (1) (2001): 63–67.

27. Hinwood, A.L., M.R. Sim, N. de Klerk et al. Are 24-hr urine samples and creatinine adjustment required for analysis of inorganic arsenic in urine in population studies? *Environmental Research* 88 (3) (2002): 219–224.

28. Nermell, B., A.L. Lindberg, M. Rahman et al. Urinary arsenic concentration adjustment factors and malnutrition. *Environmental Research* 106 (2) (2008): 212–218.

29. Kaise, T., S. Watanabe, K. Itoh. The acute toxicity of arsenobetaine. *Chemosphere* 14 (9) (1985): 1327–1332.

30. Francesconi, K.A., R. Tanggaard, C.J. McKenzie et al. Arsenic metabolites in human urine after ingestion of an arsenosugar. *Clinical Chemistry* 48 (1) (2002): 92–101.

31. Le, X.C., W.R. Cullen, K.J. Reimer. Human urinary arsenic excretion after one-time ingestion of seaweed, crab and shrimp. *Clinical Chemistry* 40 (4) (1994): 617–624.

32. Francesconi, K.A., D. Kuehnelt. Determination of arsenic species: A critical review of methods and applications, 2000–2003. *Analyst* 129 (5) (2004): 373–395.

33. Landsberger, S., G. Swift, J. Neuhoff. Nondestructive determination of arsenic in urine by epithermal neutron activation analysis and Compton suppression. *Biological Trace Element Research* 26–27 (1990): 27–32.

34. Lin, S.M., M.H. Yang. Arsenic, selenium, and zinc in patients with Blackfoot disease. *Biological Trace Element Research* 15 (1) (1988): 213–221.

35. Zeisler, R., E.A. Mackey, G.P. Lamaze et al. NAA methods for determination of nanogram amounts of arsenic in biological samples. *Journal of Radioanalytical and Nuclear Chemistry* 269 (2) (2006): 291–296.

36. Normandin, L., P. Ayottte, P. Levallois et al. Biomarkers of arsenic exposure and effects in a Canadian rural population exposed through groundwater consumption. *Journal of Exposure Science and Environmental Epidemiology* 24 (2) (2014): 127–134.

37. Ahmed, M., Z. Fatmi, A. Ali. Correlation of arsenic exposure through drinking ground water and urinary arsenic excretion among adults in Pakistan. *Journal of Environmental Health* 76 (6) (2014): 48–54.

38. Cleland, B., A. Tsuchiya, D.A. Kalman et al. Arsenic exposure within the Korean Community (United States) based on dietary behavior and arsenic levels in hair, urine, air, and water. *Environmental Health Perspectives* 117 (4) (2009): 632–638.

39. Pandey, P.K., S. Yadav, M. Pandey. Human arsenic poisoning issues in Central-East Indian locations: Biomarkers and biochemical monitoring. *International Journal of Environmental Research and Public Health* 4 (1) (2007): 15–22.

40. Adair, B.M., E.E. Hudgens, M.T. Schmitt et al. Total arsenic concentrations in toenails quantitated by two techniques provide a useful biomarker of chronic arsenic exposure in drinking water. *Environmental Research* 101 (2) (2006): 213–220.

41. Šlejkovec, Z., I. Falnoga, W. Goessler et al. Analytical artefacts in the speciation of arsenic in clinical samples. *Analytica Chimica Acta* 607 (1) (2008): 83–91.

42. Hall, M., Y. Chen, H. Ahsan et al. Blood arsenic as a biomarker of arsenic exposure: Results from a prospective study. *Toxicology* 225 (2–3) (2006): 225–233.

43. Mok, W.M., C.M. Wai. Determination of arsenic and antimony in biological materials by solvent extraction and neutron activation. *Talanta* 35 (3) (1988): 183–186.
44. Mandal, B.K., Y. Ogra, K. Anzai et al. Speciation of arsenic in biological samples. *Toxicology and Applied Pharmacology* 198 (3) (2004): 307–318.
45. Mandal, B.K., K.T. Suzuki, K. Anzai. Impact of arsenic in foodstuffs on the people living in the arsenic affected areas of West Bengal, India. *Journal of Environmental Science and Health Part A* 42 (12) (2007): 1741–1752.
46. Valenzuela, O.L., V.H. Borja-Aburto, G.G. Garcia-Vargas et al. Urinary trivalent methylated arsenic species in a population chronically exposed to inorganic arsenic. *Environmental Health Perspectives* 113 (3) (2005): 250–254.
47. Rivera-Núñez, Z., J.R. Meliker, J.D. Meeker et al. Urinary arsenic species, toenail arsenic and arsenic intake estimates in a Michigan population with low levels of arsenic in drinking water. *Journal of Exposure Science and Environmental Epidemiology* 22 (2) (2012): 183–190.
48. Heinrich-Ramm, R., S. Mindt-Prüfert, D. Szadkowski. Arsenic species excretion after controlled seafood consumption. *Journal of Chromatography B* 778 (1–2) (2002): 263–273.
49. Cui, J., J. Shi, G. Jiang et al. Arsenic levels and speciation from ingestion exposures in Shanxi, China: Implications for human health. *Environmental Science and Technology* 47 (10) (2013): 5417–5424.
50. Fu, S., J. Wu, Y. Li et al. Urinary arsenic metabolism in a Western Chinese population exposed to high-dose inorganic arsenic in drinking water: Influence of ethnicity and genetic polymorphisms. *Toxicology and Applied Pharmacology* 274 (1) (2014): 117–123.
51. Devesa, V., L.M. Del Razo, B. Adair et al. Comprehensive analysis of arsenic metabolites by pH specific hydride generation atomic absorption spectrometry. *Journal of Analytical Atomic Absorption Spectrometry* 19 (11) (2004): 1460–1467.
52. Falk, K., H. Emons. Speciation of arsenic compounds by ion-exchange HPLC-ICP-MS with different nebulizers. *Journal of Analytical Atomic Spectrometry* 15 (6) (2000): 643–649.
53. Schmeisser, E., W. Goessler, N. Kienzl et al. Volatile analytes formed from arsenosugars: Determination by HPCL-HG-ICPMS and implications for arsenic speciation analyses. *Analytical Chemistry* 76 (2) (2004): 418–423.
54. Lindberg, A.-L., W. Goessler, M. Grander et al. Evaluation of the three most commonly used analytical methods for determination of inorganic arsenic and its metabolites in urine. *Toxicology Letters* 168 (3) (2007): 310–318.
55. WHO. *Guidelines for Drinking-Water Quality*. (Geneva, Switzerland: World Health Organization, 4th edn., 2011) p. 315.
56. Agusa, T., P.T.K. Trang, V.M. Lan et al. Human exposure to arsenic from drinking water in Vietnam. *Science of the Total Environment* 488–489 (2014): 562–569.
57. Buchet, J.P., R. Lauwerys, H. Roels. Comparison of the urinary excretion of arsenic metabolites after a single oral dose of sodium arsenite, monomethylarsonate, or dimethylarsinate in man. *International Archives of Occupational and Environmental Health* 48 (1) (1981): 71–79.
58. Marchiset-Ferlay, N., C. Savanovitch, M.-P. Sauvant-Rochat. What is the best biomarker to assess arsenic exposure via drinking water? *Environment International* 39 (1) (2012): 150–171.
59. Blekastad, V., J. Jonsen, E. Steinnes et al. Concentrations of trace elements in human blood serum from different places in Norway determined by neutron activation analysis. *Acta Medica Scandinavica* 216 (1) (1984): 25–29.
60. Heydorn, K. Environmental variation of arsenic levels in human blood determined by neutron activation analysis. *Clinica Chimica Acta* 28 (2) (1970): 349–357.

61. Katiyar, S., D. Singh. Prevalence of arsenic exposure in population of Ballia district from drinking water and its correlation with blood arsenic level. *Journal of Environmental Biology* 35 (3) (2014): 589–594.

62. Sukumar, A. Factors influencing levels of trace elements in human hair. *Reviews in Environmental Contamination and Toxicology* 175 (2002): 47–78.

63. Hindmarsh, J.T. Caveats in hair analysis in chronic arsenic poisoning. *Clinical Biochemistry* 35 (1) (2002): 1–11.

64. Mandal, B.K., Y. Ogra, K.T. Suzuki. Speciation of arsenic in human and nail and hair from arsenic-affected area by HPLC-inductively coupled argon plasma mass spectrometry. *Toxicology and Applied Pharmacology* 189 (2) (2003): 73–83.

65. Gellein, K., S. Lierhagen, P.S. Brevik. Trace element profiles in single strands of human hair determined by HR-ICP-MS. *Biological Trace Element Research* 123 (1–3) (2008): 250–260.

66. Forshufvud, S., H. Smith, A. Wassen. Arsenic content of Napoleon I's hair probably taken immediately after his death. *Nature* 192 (4798) (1961): 103–105.

67. Lewin, P.K., R.G.V. Hancock, P. Voynovich. Napoleon Bonaparte—No evidence of chronic arsenic poisoning. *Nature* 299 (5884) (1982): 627–628.

68. Jones, D.E.H., K.W.D. Ledingham. Arsenic in Napoleon's wallpaper. *Nature* 299 (5884) (1982): 626–627.

69. Bass, D.A., D. Hickok, D. Quig et al. Trace element analysis in hair: Factors determining accuracy, precision and reliability. *Alternative Medicine Review* 6 (5) (2001): 472–481.

70. Harkey, M.R. Anatomy and physiology of hair. *Forensic Science International* 63 (1–3) (1993): 9–18.

71. Hanh, H.T., K.-W. Kim, S. Bang et al. Community exposure to arsenic in the Mekong River delta, Southern Vietnam. *Journal of Environmental Monitoring* 13 (7) (2011): 2025–2032.

72. Concha, G., B. Nermell, M. Vahter. Spatial and temporal variation in arsenic exposure via drinking-water in northern Argentina. *Journal of Health, Population, and Nutrition* 24 (3) (2006): 317–326.

73. Phan, K., S. Sthiannopkao, K.-W. Kim. Surveillance on chronic arsenic exposure in the Mekong River basin of Cambodia using different biomarkers. *International Journal of Hygiene and Environmental Health* 215 (1) (2011): 51–58.

74. Samanta, G., R. Sharma, T. Roychowdhury et al. Arsenic and other elements in hair, nails and skin-scales of arsenic victims in West Bengal, India. *Science of the Total Environment* 326 (1–3) (2004): 33–47.

75. Choudhury, T.R., M. Ali, S.A. Rahin et al. Trace elements in the hair of normal and chronic arsenism people. *Global Advanced Research Journal of Environmental Science and Toxicology* 2 (7) (2013): 163–173.

76. Poklis, A., J.J. Saady. Arsenic poisoning: Acute or chronic? Suicide or murder? *American Journal of Forensic Medicine and Pathology* 11 (3) (1990): 226–232.

77. Hinwood, A.L., M.R. Sim, D. Jolley et al. Hair and toenail arsenic concentrations of residents living in areas with high environmental arsenic concentrations. *Environmental Health Perspectives* 111 (2) (2003): 187–193.

78. Vahter, M., E. Marafante, L. Dencker. Metabolism of arsenobetaine in mice, rats and rabbits. *Science of the Total Environment* 30 (1983): 197–211.

79. Harkins, D.K., A.S. Susten. Hair analysis: Exploring the state of the science. *Environmental Health Perspectives* 111 (4) (2003): 576–578.

80. Harrington, J.M., J.P. Middaugh, D.L. Morse et al. A survey of a population exposed to high concentrations of arsenic in well water in Fairbanks, Alaska. *American Journal of Epidemiology* 108 (5) (1978): 377–385.

81. Slotnick, M.J., J.O. Nriagu. Validity of human nails as a biomarker of arsenic and selenium exposure: A review. *Environmental Research* 102 (1) (2006): 125–139.

82. Lander, H., P.R. Hodge, C.S. Crisp. Arsenic in the hair and nails: Its significance in acute arsenical poisoning. *Journal of Forensic Medicine* 12 (1965): 52–67.

83. Garland M., J.S. Morris, B.A. Rosner et al. Toenail trace element levels as biomarkers: Reproducibility over a 6-year period. *Cancer Epidemiology, Biomarkers and Prevention* 2 (5) (1993): 493–497.

84. Calderon, R.L., E.E. Hudgens, C. Carty et al. Biological and behavioral factors modify biomarkers of arsenic exposure in a U.S. population. *Environmental Research* 126 (2013): 134–144.

85. Cottingham, K.L., R. Karami, J.F. Gruber. Diet and toenail arsenic concentrations in a New Hampshire population with arsenic-containing water. *Nutrition Journal* 12 (2013): 149.

86. Karagas, M.R., T.D. Tosteson, J. Blum et al. Measurement of low levels of arsenic exposure: A comparison of water and toenail concentrations. *American Journal of Epidemiology* 152 (1) (2000): 84–90.

87. Nichols, T.A., S.S. Morris, M.M. Mason et al. The study of human nails as an intake monitor for arsenic using neutron activation analysis. *Journal of Radioanalytical and Nuclear Chemistry* 236 (1–2) (1998): 51–57.

88. Wilhelm, M., B. Pesch, J. Wittsiepe et al. Comparison of arsenic levels in fingernail with urinary As species as biomarkers of arsenic exposure in residents living close to a coal-burning power plant in Prievidza District, Slovakia. *Journal of Exposure Analysis and Environmental Epidemiology* 15 (1) (2005): 89–98.

89. Chen, K.L.B., C.J. Amarasiriwardena, D.C. Christiani. Determination of total arsenic concentrations in nails by inductively coupled plasma mass spectrometry. *Biological Trace Element Research* 67 (2) (1999): 109–125.

90. Sekhar, C., N.S. Chary, C.T. Kamala et al. Determination of total arsenic concentration in clinical samples for epidemiological studies using ICP-MS. *Atomic Spectroscopy* 23 (5) (2002): 170–175.

91. Slotnick, M.J., J.R. Meliker, G.A. AvRuskin et al. Toenails as a biomarker of inorganic arsenic intake from drinking water and foods. *Journal of Toxicology and Environmental Health Part A* 70 (2) (2007): 148–158.

92. Beane-Freeman, L.E., L.K. Dennis, C.F. Lynch et al. Toenail arsenic content and cutaneous melanoma in Iowa. *American Journal of Epidemiology* 160 (7) (2004): 679–687.

93. Chiou, H.Y., Y.M. Hseuh, L.L. Hsieh et al. Arsenic methylation capacity, body retention, and null genotypes of glutathione *S*-transferase M1 and T1 among current arsenic-exposed residents in Taiwan. *Mutation Research* 386 (3) (1997): 196–207.

94. Martinez, V., A. Creus, W. Venegas et al. Micronuclei assessment in buccal cells of people environmentally exposed to arsenic in northern Chile. *Toxicology Letters* 155 (2) (2005): 319–327.

95. Button, M., G.R.T. Jenkin, C.F. Harrington et al. Human toenails as a biomarker of exposure to elevated environmental arsenic. *Journal of Environmental Monitoring* 11 (3) (2009): 610–617.

96. Ng, J.C., J.P. Wang, B. Zheng et al. Urinary porphyrins as biomarkers for arsenic exposure among susceptible populations in Guizhou province, China. *Toxicology and Applied Pharmacology* 206 (2) (2005): 176–184.

97. Basu, A., J. Mahata, S. Gupta et al. Genetic toxicology of a paradoxical human carcinogen, arsenic: A review. *Mutation Research* 488 (2) (2001): 171–194.

98. Basu, A., P. Ghosh, J.K. Das et al. Micronuclei as biomarkers of carcinogen exposure in populations exposed to arsenic through drinking water in West Bengal, India: A comparative study in three cell types. *Cancer Epidemiology, Biomarkers and Prevention* 13 (5) (2004): 820–827.

99. Ghosh, P., M. Banerjee, S. De Chaudhuri et al. Increased chromosomal aberration frequencies in the Bowen's patients compared to noncancerous individuals exposed to arsenic. *Mutation Research* 632 (1–2) (2003): 104–110.

100. Ghosh, P., A. Basu, J. Mahata et al. Cytogenetic damage and genetic variants in the individuals susceptible to arsenic induced cancer through drinking water. *International Journal of Cancer* 118 (10) (2006): 2470–2478.
101. Mahata, J., A. Basu, S. Ghoshal et al. Chromosomal aberrations and sister chromatid exchanges in individuals exposed to arsenic through drinking water in West Bengal, India. *Mutation Research* 534 (1–2) (2003): 133–143.
102. Pavia, L., V. Martinez, M.E. Gonsebatt et al. Sister chromatid exchange analysis in smelting plant workers exposed to arsenic. *Environmental and Molecular Mutagenesis* 47 (4) (2006): 230–235.
103. Kitchin, K.T., S. Ahmad. Oxidative stress as a possible mode of action for arsenic carcinogenesis. *Toxicology Letters* 137 (1–2) (2003): 3–13.
104. De Vizcaya-Ruiz, A., O. Barbier, R. Ruiz-Ramos et al. Biomarker of oxidative stress and damage in human populations exposed to arsenic. *Mutation Research* 674 (1–2) (2009): 85–92.
105. Chen, Y., F. Parvez, M. Gamble et al. Arsenic exposure at low-to-moderate levels and skin lesions, arsenic metabolism, neurological functions, and biomarkers for respiratory and cardiovascular diseases: Review of recent findings from the Health Effects of Arsenic Longitudinal Study (HEALS) in Bangladesh. *Toxicology and Applied Pharmacology* 239 (2) (2009): 184–192.
106. Jomova, K., Z. Jenisova, M. Feszterova et al. Arsenic: Toxicity, oxidative stress and human disease. *Journal of Applied Toxicology* 31 (2) (2011): 95–107.
107. Manna, P., M. Sinha, P.C. Sil. Arsenic-induced oxidative myocardial injury: Protective role of arjunolic acid. *Archives of Toxicology* 82 (3) (2008): 137–149.
108. States, J.C., S. Srivastava, Y. Chen et al. Arsenic and cardiovascular disease. *Toxicological Sciences* 107 (2) (2009): 312–323.
109. Chakraborty, A.K., K.C. Saha. Arsenical dermatosis from tubewell water in West Bengal. *Indian Journal of Medical Research* 85 (1987): 326–334.
110. Tseng, W.P., H.M. Chu, S.W. How et al. Prevalence of skin cancer in an endemic area of chronic arsenicism in Taiwan. *Journal of the National Cancer Institute* 40 (3) (1968): 453–634.
111. Chen, C.-J., L.-I. Hsu, C.-H. Wang et al. Biomarkers of exposure, effect, and susceptibility of arsenic-induced health hazards in Taiwan. *Toxicology and Applied Pharmacology* 206 (2) (2005): 198–206.
112. Martinez, V.D., E.A. Vucic, D.D. Becker-Santos et al. Arsenic exposure and the induction of human cancers. *Journal of Toxicology* 2011 (2011): 1–13.
113. Antonelli, R., K. Shao, D.J. Thomas et al. *AS3MT*, *GSTO* and *PNP* polymorphisms: Impact on arsenic methylation and implications for disease susceptibility. *Environmental Research* 132 (2014): 156–167.
114. McClintock, T.R., Y. Chen, J. Bundschuh et al. Arsenic exposure in Latin America: Biomarkers, risk assessments and related health effects. *Science of the Total Environment* 429 (2012): 76–91.
115. Moore, L.E., S. Karami, C. Steinmaus et al. Use of OMIC technologies to study arsenic exposure in human populations. *Environmental and Molecular Mutagenesis* 54 (7) (2013): 589–595.
116. Hegedus, C.M., C.F. Skibola, M. Warner et al. Decreased urinary beta-defensin-1 expression as a biomarker of response to arsenic. *Toxicological Sciences* 106 (1) (2008): 74–82.

13 Chronic Arsenicosis
Clinical Features and Diagnosis

Anupam Das, Rudrajit Paul, and Nilay Kanti Das

CONTENTS

13.1 INTRODUCTION

The element arsenic that claimed many lives in the past due to its potential use as a homicidal poison has started claiming lives since the last 100 years in its new avatar, *chronic arsenicosis*. The menace was identified in the beginning of twentieth century in Argentina (1917) and since then cases were reported from Chile (1962), China/Taiwan (1968), and Southeast Asia since the last two decades of the last century (in 1983 from India, West Bengal, 1995 from Bangladesh, and 1996 from Taiwan). Since then, other countries have joined the parade including Pakistan, Myanmar, Afghanistan, and Cambodia [1] and in India, apart from West Bengal, other states including Assam, Manipur, Bihar, Jharkhand, Uttar Pradesh, Andhra Pradesh, and Chhattisgarh also are reported to be in the grip of this deadly disease [2]. The importance of groundwater in causing this chronic multisystem disorder came to the notice of the medical fraternity since the report by Dr K. C. Saha in 1982 from the School of Tropical Medicine, Kolkata [3]. Groundwater remains the key source of arsenic contamination in Southeast Asia, though surface water (especially in Latin American countries), organic sea food (in Canada, Japan, etc.), bioaccumulation (from crops, meat, milk product, etc.), and industrial poisoning (e.g., smelting and microelectronic industry) are also potential sources for arsenic exposure.

In contrast with acute poisoning with arsenic, chronic arsenicosis arises from the intake of arsenic in small amounts over a prolonged period, like what happens with drinking of water contaminated with arsenic. The World Health Organization (WHO) has defined chronic arsenicosis as a "chronic health condition arising from prolonged ingestion (not less than 6 months) of arsenic above a safe dose, usually manifested by characteristic skin lesions, with or without involvement of internal organs" [4]. The definition emphasizes on the role of cutaneous features in the diagnosis though the tale-tell signs provided by systemic involvement are also of huge importance in arriving at a diagnosis.

The WHO guideline value for arsenic in drinking water is 0.01 mg/L and that is considered a safe dose and presently regarded as the *maximum permissible limit* of arsenic in consumed water. It needs to be highlighted that *safe dose* is quantified by considering the carcinogenic potential of arsenic; thus, what is defined as *safe* may not be safe for preventing the nonmalignant cutaneous and systemic manifestations. The WHO *maximum permissible limits* can however act as guidelines for national authorities or setting up their own limits considering local, environmental, social, economic, and infrastructural conditions. Technically speaking, the WHO-recommended *maximum permissible value for carcinogenic substances* is usually related to *acceptable health risk*, which is defined as that lifetime risk for cancer that equals 10–5 (that is, 1 person in 100,000) [5]. However, in the case of arsenic, this risk would mean a standard as low as 0.00017 mg/L [6]. This can happen with the ability of people to pay for safe water and the availability of water treatment technology. Some countries like Australia have lowered the national standard below the WHO limits to 0.007 mg/L, while European Union, Japan, United States, and Vietnam operates at WHO standard of 0.01 mg/L [5]. Unfortunately, developing countries and those countries that contribute to the maximum global burden of chronic arsenicosis, like Bangladesh, India, China, and Myanmar consider 0.05 mg/L as the *maximum permissible limit* due to inadequate testing facilities for lower concentrations.

Thus, the role of clinical manifestations especially for resource-poor developing countries, need not to be overemphasized. Clinical examination can provide important clue for early diagnosis of arsenicosis. This can be of extreme significance to the patient because it can halt the progress of the disease and prevent the development of malignancies.

13.2   DIAGNOSTIC APPROACH

The present algorithmic approach (Figure 13.1) for the diagnosis of chronic arsenicosis highlights that the condition can be diagnosed as *clinically confirmed* by an arsenic expert even without laboratory confirmation of raised arsenic content in drinking water, hair, or nail [7]. This algorithmic approach allows for further arsenic estimation if resources permit and a *clinically confirmed* case is further classified as *clinically and laboratory confirmed* if laboratory results favor the diagnosis. The algorithm even allows the provision of making a diagnosis of arsenicosis in the absence of an arsenic expert by laboratory estimation of arsenic content in *probable cases* and labeling it as laboratory-confirmed cases on further laboratory estimation [8].

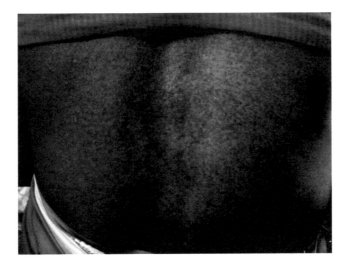

FIGURE 13.1 (**See color insert.**) Middle-aged lady showing raindrop pigmentation and leucomelanosis on the back.

13.2.1 DIAGNOSIS

The diagnosis relies on both laboratory and clinical criteria as detailed in the following.

13.2.1.1 Laboratory Criteria

Detection of arsenic in the drinking water, hair, and nails is the present day laboratory criteria for labeling a patient as *laboratory-confirmed* case of chronic arsenicosis.

1. *Water*: Consumption of drinking water with an arsenic concentration more than the prevailing national standard for at least 6 months is essential for the establishment of elevated exposure to arsenic. The maximum permissible limit of arsenic in drinking water as per the recent guideline of WHO is 0.01 mg/L. In India and Bangladesh, maximum permissible limit is 0.05 mg/L.
2. *Nails and hairs*: Both nails and hairs provide circumstantial evidence of arsenic exposure within the preceding 9 months. A hair arsenic concentration of 1 mg/kg is a marker of significant exposure; however, a concentration of 1.5 mg/kg in nail is required to correlate with a diagnosis of chronic arsenicosis.

13.2.2 CLINICAL FEATURES OF CHRONIC ARSENICOSIS

The clinical manifestations in chronic arsenicosis are broadly divided into, *cutaneous* and *extracutaneous/systemic* manifestation. The cutaneous manifestations are regarded as diagnostic marker for arsenicosis according to the operational definition of WHO, whereas the systemic manifestations are among the leading cause of morbidity and mortality associated with this deadly disease. The systemic manifestations can also serve as important tell-tale signs and can aid in the diagnosis too.

13.2.2.1 Cutaneous Manifestations of Chronic Arsenicosis

Skin lesions are the commonest and earliest manifestation in arsenicosis patients. The importance of a meticulous clinical examination cannot be overemphasized for the diagnosis of chronic arsenic poisoning. In a study [9], 3,695 (20.6%) of 18,000 persons in Bangladesh and 8,500 (9.8%) of 86,000 persons in West Bengal living in arsenic-affected districts showed dermatological features of arsenicosis [10,11]. The cutaneous features can be broadly divided into three groups; pigmentary changes (arsenical dyschromatosis), keratotic lesions, and premalignant and malignant lesions. The earliest and commonest clinical feature is melanosis [12,13]. However, the most sensitive marker is plantar keratosis [14].

13.2.2.1.1 Pigmentary Changes/Arsenical Dyschromatosis

Chronic arsenicosis leads to four basic patterns of pigmentary changes [3,15]:

1. Raindrop pattern (spotty pigmentation appearing on normal skin) (Figures 13.1 through 13.6)
2. Diffuse pigmentation (which is most intense on the trunk and patchy mostly localized to skin folds)
3. Leucomelanosis pattern (depigmented macules over normal skin or hyperpigmented background) (Figures 13.1, 13.7, and 13.8)
4. Mucosal pigmentation (Blotchy pigmentation on the undersurface of tongue or buccal mucosa).

The closest clinical differential diagnoses of these dyspigmentations are enumerated in Table 13.1.

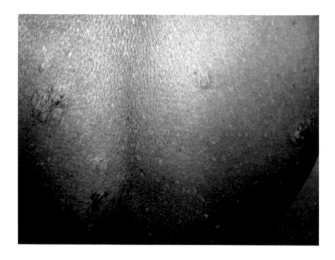

FIGURE 13.2 (See color insert.) Elderly gentleman having lesions of Bowen's disease over a background of raindrop pigmentation.

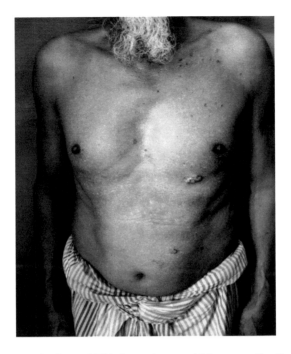

FIGURE 13.3   **(See color insert.)** Elderly gentleman with long-standing features of chronic arsenicosis, eventually developing squamous cell carcinoma on the trunk.

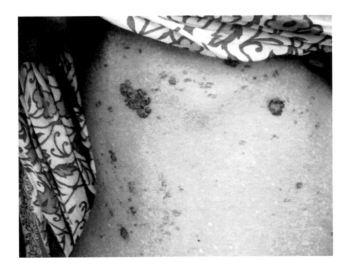

FIGURE 13.4   **(See color insert.)** Young lady with pigmentary changes, Bowen's disease, and basal cell carcinoma on the trunk.

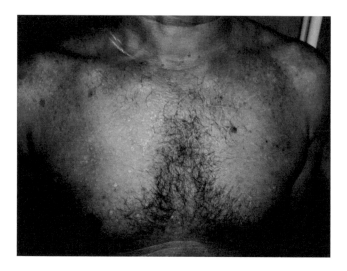

FIGURE 13.5 (See color insert.) Development of Bowen's disease on a background of raindrop pigmentation in an elderly male.

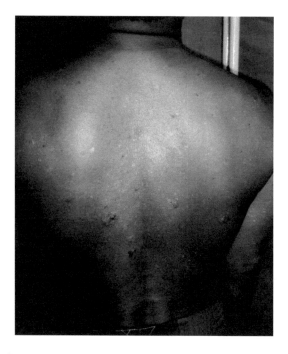

FIGURE 13.6 (See color insert.) Younger brother of the patient shown in Figure 13.5; note the lesions of Bowen's disease on the trunk.

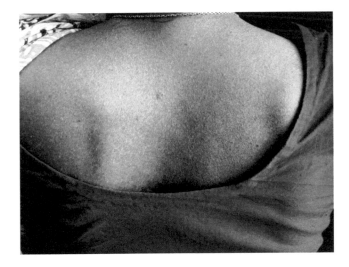

FIGURE 13.7 **(See color insert.)** A young lady with pigmentary changes suggestive of chronic arsenicosis. Note the leucomelanosis pattern.

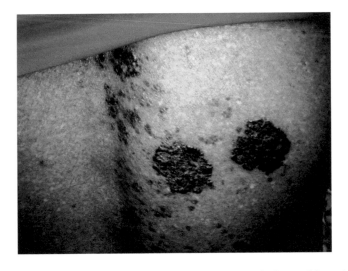

FIGURE 13.8 **(See color insert.)** A middle-aged female developing multicentric Bowen's disease, with a long-standing history of pigmentary changes.

13.2.2.1.2 *Hyperkeratosis*

They can be mild, moderate, or severe depending on the thickness and size of the keratotic lesions (papules, plaques, and nodules). Arsenical keratosis appears as diffuse thickening involving palms and soles, alone or in combination with nodules usually symmetrically distributed. The nodular forms are encountered most frequently on the thenar and lateral borders of palms, on roots or lateral surfaces of

TABLE 13.1

Mimickers of Pigmentary Changes in Chronic Arsenicosis/Arsenical Dyspigmentation

Pigmentation Abnormality in Chronic Arsenicosis	Differential Diagnosis	Salient Points for Differentiation
Diffuse or patchy pigmentation	Lichen planus pigmentosus	More on sun-exposed areas
	Addison's disease	Sun-exposed areas, on sites of trauma, scars, or chronic pressure, in the palmar creases and on nipples, areolae, axillae, perineum, and genitalia
	Vitamin B12 deficiency	More pronounced on axilla and palmoplantar areas
	Alkaptonuria	Family history present, bilateral scleral pigmentation (Osler's sign)
	Wilson's disease	Kayser–Fleischer ring in eyes, liver disease, and neuropsychiatric symptoms
	Chronic renal failure	Intractable pruritus, perforating disorders along with other systemic features
	Hemochromatosis	Generalized, but more pronounced in sun-exposed areas, genitalia, and scars
	Congenital causes	Onset since birth and family history present
	Acanthosis nigricans	Obesity, velvety corrugated plaques found
	Drug-induced	Definite history of drug intake present
	Universal acquired melanosis	Diffuse jet-black pigmentation since infancy
Raindrop pattern	Pityriasis versicolor	Perifollicular hypopigmentation with scales
	Idiopathic guttate hypomelanosis	First on the legs. Later, it may spread to other sun-exposed areas. The face is not involved
	Xeroderma pigmentosum	Consanguineous parentage, photosensitivity, family history present
	Reticulate acropigmentation of Kitamura	Palmar pits and distortion of dermatoglyphics is hallmark
	Dyschromatosis symmetrica hereditaria	Hypo- and hyperpigmented macules symmetrically on limbs and face
	Dyschromatosis universalis herediataria	Hypo- and hyperpigmented macules throughout
	Dowling-Degos disease	Perioral pitted scars and acneiform scars in neck, axilla, and inframammary crease
	Antimalarial-induced	Drug history present
	Biliary cirrhosis	Other systemic features including jaundice present
	Dyskeratosis congenita	Associated with leukokeratosis and nail dystrophy

(Continued)

TABLE 13.1 (*Continued*)
Mimickers of Pigmentary Changes in Chronic Arsenicosis/Arsenical Dyspigmentation

Pigmentation Abnormality in Chronic Arsenicosis	Differential Diagnosis	Salient Points for Differentiation
	Familial gigantic melanocytosis	Family history present
	Haber's syndrome	Rosacea-like eruption with keratotic plaques and pitted scars
	Franceschetti–Jadassohn–Naegeli syndrome	Associated with palmoplantar keratoderma and hypohidrosis
Mucosal pattern	Lichen planus	Skin lesions may be found, whitish streaks on mucosa
	Metabolic diseases (hemochromatosis, alkaptonuria, pellagra, porphyria cutanea tarda, Addison's disease, Gaucher's disease, Niemann Pick's disease)	Clinical correlation with systemic manifestations helps in ruling out the metabolic causes
Leucomelanosis pattern	Melanoma	Chronic sun exposure, lesions can be differentiated on dermoscopy
	Idiopathic guttate hypomelanosis	The face is not involved
	Post kala-azar dermal Leishmaniasis	History of Kala-azar, persistent erythema, and in duration on face, photosensitivity
	Onchocerciasis	Leopard skin, *mal morado*, lizard skin, hanging groin
	Pinta	Hyperpigmented lesions of varying colors

fingers, and soles, heels, and toes of feet. Such small nodules may coalesce to form large verrucous lesions. The nodular form may also occur in the dorsum of the hands and feet and other parts of the body. *Mild* variety is characterized by indurated, grit-like skin with invisible papules less than 2 mm in size (Figures 13.9 and 13.10). In the *moderate* variety, the lesions are elevated, punctuate, and wart-like. They are >2–5 mm in size that are readily visible (Figures 13.11 through 13.13). *Severe keratosis* (Figures 13.13 and 13.14) is characterized by keratotic elevations that are greater than 5 mm in size or that have become confluent [15]. At times, these are confluent and diffuse developing cracks and fissures[12]. Primary sites of involvement include palms and soles; however, dorsa of the extremities and trunk may also be affected [16].

The mimickers of keratotic lesions on palms and soles are tabulated in Table 13.2.

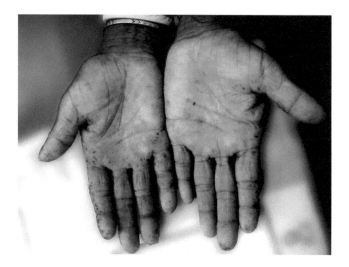

FIGURE 13.9   **(See color insert.)** Note the grit-like skin with minute papules on the palms.

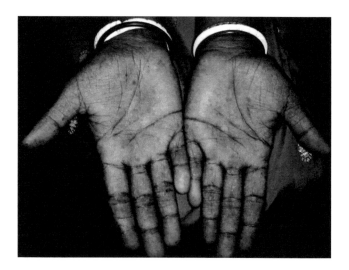

FIGURE 13.10   **(See color insert.)** Note the grit-like skin with minute papules on the palms.

13.2.2.1.3   Premalignant and Malignant Skin Lesions

Skin is thought to be perhaps the most sensitive site for arsenic-induced malignancies, mainly squamous cell carcinoma and basal cell carcinoma (Figures 13.15 and 13.16). The working group of International Agency for Research on Cancer evaluated data from ecological studies, cohort studies, and case–control studies from many countries and observed that arsenic was potentially carcinogenic for skin cancer. Malignancies appear on the hyperkeratotic areas, as well as on nonkeratotic areas of the trunk, extremities, or head [17–20]. The lesions can frequently multiple and involve covered areas of the body in contrast with the nonarsenical ones where a

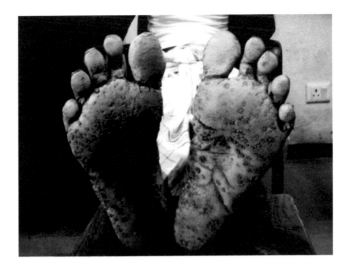

FIGURE 13.11 **(See color insert.)** Characteristic lesions of moderate arsenic keratosis on the soles of a middle-aged gentleman.

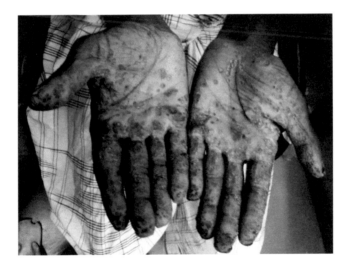

FIGURE 13.12 **(See color insert.)** Palms of the same patient as shown in Figure 13.11; note the punctuate and wart-like lesions.

solitary lesion on sun-exposed areas is the rule [21]. It has been suggested that human papilloma virus (HPV) infection could constitute an additional risk factor for the development of nonmelanoma skin cancer in humans chronically exposed to arsenic. There are also published reports of Merkel cell carcinoma [22,23]. Among the sites of development of malignancies, skin is the most sensitive site [24]. Malignant transformation is clinically suspected when there is sudden increase in size, cracks, and fissures, bleeding on top of keratotic lesions. It is proposed that patients with

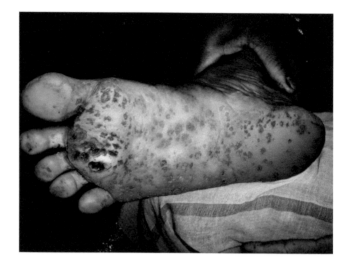

FIGURE 13.13 **(See color insert.)** Moderate to severe arsenic keratosis on the plantar surface of the feet of a middle-aged female. Note the confluence of small papules, which defines severe keratosis.

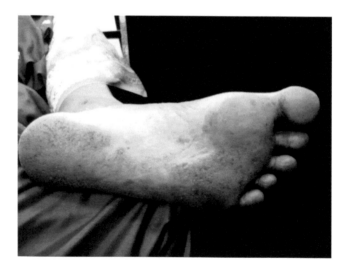

FIGURE 13.14 **(See color insert.)** Severe arsenic keratosis on the sole.

documented arsenic-induced Bowen's disease (squamous cell carcinoma in situ) should undergo a vigorous and aggressive screening for long-term complications, especially the development of subsequent visceral malignancies (small cell carcinoma of lungs, etc.) [25]. A study demonstrated an increased risk of melanoma in persons with elevated toe-nail arsenic concentrations, raising the issue relating to the role of arsenic in the development of melanoma [26].

TABLE 13.2
Mimickers of Keratotic Skin Lesions in Chronic Arsenicosis

Differential Diagnosis	Salient Points for Differentiation
Pitted keratolysis	Associated with malodor and hyperhidrosis
Darier's disease	Dirty warty papules on seborrhoeic areas, family history present, longitudinal leukonychia and erythronychia
Acrokeratosis verruciformis of Hopf	Warty papules on dorsum of hand, family history present
Cowden's syndrome	Oral and skin papilloma, trichilemmomas, gastrointestinal polyps, Lhermitte–Duclos disease
Gorlin's syndrome	Multiple basal cell cancers, odontogenic cysts of jaw
Reticulate acropigmentation of Kitamura	Reticulate pigmentation on extremities, family history present
Porokeratotic eccrine ostial and dermal duct nevus	Multiple linear punctuate pits with comedo-like plugs on palms and soles or keratotic plaques and papules that resemble linear VEN on other areas
Keratosis punctata of palmar creases	Characteristic involvement of crease only
Punctate palmoplantar keratoderma	Family history present

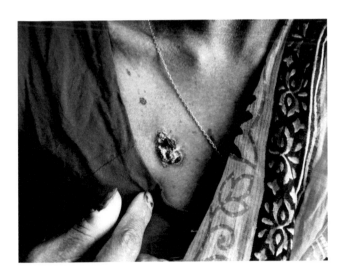

FIGURE 13.15 (See color insert.) Development of squamous cell carcinoma in a patient with a history of pigmentary changes for 15 years.

13.2.2.1.3.1 Role of Skin Biopsy and Histology Classical histology of keratotic papulonodules shows hyperkeratosis (thickening of *stratum corneum*/topmost layer of epidermis), parakeratosis (presence of nucleated cells in *stratum corneum*), acanthosis (thickening of the epidermis), and papillomatosis (elongation of the dermal papillae). Presence of dysplastic keratinocytes, mitotic figures, and wind-blown appearance of shot-gun appearance of cells are alarming signs of impending malignancy. Arsenical hyperkeratosis has been subdivided into benign type A and

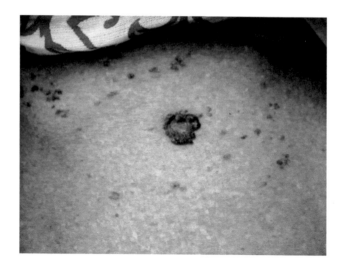

FIGURE 13.16 **(See color insert.)** Note the characteristic rolled-out border of the lesion, which is diagnostic of basal cell carcinoma.

malignant type B according to the absence or presence of cellular atypia [19]. Hence, skin histology is invaluable in differentiating a malignant and nonmalignant lesion, thus guiding the therapeutic strategy.

There is paucity of reports regarding the types and patterns of histopathological changes in skin lesions of chronic arsenicosis. In a study from Bangladesh, hyperkeratotic lesions of 70 patients with chronic arsenicosis were compared with 20 controls [27]. Significant findings included hyperkeratosis (100%), parakeratosis (97%), acanthosis (95.7%), and papillomatosis (74%). The results were found to be significantly more ($P < 0.001$) in the patients than in controls. Basal cell pigmentation was found in 42.8% ($P > 0.05$) and dysplasia and malignant changes in 7% ($P > 0.1$). Another study from Bangladesh documented hyperkeratosis, parakeratosis, acanthosis, papillomatosis, hypergranulosis, and dysplastic changes to be the most important and constant findings. However, basal pigmentation and dermal changes were found to be inconstant features. Another study focusing on the neoplastic manifestations of arsenicosis revealed precancerous skin lesions in 6.6% and cancerous lesions in 0.8% patients [28,29].

Another study was carried out in West Bengal, India. Histopathology of 20 cases of arsenicosis revealed hyperkeratosis in 16 (80%), acanthosis in 6 (30%), parakeratosis in 2 (10%), elongation of rete ridges in 2 (10%), basal cell epithelioma in 2 (10%), squamous cell carcinoma in 1 (5%), and normal pattern in 4 (20%) patients [30]. Masson Fontana's stain from the melanotic lesions of the patients revealed hyperpigmentation of basal layer in 6 (30%), pigment dropout in dermis in 4 (20%), and normal pigmentation pattern in 14 (70%) patients.

13.2.2.2 Extracutaneous/Systemic Manifestations

Apart from cutaneous manifestations, arsenicosis causes symptoms related to the involvement of the lungs, gastrointestinal system, liver, spleen, genitourinary system, hemopoietic system (or bone marrow), eyes, nervous system, and cardiovascular system; along with constitutional symptoms like generalized weakness, anorexia, and weight loss. These symptoms are classified into major and minor depending on the relative frequency of affection of certain organ system in chronic arsenicosis. Respiratory, hepatobiliary, cardiovascular, and nervous systems are the major systems that are involved.

13.2.2.2.1   Respiratory System

Prolonged arsenic exposure may result in ulceration of respiratory mucosa. Sometimes, nasal septal perforation can also occur. However, this mainly occurs in smelter workers on exposure to arsenic fumes. They can also have laryngitis, tracheobronchitis, bronchitis, and rhinitis. All these can lead to chronic cough and hoarseness of voice [31].

Groundwater arsenic toxicity may cause chronic asthmatic bronchitis. This is more common in those with skin lesions. Both *restrictive* and *obstructive lung diseases* have been reported with environmental arsenic toxicity [32]. *Chronic obstructive pulmonary disease (COPD), interstitial lung disease (ILD)*, (Figure 13.17) or *bronchiectasis* may result from arsenic ingestion. Sometimes, there may be combined features also. In a study from India, the prevalent odds ratio (OR) for chronic cough in arsenic exposed persons varied from 5 to 7 [33]. Necropsy studies have shown increased deposition of arsenic in the lung of some of these patients.

13.2.2.2.2   Hepatobiliary

Chronic environmental arsenic exposure can cause hepatomegaly. When biopsied, many of these patients show noncirrhotic portal fibrosis [34]. Another study showed incomplete septal cirrhosis, a type of macronodular cirrhosis in arsenic exposed patients. In these patients, there is high incidence of portal hypertension

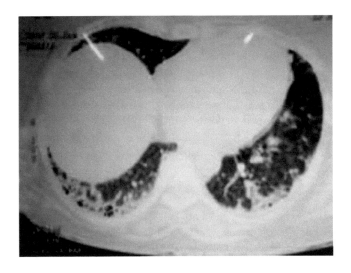

FIGURE 13.17   CT scan photograph showing interstitial lung disease.

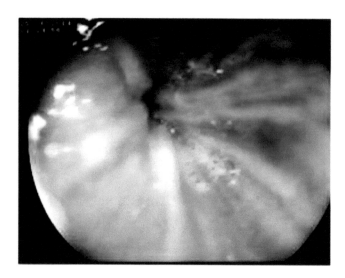

FIGURE 13.18   **(See color insert.)** Esophageal varices due to noncirrhotic portal fibrosis as seen in upper GI endoscopy.

and variceal bleeding. Arsenic can also give rise to hepatic malignancy. There is limited evidence of angiosarcoma of liver as a late complication of arsenic exposure. Earlier, most of the cases of liver damage from arsenic were reported after use of arsenic-containing medications like Fowler's solution or herbal medicines. But now, the use of these medicines has declined considerably; hence environmental arsenic constitutes the majority of cases of arsenic exposure [35].

Acute arsenic exposure may cause hepatitis mimicking a viral hepatitis. Chronic arsenic exposure may also cause fatty liver. It may also cause hepatic mitochondrial damage [36]. In mouse experiments, exposure to arsenic has been shown to cause vascular remodeling, endothelial cell capillarization, and vascularization of peribiliary capillary plexus. In addition there is also constriction of hepatic arterioles and increased sinusoidal PECAM-1 and laminin-1 protein expression. All these pathogenic mechanisms probably also contribute to liver damage in humans exposed to arsenic [37]. In these animal experiments, despite the demonstrated liver changes, the arsenic level in liver was not very high. This means that the hepatic damage can occur without actual arsenic deposition in liver. Thus, these vascular changes may be a more sensitive marker of arsenic-induced liver damage than measuring arsenic level in liver biopsy specimens. Patients may present with variceal bleeding or with enlarged tender liver (Figure 13.18).

13.2.2.2.3 Cardiovascular System

Chronic arsenic exposure can have adverse effects on the heart. In some studies, arterial thickening (even at young age), myocardial damage, and cardiac arrhythmias have been linked to chronic arsenic exposure [38]. It can also cause chronic vascular insufficiency, especially of the extremities. The rheology of flowing blood is altered in arsenic toxicity due to altered RBC membranes and altered microviscosity. The *Blackfoot disease* of Taiwan is caused by excess arsenic in drinking water [39]. This is an endemic peripheral vascular disease of Taiwan southwest coast where well water contains high arsenic levels. The patients affected with this disease have high chance of death from cardiovascular causes. Arsenic exposure may also cause *Raynaud's phenomenon* and *digital dry gangrene*. Myocardial biopsy in arsenic toxicity has shown cytoplasmic vacuoles and myofibrillar disruption. Even myocardial infarction in children has been linked to arsenic in drinking water.

Studies from Bangladesh and the United States have found relations between arsenic exposure and *hypertensive heart disease*. This occurs in a dose-dependent manner. The hypertension may lead to *ischemic heart disease* too [40]. In general, people exposed to arsenic have more incidence of cardiovascular disease. A type of *cardiomyopathy* is also said to occur due to arsenic (Figure 13.19). In fact,

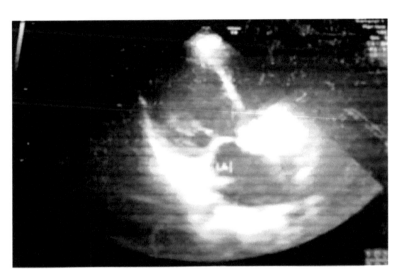

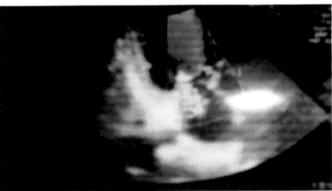

FIGURE 13.19 **(See color insert.)** Echocardiography showing arsenic-induced dilated cardiomyopathy.

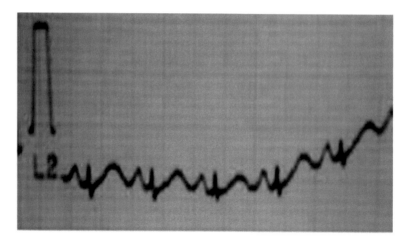

FIGURE 13.20   Photograph of ECG showing QT prolongation.

an outbreak of cardiomyopathy following beer ingestion in 2007 was linked to arsenic contamination of the alcoholic beverage. In vitro studies have found that exposure of endothelial cells to arsenic and ethanol together result in endothelial gene induction and stimulation of remodeling pathways [41]. This does not occur with either agent alone. Thus, arsenical contamination of alcohol can have deleterious effects on patients.

ECG changes in arsenic toxicity include QT prolongation (Figure 13.20), T-wave flattening, and atypical multifocal ventricular tachycardia [42]. The QT interval prolongation may precede ventricular tachycardia.

Echocardiographic study of mice has shown that concentric left ventricular hypertrophy (LVH) is common in arsenic-treated mice. This may be a result of hypertension induced by arsenic or may be an independent effect of arsenic on heart [43].

It needs to be mentioned that the maximum permissible limit of arsenic in drinking water (0.01 mg/L) is set according to the studies on carcinogenic potential of arsenic, but cardiovascular effects of arsenic occur at even lower arsenic levels.

13.2.2.2.4 Nervous System

Arsenic-induced neuropathy initially starts with a peripheral sensory neuropathy in gloves-and-stockings distribution. Later, motor neuropathy also develops [44]. Whether only arsenic is responsible for this neuropathy or other environmental toxins like lead or mercury also contribute to this presentation is still debatable [45]. There is axonopathy. The neuropathy is painful with paresthesia and muscle cramps. Sometimes, there may be ascending motor paralysis mimicking Guillain–Barré syndrome. Usually, there is no cranial neuropathy.

Arsenic inhibits the Krebs' cycle. This may be one of the mechanisms of axonal injury. Sometimes there may be only nerve conduction velocity study (NCV) abnormalities and no clinical signs. The recovery from this disease is often incomplete.

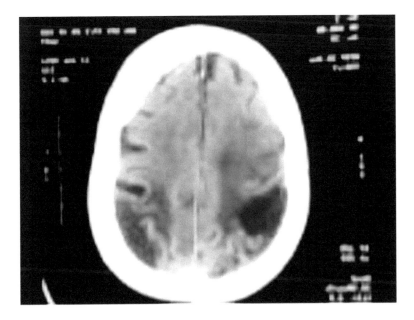

FIGURE 13.21   CT scan showing cerebral infarction.

There may also be cognitive impairment in the form of behavior changes, memory loss, and confusion [46]. Fortunately, the changes often return to normal after withdrawal of the arsenic source. Arsenic exposure has also been shown to affect the intellect of children in a study from Bangladesh [47]. A similar study from Taiwan reported similar neurodevelopmental effects in adolescents [48]. Headache and loss of libido has also been reported in adults. These patients may be sensitive to the effect of alcohol even in small amounts. Once a case was reported from India where asymmetrical phrenic nerve paralysis was seen in arsenic poisoning [49]. An increased prevalence of CVA, especially cerebral infarction, has also been reported in chronically arsenic-exposed persons (Figure 13.21).

13.2.2.2.5 Gastrointestinal System
The gastrointestinal (GI) system effects of arsenic are usually the result of acute ingestion. There will be nausea, vomiting, diarrhea, esophageal varices (Figure 13.21), and abdominal colics. There is increased GI permeability and profuse fluid loss, leading to collapse. There may be a type of hemorrhagic pangastritis and the necrosis of the mucosa may extend to other parts of small intestine too. In extreme cases, there will be gut perforation [50].

Some other clinical signs include burning lips, odynophagia, and thirst [51]. Chronic environmental arsenic exposure also causes intestinal epithelial cell damage but the acute effects are usually not seen.

Arsenic exposure has also been shown to cause gastric carcinoma in experimental animals. The human gut bacteria may act on the ingested arsenic and change it to toxic compounds called monomethylarsonous acid and monomethylmonothioarsonic acid. These may be carcinogenic [52]. Arsenic exposure also causes induction of hUGT-1 gene expression in intestinal epithelial cells. This may alter the xenobiotic metabolism and may be one of the mechanisms of intestinal injury.

13.2.2.2.6 Bone Marrow

Chronic arsenic toxicity has a number of effects on bone marrow. It can present as refractory pancytopenia. In the WBC cell line, there is neutropenia with relative lymphocytosis. If the environmental exposure is curtailed at an early stage, pancytopenia is reversible [53]. The anemia is usually normochromic and normocytic and basophilic stippling may be seen in peripheral smear. There may be mild eosinophilia and karyorrhexis in peripheral blood. Acute arsine gas toxicity may cause intravascular hemolysis. This is due to rapid depletion of intra-RBC glutathione (GSH). If there is hemolysis, there may be macrocytosis and polychromasia. Chronic arsenic exposure also causes immunosuppression.

Bone marrow pathological examination in chronic arsenic toxicity may show nonspecific mild changes. There may also be megaloblastoid changes with nuclear atypia [54]. However, in the arsenic toxicity endemic regions like India and Bangladesh, malnutrition is also an important cause of anemia. Whether there is some interaction between arsenic exposure and malnutrition to cause bone marrow suppression is still unknown.

13.2.2.2.7 Fertility

Environmental or industrial chronic arsenic exposure has been linked to spontaneous abortions and congenital malformations [55]. It is thus considered a teratogen. However, in the reported cases or teratogenicity, there were often other heavy metal exposures too. Thus, how far only arsenic contributes to the fetal damage is unknown. Most of the reported cases were after industrial exposure. Environmental arsenicosis is often difficult to quantify and thus, its exact impact on fertility is still unknown.

Animal experiments on teratogenicity of arsenic gave inconsistent results [56]. Earlier, some animal experiment data reported exencephaly after arsenic exposure. But these studies have been criticized and modern experiments have not been able to reproduce the teratogenic effects of the earlier studies.

Arsenic compounds inhibit DNA repair and thus leads to genotoxic effects. It is seen that trivalent arsenic compounds are more toxic than pentavalent ones. These chromosomal defects are transmissible to the next generations [57].

13.2.2.2.8 Kidney

Arsenic exposure may cause acute tubular necrosis. Later, cortical necrosis may also occur and this may lead to chronic renal failure. Sometimes, glomerular damage may also occur and this may lead to proteinuria. Mitochondrial damage of proximal tubular cells may cause electrolyte abnormalities [58].

Renal damage in arsenic exposure may also occur secondary to hemolysis, cardiac failure, or hepatic damage.

Renal cell cancer has been found in patients chronically exposed to arsenic. Studies from Taiwan have found renal pelvis transitional cell carcinoma in arsenic-exposed patients [59].

Besides chronic arsenic exposure, acute arsine gas toxicity may also cause tubular necrosis. However, kidneys are usually resistant to low levels of arsenic exposure. Kidneys are affected at higher arsenic doses compared to other organs like skin or heart.

13.2.2.2.9 Teeth

Arsenic can damage the periodontal tissue. But this occurs mainly with the arsenic-containing pastes earlier used by some dentists. Sometimes, necrosis of jaw and oro-antral fistula can also occur with arsenic toxicity. The gum and surrounding alveolar bones are devitalized by repeated arsenic exposure [60].

13.2.2.2.10 Eye

Arsenic only rarely affects the eye. Optic damage occurs as a part of neuropathy [61]. The papillomacular bundle is damaged due to axonal energy depletion. Arsenic replaces the phosphorus in some mitochondrial enzymes. This causes impaired oxidative phosphorylation [62]. Thus, reactive oxygen species are accumulated in the neuronal axons and this leads to progressive axonal damage. This leads to impaired central vision. Later, optic atrophy can also develop rarely (Figure 13.22).

Exposure to arsine gas may cause conjunctivitis.

13.2.2.2.11 Endocrine System

Some studies have found evidence of diabetogenic effect of long-term arsenic exposure [63]. In studies from Bangladesh and China, the relative risk for diabetes in arsenic-exposed patients was as high as 2.52. But this data has not been reproduced from other countries. However, the relation between current diabetes epidemic and arsenic exposure in countries like India is a topic of intense research.

Arsenic may cause diabetes by various mechanisms. There may be altered gene transcription, inflammation and induction of lipid peroxidation. In some animal experiments, selenium and arsenites have been found to have opposing effects on glucose metabolism [64]. Thus, low selenium along with high arsenic may be the causative factor in induction of diabetes. In in vitro studies, arsenic-inhibited insulin

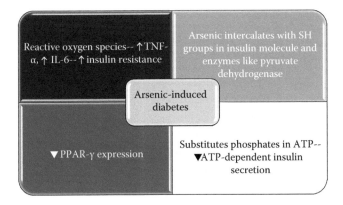

FIGURE 13.22 Pictorial representation of pathomechanism of arsenic-induced diabetes.

mediated glucose uptake in adipocytes. Figure 13.22 depicts the pathogenic mechanisms of arsenic-induced diabetes.

Apart from diabetes, arsenic may also affect the function of thyroid and adrenal glands. But these are only found in animal studies. Human data are still lacking in this field.

13.2.2.2.12 Bone

Arsenic may also affect bone formation by changes in osteoblast differentiation and function. It may cause decreased BMD and trabecular bone volume.

In vitro studies have found that arsenic causes apoptosis of cultured osteoblasts and mitochondrial injury [65]. There is increased endoplasmic reticulum (ER) stress and this may be responsible for the cell death.

13.2.2.2.13 Systemic Malignancy

Millions of people are exposed to arsenic through various routes. Thus, the carcinogenic potential of arsenic is of growing concern all over the world. Different studies have found the link of arsenic exposure with cancers of lung, liver, kidney, urinary bladder, and stomach [66]. The exact pathogenesis of arsenic-induced malignancies is not known. But potential mechanisms include inhibition of DNA repair, activation of c-myc, abnormal DNA methylation, and action of arsenic as cocarcinogen. The biochemical action of arsenic in inducing malignancy is shown in the following chart (Figure 13.23):

Workers engaged in spraying arsenic-containing insecticides have high odds of getting lung cancer. The risk of getting a malignancy from arsenic exposure from drinking water is considered similar to tobacco smoke exposure or domestic radon exposure. Usually the malignancies are of squamous cell type [67]. Rare malignancies like angiosarcoma of liver have also been reported.

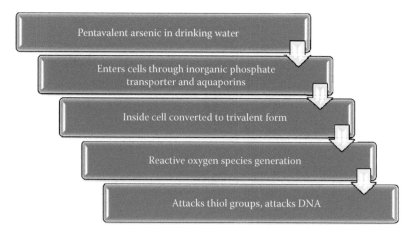

FIGURE 13.23 (**See color insert.**) Possible mechanism of development of malignancies in a long-standing case of arsenicosis.

13.3 CONCLUSION

To conclude, arsenic poisoning continues to haunt mankind since decades. A timely diagnosis of the condition can prove to be of immense help in combating this terror. For this purpose, clinicians need to be aware of the clinical features suggestive of arsenicosis and the associated premalignant conditions and a subsequent multidisciplinary approach is warranted to manage the conditions arising out of exposure to arsenic.

REFERENCES

1. IARC. *Some Drinking-Water Disinfectants and Contaminants, Including Arsenic*. Monographs on the Evaluation of Carcinogenic Risks to Humans, Vol. 84. WHO, Lyon, France, 2004.
2. Ghosh P, Roy C, Das NK, Sengupta SR. Epidemiology and prevention of chronic arsenicosis: An Indian perspective. *Indian J Dermatol Venereol Leprol* 2008;74:582–593.
3. Saha KC. Melanokeratosis from arsenic contaminated tube well water. *Indian J Dermatol* 1984;29:37–46.
4. World Health Organization. Arsenicosis case-detection, management and surveillance. Report of a Regional Consultation. WHO Regional Office for South-East Asia, New Delhi, India, June 2003.
5. Talbi A, Kemper K, Minnatullah K, Foster S, Tuinhof A. An overview of current operational response to the arsenic issue in South East Asia, Vol. II. Technical report. World Bank, Washington, DC.
6. Ahmed MF. *Arsenic Contamination: Bangladesh Perspective*. Bangladesh University of Engineering & Technology, ITN-Bangladesh, Dhaka, Bangladesh, 2003.
7. Caussy D (ed.). *A Field Guide for Detection, Management and Surveillance of Arsenicosis Cases*. World Health Organization, Regional Office of South-East Asia, New Delhi, India, 2005.
8. Das NK, Sengupta SR. Arsenicosis: Diagnosis and treatment. *Ind J Dermatol Venereol Leprol* 2008;74:571–581.
9. Khan MM, Sakauchi F, Sonoda T, Washio M, Mori M. Magnitude of arsenic toxicity in tube-well drinking water in Bangladesh and its adverse effects on human health including cancer: Evidence from a review of the literature. *Asian Pac J Cancer Prev* 2003;4:7–14.
10. Mandal NK, Biswas R. A study on arsenical dermatosis in rural community of West Bengal. *Indian J Public Health* 2004;48:30–33.
11. Rahman MM, Chowdhury UK, Mukherjee SC et al. Chronic arsenic toxicity in Bangladesh and West Bengal, India—A review and commentary. *J Toxicol Clin Toxicol* 2001;39:683–700.
12. Saha KC. Diagnosis of arsenicosis. *J Environ Sci Health A: Tox Hazard Subst Environ Eng* 2003;38:255–272.
13. Milton AH, Hasan Z, Rahman A, Rahman M. Non-cancer effects of chronic arsenicosis in Bangladesh: Preliminary results. *J Environ Sci Health A: Tox Hazard Subst Environ Eng* 2003;38:301–305.
14. Kadono T, Inaoka T, Murayama N et al. Skin manifestations of arsenicosis in two villages in Bangladesh. *Int J Dermatol* 2002;41:841–846.
15. Sengupta SR, Das NK, Datta PK. Pathogenesis, clinical features and pathology of chronic arsenicosis. *Ind J Dermatol Venereol Leprol* 2008;74:559–570.
16. Guha Mazumder DN, Das Gupta J, Santra A, Pal A, Ghose A, Sarkar S, Chattopadhaya N, Chakraborti D. Non-cancer effects of chronic arsenicosis with special reference to liver damage. In C.O. Abernathy, R.L. Calderon, W.R. Chappell (eds.), *Arsenic: Exposure and Health Effects*. Chapman & Hall, London, U.K., 1997, pp. 112–123.

17. Yoshida T, Yamauchi H, Fan Sun G. Chronic health effects in people exposed to arsenic via the drinking water: Dose–response relationships in review. *Toxicol Appl Pharmacol* 2004;198:243–252.
18. IARC. *Some Drinking-Water Disinfectants and Contaminants, Including Arsenic.* Monographs on the Evaluation of Carcinogenic Risks to Humans, Vol. 84. WHO, Lyon, France, 2004, pp. 68–70, 223–226.
19. Yeh S. Skin cancer in chronic arsenicism. *Hum Pathol* 1973;4:469–485.
20. Sommers SC, Mcmanus RG. Multiple arsenical cancers of skin and internal organs. *Cancer* 1953;6:347–359.
21. Zaldivar R, Prunes L, Ghai G. Arsenic dose in patients with cutaneous carcinomata and hepatic hemangio-enothelioma after environmental and occupational exposure. *Arch Toxicol* 1981;47:145–54.
22. Tseng WP. Effects and dose–response relationships of skin cancer and blackfoot disease with arsenic. *Environ Health Perspect* 1977;19:109–119.
23. Rosales-Castillo JA, Acosta-Saavedra LC, Torres R et al. Arsenic exposure and human papillomavirus response in non-melanoma skin cancer Mexican patients: A pilot study. *Int Arch Occup Environ Health* 2004;77:418–423.
24. Lien HC, Tsai TF, Lee YY, Hsiao CH. Merkel cell carcinoma and chronic arsenicism. *J Am Acad Dermatol* 1999;41:641–643.
25. Lee L, Bebb G. A case of Bowen's disease and small-cell lung carcinoma: Long-term consequences of chronic arsenic exposure in Chinese traditional medicine. *Environ Health Perspect* 2005;113:207–210.
26. Beane Freeman LE, Dennis LK, Lynch CF, Thorne PS, Just CL. Toenail arsenic content and cutaneous melanoma in Iowa. *Am J Epidemiol* 2004;160:679–687.
27. Sikder MS, Rahman MH, Maidul AZ, Khan MS, Rahman MM. Study on the histopathology of chronic Arsenicosis. *J Pakistan Assoc Derma* 2004;14:205–209.
28. Dhar RK, Biswas BK, Samanta G et al. Groundwater arsenic calamity in Bangladesh. *Curr Sci* 1997;73:48–59.
29. Ahmad SA, Sayed MHS, Khan MH et al. Arsenicosis: Neoplastic manifestation. *J Prevent Soc Med* 1998;17:110–115.
30. Nath T. Study of cutaneous manifestations in chronic arsenicosis. Master's thesis. University of Calcutta, Calcutta, India, 1998, p. 60.
31. Saha KC. Chronic arsenical dermatoses from tube-well water in West Bengal during 1983–87. *Indian J Dermatol* 1995;40:1–12.
32. Mazumder DN, Das-Gupta J, Santra A et al. Chronic arsenic toxicity in West Bengal— The worse calamity in the world. *J Indian Med Assoc* 1998;96:4–7, 18.
33. Guha Mazumder DN. Arsenic and non-malignant lung disease. *J Environ Sci Health A: Tox Hazard Subst Environ Eng* 2007;42:1859–1867.
34. Santra A, Das Gupta J, De BK et al. Hepatic manifestations in chronic arsenic toxicity. *Indian J Gastroenterol* 1999;18:152–155.
35. Franklin M, Bean W, Harden RC. Fowler's solution as an ecological agent in cirrhosis. *Am J Med Sci* 1950;219:589–596.
36. Santra A, Chowdhury A, Ghatak S, Biswas A, Dhali GK. Arsenic induces apoptosis in mouse liver is mitochondria dependent and is abrogated by N-acetylcysteine. *Toxicol Appl Pharmacol* 2007;220:146–155.
37. Straub AC, Clark KA, Ross MA et al. Arsenic-stimulated liver sinusoidal capillarization in mice requires NADPH oxidase-generated superoxide. *J Clin Invest* 2008;118(12):3980–3989.
38. Zaldivar R. Arsenic contamination of drinking water and foodstuffs causing endemic chronic arsenic poisoning. *Beitr Pathol* 1974;151:384–400.
39. Tseng WP. Blackfoot disease in Taiwan: A 30-year follow-up study. *Angiology* 1989;40:547–558.

40. Rahman M, Tondel M, Ahmad SA et al. Hypertension and arsenic exposure in Bangladesh. *Hypertension* 1999;33:74–78.
41. Klei LR, Barchowsky A. Positive signaling interactions between arsenic and ethanol for angiogenic gene induction in human microvascular endothelial cells. *Toxicol Sci* 2008;102:319–327.
42. Barbey JT, Pezzullo JC, Soignet SL. Effect of arsenic trioxide on QT interval in patients with advanced malignancies. *J Clin Oncol* 2003;21:3609–3615.
43. Sanchez-Soria P, Broka D, Monks SL, Camenisch TD. Chronic low-level arsenite exposure through drinking water increases blood pressure and promotes concentric left ventricular hypertrophy in female mice. *Toxicol Pathol* 2012;40:504–512.
44. Murphy MJ, Lyon LW, Taylor JW. Subacute arsenic neuropathy: Clinical and electrophysiological observations. *J Neurol Neurosurg Psychiatry* 1981;44(10):896–900.
45. Tseng HP, Wang YH, Wu MM, The HW, Chiou HY, Chen CJ. Association between chronic exposure to arsenic and slow nerve conduction velocity among adolescents in Taiwan. *J Health Popul Nutr* 2006;24(2):182–189.
46. Schenk VW, Stolk PJ. Psychosis following arsenic (possibly thalium) poisoning. *Psychiatr Neurol Neurochir* 1967;70:31–37.
47. Morton WE, Caron GA. Encephalopathy: An uncommon manifestation of workplace arsenic poisoning? *Am J Ind Med* 1989;15:1–5.
48. Tsai SY, Chou HY, The HW, Chen CM, Chen CJ. The effects of chronic arsenic exposure from drinking water on neurobehavioral development in adolescence. *Neurotoxicology* 2003;24:747–753.
49. Bansal SK, Haldar N, Dhand UK et al. Phrenic neuropathy on arsenic poisoning. *Int J Dermatol* 1991;30:304–306.
50. Tay CH, Seah CS. Arsenic poisoning from anti-asthmatic herbal preparations. *Med J Aust* 1975;2:424–428.
51. Campbell JP, Alvares JA. Acute arsenic intoxication. *Am Fam Phys* 1989;40:93–97.
52. Katsnel'son BA, Neizvestnova EM, Blokhin VA. Induction of stomach cancer by the chronic action of arsenic. *Vopr Onkol* 1986;32:68–73.
53. Islam LN, Nabi AH, Rahman MM, Khan MA, Kazi AI. Association of clinical complications with nutritional status and the prevalence of leukopenia among arsenic patients in Bangladesh. *Int J Environ Res Public Health* 2004;1:74–82.
54. Lerman BB, Ali N, Green D. Megaloblastic dyserythropoietic anaemia following arsenic ingestion. *Ann Clin Lab Sci* 1980;10:515–517.
55. Norstrom S, Beckman L, Nordenson I. Occupational and environmental risks in and around a smelter in Northern Sweden. V. Spontaneous abortion among female employees and decreased birth weight in their offspring. *Hereditas* 1979;90:291–296.
56. Okui T, Fujiwara Y. Inhibition of human excision DNA repair by inorganic arsenic and the co-mutagenic effect in V79 Chinese hamster cells. *Mutat Res* 1986;172:69–76.
57. Hood RD. Effects of sodium arsenite on fetal development. *Bull Environ Contam Toxicol* 1972;7:216–222.
58. Lewis DR, Southwick JW, Ouellet Hellstrom R et al. Drinking water arsenic in Utah: A cohort mortality study. *Environ Health Perspect* 1999;107:359–365.
59. Squibb KS, Fowler BA. The toxicity of arsenic and its compounds. In B.A. Fowler (ed.), *Biological and Environmental Effects of Arsenic*. Elsevier, New York, 1983, pp. 233–269.
60. Garip H, Salih IM, Sener BC, Göker K, Garip Y. Management of arsenic trioxide necrosis in the maxilla. *J Endod* 2004;30:732–736.
61. Kesler A, Pianka P. Toxic optic neuropathy. *Curr Neurol Neurosci Rep* 2003;3:410–414.
62. Thery JC, Jardin F, Massy N, Massy J, Stamatoullas A, Tilly H. Optical neuropathy possibly related to arsenic during acute promyelocytic leukemia treatment. *Leuk Lymphoma* 2008;49:168–170.

63. Chen YC, Lin-Shiau SY, Lin JK. Involvement of reactive oxygen species and caspase 3 activation in arsenite-induced apoptosis. *J Cell Physiol* 1998;177:324–333.

64. Das PM, Sadana JR, Gupta RK, Kumar K. Experimental selenium toxicity in guinea pigs: Biochemical studies. *Ann Nutr Metab* 1989;33:57–63.

65. Tang CH, Chiu YC, Huang CF, Chen YW, Chen PC. Arsenic induces cell apoptosis in cultured osteoblasts through endoplasmic reticulum stress. *Toxicol Appl Pharmacol* 2009;241:173–181.

66. Fishbein, L. Perspectives of carcinogenic and mutagenic metals in biological samples. *Int J Environ Anal Chem* 1987;28:21–69.

67. Hubaux R, Becker-Santos DD, Enfield KSS et al. Molecular features in arsenic-induced lung tumors. *Mol Cancer* 2013;12:20.

14 Management of Chronic Arsenicosis

An Overview

Piyush Kumar, Rudrajit Paul,
Amrita Sil, and Nilay Kanti Das

CONTENTS

14.1 INTRODUCTION

The disease that has emerged as an epidemic in the recent past due to intake of water contaminated with arsenic above the maximum permissible limit has become the leading cause of morbidity and subsequent mortality in arsenic-affected countries. Though arsenicosis is a global concern, the majority of sufferers are from Bangladesh, India (from state of West Bengal), and China [1]. Arsenicosis manifests itself as melanosis, keratosis, and malignant and premalignant skin lesions; though it can affect almost all the internal organs (the major organ systems affected being pulmonary, hepatobiliary, gastrointestinal, vascular, and nervous systems) [2]. Till date, there is no effective cure for arsenicosis and the management relies mostly on prevention and supportive therapy. It is noteworthy that some of the clinical manifestations are irreversible and a patient of arsenicosis continues to have deterioration of clinical symptoms in spite of stoppage of drinking of arsenic-contaminated water [2]. This highlights the fact that all patients of arsenicosis should be under regular follow-up even after safe-drinking water is made available to them. This chapter will focus on the various treatment options that are available for this deadly disease and highlight on the recent advances in management strategies.

14.2 MANAGEMENT OF CUTANEOUS LESIONS

Involvement of skin is an early feature of chronic arsenicosis and skin manifestations are so characteristic that they serve as clues for clinical diagnosis. However, different studies have noted different prevalence of skin involvement in chronic arsenicosis. In one large scale study, 3,695 (20.6%) of 18,000 persons in Bangladesh and 8,500 (9.8%) of 86,000 persons in West Bengal living in arsenic-affected districts were found to show dermatological features of arsenicosis [3]. However, another study by Kadono et al. has reported skin involvement in more than 50% of villagers [4]. The various skin manifestations are pigmentary changes, hyperkeratosis of palm and sole, and premalignant and malignant lesions. In general, management of chronic arsenicosis is disappointing.

14.2.1 MANAGEMENT OF PIGMENTARY CHANGES

The classical findings are diffuse hyperpigmentation and discrete hypopigmented/depigmented macules (raindrop pigmentation). Till date there is no effective treatment for these features. Chelating agents like BAL, penicillamine, and DMSA/DMPS have reported clearing of melanosis [5] and achieve significant improvement in 1–2 months. In another study, spirulina extract (250 mg) and zinc (2 mg) twice daily for 16 weeks have been shown to be useful for the treatment of chronic arsenic poisoning with melanosis [6]. Supplementation with vitamin E and selenium, either alone or in combination, slightly improved skin lesion status, although the improvement was not statistically significant [7].

14.2.2 MANAGEMENT OF HYPERKERATOSIS

Arsenical keratoses are precancerous lesions found in patients with chronic arsenicosis. The mechanisms of arsenic-induced keratoses and malignancy are not fully understood. Arsenic reacts with the sulfhydryl groups in certain tissue proteins and subsequently affects many different enzymes that are essential to cellular metabolism. Arsenic has been found to cause chromosomal mutations, chromosomal breaks, sister chromatid exchanges, and mutations in p53.

The lesions typically begin as small, 2–10 mm, punctate, yellow, keratotic papules most commonly seen on the palms and soles in areas of constant pressure or repeated trauma. In long-standing and severe cases, the lesions can be found on more widespread body areas such as the trunk, extremities, eyelids, and genitalia. The lesions tend to persist for many years, and progression to invasive SCC is known.

Available localized treatment options include paring, keratolytics, cryosurgery, curettage with or without electrocautery, CO_2 laser treatment, topical chemotherapy with 5-FU, and surgical excision. PDT with ALA has also been used to treat these lesions [8]. Oral retinoids may be useful in reducing hyperkeratosis [9].

14.2.2.1 Keratolytics

World Health Organization (WHO) recommends use of 5%–10% salicylic acid and 10%–20% urea-based ointments. For severe cases, higher concentration of keratolytics may be used [10].

14.2.2.2 Oral Retinoids

Oral retinoids by the virtue of their antikeratinizing effects in other disorders of keratinization have been tried in arsenical hyperkeratosis. They have been shown to be effective in causing regression of arsenical keratosis [11,12]. Also, retinoids are known to have effects on the expression of genes that control cell differentiation, proliferation, and induction of apoptosis and hence, they offer a significant promise

---

TABLE 14.1

Common Adverse Effects of Oral Retinoid Therapy

Mucocutaneous	Dryness of the lips, skin, and mucous membranes, photosensitivity, *Staphylococcus aureus* colonization	Use of emollients and moisturizers and avoidance of sun exposure are recommended
Ocular	Blepharoconjunctivitis, alterations in visual function, mainly poor night vision; excessive glare sensitivity; and changes in color perception are common	
Hair and nail	Diffuse idiopathic skeletal hyperostosis syndrome-like bone changes and calcification of tendons and ligaments, premature epiphyseal closure, muscle pain and cramps	
Neuropsychiatric	Increased intracranial pressure such as headache, nausea, and vomiting; depression with suicide attempts	
Metabolic	Hypothyroidism, pancreatitis, hyperlipidemia	Discontinuation of therapy is required if the triglyceride level reaches 800 mg/dL
Liver toxicity	Elevations in serum transaminase levels	Transaminase elevations of more than three times the upper normal range should lead to discontinuation of retinoid therapy
Hematologic toxicity	Neutropenia (with bexarotene, uncommon with other retinoids). Isotretinoin-induced fibrinolysis leading to bleeding complications	
Renal adverse effects	Not significant, safe for patients with end-stage kidney disease	

Source: Patton, T.J. and Ferris, L.K., Systemic retinoids, In *Comprehensive Dermatologic Drug Therapy*, Wolverton, S.E. (ed.), 3rd edition, Elsevier Saunders, Edinburgh, 2013, pp. 252–268.

---

in the chemoprevention of arsenic-related cancers [13]. However, prohibitive costs and serious adverse effects on long-term use are limiting factors for community-based management of arsenical keratosis and chemoprevention of arsenic-related malignancies.

Retinoids are known teratogens; hence, initiation of therapy in women of child-bearing age requires negative result on a pregnancy test and couples must practice effective contraception during treatment and for 1 month (2 months in some countries) after the completion of therapy. In men, retinoid therapy does not appear to produce abnormalities in spermatogenesis, sperm morphology, or sperm motility. However, men who are actively trying to father children are asked to avoid systemic retinoid therapy. Retinoid-induced birth defects include auditory, cardiovascular, craniofacial, ocular, axial and acral skeletal, central nervous system (hydrocephalus and microcephaly), and thymus gland abnormalities. The common adverse effects are summarized in Table 14.1.

14.2.2.3 Paring, Topical Chemotherapy with 5-FU, Cryosurgery, Curettage with or without Electrocautery, and CO_2 Laser Treatment

These are locally destructive methods [15,16]. Among these, paring can be done by patients too. In fact, it is advisable to educate patients about regular examination of skin and paring of hyperkeratotic lesions themselves. 5-Fluorouracil (5-FU) is a cytotoxic drug that blocks DNA synthesis by inhibiting thymidylate synthetase and is commercially available as 1% as well as 5% creams in India. It is interesting to note that 5-FU acts selectively to cause cellular destruction in actinically damaged cells but not in normal skin when used for treatment of actinic keratosis. However, whether it exerts similar selective cytotoxic effects in arsenical keratosis is not known. Also, it is less effective in treating arsenical keratosis than actinic keratosis. Irritation during treatment is common and may require reduction or interruption of treatment. The addition of emollients is often helpful to counteract irritation caused by 5-FU. Apart from providing ease of topical application, another advantage of 5-FU is its usefulness in the common neoplasms in chronic arsenicosis patients, that is, Bowen's disease, superficial basal cell carcinoma, and small squamous cell carcinoma.

Cryosurgery (using liquid nitrogen), electrocautery, and CO_2 laser are office-based treatment modalities and require a trained dermatologist. The objective of these methods is to cause selective necrosis of tissue, the extent of which depends on the type of lesion. Cryosurgery achieves that by freezing the lesion by liquid nitrogen (with $-196°C$ boiling point). Electrocautery does the same with the help of heat, generated by passing electrical current to living tissue, which is a poor conductor of electricity. CO_2 laser works on the principle of *selective photother-molysis*. Light of a particular wavelength (10,600 nm in case of CO_2 laser) targets a particular chromophore (water in the case of CO_2 laser), resulting in the generation of heat energy in the vicinity of chromophores and thus, destroying the tissue in question. All these three modalities share some common advantages—outdoor-based procedures, safe and relatively simple procedure, no hospitalization needed, no general anesthesia needed, suitable for treatment of any area of body, operative suite not required, no restriction of work or sports, useful in pregnancy and for poor surgical risk patients, and excellent for the very elderly patients. Cryosurgery and electrosurgery are also useful in Bowen's disease (carcinoma in site) and basal as well as squamous cell carcinomas.

There are few contraindications for these procedures and are specific to treatment modality. Cryosurgery should be avoided in patients with cold urticaria, cold intolerance, cryofibrinogenemia, or cryoglobulinemia. Also, it should be avoided in certain anatomical areas such as corners of the mouth, the vermilion margin of the lips, eyebrows, inner canthi, the free margin of the ala nasi, and the auditory canal, since scarring or retraction of the tissue can occur. Pain, edema, and blistering are early complications, managed conservatively. On the other hand, hypopigmentation may occur posttreatment in dark-skinned patients as melanocytes are more sensitive to cold that keratinocytes. There are no absolute contraindications to the use of electrosurgery, but it should be avoided in patients with pacemakers and defibrillator. The immediate complications are pain, swelling, and bleeding. Scarring may be noted after healing. There is virtually no contraindication for CO_2 laser, but the cost is prohibitive and dermatologists need a separate training for operating CO_2 lasers.

14.2.3 MANAGEMENT OF CANCEROUS AND PRECANCEROUS SKIN LESIONS

Squamous and basal cell carcinoma and Bowen's disease (squamous cell carcinoma in situ) are the conditions that are the major cause of mortality in arsenocosis. Study on the neoplastic manifestations of arsenicosis revealed precancerous skin lesions in 6.6% and cancerous lesions in 0.8% of the patients [17].

Bowen's disease is asymptomatic, slow-growing, and has got a favorable prognosis if treated effectively and adequately; else it can progress to squamous cell carcinoma and lead to the morbidity and subsequent mortality. In chronic arsenicosis, Bowen's disease is multicentric in nature and can relentlessly progress even if the person stops drinking arsenic-contaminated water.

14.2.3.1 Treatment of Bowen's Disease

There are a range of treatment options available for Bowen's disease including surgical interventions such as surgical excision or Mohs micrographic surgery, destructive therapies such as cryotherapy, light-based therapy such as photodynamic therapy, or topical therapies such as 5% flurouracil (5FU) or 5% imiquimod [18]. Though surgical excision remains the gold standard, noninvasive treatment options are a need of the present day. The treatment options need to be developed keeping in mind the cost of therapy, easy accessibility of treatment option, and those that could be offered an ambulatory treatment, since the sufferers of chronic arsenicosis reside at faraway places from the district headquarters (where the surgical procedure can be performed) and they belong to lower socioeconomic status with poor purchasing ability.

14.2.3.1.1 Surgical Excision

Surgical excision remains the gold standard for the management of Bowen's disease and offers the advantage of histologically examining the resection margin. As Bowen's disease is not an invasive disease, minimization of healthy tissue excision is desirable. Study has shown that a hypothetical reduction of the safety margin from

5 mm to 4 or 3 mm decreases the complete excision rate from 94.4% to 87% and 74.1%, respectively. The safe margin of resection remains a debate till date [19].

14.2.3.1.2 Mohs Micrographic Surgery

Mohs micrographic surgery (MMS), or Mohs surgery, is a newer surgical technique to treat various skin cancers. In this procedure, cutaneous tumors are excised at a 45° angle with mapping and frozen section microscopy is done for identification of residual cancer. Thus, this method provides complete histopathologic control of the surgical margins, and hence, it offers the advantage of lowest recurrence rate with maximal preservation of uninvolved tissue. Mohs surgery has become the treatment of choice for most skin cancers on the head and neck as well as for recurrent or histologically aggressive lesions. It is also particularly suitable for areas with poor healing or areas where scarring may result in significant cosmetic disfiguring [20]. MMS has shown an overall 5-year recurrence rate of 6.3% (3% for primary tumors and 9% for recurrent tumors) [21]. The main disadvantages are the requirement of trained surgeons and pathologists and specialized setup.

14.2.3.1.3 Imiquimod 5% Cream

Daily application of imiquimod 5% cream overnight (for 8 h) for 16 weeks has shown promise in complete clearance of lesion in 60% of cases [22]. Treatment may have to be stopped for 5 days on in the event of a severe inflammatory reaction or if it is uncomfortable for the participant.

14.2.3.1.4 5% Flurouracil Cream

5 FU cream rubbed to the lesions once daily for 1 week and then twice daily for 3 weeks is another therapeutic option that has shown optimistic results (complete clearance in 72.2% cases) [23] in different randomized controlled trial.

14.2.3.1.5 Cryotherapy

Cryotherapy is a simple, quick, and effective method for treating Bowen's disease. Additionally, it does not require general anesthesia or hospitalization and thus offers the advantage of accessibility in the outpatient setting [21]. A single 30 s freeze thaw cycle was shown to achieve a clearance rate of 100% and recurrence rate of 08%, with follow-up periods ranging from 6 months to 5 years [24]. Lower freezing time has not achieved similar success rate. One study with 20 s freeze thaw cycle has reported clearance rates of 68% after one treatment and 86% after two treatment sessions [25]. Cryosurgery is useful in low-risk situations and for patients who prefer to avoid surgery or cumbersome topical treatment, but poor healing and hypopigmented scarring, particularly in poorly vascularized areas, may complicate the procedure.

14.2.3.1.6 Photodynamic Therapy

Photodynamic therapy (PDT) involves the use of photochemical reactions mediated through the interaction of photosensitizing agents (5-aminolevulinic acid, ALA and methylaminolevulinate, MAL), light, and oxygen for the treatment of malignant or benign diseases. In this process, the photosensitizer is administered to the patient by topical, oral, or intravenous route. The photosensitizer is preferentially taken up by hyperproliferating cells. In next step, these photosensitizers are activated by light of a particular wavelength (Blu-U device with 405–420 nm for ALA and red light with 635 nm for MAL) in the presence of oxygen, resulting in targeted destruction of tissue [26]. MAL-PDT has shown excellent clinical clearance rates of 88%–100%, reported 3 months after one cycle [21]. ALA-PDT too has shown similar success rate of 80%–100%. PDT is an office-based procedure without requiring hospitalization and has superior cosmetic results compared with cryosurgery. PDT may be particularly appropriate for large lesions (>3 cm diameter), with two treatments of MAL-PDT, 1 week apart, clearing 96% (22/23) of lesions at 3 months [27].

14.2.3.1.7   Curettage and Electrosurgery

Curettage and electrocautery has been considered one of the simplest, least expensive, safest, and most effective treatment modality for dealing with Bowen's disease. High cure rates may be achieved with good technique and equipment; however, recurrence is known and varies from 2% to 10% [21]. This method has shown success rate similar to excision and cryotherapy in good hands. Also, this method is superior to cryotherapy in terms of pain, healing, and recurrence rate.

14.2.3.1.8   Laser

CO_2 laser has been used for the treatment of Bowen's disease involving challenging treatment sites including the digits and genitalia. However, the experience is limited and only case reports and small case series are available in medical literature. CO_2 laser in super-pulsed mode (2 W/cm^2) to treat Bowen's disease in 44 patients has shown clearance in 86% of patients after 1 treatment, with all but one of the remaining lesions cleared after a total of 2–4 treatments. However, recurrence is known and one study has documented a recurrence rate of 7% over a follow-up period of 8–52 months [28]. Deep follicular epithelium is typically spared with CO_2 laser; hence, there is the definitive risk for failure of complete clearance and recurrence. Long-pulsed 810 nm diode laser following three passes of CO_2 laser treatment and skin biopsies confirming ablation of the follicular epithelium have been tried with good results [21].

The treatment strategy for the management of Bowen's disease has been summarized in Table 14.2.

14.2.3.2   Management of Malignancies

Chronic arsenicosis patients have higher risk of developing various neoplasms as compared with general population. Patients usually develop cancers 10–20 years after the appearance of symptoms or skin lesions attributable to arsenicosis. The lesions, especially big ones, may be excised and allowed to heal by primary (with or without skin grafting, depending on the size of defect) or secondary intention. Mohs surgery is preferred over certain sites like periorbital, perinasal, periauricular, and perioral areas and on fingers, toes, and genitals. Histologic examination of regional lymph nodes may be indicated, depending on the clinical stage. Oncosurgeons should be consulted for management of such patients.

TABLE 14.2

Summary of Treatment of Bowen's Disease (Squamous Cell Carcinoma In Situ)

Lesion Characteristics	Preferred Treatment Modalities
Small, single/few, good healing sites	Curettage, cryotherapy, topical 5-FU, topical imquimod, excision, photodynamic therapy
Large, single, good healing sites	Photodynamic therapy, topical 5-FU, topical imiquimod, cryotherapy, excision
Multiple, good healing sites	Topical 5-FU, topical imiquimod, cryotherapy, curettage, PDT, excision
Small, single/few, poor healing site	Topical 5-FU, topical imiquimod, cryotherapy, curettage, PDT, excision (Mohs micrography surgery preferred)
Large, single, poor healing site	PDT, topical 5-FU, topical imiquimod, Mohs micrographic surgery
Facial	Topical 5-FU, topical imiquimod, curettage, excision, PDT
Digital	Topical 5-FU, topical imiquimod, excision, PDT, radiotherapy
Nail bed	Excision, PDT
Penile	Topical 5-FU, topical imiquimod, PDT, radiotherapy, laser
Lesions in immunocompromised patients	Cryotherapy, curettage, PDT, excision

Source: Morton, C.A. et al., *Br. J. Dermatol.*, 170, 245, 2014.

14.3 MANAGEMENT OF SYSTEMIC MANIFESTATIONS

For chronic arsenicosis, in principle, two most effective therapies are removal of the source of arsenic and chelation (*vide infra*). However, many of the systemic effects are permanent and produce morbidity even long after the arsenic exposure is terminated. Supportive therapy has to be offered in all patients with chronic arsenicosis by way of management of different systemic manifestations that can arise in the course of the disease.

14.3.1 CARDIOVASCULAR DISEASES

14.3.1.1 Blackfoot Disease

Blackfoot disease is an extreme example of peripheral vascular disease, but other milder varieties are also quite common in chronic arsenicosis. In severe cases, surgical amputation may be needed to remove the gangrenous portion of limbs [29]. However, if the exposure continues, further amputation of the higher levels of limbs may be needed at a later stage.

General treatments for peripheral vascular diseases, which include aspirin, walking rehabilitation exercise, cilostazole, or pentoxifylline, are sometimes used too [30]. However, there is no definite trial on the specific benefit of these drugs in arsenic exposure. ACE inhibitors, clopidogrel, and aggressive lipid lowering are also said to be beneficial for the condition [31].

It should be remembered that a major cause of death in patients with Blackfoot disease is cardiovascular problems. Thus, associated conditions like coronary artery disease or carotid plaques should also be managed [29]. Smoking cessation is beneficial in stopping the progression of the disease. Other contributing factors like diabetes or vasculitis may be present simultaneously and needs specific therapy.

14.3.1.2 ECG Changes

The ECG changes in arsenicosis may be innocuous findings or they may be a harbinger of cardiomyopathy. Exposure to arsenic trioxide during chemotherapy has been reported to be associated with ventricular arrhythmia [32]. In those cases, anti-arrhythmic therapy or implantable defibrillators may be needed. However, relation of environmental arsenic exposure to cardiac arrhythmia is difficult to diagnose. A recent study has found relation of arsenic exposure with QT dispersion and long-term cardiovascular mortality [33]. But apart from stopping arsenic exposure, no other definite therapy for the condition is known.

The ECG changes are said to be due to oxidative stress but no definite therapy exists to counter that effect. Recently, some experimental therapies have shown promise but they are far from convincing till now [34].

14.3.1.3 Hypertension

Arsenic-induced hypertension has been especially reported from Bangladesh [35]. A dose–response relationship has also been found between levels of arsenic exposure and the degree of hypertension [36]. There is no definite study on the treatment of this condition. As such, existing treatment guidelines for hypertension also apply to arsenic-induced cases.

14.3.1.4 Cardiomyopathy

This is still a controversial entity. There is no definite literature on the specific management of the condition. However, standard procedures like diuretics or mechanical support like intra-aortic balloon counterpulsation (IABP) are used. In severe cases of heart failure, cardiac transplantation may be needed.

14.3.2 RESPIRATORY DISEASES

14.3.2.1 Interstitial Lung Disease

This is a progressive condition. For this, respiratory rehabilitation is needed. Home oxygen support may be needed is advanced stages. Lung transplantation is another alternative if facilities are available. Steroids or other immunosuppressive drugs are generally useless in this setting.

14.3.2.2 Obstructive Lung Disease

Smoking cessation should be advised. Inhaled drugs like salbutamol may be used for symptom relief. The entity of arsenic-induced obstructive lung disease is a relatively new finding and thus, it is not well characterized [37]. Standard protocols for COPD patients should be followed. However, chronic cough may be a troublesome symptom and may need special care.

An interesting finding in recent years is that exposure to arsenic in utero may give rise to lung problems in later life [38]. Thus, early environmental changes like changing drinking water source of the pregnant woman are warranted.

14.3.3 Hepatic Diseases

14.3.3.1 Noncirrhotic Portal Fibrosis

Noncirrhotic portal fibrosis is a rare cause of portal hypertension. It is usually found in young adults or middle-aged women [39]. Liver function is usually preserved. Management centers round treatment and prophylaxis for portal hypertension and bleeding [39]. This includes endoscopic variceal ligation, octreotide, terlipressin for acute management, and propranolol or nitrates for chronic management. Surgical management is advocated if medical and/or endoscopic management fails [40]. The data on surgical management of NCPF is less than EHPVO; however, benefit has been documented. Earlier, central shunt surgery was done but that led to complications like ascites and hepatic encephalopathy [41]. However, now selective shunts like distal splenorenal shunts (DSRS) are done, which are said to have lesser complication rates [42].

The patients need regular follow up at 6 monthly intervals or as the clinical state demands. If endoscopic variceal ligation had been done, more frequent follow up including repeat endoscopy may be needed [39]. In refractory bleeding, esophagogastric devascularization procedures like sugiura procedure may be needed [43].

Although liver function usually remains normal in NCPF, sometimes progressive liver decompensation can develop. In those cases, appropriate management of conditions like hepatic encephalopathy is needed. In extreme cases, liver transplantation is needed [44]. Newer therapies for NCPF include partial splenic embolization (especially if there is large splenomegaly with/without hypersplenism) and balloon-occluded retrograde transvenous obliteration (BRTO) [45].

14.3.3.2 Fatty Liver

This condition is usually asymptomatic and is incidentally diagnosed by imaging studies. Occasionally, patients may present with hepatomegaly [46]. There is no specific treatment. Weight loss and dietary control may help in checking progression. However, sudden severe weight loss may actually aggravate the condition [47].

Other therapies that have been tried for the condition include metformin, atorvastatin, antioxidants like vitamin E, ursodeoxycholic acid (UDCA), and orlistat. Of these, once there was much enthusiasm about UDCA. But now, it has not been found to be as beneficial as previously claimed [48]. Newer experimental therapies tried in fatty liver include betaine and losartan [49].

Another exciting development in treatment of fatty liver is the use of probiotics for normalizing liver enzymes. It has been tried in nonalcoholic fatty liver disease and the results are promising [50]. However, more data is needed before this can be recommended.

14.3.3.3   Angiosarcoma

This is a very rare and very aggressive malignancy of liver [51]. The initial clinical features are often nonspecific and this makes early diagnosis difficult. Mortality rate is very high. Till now, there is no definite guideline on management of the condition. Even liver transplantation is not always protective [52].

Hepatic resection with or without adjuvant therapy is the mainstay of therapy [53]. Chemotherapy is still not standardized and some people have used adriamycin or methotrexate [54]. However, the survival rate is still very poor in almost all the reported cases [55]. A rare complication of angiosarcoma is spontaneous rupture with intraperitoneal hemorrhage. This is usually uniformly fatal.

Hepatic arterial infusion chemotherapy is a newer option for treatment. Although data is very sparse, preliminary results show promise [56].

14.3.4   BONE MARROW DISEASES

14.3.4.1   Anemia

This is often insidious in onset. The regions of the world where arsenicosis is found are also heavily affected by malnutrition. Thus, it is difficult to deduce the exact role of arsenic in anemia. However, change of water source and nutritional support generally leads to resolution in many cases.

In extreme cases, RBC exchange transfusion may be needed [57].

Sometimes macrocytosis is also found in arsenic toxicity. Stopping the arsenic exposure generally leads to resolution of the condition [58].

Basophilic stippling also does not require separate treatment. Stopping the exposure to heavy metal leads to disappearance of this entity [59].

14.3.4.2   Pancytopenia

Blood product support is needed in severe cases. For refractory pancytopenia, bone marrow transplantation is needed.

14.3.5   RENAL DISEASES

14.3.5.1   Tubular Necrosis

In acute kidney injury, hemodialysis support may be needed. Also, treatment of other complications like electrolyte abnormalities or anemia requires a multidisciplinary approach. In many cases, if the exposure is removed, the kidneys slowly return to normal. However, in some cases, they may progress to chronic renal disease. In that case, renal replacement therapy may be needed.

14.3.5.2   Carcinoma

Tumors are primarily renal cell carcinoma (RCC). This often remains silent in the early stage and presents with a large mass with metastasis or vascular invasion later. Thus treatment is often unsatisfactory. Current therapies for this malignancy are surgery, radiotherapy, chemotherapy and biological, singly or in various combinations. Newer biological used for the treatment of RCC include sorafenib, sunitinib, and bevacizumab. Pazopanib has also been used recently [60]. Recently, NICE guidelines have also approved this drug in selected cases [61].

14.3.6 Nervous System Diseases

14.3.6.1 Peripheral Neuropathy

Arsenic induced peripheral neuropathy may be progressive and may even cause quadriplegia and respiratory failure [62]. In those cases, ventilator support and intensive care may be needed. Treatment with dimercaptosuccinic acid leads to resolution of symptoms; sometimes iv therapy may be needed [62]. Residual weakness may remain in some cases.

For the neuropathic pain, aspirin or codeine may be used [63].

Early diagnosis and removal from arsenic source usually leads to good recovery. But advanced cases may be progressive even after the removal of exposure [64].

14.3.6.2 Dementia

Arsenic exposure has been linked to cognitive dysfunction. Animal model experiments have shown significant alterations in the brain in arsenic exposed mice [65]. Also in humans, arsenic exposure has been shown to cause increased deposition of β-amyloid in the brain [66]. This may be responsible for memory loss and other cognitive impairments.

Oxidative stress is hypothesized to be the main pathophysiology behind cerebral dysfunction in chronic arsenicosis. Herbal preparations like curcumin are said to be protective in this condition, although further studies are needed [67]. In a study from the United States, environmental exposure to heavy metals like manganese and cadmium along with arsenic was found to affect the neurodevelopmental parameters in children [68]. Thus, how far only arsenic contributes to this effect is difficult to elucidate.

Another new therapy advocated in this condition is S-adenosyl methionine [69]. It is said to counteract the folate deprivation that potentiates arsenic toxicity. However, as of now, there is no definitive therapy for arsenic-induced dementia, except removal of the source of exposure.

14.3.6.3 Cognitive Impairment in Children

Similar to dementia, this has no specific treatment. New research has shown promise with therapies like selenium [70]. Zinc has also been shown to protect against arsenic-induced neuronal apoptosis [71]. However, except chelator therapy, none of these are approved till now.

14.3.6.4 Cerebrovascular Accident

Management of ischemic stroke due to arsenicosis is similar to stroke from other causes. Secondary prevention with aspirin and/or statin is used. Physiotherapy and rehabilitation are also needed.

Arsenic is said to be a frequent contaminant in recreational drugs like cocaine. This combination may be responsible for cerebral thrombangitis obliterans [72].

14.3.7 MALIGNANCY

Malignancy involving lung, urinary bladder, and stomach is common in chronic arseniocosis patients. The detailed discussion on management of visceral malignancy is beyond the scope of this chapter.

14.3.8 ARSENIC-INDUCED DIABETES

In arsenic-induced secondary diabetes, therapy may be difficult. Arsenic is said to hinder the action of insulin [73]. This is done by alterations in the ATP-dependent insulin secretion mechanism. Also, oxidative stress induced by arsenic may cause beta-cell dysfunction. There is also resistance to insulin action at tissue level.

Recently, nanoparticle-coated insulin has been found to counter this effect of arsenic [74]. This may be a potential therapeutic option in future.

Some other recent researches in arsenic-induced diabetes include spirulina [75], garlic, and vitamin E. But none of them have shown definite human benefit till now.

14.4 ROLE OF CHELATORS

The word chelation is believed to have its origin in Greek word *chele* that means claw of a lobster. Chelating agents are organic or inorganic compounds capable of binding metal ions to form complex ring-like structure called *chelates* and act by converting metals to a chemically inert form and by enhancing sequestration, transportation, and excretion [76]. Metal-binding proteins, including metallothioneins and glutathione are potent chelators for heavy metals and are central to the natural response of the body to toxic heavy metals. No wonder, glutathione acts as a biomarker for toxic metal overload [77]. The first experimental use of a chelator against metal poisoning was the use of citrate as an antidote toward acute lead intoxication in 1941 [78]. Soon, during World War II 2,3-dimercaptopropanol (BAL) was developed as an experimental antidote against arsenic-based war gases. It has been used clinically since 1949 in arsenic, cadmium, and mercury poisoning; however, its high toxicity and the high frequency of various side effects have limited its widespread use. Also, BAL increased the toxicity of cadmium and lead in animal experiments [78]. Later on, dimercaptosuccinic acid (DMSA or succimer) and dimercaptopropane sulfonate (DMPS) were added to armamentarium of chelators. These chelators had greater water solubility and were administered as oral, intravenous, suppository, or transdermal preparations. However, these chelators lacked specificity and increased excretion of essential minerals was undesirable effect. In one study, use of DMPS resulted in significantly increased excretion of copper, selenium, zinc, and magnesium, necessitating replenishment of these essential minerals before and after treatment [79]. Calcium disodium ethylenediaminetetraacetic acid ($CaNa_2EDTA$), another chelator that has been extensively used in childhood lead poisoning since 1950s, has been found to be responsible for three deaths due to increased excretion of calcium, resulting in cardiac arrest [77].

TABLE 14.3
Features of Ideal Chelating Agents

Greater affinity to toxic heavy metal, less affinity to essential minerals
Low toxicity
Same distribution as metal
Forms nontoxic complexes with toxic heavy metal
Ability to compete with natural chelators (e.g., glutathione)
Rapid elimination of toxic heavy metal
High water solubility
Does not lead to redistribution of toxic metals

TABLE 14.4
Common Chelating Therapy Used in Heavy Metal Poisoning

Metal	Chelator
Arsenic	Dimercaprol, succimer
Cadmium	None
Lead	Succimer, Ca-EDTA
Mercury	Dimercaprol, succimer, *N*-acetylpenicillamine
Thallium	Prussian blue
Iron	Desferioxamine
Copper	D-Penicillamine

Clinical experience with chelators led to better understanding of ideal chelators. Another major drawback is redistribution of toxic metals. For example, BAL, apart from rapid mobilization of arsenic from the body, causes a significant increase in brain arsenic [80]. Similarly, treatment with DMSA has caused an elevation of mercury in motor axons, likely due to redistribution of mercury nonneural tissues such as the kidneys and liver [76]. The characteristics of ideal chelators have been summarized in Table 14.3 and the common chelators used for heavy metal toxicity are enumerated in Table 14.4.

In arsenicosis, chelation therapy is considered to be the specific therapy by many since in principle it can reduce the arsenic stores in the body thus has the potential of relieving the systemic clinical manifestations reducing subsequent cancer risk. Many agents are being tried in arsenicosis (Table 14.5) and there were many studies that were conducted to evaluate the efficacy of specific chelation therapy in arsenicosis that revealed conflicting reports regarding its usefulness. The study evaluating DMSA (dimercaptosuccinic acid) did not yield better efficacy than control subjects treated with placebo [81]. But in another single blind placebo control trial, DMPS (dimercaptopropane sulfonate) was shown to result in significant improvement of

TABLE 14.5

Chelators Used in Arsenic Toxicity

Name	Route of Use	Dose	Cost	Availability in India	Pregnancy Category
BAL	im	10–12 mg/kg/day divided q6 hr deep IM ×2 days	Rs. 105/100 mg	Yes	C
Succimer (DMSA)	Oral	750–1000 mg/m²/ day	$55.90 per 45 capsules, 100 mg each	Available online	C
Penicillamine	Oral	750 mg/day	Rs. 160/10 capsules of 250 mg each	Yes	D

clinical score in comparison to placebo [82]. DMPS was also shown to increase the urinary excretion of arsenic during the period of chelation therapy. The use of this drug is limited by its exorbitant cost, lack of easy availability, and long-term prospective trial demonstrating the efficacy and safety. Owing to these limitations chelation therapy is not yet a recommended form of therapy in arsenicosis. Considering the inadequacies of conventional chelators, newer chelators have been developed and have been tried against cases of experimental heavy metal poisoning. Monoisoamyl DMSA (MiADMSA), monomethyl DMSA (MmDMSA), and monocyclohexyl DMSA (MchDMSA) are some of the newer chelators that are more effective and has better safety profile. However, extensive studies are required to reach at a final conclusion [78].

Another, newer approach in treating heavy metal poisoning is use of two chelators, which act differently. This approach is based on assumption that different chelating agents are likely to mobilize toxic metals from different tissue compartments, leading to efficient excretion of toxic metal and has found some positive support from animal studies. It has been shown that coadministration of DMSA and MiADMSA at lower dose was effective in reducing arsenic-induced oxidative stress, depleting arsenic from blood and soft tissues and was also able to repair DNA damage caused following arsenic exposure [83]. There is renewed interest in chelators in recent times with the animal study on guinea pigs with the use of different protocol based on combination of two chelators, DMSA and MiADMSA, showing effectiveness in mitigating arsenic-induced neuronal apoptosis by way of preventing Ca^{2+} influx through L-type calcium channel [84].

Coadministration of dietary nutrients like vitamins, essential metals, antioxidants, or an amino acid like methionine with a chelating agent leads to many beneficial effects including better clinical recovery as well as increased excretion of toxic metal. Mishra et al. has shown that combined administration of quercetin and monoisoamyl 2,3-dimercaptosuccinic acid (MiDMSA) led to a rapid mobilization of arsenic [85]. Based on experiments, it appears that combination therapy is a better approach to treat cases of metal poisoning. However, clinical experience is limited and more number of studies is required in this area.

14.4.1 Dimercaprol (BAL)

Arsenic and its compounds attack the thiol group of body enzymes and inhibit their activity. The dimercaprol has two –SH groups that compete with body enzymes to bind to arsenic and thus prevent the enzyme-inhibiting action. The structure of BAL is

$$CH_2-OH$$
$$|$$
$$CH-SH$$
$$|$$
$$CH_2-SH$$

After chelating arsenic (As) it forms

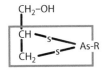

However, this has a narrow therapeutic range and often causes severe toxicity. The main toxicities reported are

- Local injection site reaction
- Chest tightness
- Tingling of hands and feet
- Chest pain
- Anemia in G6-PD deficiency
- Nephropathy

The exact mechanism of toxicity is not known but it is thought to be due to the inhibition of enzymes and increased capillary permeability. The safe limit is 5 mg/kg/dose and most side effects are transient. There is no reported carcinogenic potential but it is said to be teratogenic in mice [86]. It is a pregnancy category C drug and should be avoided in pregnancy if possible.

Immediately after injection, many patients develop local pain, tingling, and a transient rash. Later, sterile abscess can develop at injection site. The drug should always be given by deep im injection and never by sc or iv route. Nausea and vomiting may also occur. There may be hyperpyrexia, especially in children. If it comes in contact with eye, it may cause irritation and red eye. Repeated high doses may cause cardiovascular collapse. Usually there is transient hypertension and tachycardia after the injection and this is proportional to the dose administered. After injection, sometimes a constricting sensation in chest or thorax or muscle spasms can also occur. Other minor side effects immediately after injection include urticaria, abdominal pain, and rhinorrhoea. In children, sometimes convulsions and coma has also been reported after repeated high doses.

A peculiar symptom sometimes reported after BAL injection is a burning sensation of the penis. But the exact cause of this is not known.

Dimercaprol can also cause a sulfurous smell in breathe like rotten eggs due to mercaptans.

In G-6PD-deficient patients, BAL can cause hemolysis. However, this effect is infrequently reported [87]. Another hematological side effect is transient decrease in polymorphonuclear leukocytes, especially in children.

Dimercaprol is contraindicated in those with hepatic failure. It can cause fatty liver and sometimes, hepatitis. It can also cause acute renal failure in some cases.

BAL may cause metabolic acidosis with increased lactate and blood sugar. But after some time, there may also be hypoglycemia and glycosuria. The exact mechanism of hyperglycemia is not known but it is said to interfere with the action of insulin by attacking the disulfide bridges. Also, the surge of adrenalin after the injection may be responsible for this glycemic effect.

There is no specific toxicological treatment for BAL poisoning, but symptomatic therapy is given.

Another effect of dimercaprol is that it causes redistribution of arsenic to brain and testes. But the exact effect of this redistribution is not known.

14.4.2 SUCCIMER

This is a water-soluble analogue of BAL. its chemical formula is meso-2,3-dimercaptosuccinc acid (DMSA). Toxicological studies in mice showed it to be more effective than BAL in arsenic toxicity [88].

This is also an organosulfur compound and the mesoisomer acts as chelator. It is useful in arsenic as well as antimony and mercury poisoning.

The most important side effect of succimer is a transient increase in transaminase. But this is not always clinically significant [89]. A skin reaction can also occur. However, this is an oral drug. Thus, injection-related side effects like those in BAL will not occur.

Other minor side effects include

- Nausea
- Headache
- Proteinuria
- Pruritus
- Dizziness
- Metallic taste in mouth
- Anorexia
- Eosinophilia

However, side effects are much less compared to BAL. Dermatological effects reported include mild herpetic or papular rash or mucous eruptions [90]. An adult patient took intentional overdose (4 g) of succimer in 2001. However, the only reported effects were a sensation of jitteriness and palmer keratosis that resolved within 4 weeks. In some cases, chronic use of DMSA caused candidiasis, although the cause of this is not known [91]. To avoid this, a new congener, called DMPS, dimercaptopropane sulfonic acid (dimerval), has been used in recent times.

This is a new drug and availability is limited. But human studies have shown increased urinary excretion of arsenic compounds after using DMPS [92]. This drug is available in both oral and parenteral forms. It is not freely available in United States.

14.4.3 PENICILLAMINE

This is a new addition to arsenic chelation therapy. This also acts by active –SH group. This is also an oral drug and thus does not have injection-related side effects. However, this drug has a lot of other side effects.

1. Cutaneous: Urticaria, pemphigoid, lupus like lesions, macular rash
2. Hematological: Aplastic anemia, agranulocytosis, leukopenia
3. Nephropathic: Membranous glomerulonephritis causing nephrotic syndrome
4. Gastrointestinal: Nausea, vomiting, loss of taste
5. Hepatic: Hepatitis
6. Secondary myasthenia gravis, myopathy
7. Hypersensitivity
8. Elastosis perforans serpiginosa, anetoderma
9. Reversible optic neuritis
10. Secondary Goodpasture's syndrome and secondary dermatomyositis
11. Alopecia

Due to these side effects, this drug is not very popular in treatment.

Thus, the chelators used in arsenic toxicity have some side effects. They should be used with caution and regular monitoring of adverse reactions is mandatory.

14.4.4 RECENT ADVANCES

The search for ideal agent for treatment of arsenicosis is continuing and many different management strategies would emerge in the future taking lead from the recent advances that are taking place in the management of arsenicosis. The findings are detailed in the following.

14.4.4.1 Nutritional Status and Arsenicosis

Bioaccumulation of arsenic in humans is a complex process that is influenced by factors such as environmental quality, age, gender, nutrition, speciation, and binding nature [93]. Nutrition, among these factors, has interested researchers a lot as it has huge therapeutic implications. In their study, Samal et al. have concluded that participants with low nutrition (<1000 kcal) exhibited a higher probability of keratosis and melanosis than those with adequate nutrition (>1000 kcal). They have found that the odd ratios for low nutrition in case of keratosis and melanosis are 1.21 (95% CI: 0.78–1.85) and 2.53 (95% CI: 1.68–3.81), respectively. However, the odds ratios were low for participants with adequate nutrition and the odd ratios for keratosis and melanosis are 0.78 (95% CI: 0.46–1.32) and 0.95 (95% CI: 0.57–1.58), respectively. Similar observations have been made by Roychowdhury et al. [94], Chakraborty et al. [95], and Mondal et al. [96]. The importance of adequate nutrition can be judged by the fact that a group of people in Alaska did not show any arsenicosis-related skin lesions, despite consuming high concentration of arsenic in drinking water [97]. Similar protective effects of adequate nutrition have been documented in India too by various researchers. Das et al. have noted that the inhabitants consuming nutritious food show almost no arsenic skin lesions in some arsenic-affected villages of West Bengal [98]. In their study, Roychowdhury et al. documented that the children in Baruipara block, South 24 Parganas district of West Bengal, used to consume highly contaminated water, but did not show any such skin lesions. On the other hand, many children from Murshidabad district, West Bengal, suffer from arsenicosis-related melanosis, despite consuming the same amount of arsenic-contaminated water [99]. It seems that people can tolerate arsenic up to a certain range although they have a high level of arsenic in their hair, nail, and urine samples.

14.4.4.2   Effect of Vitamin C (Ascorbic Acid)

The administration of ascorbic acid has been shown to normalize various biomarkers of arsenicosis. Rana et al. in their study have noted normalization of levels of AST ($p < 0.05$), ALT ($p < 0.05$), ALP, ACP ($p < 0.05$), and LDH ($p < 0.05$) activities in the plasma after 12 weeks of arsenic plus ascorbic acid coexposure in rats [100]. Ascorbic acid has also shown beneficial effects on lipid peroxidation induced by arsenic in erythrocytes as demonstrated by malondialdehyde assay (MDA). Cotreatment with ascorbic acid was found to be associated with significant ($p < 0.05$) decrease in erythrocytes MDA level compared to arsenic-treated rats. Similar beneficial effects have been noted for increased protein carbonyl (PC), decreased activity of antioxidant enzymes superoxide dismutase (SOD) and catalase (CAT), elevated nitric oxide production, and down regulated the mRNA expression of SOD2 in erythrocytes, associated with arsenicosis. The changes described in erythrocytes are considered sensitive indicators of oxidative reactions, in part due to role of RBCs in feneration of free radical species [101]. These findings suggest that the ascorbic acid is effective in mitigating arsenic induced oxidative damage in rat, which could be attributed to its antioxidant nature, free radical scavenging, and metal chelating properties [100].

14.4.4.3   Selenium and Arsenicosis

The identification of seleno-*bis*(*S*-glutathionyl) arsinium ion (As/Se/GSH) has provided the most direct link between dietary selenium and arsenic toxicity. This As/Se/GSH complex appears to be an excretory route for excess ingested arsenic, not Selenium, via the hepatobiliary route and fecal elimination [102]. Gailer et al. have demonstrated this complex in the bile of the rabbit [103]. Hence, the absence of a low nonbioavailable dietary selenium intake would appear to adversely affect arsenic excretion. The same has been demonstrated in the mouse by Kenyon et al. in 1997 [104]. Selenium appears to counteract arsenic toxicity by the formation of insoluble complexes without the involvement of reduced glutathione (GSH) [105]. Arsenic hemiselenide (As_2Se) is the most likely complex formed in tissues between arsenic and selenium. These insoluble complexes get deposited in tissues (thus annihilating its toxicity) and might as well contribute to the darkening of skin as this complex is black in color [106]. Glutathione peroxidases and the other selenoproteins, like thioredoxin reductase, are selenium dependent and hence, selenium depletion, by decreased levels of antioxidants enzymes, offers no protection against free radical damage [107]. Wang et al. have documented clinical beneficial effects of selenium supplementation in chronic arsenicosis patients [108]. They have shown that 14 months of selenium supplementation was associated with reduction in the severity of symptoms of arsenicosis by 75%. Also, people with arsenicosis taking selenium supplements showed no deterioration of symptoms. Arsenic in blood, urine, and hair too showed decline with selenium treatment over time in comparison with control subjects [109].

14.4.4.4   Sodium Thiosulfate

Sodium thiosulfate, an exogenous source of sulfur, was first used for the treatment of arsenical dermatosis by Ravaut in 1920 and since then, many researchers have *attested its value in shortening the duration and lessening the severity of arsenical eruptions.* However, randomized control trials are still lacking [110]. The exact mechanism of action of sodium thiosulfate in arsenical dermatosis is unknown, but it is believed to increase in the elimination of arsenic through urine and feces after oral or intravenous administration [111]. It is postulated that the sulfate moiety may react with and chelate arsenic, allowing its removal. In their study on cattle, Ghosh et al. have documented increased excretion of arsenic in urine and feces [112]. They have found that sodium thiosulfate in a dose of 20 and 40 g increases the excretion of arsenic from urine significantly ($p < 0.05$) from the control value. However, the urinary excretion of arsenic does not appear to be dose dependent as the excretion of arsenic concentration through urine was not significant between the cattle treated with 20 and 40 g of sodium thiosulfate. Similarly, increased excretion of arsenic through feces was noted. The decrease in arsenic content of hair was less, and only cattle treated with sodium thiosulfate at 40 g showed a significant ($p < 0.05$) decrease of the arsenic content of the hair from the basal value. The usefulness of sodium thiosulfate in acute arsenicosis has been also documented by Bertin et al. in their study [113]. However, there are no recent human studies.

14.4.4.5   Zinc

The evidence of the role of zinc supplementation in the treatment of arsenicosis is not strong. However, it has been shown that zinc supplementation, either alone or in combination with monoisoamyl dimercaptosuccinic acid (DMSA) during and after arsenic exposure resulted in more pronounced elimination of arsenic in male mice [114].

14.4.4.6   Vitamin E

The role of vitamin E in reducing the arsenic burden is limited, but it does offer some beneficial effects because of its antioxidant properties. It has been shown to be useful in the restoration of altered biochemical pathways, particularly related to heme biosynthesis and oxidative injury [115].

14.4.4.7 Other Antioxidants

Coadministration of N-acetylcysteine, melatonin, α-lipoic acid, and taurine has been used as adjuvant along with chelators in heavy metal poisoning such as lead and arsenic in animals and have shown some protective effects from oxidative injury. Though, they have no role in decreasing the burden of arsenic in tissues [116].

14.4.4.8 Plant Products

As obvious from earlier discussions, conventional chelating agents may have toxic side effects and have some disadvantages. Coadministration of various plants extracts either during exposure or treatment with chelators has shown some benefit in arsenic-induced hematological, renal, and hepatic disorders in laboratory animals. Fruits and leaves of *Hippophae rhamnoides* (rich in antioxidants vitamin A, C, and E; carotenoids; and organic acids) [116], and leaves of *Psidium guajava* (rich in flavonoids and polyphenols) [117] are of special interest. In their study, Gupta et al. have found that aqueous fruit extract of *H. rhamnoides* (Seabuckthorn) [116] extracted at room temperature had a favorable effect on various arsenic-induced altered biochemical variables in albino mice. The authors concluded that "fruit extract of *H. rhamnoides* has a significant protective role against arsenic-induced oxidative injury" [118]. Tandon et al. have studied the effects of leaves of *P. guajava* on arsenic-induced lipid peroxidation in erythrocytes, level of reduced glutathione (GSH), superoxide dismutase (SOD), and catalase (CAT) activities. Their findings suggest that *P. guajava* leaves may have a role in protecting from the oxidative damage by ROS (reactive oxygen species) and in preventing the arsenic-induced alterations in the oxidative stress markers value. Also, it reduces the arsenic levels in blood and tissues and showed metal chelation property [117]. Another medicinal plant *Bacopa monnieri* (Brahmi, waterhyssop) has shown significant antioxidant properties [119]; however, it has not been studied in arsenicosis.

14.5 CONCLUSION

The management of arsenicosis has a long way to go before a standard of care would emerge. The relentless efforts of the researchers along with the help from governmental and nongovernmental funding agencies would help the mankind find the optimum agent to put an end to this menace.

REFERENCES

1. Ghosh P, Roy C, Das NK, Sengupta SR. Epidemiology and prevention of chronic arsenicosis: An Indian perspective. *Indian J Dermatol Venereol Leprol* 2008;74:582–593.
2. Sengupta SR, Das NK, Datta PK. Pathogenesis, clinical features and pathology of chronic arsenicosis. *Indian J Dermatol Venereol Leprol* 2008;74:559–570.
3. Rahman MM, Chowdhury UK, Mukherjee SC et al. Chronic arsenic toxicity in Bangladesh and West Bengal, India—A review and commentary. *J Toxicol Clin Toxicol* 2001;39:683–700.

4. Kadono T, Inaoka T, Murayama N et al. Skin manifestations of arsenicosis in two villages in Bangladesh. *Int J Dermatol* 2002;41:841–846.
5. Saha JC, Dikshit AK, Bandyopadhyay M, Saha KC. A review of arsenic poisoning and its effects on human health. *Crit Rev Environ Sci Technol* 1999;29(3):281–313.
6. Misbahuddin M, Islam AZ, Khandker S, Ifthaker-Al-Mahmud, Islam N, Anjumanara. Efficacy of spirulina extract plus zinc in patients of chronic arsenic poisoning: A randomized placebo-controlled study. *Clin Toxicol* (Phila) 2006;44(2):135–141.
7. Verret WJ, Chen Y, Ahmed A, Islam T, Parvez F, Kibriya MG, Graziano JH, Ahsan H. A randomized, double-blind placebo-controlled trial evaluating the effects of vitamin E and selenium on arsenic-induced skin lesions in Bangladesh. *J Occup Environ Med* October 2005;47(10):1026–1035.
8. Schwartz RA. Arsenic and the skin. *Int J Dermatol* 1997;36:241–250.
9. Das NK, Sengupta SR. Arsenicosis: Diagnosis and treatment. *Indian J Dermatol Venereol Leprol* 2008;74:571–581.
10. A field guide for detection, management and surveillance of arsenicosis cases. In: Caussy, D. (ed.), World Health Organization, Regional Office of South-East Asia, New Delhi, India, 2005. Available from http://apps.searo.who.int/PDS_DOCS/B0301.pdf.
11. Thiaprasit M. Chronic cutaneous arsenism treated with aromatic retinoid. *J Med Assoc Thailand* 1984;67:93–100.
12. Biczo Z, Berta M, Szabo M, Nagy GY. Traitement de Iíarsenicisme chronique par Iíetretinate. *La Presse Medicale* 1986;15:2073.
13. Miller WH. The emerging role of retinoids and retinoic acid metabolism blocking agents in the treatment of cancer. *Cancer* 1998;83:1471–1482.
14. Patton TJ, Ferris LK. Systemic retinoids. In *Comprehensive Dermatologic Drug Therapy*. Wolverton SE, ed. 3rd edition, Elsevier Saunders, Edinburgh, 2013, pp. 252–268.
15. Krüger-Corcoran D, Vandersee S, Stockfleth E. Precancerous tumors and carcinomas in situ of the skin. *Internist (Berl)* 2013 Jun; 54(6):671–682.
16. Hoover WD 3rd, Jorizzo JL, Clark AR, Feldman SR, Holbrook J, Huang KE. Efficacy of cryosurgery and 5-fluorouracil cream 0.5% combination therapy for the treatment of actinic keratosis. *Cutis* 2014 Nov; 94(5):255–259.
17. Ahmad SA, Sayed MHS, Khan MH et al. Arsenicosis: Neoplastic manifestation. *J Prevent Soc Med* 1998;17:110–115.
18. Bath-Hextall FJ, Matin RN, Wilkinson D, Leonardi-Bee J. Interventions for cutaneous Bowen's disease. *Cochrane Database Syst Rev* 2013;Issue 6:Art. No. CD007281. doi: 10.1002/14651858.CD007281.pub2.
19. Westers-Attema A, van den Heijkant F, Lohman BG, Nelemans PJ, Winnepenninckx V, Kelleners-Smeets NW, Mosterd K. Bowen's disease: A six-year retrospective study of treatment with emphasis on resection margins. *Acta Derm Venereol* July 2014;94(4):431–435.
20. Jiang SIB, Kim SS. Mohs surgery. Available from http://emedicine.medscape.com/article/2212475-overview#a0101. [Accessed on April 26, 2014].
21. Morton CA, Birnie AJ, Eedy DJ. British association of dermatologists' guidelines for the management of squamous cell carcinoma in situ (Bowen's disease). *Br J Dermatol* 2014;170:245–260.
22. Patel GK, Goodwin R, Chawla M et al. Imiquimod 5% cream monotherapy for cutaneous squamous cell carcinoma in situ (Bowen's disease): A randomised, double-blind placebo-controlled trial. *J Am Acad Dermatol* 2006;54(6):1025–1032.
23. Morton C, Horn M, Leman J et al. Comparison of topical methyl aminolevulinate photodynamic therapy with cryotherapy or fluorouracil for treatment of squamous cell carcinoma in situ: Results of a multicenter randomized trial. *Arch Dermatol* 2006;142(6):729–735.

24. Holt PJ. Cryotherapy for skin cancer: Results over a 5-year period using liquid nitrogen spray cryosurgery. *Br J Dermatol* 1988;119:231–240.

25. Cox NH, Dyson P. Wound healing on the lower leg after radiotherapy or cryotherapy of Bowen's disease and other malignant skin lesions. *Br J Dermatol* 1995;133:60–65.

26. Rao J, Bissonnette R, Taylor CR. Photodynamic therapy for the dermatologist. Available from http://emedicine.medscape.com/article/1121517-overview#a1 [Accessed on March, 2014].

27. Lopez N, Meyer-Gonzalez T, Herrera-Acosta E et al. Photodynamic therapy in the treatment of extensive Bowen's disease. *J Dermatol Treat* 2012;23:428–430.

28. Dave R, Monk B, Mahaffey P. Treatment of Bowen's disease with carbon dioxide laser. *Lasers Surg Med* 2003;32:335.

29. Tseng W. Blackfoot disease in Taiwan: A 30-year follow-up study. *Angiology* 1989;40:547–558.

30. Money SR, Herd JA, Isaacsohn JL et al. Effect of cilostazol on walking distances in patients with intermittent claudication caused by peripheral vascular disease. *J Vasc Surg* 1998;27:267–274.

31. Regensteiner JG, Hiatt WR. Current medical therapies for patients with peripheral arterial disease: A critical review. *Am J Med* 2002;112:49–57.

32. Ohnishi K, Yoshida H, Shigeno K et al. Prolongation of the QT interval and ventricular tachycardia in patients treated with arsenic trioxide for acute promyelocytic leukemia. *Ann Intern Med* 2000;133:881–885.

33. Wang C, Chen C, Hsiao CK et al. Arsenic-induced QT dispersion is associated with atherosclerotic diseases and predicts long-term cardiovascular mortality in subjects with previous exposure to arsenic: A 17-year follow-up Study. *Cardiovasc Toxicol* 2010;10:17–26.

34. Kaur T, Goel RK, Balakumar P. Effect of rosiglitazone in sodium arsenite-induced experimental vascular endothelial dysfunction. *Arch Pharm Res* 2010;33:611–618.

35. Rahman M, Tondel M, Ahmad SA, Chowdhury IA, Faruquee MH, Axelson O. Hypertension and arsenic exposure in Bangladesh. *Hypertension* 1999;33(1):74–78.

36. Abhyankar LN, Jones MR, Guallar E, Navas-Acien A. Arsenic exposure and hypertension: A systematic review. *Environ Health Perspect* 2012;120:494–500.

37. Parvez F, Chen Y, Yunus M et al. A prospective study of arsenic induced respiratory symptoms and chronic obstructive pulmonary disease (COPD): Findings from Health Effects of Arsenic Exposure Longitudinal Study (HEALS) in Bangladesh. *Epidemiology* 2009;20:S82–S83.

38. Ramsey KA, Larcombe AN, Sly PD, Zosky GR. In utero exposure to low dose arsenic via drinking water impairs early life lung mechanics in mice. *BMC Pharmacol Toxicol* 2013;14:13.

39. Khanna R, Sarin SK. Non-cirrhotic portal hypertension—Diagnosis and management. *J Hepatol* 2014;60:421–441.

40. Mitra SK, Rao KLN, Narasimhan KL et al. Side-to-side lienorenal shunt without splenectomy in noncirrhotic portal hypertension in children. *J Pediatr Surg* 1993;28:398–401.

41. Pal S, Radhakrishna P, Sahni P, Pande GK, Nundy S, Chattopadhyay TK. Prophylactic surgery in non-cirrhotic portal fibrosis: Is it worthwhile? *Indian J Gastroenterol* 2005;24:239–242.

42. Sarin SK, Agarwal SR. Extrahepatic portal vein obstruction. *Semin Liver Dis* 2002;22:43–58.

43. Gupta AK, George J, Giri P, Gupta R. Non-cirrhotic portal fibrosis. Association of physicians of India. 2010. Available online from http://www.apiindia.org/pdf/medicine_ update_2010/ge_and_hepatology_01.pdf [Accessed October 18, 2014].

44. Krasinskas AM, Eghtesad B, Kamath PS, Demetris AJ, Abraham SC. Liver transplantation for severe intrahepatic noncirrhotic portal hypertension. *Liver Transpl* 2005;11:627–634.

45. Nijhawan S, Katiyar P. Non-cirrhotic portal fibrosis. API update of Medicine 2012. Available online from http://www.apiindia.org/pdf/medicine_update_2012/miscellaneous_04.pdf [Accessed October 17, 2014].

46. Sanyal AJ. American Gastroenterological Association. AGA technical review on nonalcoholic fatty liver disease. *Gastroenterology* 2002;123:1705–1725.

47. Andersen T, Gluud C, Franzmann MB et al. Hepatic effects of dietary weight loss in morbidly obese subjects. *J Hepatol* 1991;12:224–229.

48. Lindor KD, Kowdley KV, Heathcote EJ et al. Ursodeoxycholic acid for treatment of nonalcoholic steatohepatitis: Results of a randomized trial. *Hepatology* 2004;39:770–778.

49. Yokohama S, Yoneda M, Haneda M et al. Therapeutic efficacy of an angiotensin II receptor antagonist in patients with nonalcoholic steatohepatitis. *Hepatology* 2004;40:1222–1225.

50. Loguercio C, Federico A, Tuccillo C. Beneficial effects of a probiotic VSL#3 on parameters of liver dysfunction in chronic liver diseases. *J Clin Gastroenterol* 2005;39:540–543.

51. Mani H, Van Thiel DH. Mesenchymal tumors of the liver. *Clin Liver Dis* 2001;5:219–257.

52. Maluf D, Cotterell A, Clark B, Stravitz T, Kauffman HM, Fisher RA. Hepatic angiosarcoma and liver transplantation: Case report and literature review. *Transplant Proc* 2005;37:2195–2199.

53. Duan XF, Li Q. Primary hepatic angiosarcoma: A retrospective analysis of 6 cases. *J Dig Dis* 2012;13:381–385.

54. Dannaher CL, Tamburro CH, Yam LT. Chemotherapy of vinyl chloride-associated hepatic angiosarcoma. *Cancer* 1981;47:466–469.

55. Zheng Y, Zhang X, Zhang J et al. Primary hepatic angiosarcoma and potential treatment options. *J Gastroenterol Hepatol* 2014;29:906–911.

56. Huang NC, Wann SR, Chang HT, Lin SL, Wang JS, Guo HR. Arsenic, vinyl chloride, viral hepatitis, and hepatic angiosarcoma: A hospital-based study and review of literature in Taiwan. *BMC Gastroenterol* 2011;11:142.

57. Valbonesi M, Bruni R. Clinical application of therapeutic erythrocytapheresis. *Transfus Sci* 2000;22:183–194.

58. Heaven R, Duncan M, Vukelja SJ. Arsenic intoxication presenting with macrocytosis and peripheral neuropathy, without anemia. *Acta Haematol* 1994;92:142–143.

59. Miwa S, Ishida Y, Takegawa S, Urata G, Toyoda T. A case of lead intoxication: Clinical and biochemical studies. *Am J Hematol* 1981;11(1):99–105.

60. Motzer RJ, Hutson TE, Cella D et al. Pazopanib versus sunitinib in metastatic renal-cell carcinoma. *N Engl J Med* 2013;369:722–731.

61. Gupta S, Spiess PE. The prospects of pazopanib in advanced renal cell carcinoma. *Ther Adv Urol* 2013 Oct; 5(5):223–232.

62. Wax PM, Thornton CA. Recovery from severe arsenic-induced peripheral neuropathy with 2,3-dimercapto-1-propanesulphonic acid. *J Toxicol Clin Toxicol* 2000;38(7):777–780.

63. Heyman A, Pfeiffer JB, Willett RW et al. Peripheral neuropathy caused by arsenical intoxication—A study of 41 cases with observations on the effects of Bal (2,3-dimercaptopropanol). *N Engl J Med* 1956;254:401–409.

64. Mukherjee SC, Rahman MM, Chowdhury UK et al. Neuropathy in arsenic toxicity from groundwater arsenic contamination in West Bengal, India. *J Environ Sci Health* 2003;38:165–169.

65. Wang Y, Li S, Piao F, Hong Y, Liu P, Zhao Y. Arsenic down-regulates the expression of Camk4, an important gene related to cerebellar LTD in mice. *Neurotoxicol Teratol* 2009;31:318–322.

66. Dewji NN, Do C, Bayney RM. Transcriptional activation of Alzheimer's β-amyloid precursor protein gene by stress. *Mol Brain Res* 1995;33:245–253.

67. Yadav RS, Shukla RK, Sankhwar ML et al. Neuroprotective effect of curcumin in arsenic-induced neurotoxicity in rats. *Neurotoxicology* 2010;31:533–539.

68. Wright RO, Amarasiriwardena C, Woolf AD et al. Neuropsychological correlates of hair arsenic, manganese, and cadmium levels in school-age children residing near a hazardous waste site. *Neurotoxicology* 2006;27:210–216.

69. Dubey M, Shea TB. Potentiation of arsenic neurotoxicity by folate deprivation: Protective role of S-adenosyl methionine. *Nutr Neurosci* 2007;10:199–204.

70. Roy S, Chattoraj A, Bhattacharya S. Arsenic-induced changes in optic tectalhistoarchitecture and acetylcholinesterase-acetylcholine profile in Channapunctatus: Amelioration by selenium. *Comp Biochem Physiol C: Toxicol Pharmacol* 2006;144:16–24.

71. Milton AG, Zalewski PD, Ratnaike RN. Zinc protects against arsenic-induced apoptosis in a neuronal cell line, measured by DEVD-caspase activity. *Biometals* 2004;17:707–713.

72. Noël B. Vascular complications of cocaine use. *Stroke* July 2002;33:1747–1748.

73. Tseng CH. The potential biological mechanisms of arsenic-induced diabetes mellitus. *Toxicol Appl Pharmacol* 2004;197(2):67–83.

74. Samadder A, Das J, Das S, De A, Saha SK, Bhattacharyya SS, Khuda-Bukhsh AR. Poly(lactic-co-glycolic) acid loaded nano-insulin has greater potentials of combating arsenic induced hyperglycemia in mice: Some novel findings. *Toxicol Appl Pharmacol* 2013;267:57–73.

75. Kulshreshtha A, Zacharia AJ, Jarouliya U, Bhadauriya P, Prasad GBKS, Bisen PS. Spirulina in health care management. *Curr Pharm Biotechnol* 2008;9:400–405.

76. Flora SJS, Pachauri V. Chelation in metal intoxication. *Int J Environ Res Public Health* 2010;7(7):2745–2788.

77. Sears ME. Chelation: Harnessing and enhancing heavy metal detoxification—A review. *Sci World J* 2013;2013:219840.

78. Flora SJ, Mittal M, Mehta A. Heavy metal induced oxidative stress & its possible reversal by chelation therapy. *Indian J Med Res* 2008;128(4):501–523.

79. Torres-Alanís O, Garza-Ocañas L, Bernal MA, Piñeyro-López A. Urinary excretion of trace elements in humans after sodium 2,3-dimercaptopropane-1-sulfonate challenge test. *J Toxicol Clin Toxicol* 2000;38(7):697–700.

80. Hoover TD, Aposhian HV. BAL increases the arsenic-74 content of rabbit brain. *Toxicol Appl Pharmacol* 1983;70:160–162.

81. Guha Mazumder DN, Ghoshal UC, Saha J et al. Randomized placebo-controlled trial of 2,3-dimercaptosuccinic acid in therapy of chronic arsenicosis due to drinking arsenic-contaminated subsoil water. *J Toxicol Clin Toxicol* 1998;36:683–690.

82. Guha Mazumder DN, De BK, Santra A et al. Randomized placebo-controlled trial of 2,3-dimercapto-1-propanesulfonate (DMPS) in therapy of chronic arsenicosis due to drinking arsenic-contaminated water. *J Toxicol Clin Toxicol* 2001;39:665–674.

83. Bhadauria S, Flora SJS. Response of arsenic induced oxidative stress, DNA damage and metal imbalance to combined administration of DMSA and monoisoamyl DMSA during chronic arsenic poisoning in rats. *Cell Biol Toxicol* 2007;23:91–104.

84. Pachauri V, Mehta A, Mishra D, Flora SJ. Arsenic induced neuronal apoptosis in guinea pigs is Ca^{2+} dependent and abrogated by chelation therapy: Role of voltage gated calcium channels. *Neurotoxicology* March 2013;35:137–145.

85. Mishra D, Flora SJ. Quercetin administration during chelation therapy protects arsenic-induced oxidative stress in mice. *Biol Trace Elem Res* May 2008;122(2):137–147.

86. Dimercaprol. INCHEM. 1990. Available online from http://www.inchem.org/documents/pims/pharm/dimercap.htm#SubSectionTitle:7.1.1 Toxicodynamics [Cited May 16, 2014].

87. Janakiraman N, Seeler RA, Royal JE, Chen MF. Hemolysis during BAL chelation therapy for high blood lead levels in two G6PD deficient children. *Clin Pediatr* 1978;17:485–487.

88. Reichl RX, Kreppel H, Forth W. Pyruvate and lactate metabolism in livers of guineapigs perfused with chelating agents after repeated treatment with As_2O_3. *Arch Toxicol* 1991;65:235–238.

89. Bradberry S, Vale A. Dimercaptosuccinic acid (succimer; DMSA) in inorganic lead poisoning. *Clin Toxicol* (Phila) 2009;47:617–631.

90. LundbeckInc. Chemet® prescribing information, 2009. http://www.lundbeckinc.com/USA/products/hospital/chemet/USA-CHE-PI-Web-CC-May-2009.pdf viewed [February 15, 2010].

91. Buchwald AL. Intentional overdose of dimercaptosuccinic acid in the course of treatment for arsenic poisoning. *J Toxicol Clin Toxicol* 2001;39:113–114.

92. Aposhian HV, Arroyo A, Cebrian ME et al. DMPS-arsenic challenge test. I: Increased urinary excretion of monomethylarsonic acid in humans given dimercaptopropane sulfonate. *J Pharmacol Exp Ther* 1997;282:192–200.

93. Samal AC, Kar S, Maity JP, Santra SC. Arsenicosis and its relationship with nutritional status in two arsenic affected areas of West Bengal, India. *J Asian Earth Sci* 2013; 77:303–310.

94. Roychowdhury T, Uchino T, Tokunaga H, Ando M. Survey of arsenic in food composites from an arsenic-affected area of West Bengal, India. *Food Chem Toxicol* 2002;40:1611–1621.

95. Chakraborty D, Rahaman MM, Paul K, Chowdhury UK, Sengupta MK, Lodh D, Chanda CR, Saha KC, Mukherjee SC. Arsenic calamity in the Indian subcontinent what lessons have been learned? *Talanta* 2002;58:3–22.

96. Mondal NK, Roy P, Das B, Datta JK. Chronic arsenic toxicity and it's relation with nutritional status: A case study in Purabasthali-II, Burdwan, West Bengal, India. *Int J Environ Sci* 2011;2(2):1103–1118.

97. Harrington JM, Middaugh JP, Morse DL, Housworth J. A survey of a population exposed to high concentrations of arsenic in well water in Fairbanks, Alaska. *Am J Epidemiol* 1978;108:377–385.

98. Das D, Chatterjee A, Mandal BK, Samanta G, Chakraborti D, Chanda B. Arsenic in ground water in six districts of West Bengal, India: The biggest arsenic calamity in the world, Part 2. Arsenic concentration in drinking water, hair, nails, urine, skin-scale and liver tissue (Biopsy) of the affected people. *Analyst* 1995;120:917–924.

99. Roychowdhury T, Mandal BK, Samanta G et al. Arsenic in groundwater in six districts of West Bengal, India: The biggest arsenic calamity in the world: The status report up to August, 1995. In: Abernathy, C.O., Calderon, R.L., Chappell, W.R. (eds.), *Arsenic: Exposure and Health Effects*. Chapman & Hall, London, U.K., 1997, pp. 93–111.

100. Rana T, Bera AK, Das S, Pan D, Bandyopadhyay S, Bhattacharya D, De S, Sikdar S, Das SK. Effect of ascorbic acid on blood oxidative stress in experimental chronic arsenicosis in rodents. *Food Chem Toxicol* 2010;48:1072–1077.

101. Sato Y, Kanazawa S, Sato K, Suzuki Y. Mechanism of free radical induced hemolysis of human erythrocytes: II. Hemolysis by lipid soluble radical initiator. *Biol Pharm Bull* 1998;21:250–256.

102. Gailer J, George GN, Pickering IJ, Prince RC, Ringwald SC, Pemberton JE, Glass RS, Younis HS, DeYoung DW, Aposhian V. A metabolic link between arsenite and selenite: The seleno-bis(S-glutathionyl) arsinium ion. *J Am Chem Soc* 2000;122:4637–4639.

103. Gailer J, George GN, Pickering IJ, Prince RC, Younis HS, Winzerling JJ. Biliary excretion of [(GS)(2)AsSe] intravenous injection of rabbits with arsenite and selenate. *Chem Res Toxicol* 2002;15:1466–1471.
104. Kenyon EM, Hughes MF, Levander OA. Influence of dietary selenium on the disposition of arsenate in the female B6C3F1 mouse. *J Toxicol Environ Health* 1997;51:279–299.
105. Csanaky I, Gregus Z. Effect of selenite on the deposition of arsenate and arsenite in rats. *Toxicology* 2003;186:33–50.
106. Berry JP, Galle P. Selenium–arsenic interaction in renal cells: Role of lyososomes. Electron microprobe study. *J Submicrosc Cytol Pathol* 1994;26:203–210.
107. Spallholz JE, Boylan LM, Rhaman MM. Environmental hypothesis: Is poor dietary selenium intake an underlying factor for arsenicosis and cancer in Bangladesh and West Bengal, India? *Sci Total Environ* 2004;323:21–32.
108. Wang W, Yang L, Hou S, Tan S, Li H. Effects of selenium supplementation on arsenism: An intervention trial in inner mongolia. *Environ Geochem Health* 2002;24:359–374.
109. Wang W, Yang L, Hou S, Tan S, Li H. Prevention of endemic arsenism with selenium. *Curr Sci* 2001;81:1215–1219.
110. Ayres Jr S, Anderson NP. Sodium thiosulfate and the elimination of arsenic. *JAMA* 1938;110(12):886–887.
111. Ayres Jr S, Anderson NP. Sodium thiosulfate and the elimination of arsenic. *JAMA* 1938;110(12):886–887.
112. Ghosh CK, Datta BK, Biswas S, Maji C, Sarkar S, Mandal TK, Majumder D, Chakraborty AK. Chronic arsenicosis of cattle in West Bengal and it's possible mitigation by sodium thiosulfate. *Toxicol Int* 2011;18:137–139.
113. Bertin FR, Baseler LJ, Wilson CR, Kritchevsky JE, Taylor SD. Arsenic toxicosis in cattle: Meta-analysis of 156 cases. *J Vet Intern Med* July–August 2013;27(4):977–981.
114. Modi M, Pathak U, Kalia K, Flora SJ. Arsenic antagonism studies with monoisoamyl DMSA and zinc in male mice. *Environ Toxicol Pharmacol* January 2005;19(1):131–138.
115. Kannan GM, Flora SJ. Chronic arsenic poisoning in the rat: Treatment with combined administration of succimers and an antioxidant. *Ecotoxicol Environ Saf* May 2004;58(1):37–43.
116. Kalia K, Flora SJ. Strategies for safe and effective therapeutic measures for chronic arsenic and lead poisoning. *J Occup Health* January 2005;47(1):1–21.
117. Tandon N, Roy M, Roy S, Gupta N. Protective effect of *Psidium guajava* in arsenic-induced oxidative stress and cytological damage in rats. *Toxicol Int* 2012;19:245–249.
118. Gupta R, Flora SJ. Protective effects of fruit extracts of *Hippophae rhamnoides* L. against arsenic toxicity in Swiss albino mice. *Hum Exp Toxicol* June 2006;25(6):285–295.
119. Kapoor R, Srivastava S, Kakkar P. *Bacopa monnieri* modulates antioxidant responses in brain and kidney of diabetic rats. *Environ Toxicol Pharmacol* 2009 Jan; 27(1):62–69.

Section IV

Remediation of Arsenic by Nutraceuticals and Functional Food

15 Influence of Nutraceuticals in Combating Arsenic Toxicity

Nidhi Dwivedi, Vidhu Pachauri, and S.J.S. Flora

CONTENTS

15.1 INTRODUCTION

Arsenic is number one on the Agency for Toxic Substances and Disease Registry's (ATSDR) *Top 20 List*. The delayed health effects of exposure to arsenic, the lack of common definitions and of local awareness, as well as poor reporting in affected areas are major problems in determining the extent of the arsenic-in-drinking-water problem. More than 20 arsenic compounds are present in the natural environment and biological systems. Trivalent arsenic species, such as inorganic arsenite (As^{III}), monomethylarsonous acid (MMA^{III}), and dimethylarsinous acid (DMA^{III}) are more toxic as compared to pentavalent arsenic species (As^{v}). Naturally, arsenic occurs

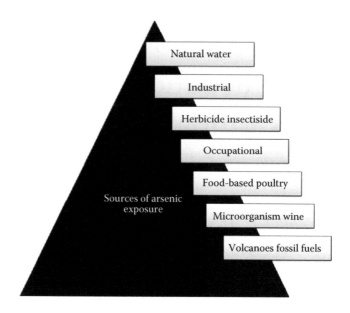

FIGURE 15.1   Sources of arsenic exposure.

in rocks, soil, metal ores (such as copper and lead), and in the form of minerals. Exposure to arsenic may also come from industrial processes (semiconductor
manufacturing, wood preservatives, metallurgy, glass clarification, and smelting
and refining of metals and ores), commercial products (pesticides, herbicides, fun-
gicides, and fire salts), and medicines (antiparasitic drugs and folk medicines) [1]
(see Figure 15.1).

Absorbance of inorganic arsenic from the gastrointestinal tract depends on the
solubility and composition of arsenicals. Experimental data from both humans as
well as animals show that over 90% of an ingested dose of dissolved inorganic arsenic is absorbed in the gastrointestinal tract, passed into blood stream and distributed
to organs/tissues after first passing through the liver. Therefore, the distribution of
arsenic taken up by the gastrointestinal tract is influenced by the metabolic activity in the liver. The metabolism of arsenicals in liver and their resulting chemical
forms in organs/tissues affect the toxicity of animals [2,3]. It is also reported that
the highest absorbance rate was found in the small intestine followed by mouth cav-
ity and stomach. Systemic toxic effects have resulted from occupational accidents
where the splashing of arsenic acid or arsenic trichloride on workers could be the
possible route for the absorption of arsenic in skin [4]. However, the absorption may
have been enhanced by epidermal lesions caused by the compounds. Arsenic has
some properties similar to phosphorous and this helps arsenic to cross the selective membrane barrier of the cell. As arsenic is structurally similar to phosphorous, it enters the host cell through phosphate ion channels, causing its absorption/
entry in the cell. Following absorption by the lungs or the gastrointestinal tract,
arsenic is transported through blood to other body parts. Inorganic arsenic is rapidly cleared from the blood both in humans and experimental animals, like mice,

rabbits, and hamsters. However, in rats, absorbed arsenic is accumulated in the red blood cells due to its irreversible binding with hemoglobin. Metabolism of arsenic is a two-step procedure: The first step is oxidation from trivalent to pentavalent or reduction from pentavalent to trivalent by a common enzyme arsenate reductase [3,5]. In the second step, reduced arsenite is detoxified from inorganic to organic compound through sequential methylation reactions. This detoxification converts inorganic arsenics to form MMA and dimethylarsinic acid (DMA) in vivo using *S*-adenosyl methionine (SAM) as the methyl donor and glutathione (GSH) as an essential cofactor [6].

Arsenic affects cellular functioning by various mechanisms of action. Arsenic, particularly the trivalent forms, binds to sulfhydryl groups, disrupts essential enzyme activity, and leads to impaired gluconeogenesis and oxidative phosphorylation. In another mechanism, arsenic inhibits the enzyme pyruvate dehydrogenase by binding to its sulfhydryl groups. Importantly, arsenic inhibits the synthesis of GSH, one of the most powerful cellular antioxidant. The toxicity of pentavalent inorganic arsenic occurs by two ways: First, via its reduction to trivalent arsenic [7] and second, due to its resemblance with inorganic phosphate it gets substituted in place of phosphate in glycolytic and cellular respiration pathways and preferentially forms ADP-arsenate, having less stable bonds in comparison with high energy compounds such as adenosine triphosphate (ATP). This *arsenolysis* causes rapid hydrolysis of these bonds, resulting in premature uncoupling oxidative phosphorylation.

Studies have confirmed the generation of various types of reactive oxygen species (ROS) during arsenic metabolism in cells [8]. A clear relation between oxidative stress and the development of arsenic-related diseases including cancers has been established. In addition to ROS, reactive nitrogen species (RNS) are also believed to be directly associated in the oxidative damage of lipids, proteins, and DNA in cells exposed to arsenic. Recent experimental studies proved that arsenic-induced free radicals generation can cause cell damage and death through activation of oxidative sensitive signaling pathways [9]. It is proposed that their formation involves intermediary arsine species.

GSH is a very effective cellular antioxidant and plays an important role in maintaining cellular redox homeostasis. On observation of these studies it is indicated that GSH possibly acts as an electron donor for the reduction of pentavalent to trivalent arsenicals and that arsenite has high affinity to GSH, rendering it unavailable for subsequent reactions.

15.2 ARSENIC-INDUCED HUMAN DISEASES

Arsenic toxicity is associated with many consequences like inhibition of proteins, generation of free radicals, formation of less stable energy yielding compound ADP–arsenate, GSH inhibition, and inhibition of phosphate. These effects result in disturbances in cellular homeostasis causing different cell malfunctions including cancer. Many recent studies have provided experimental evidences that show that arsenic exposure has been linked with various types of cancer [10], cardiovascular disease [11], diabetes [12], neurological disorders [13], and dermal effects [14]. Chronic exposure of arsenic results in the development of lesions on the skin, including hyperkeratosis and hyperpigmentation. These symptoms are often used as a diagnostic feature for arsenicosis [15].

Arsenic is carcinogenic and remains stable in environment for long time. It is mainly associated with cancers of the skin, but there are evidences for lung, bladder, liver, and kidney cancers being caused by exposure to arsenic [16]. A large number of epidemiological trials have reported that inhalation exposure to inorganic arsenic increases the risk of lung cancer. Human exposure to inorganic arsenic is associated with an increased risk of dermal malignancies [17]. Long-term arsenic exposure has been reported to cause a malignant transformation of human keratinocytes in vitro [17].

Oral exposure to arsenic causes adverse effects on the cardiovascular system. There are evidences from epidemiological studies that the cardiovascular system may also be affected by inhaling inorganic arsenic [18]. Hypertension is another severe disorder associated with increased arsenic exposure [19]. Altered antioxidant defense level causes elevated systolic blood pressure. Gastroenteritis has been reported in a number of cases of chronic arsenicosis due to drinking arsenic-contaminated water. Symptoms like nausea, diarrhoea, anorexia, and abdominal pain are also associated with chronic arsenic toxicity. Prolonged drinking of arsenic-contaminated water is associated with hepatomegaly, predominant lesion being hepatic fibrosis. Typical cutaneous signs of long-term arsenic exposure were also observed in some of the patients. There have also been case reports of liver cirrhosis following medication with inorganic arsenic compounds. Peripheral neuritis, sleep disturbances, weakness, and cognitive and memory impairment have been reported in cases of

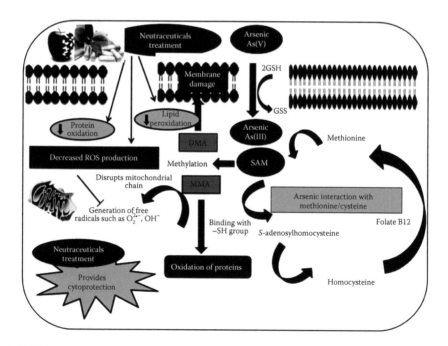

FIGURE 15.2   Plausible mechanisms of nutraceuticals against arsenic toxicity.

arsenic poisoning. Headache has been reported to occur in people drinking arsenic-contaminated water while irritability symptoms like lack of concentration, depression, sleep disorders, headache, and vertigo were reported in arsenicosis patients showing features of neuropathy in arsenic-exposed subjects (shown in Figure 15.2).

15.2.1 SYMPTOMS

Symptoms of acute arsenic poisoning are many and may be severely fatal at high doses. Severe cases of arsenic poisoning have exhibited symptoms of fever, aversion to food, abnormal liver enlargement (hepatomegaly), cardiac arrhythmia, development of dark patches on skin and other tissue (melanosis), peripheral neuropathy including sensory loss in the peripheral nervous system, gastrointestinal disorders, cardiovascular effects, and adverse effects on red blood cell formation, which can result in anemia. Chronic effects of arsenic poisoning include neurotoxic effects to the central and peripheral nervous systems. Symptoms include sensory changes, muscle sensitivity, prickling and tingling sensations (paresthesia), and muscle weakness. Liver injury is a common symptom of chronic arsenic poisoning. In people chronically exposed to toxic doses of arsenic, some cancers may be preceded by discolored skin (hyperpigmentation) and development of horny skin surfaces (hyperkeratosis). Analyses of hair, fingernails, and toenails can serve as evidences of arsenic ingestion. Such analyses are complicated by the possible presence of external arsenic contamination, particularly in a work environment in which the air and surroundings may be contaminated with arsenic. Levels of arsenic may be correlated with the growth of nails and hair so that careful analysis of segments of these materials can indicate time frames of exposure.

15.2.2 ARSENIC-INDUCED OXIDATIVE STRESS

Oxidative stress is a relatively new theory of arsenic toxicity [20–22]. Erythrocytes may be susceptible to oxidative damage due to the presence of heme-iron, polyunsaturated fatty acid (PUFA), and oxygen, which may initiate the reactions that induce oxidative changes in RBC [23]. Generation of various types of ROS during arsenic metabolism in cells has been confirmed [8]. In addition to ROS, RNS are also believed to be directly associated in the oxidative damage to lipids, proteins, and DNA in cells exposed to arsenic. Superoxide anion radicals like (O^{2-}), singlet oxygen ($^1O_2$), the peroxyl radical (ROO), nitric oxide (NO), hydrogen peroxide (H_2O_2), hydroxyl radicals [24], dimethylarsenic peroxy radicals ($[(CH_3)_2AsOO]$), and also the dimethylarsinic radical $[(CH_3)_2As]$ are generated through reactions involving arsenic. However, the exact mechanism for the generation of these reactive species is not clear, but it is proposed that their formation involves intermediary arsine species. The interaction of arsenic with glutathione and its related enzymes by changing their redox status may lead to the alterations of their biological function. Inactivation of GSH-related enzymes could have deleterious effects on the detoxification processes and other critical cellular processes involving GSH mediated redox regulation. GSH, an effective cellular antioxidant is reported to reduce or elevate during arsenic exposure [25]. These trends on arsenic exposure are dose and duration dependent.

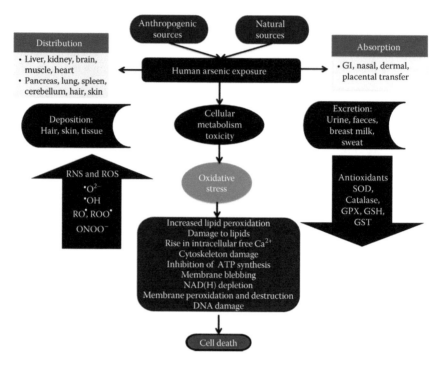

FIGURE 15.3   Arsenic-induced oxidative stress.

GSH possibly acts as an electron donor for the reduction of pentavalent to trivalent arsenicals and that arsenite has high affinity to GSH, rendering it unavailable for subsequent reactions. These effects in combination disturb cellular homeostasis causing different cell malfunctions including cancer. Many recent studies have provided experimental evidences that show that arsenic exposure has been linked with various types of cancer, cardiovascular disease, diabetes, neurological disorders, and dermal effects.

   The cell has several ways to alleviate the effects of oxidative stress, either by repairing the damage or by directly diminishing the occurrence of oxidative damage by means of enzymatic and nonenzymatic antioxidants, herbal drugs, and flavonoids. These nutraceuticals have also been shown to scavenge free radicals and ROS and can also act to overcome the oxidative stress (see Figure 15.3).

15.3   NUTRITIONAL AND FUNCTIONAL FOOD TO COMBAT ARSENIC TOXICITY

Although the seriousness of the problem of arsenicosis has been recognized, yet there is no effective treatment. Prevention might provide relief to the affected population. Thus, dietary prevention can be a better remedial action for chronic arsenic poisoning. Rich, healthy, and balanced nutrition can modulate the delayed effects of arsenic-contaminated drinking water. Nutritional factors are known to influence arsenic metabolism and poor nutritional status as reflected by a lack of various vitamins and antioxidants is thought to confer greater susceptibility to arsenic toxicity. Balanced nutrition may modulate the delayed effect of arsenic toxicity. Nutritional status is one of the critical factors that may influence arsenic toxicity and its adverse health effects. For the amelioration/remediation of toxicological effects of the arsenic, natural products obtained from plants are being positively used by researchers. Plant products without any chemical treatment have active ingredients, which may be effective against arsenic without causing any harmful adverse health effects. In the direction of preventive efforts, the present chapter attempts to provide comprehensive details of various nutraceuticals including antioxidants like vitamins, flavonoids, essential metals, and also functional food like jaggery (a natural product of sugarcane, rich in protein, carbohydrate, iron, and vitamins, especially vitamin C) and honey in preventing arsenic toxicity. We have also provided readers the available information, if these nutraceuticals can be administered during chelation treatment to provide a better therapeutic benefit against arsenic toxicity.

It is now well known that arsenic causes their toxicity by the involvement of ROS. These metals bind to macromolecules and produce different free radicals that in turn attack the building blocks of the biological systems. Thus to counteract the arsenic-induced toxic manifestations, it becomes crucial to maintain nutritional health; an effective option to minimize arsenic burden and reducing altered biochemical variables. Overload or deficiency of essential trace elements and other dietary nutrients assist arsenic absorption and supplementation of such nutrients ameliorates the toxicity. In addition to the role of micronutrients in modifying arsenic toxicity, these nutritional components can also act as adjuvant increasing the efficacy of a known chelator or by acting independently. Few reports suggest that arsenic exposure has led to the alteration in metabolism of the body stores of some essential elements and abolish the anticarcinogenic effects like selenium (Se) and copper [26,27]. Arsenic is known to induce metallothionein (MT) [28], a low-molecular-weight cysteine-rich metal-binding protein. This implies that arsenite can be detoxified by MT. Dietary antioxidants such as vitamin E, vitamin C, and vitamin A also play beneficial role in alleviating toxic effects of arsenic [22,29] (see Figure 15.4).

15.3.1 NUTRACEUTICALS AS PREVENTIVE MEASURES

15.3.1.1 Essential Metals

Micronutrients are elements required in minute quantity of less than few milligrams per day by the body, for example, iron, iodine, fluorine, zinc (Zn), copper, cobalt, chromium, manganese, molybdenum, selenium, nickel, tin, silicon, and vanadium. They also represent a class of inorganic substance and exist naturally in a variety of food. The human body needs different types of micronutrients in order to function properly; for example, a variety of metabolic and physiologic processes in the human body require micronutrients. Insufficiencies of several essential elements have been shown to aggravate the toxic effects of arsenic and supplementation of

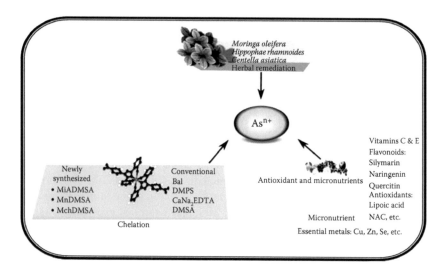

FIGURE 15.4 Preventive and therapeutic strategies against arsenic toxicity.

such micronutrients/essential metals ameliorates the toxicity. Role of essential metals in reducing the toxic burden is widespread, and supports the conclusion that ingestion of micronutrients like calcium, iron, zinc, etc., may lower the metal toxicity while it has been seen that animals raised on a diet low in these metals have a much higher metal concentration [30]. Few important micronutrients found to be effective in eliciting arsenic-induced disorders are mentioned in the following.

15.3.1.1.1 Selenium

Selenium is required as a dietary element for health and is essential for the function of various selenoenzymes, and selenoproteins acquire antioxidant property but they also possess toxic effects on excess intake. The recommended dietary allowance of selenium is 70 µg/day for adults [31]. It is an integral component of glutathione peroxidase (GPx), an antioxidant enzyme. GPx is a first established selenoenzyme and is known to be inhibited by arsenic [32]. Selenoproteins and their metabolites are critical in maintaining antioxidant/anti-inflammatory homeostasis. Recent studies suggest that thioredoxin reductase (TrxR) in liver tissue and type I iodothyronine deiodinase are potent inhibitors of arsenite [33]. The toxic effects of arsenic on these antioxidant selenoenzymes could augment oxidative stress and lead to the abnormal functioning of tissues. Selenium together with vitamin, superoxide dismutase (SOD), and catalase (CAT) neutralizes free radicals. The major important metabolic role of selenium is due to its function in the active site of thioredoxin reductase and GPx antioxidant enzymes [34,35]. Several examinations in human populations recommend a possible association between arsenic sensitivity and Se nutrition. Wuyi et al. [36] elucidated progression in arsenic-associated symptoms in an Inner Mongolian population, with a dietary supplementation of Se-enriched yeast. *Keshan Disease* is an fatal endemic cardiomyopathy found primarily in children and the only disease caused due to selenium deficiency. Severe chronic degenerative diseases in humans could be prevented by selenium supplementation when used alone or in combination [37,38]. Patients suffering from chronic arsenicosis had lower Se levels than the controls in a Taiwan population [39]. These observations imply enhanced arsenic toxicity due to low selenium intake.

15.3.1.1.1.1 Mechanism of Action   Arsenic and selenium are known metalloids with almost comparable chemical properties. Selenium and arsenic interaction would be of practical significance because the consumption of animal food, which is a major source of dietary Se, is often low in developing countries, where population is exposed to arsenicosis [40]. Selenite (Se^{4+}) or selenate (Se^{6+}), inorganic forms of selenium are reduced by glutathione (GSH) to yield selenodiglutathione (GS–Se–SG), which is further converted to hydrogen selenide (H_2Se) catalyzed by glutathione reductase. Hydrogen selenide acts as an intermediary metabolite that either serves as a precursor for the synthesis of selenocysteine or is methylated to methylselenol, dimethylselenide, and trimethylselenonium cation. Berry and Galle [41] demonstrated that Se reduced arsenic toxicity via the formation of a selenide precipitate (As_2Se) that is deposited into tissues. It has also been revealed that higher dietary Se intake may lower the risk of arsenic-associated skin lesions. Selenium promotes biliary excretion of exogenous selenium and selenite that enhances the excretion of arsenic into bile [42,43]. These observations demonstrated that arsenic augments the hepatobiliary transport of selenium and facilitates accumulation of selenium in red blood cells, which also facilitates the biliary excretion of arsenic. Glattre et al. [44] revealed interaction, accumulation, and distribution of selenium and arsenic in rat thyroid tissue. There might be competition for binding between selenium and arsenic with functional proteins, active tissue sites, and bioligands, and thus leads to the formation of metal selenide, a reversible compound that may be a possible mechanism for the reduction of available *free* arsenic ions in the body. In addition selenium may also play a role in arsenic elimination by forming a selenium–arsenic conjugate in the liver prior to excretion into the bile. Hence, this finding evaluates the hypothesis that Se facilitates the biliary elimination of As, possibly via Se–As conjugate formation using glutathione complex [45,46]. GPx is an Se-dependent enzyme and effect of arsenic on synthesis and activity of selenoenzymes has been previously reported. Arsenic may directly interact with Se and form insoluble and inactive As–Se complex [42,47] rendering it unavailable and ultimately resulting in the inhibition of GPx activity or alter the expression and synthesis of selenoproteins like GPx [48,49].

15.3.1.1.2   Zinc
Zinc is a ubiquitous trace element present in more than 70 different enzymes that function in cellular metabolism, metabolism of various macromolecules (proteins, lipids, and carbohydrates), muscle energy production, and protein synthesis. The adult human body contains approximately 1.5–2.5 g of zinc [50]. When dietary zinc is increased over requirement level, it reduces trace metal absorption. Zinc maintains insulin and blood glucose concentration and plays an essential role in the maintenance and development of the body's immune system. Zinc deficiency is more prevalent in the Middle East and may lead to retarded growth,

hypogonadism, loss of appetite, dermatitis, reduced taste acuity, delayed wound healing, impaired reproduction, altered normal cell metabolism, and poor immune function.

15.3.1.1.2.1 Mechanism of Action   Zinc as an antioxidant functions in the protection of thiol groups of proteins like DNA zinc-binding proteins (zinc fingers) against free radical attack and reduction of free radical formation through antagonism of redox-active transition metals (iron and copper) [51]. Zinc also acts as an inhibitor of nicotinamide adenine dinucleotide phosphate (NADPH) oxidase; inducer of MT ROS activates NF-κB, growth factors, antiapoptotic molecules, inflammatory cytokines, and adhesion molecules [52]. Zinc also reduces inflammatory cytokine production by upregulating zinc-finger protein, which inhibits NF-κB activation via TNF receptor associated factors (TRAF) pathway [53]. Thus zinc also functions as an anti-inflammatory agent besides an antioxidant. In vivo studies suggest that long-term absence of zinc makes an organism more prone to oxidative stress-induced injury including atherosclerosis, cancers, neurological disorders, and autoimmune diseases. It has been well known that arsenic exerts its toxicity by inhibiting ALAD activity (Zn-dependent enzyme) and causing oxidative stress condition by hindering SOD activity (a Cu/Zn-dependent enzyme). Thymidine kinase, deoxyribonucleic acid (DNA), ribonucleic acid (RNA), and polymerase are zinc-dependent enzymes that act as a catalysts in the replication and transcription of DNA during cell division [54]. Various studies suggest Zn coadministration with chelating agent as a therapeutic measure against arsenic toxicity [43,55]. Zn may alter glutathione synthesis, which is believed to be an essential cofactor for arsenic methylation. Biotransformation of arsenic is Zn dependent as it serves as a cofactor of *S*-adenosylmethionine transferase enzyme.

MT is a metal-binding protein high in cysteine-rich residues containing absence of aromatic amino acids or disulfide bonds, which is capable of chelating arsenic from in vivo sites. Maret [56] has reported that the MTs signifies a connection between cellular zinc and the redox status of the cell. Increased oxidative stress alters cellular redox homeostasis causing the release of zinc from MT as a result of sulfide/disulfide exchange. MT can bind seven atoms of zinc per molecule, which might act as a reservoir for utilization in times of zinc deficiency. Zinc exerts its antioxidant property by occupying iron (Fe)- or copper (Cu)-binding sites in DNA, proteins and lipids as Fe and Cu are considered to be redox-active metals that participate in electron transfer reactions with the consequent production of ROS capable of peroxidizing cell components. Zinc preferentially binds to negatively charged phospholipids, thereby, preventing their binding with redox-active metals. Enhanced zinc also increases the renal and hepatic contents of MT.

15.3.1.1.3 Copper

Copper, an essential trace element is a cofactor of many enzymatic reactions like cytochrome c oxidase, ascorbate oxidase, or SOD. Besides its enzymatic roles, copper is also a component of mitochondrial electron transport chain that functions in iron absorption and mobilization [8]. Azurin and plastocyanin are the blue copper proteins that participate in electron transport. Copper depletion leads to decreased potential of cells to produce SOD. It is an integral part of ceruloplasmin and is essential for the intestinal absorption of iron. Ceruloplasmin contains about 95% of the copper found in serum. Copper deficiency is coupled with reduced hemoglobin and elastin formation and abnormal amino oxidase activity. Dietary copper supplementation lowers the mortality rate and severity of anemia in in vivo study and high arsenic intake can affect copper metabolism [57].

15.3.1.1.3.1 Mechanism of Action   Copper is readily absorbed from the diet and stored in the liver. Copper is bound to serum albumin or histidine and distributed through the bloodstream for delivery to tissues or storage in the liver. After absorption, copper reaches MT pool, transported to the mitochondria for cytochrome c oxidase absorption or for delivery to Cu/Zn-SOD [58]. It has been revealed that arsenic exposure reduced copper concentrations in liver, plasma, and blood cells but elevates copper concentration in the kidney in rats. It might be due to the formation of arsenical-copper complex, which results in the accumulation of copper in the renal cortex because of altered glomerular filtration or tubular reabsorption of copper [59].

15.3.1.1.4   Iron
Iron is essential for various cellular activities, and optimum concentrations of iron in body are appreciated for cell survival. It is an integral part of many proteins and 65% of iron is bound to hemoglobin, myoglobin (10%), cytochromes, iron containing enzymes and iron storage proteins like ferritin (25%) involved in the transport and metabolism of oxygen [60]. Lesser amount of body iron circulates in the plasma as an exchangeable pool and is bound to transferrin. Iron deficiency limits oxygen supply to cells, resulting in symptoms like fatigue, poor work performance, and decreased immunity [61], whereas, in excess it may lead to severe toxicity and even cell death [61,62]. Interestingly, iron load can also get increased by consuming citrus fruits rich in ascorbic acid, which facilitates the absorption of nonheme iron in duodenum. Ingestion of food cooked in iron vessels, as well as consumption of iron-containing vitamins or nutritional supplements may also account in the loading of excess of iron in body. Arsenic poisoning depletes iron level gradually in the body, which usually begins with a negative iron balance, when iron intake does not meet the daily requirement for dietary iron [63]. This negative balance initially lowers the storage form of iron while the blood hemoglobin level, a marker of iron status, remains normal. Anemia caused by iron deficiency is an advanced stage of iron depletion. It occurs when storage sites of iron are deficient and blood levels of iron cannot meet daily needs. Blood hemoglobin levels are below normal with iron deficiency anemia [64]. A supplement of 400 ppm iron as ferrous sulfate had a marked effect in preventing the low hemoglobin level produced by metal poisoning.

15.3.1.1.4.1   Mechanism   The body's redox status is mainly dependent on an iron (and copper) redox couple [65]. The redox-couple mechanisms prevent excessive iron absorption in the proximal intestine and regulate the recycling process with rate of iron release. Cellular iron that is unused by other ferroproteins accumulates in ferritin; however, its iron-binding ability is limited [50,66]. Iron regulates the oxidative processes because it is required by cells for survival, but its overload in the body is detrimental leading the generation of ROS through the Fenton reaction.

15.3.1.2 Antioxidants

Antioxidants are enzymatic or nonenzymatic substances that inhibit or delay the oxidation of a substrate while present in minute amounts. Nutritional antioxidants act through different mechanisms and in different compartments, but are mainly free radical scavengers: They directly scavenge free radicals, lessen the peroxide concentrations, and repair oxidized membranes. These enzymes regulate the glutathione redox status of the cell for maintaining critical balance. They quench iron to decrease ROS production; via lipid metabolism, short-chain free fatty acids and cholesteryl esters neutralize ROS [67]. Typical natural antioxidants include tocopherol, ascorbic acid, flavonoids, quercetin, carotene, cinnamic acid, peptides, and phenolic compounds. Antioxidants are broadly classified into the following categories explained in Table 15.1 [68].

15.3.1.2.1 Glutathione

Glutathione is a nonenzymatic and a major thiol antioxidant. It is one of the most ubiquitous tripeptide, comprises three important amino acids glutamine, cysteine, and glycine considered as the major thiol-disulfide redox buffer of the cell. Glutathione defines its activity by the presence of sulfhydryl group on cysteine residue and glutamyl linkage through which antioxidant activity exists as diverse defense pathways. Glutathione employs glutathione-S-transferase (GST) and GPx in the detoxification of various electrophilic compounds, peroxides, and various biological reactions such as glyoxalase system, reduction of ribonucleotides to deoxyribonucleotides, etc. [69]. Deficiency of glutathione leads the cellular system at risk of oxidative damage. This tripeptide exists either in oxidized form (GSSG) or reduced form (GSH) intracellularly and maintenance of GSH:GSSG ratio in the

TABLE 15.1
Kinetically Based Classification of Antioxidants

No.	Antioxidant Category	Examples
1.	Antioxidants that break chains by reacting with peroxyl radicals having weak O–H or N–H bonds	Phenol, naphthol, hydroquinone, aromatic amines, and aminophenols
2.	Antioxidants that split chains by reacting with alkyl radicals	Quinones, nitrones, and iminoquinones
3.	Hydroperoxide decomposing antioxidants	Sulfide, phosphide, and thiophosphate
4.	Antioxidants with metal deactivation	Diamines, hydroxyl acids, and bifunctional compounds
5.	Antioxidants with cyclic chain termination	Aromatic amines, nitroxyl radical, and variable valence metal compounds
6.	Synergism among several antioxidants	Phenol sulfide; phenolic group reacts with peroxyl radical and sulfide group with hydroperoxide

body is a priority for cell survival. Glutathione has the ability to regenerate important antioxidants such as vitamin C and E, which is linked to the redox status of GSH:GSSG and are necessary for DNA repair and gene expression [70]. The sulfhydryl group of cysteine donates electron to ROS in the reduced form. During this process, GSH become highly reactive, but readily binds with another reactive GSH to form GSSG. Glutathione-related enzymes like GPx and GR recycles the conversion of GSH into GSSG and regeneration of GSH from GSSG at the expense of NADPH. GPx also converts lipid peroxides (LPs) along with hydrogen. Activities of GPx and GR are strictly regulated with the changes in GSH level. Arsenic accumulates the oxidized glutathione inside the cells and then the GSH:GSSG ration becomes a good measure of oxidative stress. GSH has the ability to chelate transition metal ions, thereby reducing their toxic effects [68,71]. Glutathione administered orally helps in replenishing the status of GSH in both animals and humans [72]. Arsenic toxicity has been documented to alter glutathione-linked pathway. Its deficiency leads to be implicated in several pathophysiological disorders. Thus, glutathione has been considered a direct and critical biomarker for the oxidative stress–induced disorders [73].

15.3.1.2.2 Vitamin C (Ascorbic Acid)

It is powerful and water soluble antioxidant function in aqueous environment of the body. They are required in tiny amounts and are classified on the basis of their biological and chemical activity. It protects low-density lipoprotein cholesterol (LDL-C), cancer, and effects of aging against free radical damage. Vitamins are necessary for body not only to maintain normal metabolic functions but also for the prevention of pathological diseases. They also function as cofactors or coenzymes in various enzymatic reactions' dietary needs of vitamin C increases due to smoking, alcohol, and antibiotics. Vitamin C together with vitamin E regenerates α-tocopherol from α-tocopherol radicals in lipoproteins and biomembrane. Recent studies elucidated that vitamin C ameliorated arsenic-induced toxicity [74]. Vitamin C plays both a preventive and therapeutic role in a number of pathologies including cancer, atherosclerosis, and viral infections. Ascorbic acid act as a vinylogous carboxylic acid and its double bond transmits electron pairs between the hydroxyl and the carbonyl. ROS oxidize ascorbate to monodehydroascorbate and dehydroascorbate and then reduced to water, while the oxidized forms of ascorbate are relatively nonreactive and do not cause cellular toxicity. Numerous studies suggested that vitamin c may influence apoptosis, gene expression, and many biological functions [68,75]. Studies also showed a reduction in markers of oxidative DNA, lipid, and protein damage after supplementation with vitamin C and in vitro and ex vivo studies revealed dose-dependent potential of vitamin C in providing resistance to lipid peroxidation [76]. The remarkable effects of vitamin C with *meso*-2,3-dimercaptosuccinic acid (DMSA) or its monoisoamyl derivative (MiADMSA) inhibited hemebiosynthesis pathway and reduced the arsenic-induced toxic effects. Combination therapy of vitamin C and MiADMSA was effective in reducing hepatic and renal arsenic burden, which justifies that vitamin C acts as an antioxidant by the formation of soluble complex. Cell transports and accumulates vitamin C through the facilitative glucose transporters that transport dehydroascorbic acid (DHA), the oxidized

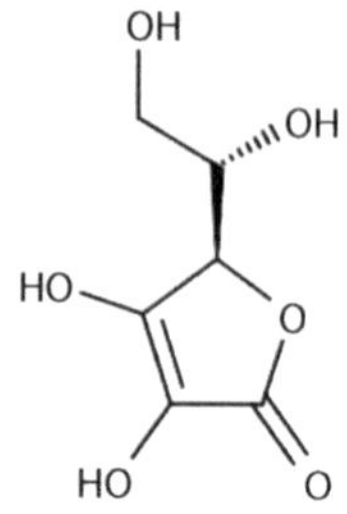

FIGURE 15.5 Chemical structure of vitamin C.

form of vitamin C, and sodium-dependent ascorbic acid transporters that transport the reduced form of vitamin C, ascorbic acid that are widely expressed at mRNA level; however, only specialized cells appear to transport vitamin C directly [68,77] (see Figure 15.5).

15.3.1.2.3 Vitamin E (α-Tocopherol)

Vitamin E is a collective name for a group of eight fat soluble compounds; tocopherol and tocotrienol, of which the most abundant form is α-tocopherol. The recommended dietary allowance of vitamin E for adult male and female is 10 and 8 mg, respectively. It breaks the propagation of reactive species chain reaction in the lipids of biological membranes [67]. It is known to be membrane-bound antioxidant that provides protection against lipid peroxidation and also against cancer, cardiovascular diseases, and diabetes. Vitamin E also maintains membrane fluidity and ion transport to preserve cell membrane. Recent evidences suggest that α-tocopherol and ascorbic acid function together in which α-tocopherol is converted to an α-tocopherol radical that can again be reduced to the original α-tocopherol form by ascorbic acid. It exerts its antioxidant effects by cooperating with vitamin C, β-carotene and selenium. The free hydroxyl group on the aromatic ring of vitamin E is responsible for its antioxidant properties. The hydrogen ion from this group is donated to the free radical, resulting in a stable free radical form of the vitamin. Vitamin E administration has been shown to restore various altered biochemical variables indicative of heme-biosynthesis pathway and oxidative stress in blood and soft tissues of Swiss albino mice [20,78]. Vitamin has the potential to prevent cellular damage by maintaining thiol group of membranous proteins and by scavenging free radicals. Vitamin E supplementation has been reported against arsenic toxicity mainly due to its antioxidative property and its location in the biological membrane to stabilize against ROS. Coadministration of vitamin E with arsenic showed increased antioxidant enzymatic activity and decreased lipid peroxidation in experimental animals [67]. Thus, nutrient supplementation of vitamins strengthens the body defense system by removing the toxic burden (see Figure 15.6).

15.3.1.2.4 Captopril

Captopril is a potent antioxidant with a potential of angiotensin-converting enzyme (ACE) inhibitor and its ability to contribute in the formation of thiols. Numerous

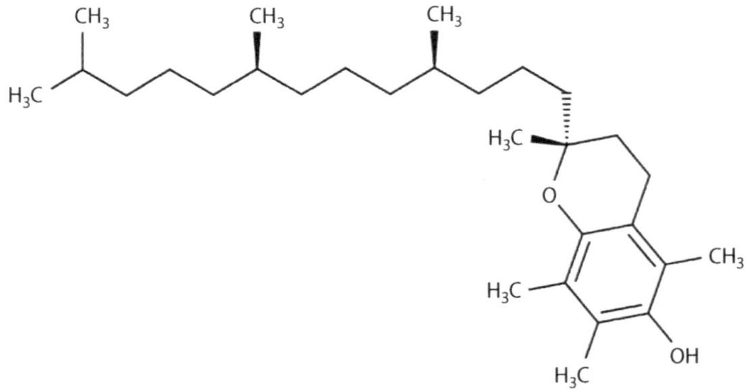

FIGURE 15.6 Chemical structure of vitamin E.

studies suggest that captopril may block the conversion of angiotensin I to angiotensin II and may also stimulate prostaglandins, bradykinin, and eNOS synthesis (inhibition of superoxide formation) [79]. Captopril functions as an antioxidant and metal chelator by quenching ROS and restoring the activities of antioxidant enzymes like SOD and GPx. Thus, the binding of captopril with toxic metal ion greatly depend upon the pH and metal/captopril ratio [80]. Captopril is also effective against type II diabetes, lowering diastolic and systolic blood pressure, and also provides protection against nephropathy.

Captopril is absorbed through the mucosa and is converted into mixed disulfides with endogenous thiol compounds [79]. Carboxyl, amidic, and mercapto are the three functional groups of captopril coordinated to soft metal ions through thiol-donating group. Captopril due to single thiol moiety quenches single free radical by donating one electron at a time. Captopril with GSH sulfhydryl nucleophile can scavenge ROS by the utilization of GSH in detoxification reactions [73,81]. Captopril exists in *cis* and *trans* isomeric forms and a similar ratio of *cis* to *trans* isomer present at physiological pH [82]. Bhatt and Flora [83] demonstrated that captopril effectively reduced the levels of lipid peroxidation in various tissues, thereby reducing the oxidative stress induced by gallium arsenide toxicity. There are various reports that support the efficacy of captopril in restoring antioxidant level against arsenic-induced toxic manifestations through the generation of oxidative stress and cellular injury to biomolecules [84]. These results can be attributed to the free radical scavenging activity of captopril [85]. Wojakowski et al. [86] have also reported a potent antioxidant nature of captopril against oxidative cellular damage (Figure 15.7).

15.3.1.2.5 Melatonin

N-acetyl-5-methoxytryptamine is an indole amine and tryptophan derivative commonly recognized as melatonin. It is mainly produced in the pineal gland and retina besides bacterial and unicellular organisms. It is identified as a potent *chronobiotic* regulating normal circadian rhythms. Endogenously synthesized melatonin releases quickly into the bloodstream and reaches CSF, saliva, and bile while orally

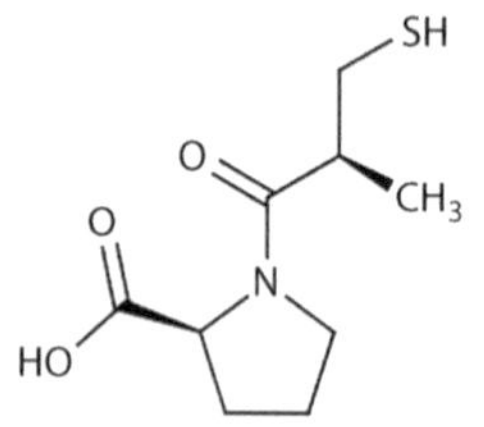

FIGURE 15.7 Chemical structure of captopril.

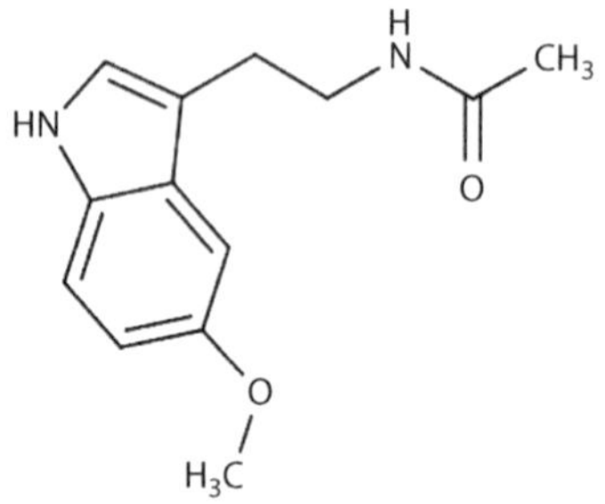

FIGURE 15.8 Chemical structure of melatonin.

ingested exogenous melatonin gets rapidly absorbed and serum levels becomes visible after 60–150 min. Melatonin detoxifies free radicals and highly reactive species derived from the oxygen and nitrogen such as hydroxyl radicals, hydrogen peroxide, peroxynitrite anions, etc. Melatonin also indirectly acts as an antioxidant by either increasing the translational levels or other antioxidant enzymes activity. Ascorbate and urate recycles melatonin that further accentuates the significance of melatonin as a free-radical quencher and an effective antioxidant [73,87] (described in Figure 15.8).

The presence of aromatic indole ring formulates melatonin to behave as a potent electron donor, which helps in repairing electrophilic radicals. Due to indole moiety, the reactivity and selectivity of melatonin is strongly affected. Free radicals, such as •OH and LOO•, may abstract a single electron from melatonin, giving rise to a melatonyl cation radical which, in turn, could donate a second electron to $O2^-$• to produce N-acetyl-N-formyl-5-methoxykynuramine (AFMK). Melatonin on reaction with H_2O_2, a precursor of OH radical, leads to oxidative cleavage of the indole ring, which is also an effective free radical scavenger. Melatonin also augments the activity of glutathione-linked enzymes that metabolize the free radicals into harmless agents. Melatonin stimulates the activity of GPx that results in the rapid conversion of hydroperoxides into water and oxygen. Additionally, melatonin also stimulates the activity of glucose-6-phosphate dehydrogenase (G-6-PD), imperatively providing reducing equivalents like NADPH for the action of GRd. Melatonin associates with quinone reductase 2, considered an important melatonin receptor for the action of detoxification of prooxidant quinines [88]. Synergistic action of melatonin with vitamin C and E shows protection from oxidative stress and also cross BBB.

The antioxidant property of melatonin is attributed to its radical scavenging activity, GSH-Px activating property, and regulatory activity for organ function. AFMK is a metabolic product of melatonin known to reduce the DNA damage [89]. However, melatonin restricts its protection in every organ due to the fact that it works through receptor-mediated mechanisms [90]. Despite its receptor-based mechanism, melatonin shows beneficial role in several studies where unwarranted free radical generation occurs. Thus, supplementation of melatonin should be employed in pathological disorders in the case of oxidative damage [91].

15.3.1.2.6 Carotenoids

Carotenoids are naturally occurring pigments such as β-carotene, lycopene, and astaxanthin, found in plants and microorganisms. Carotenoids possess antioxidant properties and inhibit various toxicological manifestations such as cancers, arthrosclerosis, age-related muscular degeneration, etc. [92]. Dietary carotenoids are present as microcomponents in fruits and vegetables as the key sources. All carotenoids also possess polyisoprenoid structure known as poliene system, joined by conjugated double bonds. These conjugated double bonds delocalize unpaired electrons responsible for the excellent potential of α-carotene to scavenge singlet oxygen without degradation and for the chemical reactivity of α-carotene with free radicals such as the peroxyl and superoxide radicals [68,93] (see Figure 15.9).

Carotenoids exert its antioxidant effects in lipid phase by scavenging free radicals and accumulate in lipophilic compartments like biomembranes or lipoproteins. Carotenoids may act as antioxidants without losing its structure and the reactions between carotenoids and free radicals such as lipid peroxidation causes further radical reactions. The lipophilicity of carotenoids also affects their absorption, distribution, and excretion in the body. Upon carotenoids ingestion, they get absorbed into the GI tract and are found unchanged in circulation and tissues. The major site of carotenoids accumulation in the body is adipose tissue [73]. The biological activity of carotenoids is mainly due to its capacity to form vitamin A within the body [94]. Excessive consumption of carotenoid-rich diet is associated with a minimal risk for several toxic insult such as various types of cancer, cardiovascular or ophthalmological diseases, etc. [95]. Sindhu et al. [96] recently reported that carotenoids exhibit antioxidant property in both in vivo and in vitro model.

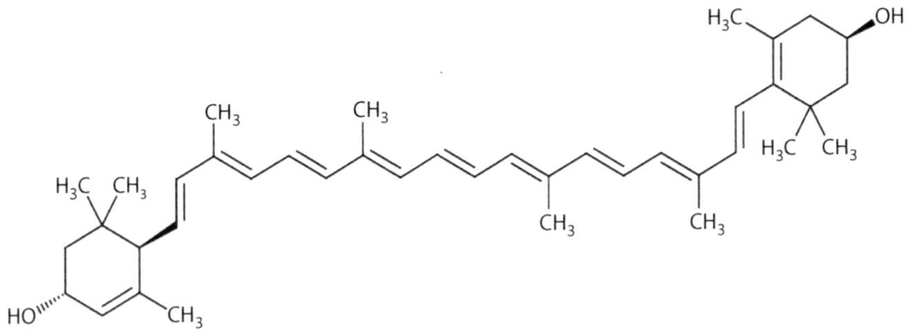

FIGURE 15.9 Chemical structure of carotenoids.

15.3.1.2.7 Taurine

Taurine (2-aminoethanesulfonic acid), is a nonessential sulfur-containing amino acid synthesized chiefly in pancreas via cysteine sulfinic pathway. It functions with glycine and γ-amino butyric acid (neuroinhibitory transmitter) and is present in several tissues like muscles, brain, heart, etc. It maintains the level of cysteine, which is an important precursor of glutathione. High water solubility and low lipophilicity of taurine represents its zwitter-ionic nature that may be attributed to the presence of sulfonate group in taurine, a strong acid that makes it completely zwitter ionic over the physiological pH range [68]. Taurine maintains various physiological functions including calcium homeostasis, bile acid conjugation, membrane stabilization, strengthening of cardiac contractility, and also crosses blood–brain barrier (BBB). Taurine increases the overexpression of peroxyredoxin-1, thioredoxin-1, and heme oxygenase, thereby depleting ROS generation. Taurine exhibit antioxidant property due to free radical quenching activity and reduces peroxidation of unsaturated membrane lipids due to oxidative stress. Taurine prevents ion leakage, water influx, and cell swelling and alters phospholipids methyltransferase enzyme activity that regulates the activity of phosphatidylethanolamine (PE) and phosphatidylcholine (PC) (constituents of membrane) due to its membrane-protective effects induced by toxic metals. Taurine has been used in the treatment of hepatic disorders, cystic fibrosis; Alzheimer's disease; cardiovascular, seizure, and ocular disorders; inflammation; and arrhythmia. It is a potent neuroprotectant and thus prevents neural oxidative damage. Administration of taurine alone significantly reduced hepatic oxidative stress in arsenic-induced toxicity in rats. Coadministration of taurine with MiADMSA led to more beneficial recovery from oxidative injury and reduction in arsenic burden facilitating the entry of chelator to the intracellular sites and thereby reducing arsenic concentration [97]. Ghosh et al. [98] recently suggested that taurine prevents arsenic-induced cardiac oxidative stress and apoptotic damage via the role of NF-κB, p38, and JNK MAPK pathway. Thus, ample evidences validated that taurine supplementation may have a beneficial property in oxidative stress condition through direct scavenging of ROS, stabilization of membrane ATPase, and inhibition of lipid peroxidation in arsenic-induced toxicity (see Figure 15.10).

15.3.1.2.8 Gallic Acid

Gallic acid (GA) is a natural antioxidant and secondary polyphenolic metabolites. It is mainly present in gallnuts, sumac, tea leaves, oak bark, etc., and is used as an

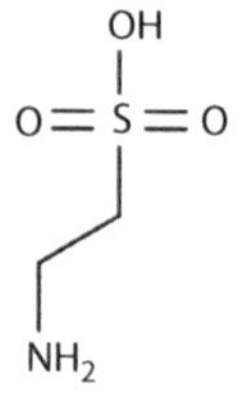

FIGURE 15.10 Chemical structure of taurine.

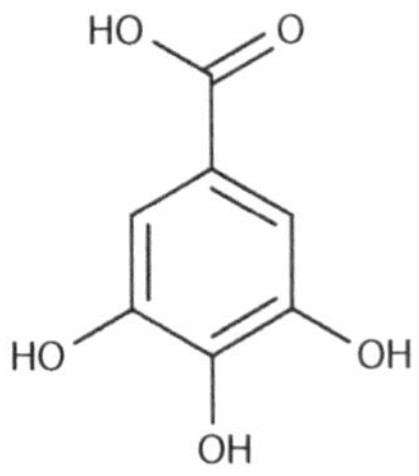

FIGURE 15.11   Chemical structure of gallic acid.

indicator for adulteration in fruit juices and alcoholic beverages. Alkyl gallate, alkyl derivative of gallic acid possess antioxidant activity and may suppress the Fenton reaction derived by superoxide radical that leads to the generation of ROS. Both gallic acid and alkyl gallate exhibit chelating properties and also reported to activate microsomal glutathione *S*-transferase through oxidative modification of the enzyme [99]. Gallic acid behaves as a potent peroxynitrite scavenger and exhibit antiproliferative activity. Gallic acid not only depends on its capacity to scavenge free radicals, but also depends on its hydrophobic property that allows it to cross biological membranes, which may be attributed to the relation between hydrophobicity and free radical scavenging activity of an antioxidant. In addition, gallic acid also exhibit dual nature as, in the presence of H_2O_2, it acts as an antioxidant and inhibit the generation of oxidizing products, whereas in the absence of H_2O_2, it acts as an oxidizing compound [100]. Gallic acid blocks the activity of tyrosine due to its phenolic structure and number of hydroxyl groups attached to the ring. Thus gallic acid exhibits its protective efficacy as ROS and RNS scavenger and chelating agent, suppresses Fenton reaction, and activates GST enzyme against arsenic-induced oxidative damage (described in Figure 15.11).

15.3.1.2.9   *Alpha-Lipoic Acid*

Alpha-lipoic acid (α-LA) (1,2-dithione-3-pentanoic acid) is a thiol-containing antioxidant with metal-chelating. Lipoic acid is a powerful antioxidant with the potential to be active in both lipids as well as in aqueous phase. α-LA is a common dietary supplement widely used in the prevention of various pathological diseases. Its main function is to increase the production of glutathione, which helps to detoxify various toxic substances in the body. Dihydrolipoic acid (DHLA) is the reduced form of lipoic acid. The molecule consists of a carboxylic and a cyclic disulfide groups. Due to an asymmetric carbon having four different attached groups, lipoic acid exists in R-enantiomer and S-enantiomer. Naturally occurring lipoic acid is the R-form, but synthetic lipoic acid (known as α-lipoic acid) is a racemic mixture of both R-form and S-form. It functions as a coenzyme in the Kreb's cycle and in the production of cellular energy (ATP) [101]. α-LA is a unique biological antioxidant because it retains protective functions in both of its reduced and oxidized forms [102] and is amphipathic in nature and may act as an antioxidant both in hydrophilic and lipophilic environments. Lipoic acid readily crosses BBB and then passes easily into the brain. It is thought to protect brain and nerve tissue by preventing free radical damage as it is taken up by all areas of the central and peripheral nervous system [103]. α-LA may also be effective in both prevention and treatment of oxidative stress in a number of models or clinical conditions, including ischemia reperfusion injury, diabetes, aging, and neurodegenerative diseases. LA is active against hydroxyl radicals and hypochlorous acid, but not against hydrogen peroxide or singlet oxygen. In general, DHLA has superior antioxidant activity than LA. By donating two hydrogen atoms, DHLA can neutralize free radicals without itself becoming a free radical. Prevention of oxygen-based damage to nerves is also a key area of clinical research on the possible use of lipoic acid. Because of its twofold interactions with both water-soluble (vitamin C) and fat-soluble (vitamin E) substances, lipoic acid has been shown to prevent deficiency of both vitamins in both human and animal studies. Other antioxidants seem to benefit equally by the presence of lipoic acid and these antioxidants include coenzyme Q, glutathione, and NADH. Lipoic acid is found in almost all food items, but slightly more so in kidney, heart, liver, spinach, broccoli, and yeast extract. Naturally occurring lipoic acid is always covalently bound and not immediately available from dietary sources. Lipoic acid is considered a superlative scavenger because of the five-membered ring in the intramolecular disulfide [104] (Figure 15.12).

The probable mechanisms through which LA shows beneficial effects against oxidative stress induced by toxic agent might be the DHLA, reduced form of LA. NADH reduces LA to DHLA, which is a strong antioxidant, to quench ROS and regenerates endogenous antioxidants including vitamin E, C, and glutathione. DHLA possess strong chelating properties to prevent free radical generation thus to reduce oxidant attacks on biomolecules [68]. LA is also the key cofactor of pyruvate dehydrogenase and α-ketoglutaric dehydrogenase enzymes and a supplementation of required LA helps to potentiate the enzymatic activities, thereby promoting and ameliorating oxidative phosphorylation and mitochondrial respiration. However, studies have shown that lipoic acid exhibit prominent antioxidizing and chelating properties that quench ROS, regenerate endogenous antioxidants, and repair oxidative damage [105]. DHLA is the reduced form of lipoic acid and is more potent than its oxidized version with respect to chelate metals and scavenging ROS [20,83]. Moreover, the ability to regenerate endogenous antioxidants and to repair oxidative damage are some of the exclusive properties of LA that makes it a good candidate for the preventive and clinical treatment of various pathophysiologies associated with arsenic toxicity.

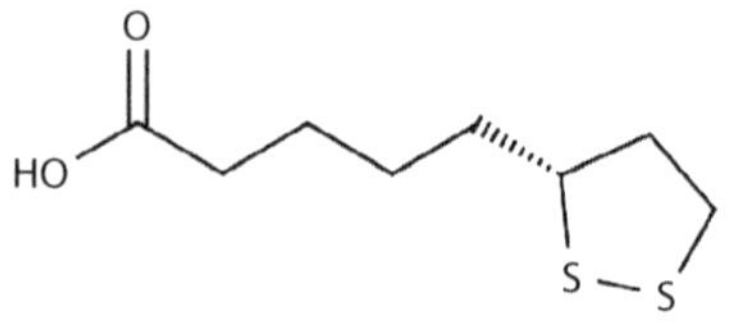

FIGURE 15.12   Chemical structure of alpha-lipoic acid.

15.3.1.2.10 N-Acetyl Cysteine

N-acetyl cysteine (NAC) is a thiol antioxidant containing acetylated variant of the cysteine residue. It is a precursor of both naturally occurring cysteine and reduced glutathione and thereby maintains intracellular GSH levels and scavenges ROS by providing the substrate for the rate-limiting step in GSH synthesis. NAC possess clinical applications against paracetamol intoxication and exhibit mucolytic ability. Isomeric forms of NAC are L and D form, of which L-NAC is active and metabolized into cysteine and GSH. It is rapidly absorbed through ingestion and undergoes first pass metabolism leading to the formation of several metabolites capable of stimulating GSH synthesis and promoting free radical detoxification. The presence of thiol group in NAC is responsible for its metabolic activity, while the acetyl-substituted amino group stabilizes the molecule against oxidation. NAC is a known metal chelator attributed by the presence of thiol groups to reduce free ROS and provide chelating site for metals. Thus, NAC has a strong ability to reverse the altered prooxidant/antioxidant balance in arsenic poisoning. NAC can readily cross the biological membrane and provide intracellular effects. Coadministration of NAC and succimer against arsenic exposure led to altered biochemical variables indicative of oxidative stress and removal of arsenic from the soft tissues. Thus the more pronounced effects of combination therapy than monotherapy could be attributed to the fact that NAC, a thiol antioxidant, provided an additional binding site for arsenic. This also led to more effective reduction of toxic effects. Recent studies suggest the preventive efficacy of Zn and NAC coadministration against arsenic-induced oxidative stress. NAC also reduced arsenic-induced liver cell injury in mice associated with the induction of oxidative stress, the perturbations in the mitochondrial redox state, and arsenic-induced apoptosis of hepatocytes [106,107] (see Figure 15.13).

The mechanism of action by which NAC exerts its antioxidant potential is the replenishment of glutathione. Additional cysteine transport may increase the GSH level in the case of decreased and increased demand of GSH despite the fact that it is impossible to administer the active form of cysteine because of its low intestinal absorption, poor water solubility, and rapid hepatic metabolism [108,109]. However, the acetyl radical of NAC linked to amine function eradicates these limitations. Since glutathione is an endogenous antioxidant and has the potential to quench ROS and RNS, it thereby prevents oxidative damage [110]. Moreover, its nucleophilic nature also entraps free radicals leading to mutagenic effects and the inhibition of nitrosation reactions. NAC does not alter the tissue distribution of micronutrients. Clinically, NAC is effectually used in various pathological conditions such as

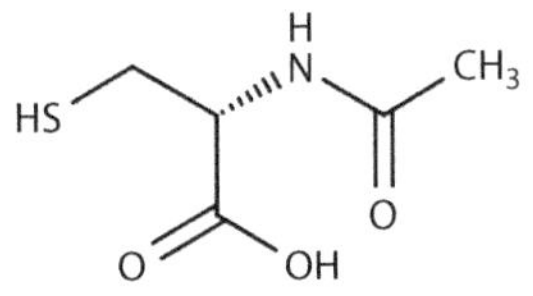

FIGURE 15.13 Chemical structure of *N*-acetylcysteine.

cancer, ischemic-reperfusion damage, cardiovascular diseases, respiratory diseases, etc. [111,112]. Thus, NAC is a justified candidate as antioxidant for preventing arsenic intoxications and can also be employed in combination therapy with other chelators to accomplish better preventive and therapeutic efficacy.

15.3.1.3 Flavonoids

Flavonoids are polyphenolic compounds or bioflavonoids and constitute one of the most commonly occurring and ubiquitous groups of plant metabolites that contributes to the protective effects against toxic manifestations. They are widely distributed in various plant parts such as seeds, bark, fruits, flowers, etc. However fruits, vegetables, and beverages prepared from plant extracts are the principal dietary sources of flavonoids. Flavonoids also regulate various physiological processes such as inflammation, mutagenesis, thrombosis, allergic reactions, and vasodilatation. Numerous reports suggest that it may possess antineoplastic property and the ability to cross placenta barrier and thus provide protective mechanism [113]. The presence of diphenylpropane moiety in flavonoid structure consists of two aromatic rings linked through three carbon atoms that together form an oxygenated heterocycle. Flavonoids consists of three phenolic rings named A, B, and C ring; depending on the oxidation level of the C ring flavonoids can be classified into flavones, flavonols, flavanones, or isoflavones. Flavonoids are potent antioxidants freely available as dietary supplements and may function as scavengers of free radical and as chelators of metal ions that induce oxidative stress. The presence of 3-hydroxy-carbonyl on the C-ring, 5-hydroxy-carbonyl on the A-ring, and 3',4'-dihydroxyl on the B-ring is responsible for the metal chelating property of flavonoids [114]. Electron donating ability of the flavonoid molecule is primarily associated with the presence of a B-ring catechol group (dihydroxylated B-ring). Flavonoids are potential cancer-preventing agents and their interaction with drug-metabolizing enzymes is the major molecular mechanism involved in its anti-cancerous property. Flavonoids also inhibit cytochrome P450 (phase I metabolizing enzymes), which is known to trigger carcinogenesis. On the other hand, flavonoids enhance the activity of glutathione-*S*-transferase (phase II metabolizing enzymes), which detoxify the carcinogens and eliminate them out of the body [115].

15.3.1.3.1 Quercetin

Quercetin (3',3,4',5,7-pentahydroxyflavone) is one of the most abundant dietary bioflavonoids, found in many fruits and vegetables, olive oil, red wine, and tea. Quercetin is capable of scavenging free radicals and reduces the cellular effects of low density lipoproteins. High consumption of flavonoid-rich food leads to reduced incidence of cancers and have biological, pharmacological, and medicinal properties including anti-inflammatory, antiallergic, antiviral, antithrombotic, antimutagenic, antineoplastic, and cytoprotective effects. Quercetin also prevents neurodegenerative disorders like age-related cognitive, motor, and mood decline and protect against oxidative stress as well as cerebral ischemic injuries. Quercetin is also reported to prevent apoptosis in several cell lines such as fibroblasts, cardiomyoblasts, and epithelial cells [116,117]. Quercetin was found to attenuate oxidative stress induced by arsenic by restoring GSH contents and ROS levels and reducing TBARS levels [24]. However, it is widely accepted that the beneficial effects of

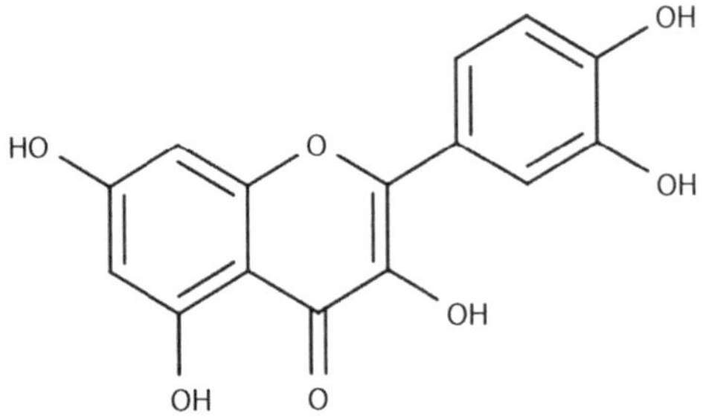

FIGURE 15.14   Chemical structure of quercetin.

quercetin are due mainly to its antioxidant properties and also due to the regulation of signaling pathway. Quercetin administration was also found to be associated
with reduced condition of oxidative stress induced by GaAs exposure [83]. Mishra
and Flora [24] have also reported that the combined treatment with quercetin and
MiADMSA was not only able to chelate arsenic from the cell but also ameliorate
oxidant levels. Dwivedi and Flora [118] provided a probable insight on the beneficial effects of quercetin intake in preventing arsenic poisoning by reducing oxidative stress. Besides, providing beneficial effects in reversing the altered biochemical
variables, quercetin intake could also be useful in enhancing endogenous antioxidant levels (described in Figure 15.14).

15.3.1.3.1.1   Mechanism   The exact mechanism of quercetin is poorly understood.
It is orally absorbed and excreted mainly through bile as sulfates and conjugates.
It offers significant protection to various toxic models of experimental liver diseases
in laboratory animals. Mishra and Flora [24] reported that quercetin significantly
attenuates oxidative damage induced by arsenic by significantly restoring GSH and
ROS levels and reducing TBARS [119]. Quercetin might exert the protective effect
against the oxidative stress associated with the generation of ROS. The protective
efficacy is due its chemical structure where the presence of multiple hydroxyl groups
contributes for its substantial antioxidant and chelating potential. A double bond
and carbonyl function in the heterocycle increases activity by forming a more stable radical through electron delocalization and conjugation. Quercetin undergoes
one or two electron oxidations to form semiquinone and quinine compounds. With
its metal-chelating property, it chelates metal ions and form five-member chelating ring, and it belongs to a special class of bidentate O,O-coordinating ligands.
Quercetin possesses three chelating sites in competition: (1) 3-hydroxychromone, (2)
5-hydroxychromone, and (3) 3′,4′-dihydroxyl groups. Quercetin is most widely used
for the detection of metals bound to flavonoid ligands. Its antiradical property is
directed to scavenge hydroxyl ion and the superoxide anion, highly reactive oxygen
species implicated in the initiation of lipid peroxidation.

15.3.1.3.2   Silymarin
Silymarin is a polyphenolic flavonoid derived from milk thistle (*Silybum marianum*),
which exhibit anticarcinogenic, cytoprotective, and anti-inflammatory effects [120].

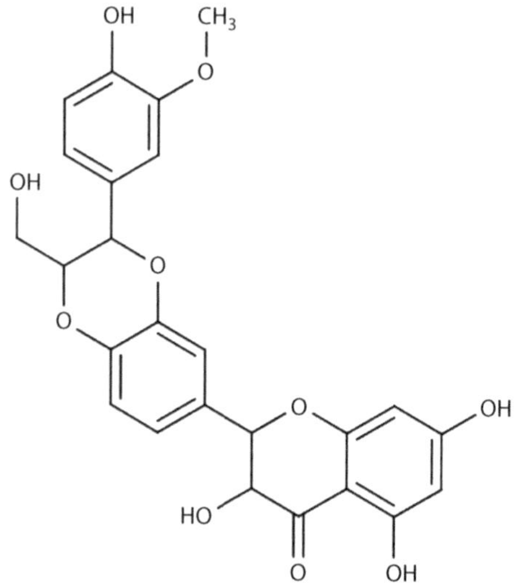

FIGURE 15.15 Chemical structure of silymarin.

Silymarin has been shown to improve the antioxidant status in blood and soft tissues [121]. Effects of the plant flavonoids, silymarin, and quercetin on arsenite-induced oxidative stress have been investigated in CHO-K1 cells [122] as well as human breast adenocarcinoma cell lines [123]. Clinical trials have shown that silymarin is safe and may be potentially effective in improving the symptoms of acute clinical hepatitis, particularly in patients with alcoholic cirrhosis [124]. It also suppresses the TNF-induced production of ROS and lipid peroxidation. Silymarin may also show neuroprotection since besides being an antioxidant it penetrates the BBB to reach CNS. It can prevent the absorption of toxins into the hepatocytes by occupying the binding sites as well as inhibiting many transport proteins at the membrane. Oral effectiveness, good safety profile, easy availability, and, most importantly, an affordable price make Silymarin a suitable candidate for the treatment of arsenic-induced toxicity (see Figure 15.15).

15.3.1.3.2.1 Mechanism of Action   The essential activity of silymarin is an antioxidant effect of its flavonolignans and of other polyphenolic substituent, which is attributed to the radical scavenging ability of both free radicals and ROS [125]. Silymarin's hepatoprotective effects are purportedly accomplished via several mechanisms like antioxidation, inhibition of lipid peroxidation, stimulation of ribosomal RNA polymerase, and subsequent protein synthesis leading to enhanced hepatocyte regeneration, enhanced liver detoxification via inhibition of phase I detoxification, enhanced glucuronidation, and protection from glutathione depletion. In few human studies, silymarin has been reported to be effective in protecting patients with acute hepatitis [124]. Jain et al. [126] reported that silymarin scavenges free radical generation by arsenic. These effects may reflect the ability of silymarin (1) to enhance the scavenging and inactivation of H_2O_2 and hydroxyl radicals; (2) to chelate with redox metals, including Fe^{2+} that catalyzes the formation of free radicals via the Fenton reactions; and (3) to reduce lipid peroxidation by induction of enzymatic and nonenzymatic antioxidants, such as GSH, SOD, and CAT [127]. Jain et al. [126] suggested the beneficial role of silymarin in reducing altered biochemical variables suggestive of oxidative stress. Silymarin may be easily incorporated into the diet or could be cosupplemented during chelation treatment and thus may afford a hepatoprotective effect against arsenite-induced cytotoxicity [126,128].

15.3.1.3.3 Naringenin

Naringenin is natural flavonoids, aglycone of naringin, and is widely distributed in citrus fruits, tomatoes, cherries, grapefruit, and cocoa. Naringenin has also been extensively investigated for its pharmacological activities including antitumor [129], anti-inflammatory, and hepatoprotective effects. Structural similarity of naringenin with quercetin raised the possibility of its use as an antioxidant/chelating agent against arsenic toxicity. Antioxidant, metal chelating, and free-radical-scavenging property of naringenin have also been reported earlier. The beneficial properties of naringenin have been related to the inhibition of LPs formation or scavenging of free radicals [128,130] (shown in Figure 15.16).

15.3.1.3.4 Catechin

Tea is the most popular and readily available beverages in the world. It is being greatly consumed and considered to be beneficial to human health as it exerts various protective mechanisms. Catechins are the main constituents of green tea leaves and are bestowed with numerous biological and pharmacological properties, which ultimately attribute to prevention and therapy for various pathological diseases. Catechins possess anti-inflammatory properties and show protection against platelet aggregation, cancer, and cardiovascular and neurological disorders. Catechin consists of eight polyphenolic flavonoid compounds such as catechin, epicatechin, gallocatechin, epigallocatechin, catechin gallate, epicatechin gallate, gallocatechin gallate, and epigallocatechin gallate [131]. The catechol group in ring A and resorcinol group in ring B are two pharmacophores that exist in catechin. In addition, it also possess a hydroxyl group (–OH) in ring C at position 3. The catechin oxidation is mainly recounted to the catechol group in ring B and three OH groups in rings A and C [132] (see Figure 15.17).

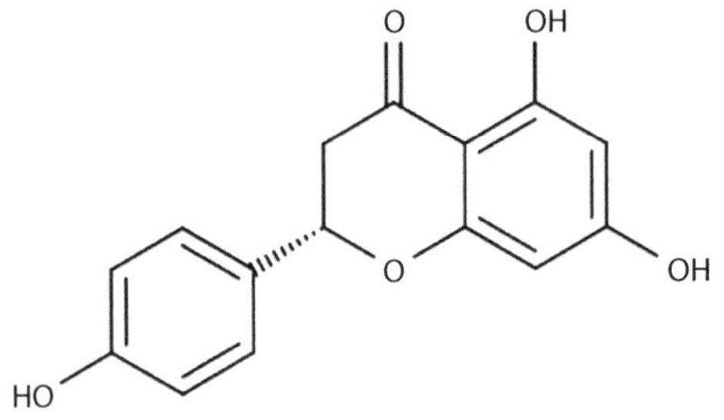

FIGURE 15.16 Chemical structure of naringenin.

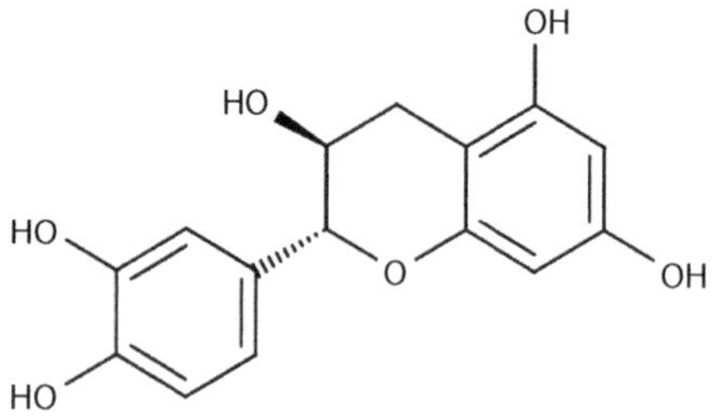

FIGURE 15.17   Chemical structure of catechin.

15.3.1.3.4.1   Mechanism   Catechin acts as a biological antioxidant with free-radical-scavenging property. This property is increased upon deprotonation as the ionization potential (IP) becomes significantly lower and as a result, the radical scavenging antioxidant activity of catechin is increased. Catechin is also known to chelate various metal ions, thereby preventing the generation of potentially damaging free radicals. Catechin oxidation results in the formation of dimerized products and stable semiquinone free radicals, another possible mechanism contributing to the scavenging and chelating activity of these flavonoids. Metal chelation by flavanols prevents lipid peroxidation by sequestering metal ions into inert complexes that becomes unable to decompose H_2O_2 and by restricting the access of metal ions toward lipid hydroperoxides. Catechin has been reported to inhibit xanthine oxidase enzyme, whose upregulation may lead to oxidative stress and gout on metal exposure. Animal studies have shown that catechin shows anticancerous protection by exhibiting chemopreventive as well as antiangiogenic and antimutagenic properties. It can also penetrate BBB and provide protection against neurodegenerative disorders including Parkinson's disease, Alzheimer's disease, and ischemic damage. Recent report demonstrates that catechins and their metabolites have various mechanisms of action through which they explore beneficial mechanisms against metal-induced oxidative stress and pathologies [73]. Despite their versatile therapeutic potential, more epidemiological and clinical studies need to be undertaken for their beneficial efficacy to be fully elucidated.

15.3.1.4   Functional Food

15.3.1.4.1   Jaggery

Nutritional/functional factors can alter the body response to environment toxicants and provide health benefits besides the basic nutrition to improve the physiological function of the body [133]. Vahter [134] suggests that nutritious deficient diet can increase the vulnerability to adverse effects of arsenic in drinking water. Jaggery is a natural sweetener prepared from sugarcane juice (*Saccharum officinarum*) without any chemicals or synthetic additives or preservatives. It is rich in protein, carbohydrate, iron, and vitamins (especially vitamin C) and minerals and has great nutritive and medicinal value. Jaggery comes under ayurveda, the Indian system of medicine and is popularly known as *Gur* in India and *Panela* in South America. Earlier studies demonstrated that jaggery has an astounding beneficial effect on pulmonary system against environmental toxicants [135]. Singh et al. [136,137] reported that jaggery can counteract the genotoxic effects induced by arsenic. Numerous epidemiological studies documented that balanced diet can amend the delayed effect of toxic manifestations of arsenic.

15.3.1.4.1.1 Mechanism of Action   Recent reports elucidates that some natural products can restore GPx and GR enzymatic activities through various biological mechanisms affected by arsenic exposure [138]. Jaggery supplementation also improves the pulmonary defense mechanisms against respiratory disorders induced by various xenobiotics [135]. Vitamin C, a nonenzymatic antioxidant is an important ingredient of jaggery that is essential for the normal functioning of the body, especially lung. The presence of vitamin C in the extracellular fluids lining the lung decreases the airway inflammation and oxidant attack [139]. Nayaka et al. [140] suggested that prolonged ingestion of jaggery has potential in minimizing the toxic effects induced by environmental toxicants. Singh et al. [137] demonstrated that jaggery supplementation improves the body defense mechanism and prevents toxicological manifestations. The exact mechanism of prevention is poorly known but the active ingredients present in jaggery plays a crucial role in preventing arsenicosis. Additionally, vitamins, minerals, carotene, and polyphenols present in jaggery maintain the redox equilibrium to diminish the cellular oxidative damage induced by arsenic within the body. Functional food can mobilize and upregulate the antioxidant potential of cells to counteract excessive free radicals.

15.3.1.4.2 Honey
In Indian tradition, honey is very popular and known not only as nutrient, but is also used in medicines. Honey has gained limelight recently as a source of antioxidant rich phenolic acids, flavonoids, glucose oxidase, CAT, ascorbic acid, carotenoid derivatives, organic acids, amino acids, and proteins. Honey is composed of saturated solution of sugars, mainly fructose and glucose. Recently, researchers have documented the protective effects of honey against cardiovascular diseases, cancer, and microbial infections. Macronutrients and micronutrients are the possible sources of energy in honey as 100 g of honey gives approximately 300 kcal and a daily consumption of 20 g provides about 3% of the recommended daily intake of energy. Honey also constitutes choline, which is essential for cardiovascular and acetylcholine functions in brain as a neurotransmitter as well as in cellular membrane integrity and repair [141–143].

15.3.1.4.2.1 Mechanism of Action   Functional food protects body cell by countering free radicals generated during toxicity. Supplementations of honey combats free radicals and contribute to better health. These oxidative reactions can lead to deleterious reactions in food products, which may cause chronic diseases and cancers. The presence of beneficial bioactive compounds in honey reflects a dietary source of natural antioxidants. Numerous studies explained the antioxidant potential of honey to scavenge ROS and enzymatic or nonenzymatic capacity of lipid peroxidation inhibition. The beneficial effects of honey to neutralize free radicals and to protect against lipid peroxidation may contribute toward reducing some inflammatory effects suggestive of oxidative stress. A greater understanding to delineate the mechanisms and factors governing the bioavailability of honey phytochemicals will be crucial in exploring the mechanisms by which honey exerts its beneficial effects on human health [142,143].

15.3.1.5 Herbal Extracts

Natural antioxidants occur in many plants and plant parts like wood, bark, stem, leaf, and pod. The traditional medicine all over the world is nowadays gaining attention from researchers for their therapeutic efficacy. Ayurvedic medicine commonly regarded as the plant medicine has been practiced since long time in the Indian subcontinent. The word Ayurveda comes from two Sanskrit words: *ayus*, meaning life or life span and *veda*, meaning knowledge or science. Ayurveda is translated as *the science of life*, which emphasizes its orientation toward prevention. A positive association has been developed between dietary supplementation with various vegetables and plant parts and the reduction of carcinogenic effects of various environmental agents including arsenic. Nowadays a great attention has been focused on natural antioxidants containing plant products because these are broadly available, relatively inexpensive, and nontoxic as compared to the synthetic drugs. Natural antioxidants occur in many plants and plant parts like wood, bark, stem, leaf, and pod [144]. Typical natural antioxidants include tocopherol, ascorbic acid, flavonoids, quercetin, carotene, cinnamic acid, peptides, and phenolic compounds. The positive results of a number of studies create an increasing research interest in the use of herbal extracts against arsenic toxicity. *Aloe vera* has been reported to possess antiulcer, antidiabetic, antioxidant, and free-radical-scavenging activity [138]. *Centella asiatica* improves learning and memory and possess antioxidant, antiulcer, and radioprotective activity [145]. It is proven recently that shelled *Moringa oleifera* seed powder has the ability to remove cadmium and arsenic from the aqueous system. Fourier transform infrared (FTIR) spectrometry highlights protein/amino acid–arsenic interactions responsible for sorption phenomenon of seed powder of *M. oleifera* [146]. Garlic has been reported to prevent arsenic-induced oxidative stress and apoptosis and reversing altered clinical variables [147].

15.3.1.5.1 Aloe vera

Aloe vera is a perennial plant belongs to a member of Liliaceae family; it is very cactus-like in its characteristics and has been known for its medicinal properties for many centuries [148]. Aloe gel is a clear gelatinous material obtained from the inner thin-walled mucilaginous parenchyma cells, while latex drains out from the pericyclic cells beneath the skin of the leaf. It has been used in the traditional medicine of many cultures and is said to be beneficial in the treatment of various disorders such as peptic ulcers and burns. In the last decade, Aloe vera has been used extensively in health drinks, topical creams, toiletries, and cosmetics. Experimental studies proved that aloe vera has antiulcer, antidiabetic, immunostimulant, anti-inflammatory, antiviral, antifungal, antibacterial, antitumor, antioxidant, and free-radical-scavenging activity. The protective value of aloe vera has also been investigated against the toxic effects of arsenic in rats and the results suggested that simultaneous administration of aloe vera with arsenic prevents arsenic-induced oxidative injury but have no effect on arsenic concentration in blood and soft tissues. So the study suggested that aloe vera has very limited preventive effects against arsenic [138]. Beneficial properties of aloe vera may be attributed to the presence of mucopolysaccharides in the inner gel of the leaf, especially acemannan (acetylated mannans) glycosides of anthraquinones and dihydroanthraquinones and acemannan and saponins. Apart from the polysaccharide components, 17 common amino acids were also detected in the free state, arginine being the most abundant, along with traces of lupeol, cholesterol, campestrol, and β-sitosterol.

15.3.1.5.2 Centella asiatica

C. asiatica (Syn: Hydrocotyl asiatica, family: Umbelliferae) is a sweetish, arid, digestible plant of tropical areas, which grows as a weed in marshy places throughout the country up to 200 m. In the Indian system of medicine ayurveda, C. asiatica has been used in various parts of India for different ailments like headache, body ache, insanity, asthma, leprosy, ulcers, eczemas, and wound healing. The plant showed CNS depressant activity, antitumor properties, and an inhibitory effect on the biosynthetic activity of fibroblast cells. C. asiatica has found to be immensely used in leprosy probably by dissolving the waxy covering of bacillus leprae. C. asiatica is used to revitalize the brain and nervous system, increase attention span and concentration, and combat aging [149]. It possesses potent antioxidant properties [150] and has been found to protect against membrane peroxidation and lipid peroxidation. Effect of C. asiatica on CNS has been reported to be beneficial in improving memory and in overcoming the negative effects of fatigue and stress. The alcoholic extract of leaves was found to produce tranquilizing effect in rats. Numerous studies documented the involvement of biogenic amines in learning and memory process. Saxena and Flora [145] reported C. asiatica to be beneficial in providing more pronounced effects as compared with chelation therapy particularly in the recovery of oxidative stress variables and restore the altered neurotransmitters (NE, DA, and AChE) levels. Gupta and Flora [150] reported that the aqueous extract of C. asiatica at the concentration of 200 and 500 mg/kg body weight showed beneficial effects following arsenic-induced toxic effects.

15.3.1.5.2.1 Mechanism of Action   Previous studies revealed that C. asiatica possess anti-inflammatory properties that might involve opioid receptors and the presence of high content of triterpenes [151]. Administration of asiaticoside, derivatives of C. asiatica caused the restoration of 40% of the damaged liver cells to normal state in rats [152]. Aqueous extracts of C. asiatica has been known for its antioxidant, cognitive-enhancing, and antiepileptic properties. It has also been reported that phenolic compounds are the major contributors to the antioxidant activities of C. asiatica [153]. Ayurvedic medicine used in the treatment of inflammation, anemia, asthma, blood disorders, bronchitis, fever, urinary discharge, and splenomegaly contains C. asiatica as principal ingredient [154]. Asiatic acid, a principal constituent of C. asiatica is a pentacyclic triterpene found to be effective as anticancer [155]. They proposed that Asiatic acid–induced apoptosis mediated through generation of ROS, alteration of Bax/Bcl-2, ratio and activation of caspase-3, but p53-independent.

15.3.1.5.3 Ocimum sanctum

Ocimum sanctum is a tropical annual herb, up to 18 in. tall and grows into a low bush and is a member of the family Lamiaceae (Labiatae). It is a well-known medicinal plant widely distributed throughout India and its role as a healing herb is well identified in tradition Indian medicine system ayurveda where it is considered to be a panacea for many diseases. Leaves of this plant are used in a variety of pathophysiological states like asthma, dysentery, dyspepsia, chronic fever, skin disease, and helminthiasis and for ring worms. It is also used as an antistressor. It has been observed that tulsi leaves exert hypocholesterolemic, hypotriglyceridemic, and hypophospholipidemic effects in the normal rabbits. Leaves of *O. sanctum* are rich in essential oils. The presence of eugenol in it in considerable amounts has been shown to possess significant antioxidant properties and efficiently inhibit lipid peroxidation [156,157]. The oil of *O. sanctum* possesses antibacterial, antifungal, antistress, immunostimulatory, anticarcinogenic, anti diabetic, and radioprotective effects. The plant also possesses anti-inflammatory and antioxidant properties. Free-radical-scavenging activity is a major mechanism by which *O. sanctum* products protect against cellular damage. One of the recent studies suggest that *O. sanctum* has the ability for faster wound healing property and is useful in the management of abnormal healing and hypertrophic scars by increasing antioxidant enzymes level, namely, SOD, CAT, and GPx [158].

15.3.1.5.4 Hippophae rhamnoides

Hippophae rhamnoides, a member of the Elaegnaceae family, is a perennial plant. *H. rhamnoides* is its botanical name of sea buckthorn and the other names are sandthorn and swallow-thorn. Recently, it has been extensively planted across much of the northern China and in other countries to prevent soil erosion and to serve as an economic resource for food and medicine products. The name is taken from its habit of growing near the sea and from the possession of many spines or thorns that are reminiscent of some buckthorn species. Fruits of sea buckthorn have been used extensively in traditional medicine in Turkey to treat constipation, skin wounds, and influenza infections. In addition to this, many medicinal effects of sea buckthorn against flu, cardiovascular diseases, mucosal injuries, and skin disorders have been suggested to be due to high contents of antioxidant substances present in this plant. The extracts of *H. rhamnoides* scavenge superoxide radicals and prevents lipid peroxidation, perhaps due to the polyphenols in the extract. In addition to the antioxidant activity, sea buckthorns also have antibacterial, antiulcer, radioprotection, and immunomodulatory activity. *H. rhamnoides* is able to reduce cytotoxicity arising due to sodium nitroprusside, nicotine, and ethanol. Recently Geetha et al. [159,160] have provided evidence in which leaf and fruit extract of this plant has been found to have potent antioxidant activity against chromium-induced oxidative stress in rats. No study, however, is available regarding the protective role of sea buckthorn against arsenic toxicity. Sea buckthorn has been shown to have a potent antioxidant activity, which is mainly attributed to its flavonoids and vitamin C content [161]. In recent years, it has gained great research interest in cancer therapy, cardiovascular diseases, gastrointestinal ulcers, skin disorders, and as a liver-protective agent (for chemical toxins) and a remedy for liver cirrhosis. The fruits are rich in vitamin C and vitamin A. Sea buckthorn is an excellent source of plant-based polyunsaturated (essential) fatty acids for maintaining good health and normal growth and development. Its seed oil naturally provides a 1:1 ratio of omega-3 to omega-6. Sea buckthorn is also an excellent source of oleic acid, an essential fatty acid known to help reduce blood cholesterol levels. These essential fatty acids are important in the prevention of heart disease, cancer, and maintaining an overall healthy immune system.

15.3.1.5.5   Allium sativum

Garlic (*Allium sativum* L., family Liliaceae), called lasun in India, is a medicinal plant which has been used for thousands of years in Indian ayurvedic medicine. It is also used with spices to give the food a special flavor and fragrance. Many beneficial health properties of garlic are attributed to organosulfur compounds, particularly to thiosulfinates. The pharmacological effects of garlic have mostly been attributed to its hypoglycemic, hypolipidimic, anticoagulant, antihypertensive, antihepatotoxic, anticancer, immune system–modulatory, antiatherosclerotic, antimicrobial, antidote (for heavy metal poisoning), and antioxidant properties [162]. In addition to this, the bulb of garlic is used as an antirheumatic and stimulant beside its use in conditions like paralysis, forgetfulness, tumor colicky pain, and chronic fever. Recent studies have demonstrated that garlic exerts its therapeutic effect by increasing nitric oxide (NO) production. It is also found to have free-radical-scavenging action and inhibits oxidative modification of low-density lipoproteins. An elucidation of the ability of garlic extract to scavenge hydroxyl radical (•OH) will be an important step in the understanding of the mechanism of the beneficial effects of garlic in various diseases. The intrinsic antioxidant activity of garlic, garlic extracts, and some garlic constituents have been widely documented in vivo and in vitro. Roy Choudhury et al. [163] also reported that the coadministration of garlic extract is able to reduce the clastogenic effects of sodium arsenite. Garlic is administered orally and topically.

15.3.1.5.5.1 Mechanism of Action   The sulfur moieties in the crude garlic extract are, therefore, able to significantly alleviate the action of arsenic on the cellular components. Roy Choudhury et al. [163] also reported that crude garlic extract significantly reduced the clastogenic effects of arsenic in animals that are directly exposed to sodium arsenite and have been given the extract in their diet for long periods. The components of the extract are probably not able to cross the transplacental barrier in appreciable amounts unlike the metal. The clastogenic effects of sodium arsenite are, therefore, reduced in the progeny from the parents given only metal and the metal together with the crude garlic extract. Another finding of the same group also suggested that simultaneous administration of mustard oil and garlic extract enhances the degree of modulating capacity against sodium arsenite clastogenicity because of the fact that fatty acids present in mustard oil and the sulfhydryl components of the garlic together act strongly as radical scavengers in target tissues and cells. This amount is well within the daily dietary intake and may well be able to protect against the toxicity of arsenic through drinking water. So, garlic could emerge as a putative preventive agent against arsenic. Another finding also suggested that simultaneous administration of mustard oil and garlic extract enhances the degree of modulating capacity against sodium arsenite clastogenicity because of the fact that fatty acids present in mustard oil and the sulfhydryl components of the garlic together act strongly as radical scavengers in target tissues and cells [164].

15.3.1.5.6 Phyllanthus emblica

Fruits of *Phyllanthus emblica* L. (a rich source of vitamin C) belong to the Euphorbiaceae family and related species have been extensively used in unani and ayurvedic systems of medicine in India for the treatment of a wide variety of diseases including scurvy, ulceration, and leucorrhoea. *P. emblica* has the ability to stimulate natural antioxidant enzyme systems including CAT, SOD, and GPx [165,166]. Dried fruit is useful in hemorrhage, diarrhea, and dysentery. In combination with iron, it is used as a remedy for anemia, jaundice, and dyspepsia. Fermented liquor prepared from the fruit is used in jaundice, dyspepsia, and cough. Amla fruits possess expectorant, cardiotonic, antioxidative, antitumor, anticancer, anti-inflammatory, antibacterial, antiviral, antiemetic, and cardiovascular protective properties. The dried fruit extract is nontoxic and inhibits emesis induced by apomorphine. the fruits are used in the treatment of leucorrhoea and atherosclerosis and also to purify low turbidity water and can be used as a coagulant in the treatment of water [167]. *P. emblica* also act as hypocholesterolemic agent and lowers cholesterol level by the unique concerted action of both inhibiting cholesterol production and enhancing cholesterol degradation.

  P. emblica appears to be an excellent antidote offering protection from a number of heavy metals. It has even been proven to prevent DNA and cell damage from arsenic poisoning [168]. This is particularly significant because environmental toxins are carcinogenic, and since *P. emblica* inhibits cellular mutation resulting from heavy-metal poisoning, it may have a role in cancer prevention. Sairam et al. [169,170] reported the antioxidant and immunomodulatory properties of the fruit extract of amla using chromium(VI) as an immunosuppressive agent, with the emphasis on lymphocytes.

15.3.1.5.7 Curcumin

Curcumin (1,7-bis(4-hydroxy-3-methoxyphenyl)-1,6-heptadiene-3,5-dione) is an active ingredient of turmeric isolated from the root of *Curcumin longa* L. and has been used for centuries in South and Southeast Asia as traditional medicine. Curcumin is known to exhibit strong antioxidant, anti-inflammatory, and anti-infective properties that is used for treating a variety of inflammatory diseases including atherosclerosis, Alzheimer's, diabetes, arthritis, and inflammatory bowel disease [171,172]. Curcumin decreased the level of LPs and increased the levels of SOD, CAT, GPx and increased maturation and cross-linking of collagen. Curcumin treatment resulted in increased formation of granulation tissue, neovascularization and enhanced biosynthesis of extracellular matrix proteins such as collagen [173] (described in Figure 15.18).

  Curcumin has been used in therapeutic activities and is associated with the suppression of inflammation, angiogenesis, tumorigenesis, and diabetes, and with therapeutic effects in diseases of the cardiovascular, pulmonary, and neurological

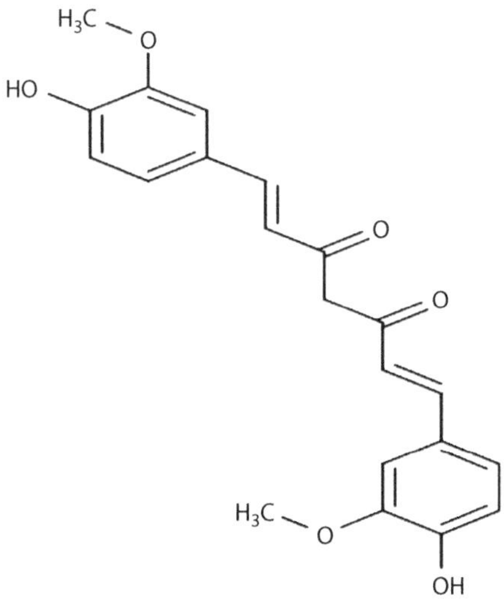

FIGURE 15.18 Chemical structure of curcumin.

system. In general, most of these effects can be attributed to the antioxidant, anti-inflammatory, and anticancer activities of curcumin. Curcumin is an effective scavenger of ROS and RNS [174]. Curcumin protective function against peroxidative damage of biomembranes, known to be a free-radical-mediated chain reaction, has mainly been attributed to the scavenging of the reactive free radicals involved in peroxidation. The scavenging properties of curcumin have also been considered to be responsible for its protective role against oxidative damage of DNA and proteins, believed to be associated with a variety of chronic diseases such as cancer, atherosclerosis, neurodegenerative diseases, and aging [175]. In addition to its direct antioxidant activity, curcumin may function indirectly as an antioxidant by inhibiting the activity of inflammatory enzymes or by enhancing the synthesis of glutathione. Curcumin has promising chemopreventive and therapeutic potential for various cancers including leukemia, lymphoma, and cancers of the gastrointestinal tract, genitourinary system, breast, ovary, head and neck, lung, and skin [174]. Besides avoiding well-established risk factors such as smoking or obesity, the chemopreventive effects of curcumin are one of the most promising approaches to reduce the risk of cancer. Curcumin likely exerts its inhibitory effect on cancer development by several mechanisms such as inhibition of carcinogen activation and stimulation of carcinogen detoxification, prevention of oxidative DNA damage, and its capacity to reduce inflammation. In several culture and in vivo models, curcumin has been shown to modulate enzymes that are involved in the metabolic activation of carcinogens (e.g., inhibition of cytochrome P450) and the detoxification and excretion of such compounds (e.g., induction of glutathione-S-transferase). Srivastava et al. [176] recently reported that arsenic-induced cholinergic deficits in the brain are linked with enhanced generation of ROS and increased apoptosis due to modulation in the expression of pro- and antiapoptotic mitochondrial proteins and these changes are protected by curcumin.

15.3.1.5.8 Arjunolic Acid

Terminalia arjuna (TA) belongs to ayurvedic and unani systems of medicine. It is an ornamental tree and its specimen is placed in the Central National Herbarium (CNH), Botanical Survey of India (BSI), West Bengal, India. The bark of TA is used since long time against cardiac disorders. Many active constituents like tannins, triterpenoid saponins (arjunic acid, arjunolic acid arjungenin, and arjunglycosides), flavonoids, ellagic acid, gallic acid, etc., are present in its bark [177]. Manna et al. [120] demonstrated the protective role of triterpenoid saponin, isolated from the bark of *T. arjuna* against arsenic-induced oxidative stress in mouse hepatocytes. Recent report suggests that arjunolic acid (AA) shows protection in restoring arsenic-induced oxidative stress and nephrotoxicity in murine brain and mouse and is probably due to its antioxidant activity [177,178] (pictured in Figure 15.19).

15.3.1.5.8.1 Mechanism of Action Arjunolic acid has one primary and two secondary hydroxyl groups that oxidized to give ketoderivatives such as vitamin C when it gets oxidized by ROS. AA also contains one carboxylic hydrogen atom that can easily be neutralized by any oxidant species. This property of AA justifies it as a potential radical scavenger and thus prevents the arsenic-induced toxic manifestations. Interestingly, the structure of AA contains polyhydroxyl groups (two of which are viscinal groups) and these two groups may form a five-membered chelate complex with arsenic and inhibits its toxic effects. Thus, AA possesses a good chelating property against arsenic-induced toxicity [120,177,178].

15.3.1.5.9 Moringa oleifera

M. oleifera is a tropical plant belonging to the family of Moringaceae; *M. oleifera* is the most widespread species, which grows quickly at low altitudes in the whole tropical belt, including arid zones. It is widely distributed in India, the Philippines, Sri Lanka, Thailand, Malaysia, Burma, Pakistan, Singapore, the West Indies, and Nigeria. It is generally known in the developing world as a vegetable (leaves, green, pods, flowers, and roasted seeds), for spices (mainly roots), for cooking and cosmetic oil (seeds) and

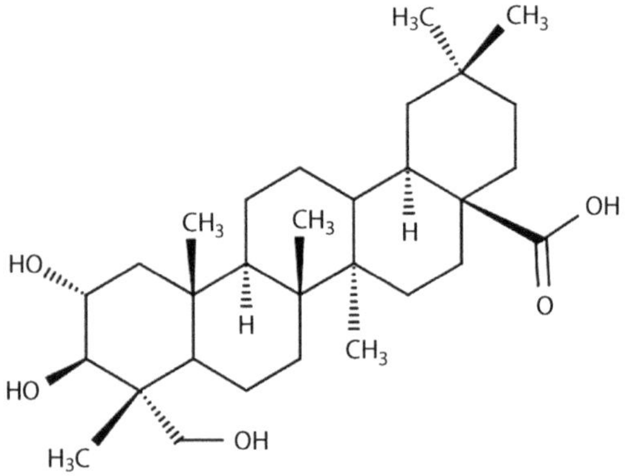

FIGURE 15.19 Chemical structure of arjunolic acid.

as a medicinal plant (all plant organs). Leaves of this plant are traditionally known for or reported to have various biological activities, including hypocholesterolemic agent, regulation of thyroid hormone status, antidiabetic agent, gastric ulcers, antitumor agent, and hypotensive agent. Besides leaves, flowers, roots, gums, fruits, and pods of *M. oleifera* are extensively used for treating inflammation, cardiovascular action, hepatic disease, and hematological and hepatorenal function. *M. oleifera* seeds possess effective coagulation and pharmacological properties and the plant are neither toxic to humans nor to animals [179]. Gupta et al. [180] documented that concomitant administration of *M. oleifera* seed powder with arsenic could significantly protect animals from oxidative stress and reduced tissue-arsenic concentration. Administration of *M. oleifera* seed powder thus could also be beneficial during chelation therapy with a thiol chelator.

15.3.1.5.9.1 Mechanism of Action   One of the strategies to achieve these goals could be by exploring the efficacy of herbal extracts. For safe and effective treatment of arsenic, two points should be achieved by herbal extracts: (1) prevent arsenic accumulation in the target tissues and (2) revert arsenic-induced biochemical alterations indicative of oxidative stress. The active ingredient present in plants is used to treat diseases. Plants are generally considered to be less toxic and free from many of the side effects that a synthetic drug may exhibit. Seeds of *M. oleifera* (SMO) contain significantly high concentration of methionine and cysteine. Due to the presence of –SH group, these amino acids stimulate GSH synthesis and provide additional binding site for arsenic that ultimately facilitate the removal of arsenic from the body. Oxidative methylation reactions in which trivalent forms of arsenic is sequentially methylated to form mono- and dimethylated products (MMA, DMA, and excretory products) using SAM as the methyl donor and GSH as an essential cofactor [6]. SAM, the universal methyl donor is synthesized from the condensation of L-methionine with ATP. L-Methionine thus might indirectly be playing an important role in the removal of arsenic from the in vivo site. Rich source of ascorbic acid in pods and leaves of *M. oleifera* is known for its antioxidant properties and its beneficial effects could be attributed to its ability to form a poorly ionized but soluble complex with toxic metal/metalloid. These components seem to offer protection and to maintain the membrane integrity of soft tissues [180] (represented in Figure 15.20).

15.3.2 NUTRACEUTICAL SUPPLEMENTATION DURING THE CHELATION OF ARSENIC

15.3.2.1 Chelation

Chelation therapy has been the mainstay therapy in metal poisoning and related disorders. The term chelate was first applied by Sir Gilbert T. Morgan and H.D.K. Drew in 1920. They suggested the term for the calliper-like groups that function as two associating units and fasten to a central atom so as to produce heterocyclic rings [181]. Greek work chelate means claw and the process of ring formation is termed chelation. Understanding the concept of chelation promises its application as drug to reduce the body burden of toxic metal. Chelating agents have (1) at least two functional groups with donor atoms capable of combining with a metal and (2) the donor

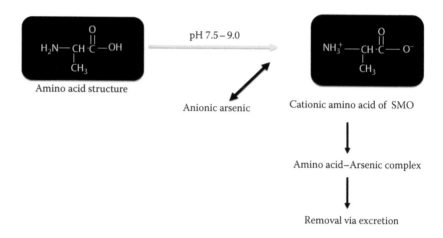

FIGURE 15.20 Possible mechanism showing interaction between arsenic and amino acid. (From Gupta, R. and Flora, S.J.S., *J. Appl. Toxicol.*, 26, 213, 2006.)

atoms that must be situated as to allow ring formation with metal atom as the closing member. However, for application in the biological system, the desired properties extend beyond simple chemical behavior. For these compounds to be of therapeutic relevance they must (1) cross through physiological barriers into compartments where a toxic metal ion is accumulated; (2) form a stable complex with the metal; (3) be able to compete with the biological chelator, if required and remove metal from the site; and (4) form chelation complex that is nontoxic and easily excrete from the site of deposition and body [182]. Chelating agents available for human use may be classified based on structural properties like polyaminocarboxylic acids, chelators with vicinal –SH groups, β-mercapto-α-aminoacids, hydroxamic acids, orthohydroxycarboxylic acids or orthodiphenols, and miscellaneous agents. Commonly known chelating agents used in chelation therapy include calcium disodium ethylenediaminetetraacetic acid (CaNa$_2$EDTA), 2,3-dimercaptopropanol also known as British anti-Lewisite (BAL), D-penicillamine, DMSA, and 2,3-dimercaptopropane-sulfonic acid (DMPS), BAL or 2,3-dimercaptopropanol (dimercaprol), one of the oldest chelating agents is indicated in arsenic poisoning following ingestion, inhalation, or absorption [183]. However, BAL is rather considered most toxic of the chelators available that restricts its application to a few acute poisoning cases. The drug has a low therapeutic index and shows brain redistribution of metal. Other disadvantages include difficult storage owing to easy oxidation and painful intramuscular mode of administration due to lipophilicity. Minor adverse drug reactions like fever, conjunctivitis, lacrimation, headache, and nausea, etc., may be accompanied by serious effects including infection, liver damage, high blood pressure and heart rate, etc. BAL is rapidly absorbed and intracellularly distributed with metabolites excreted in urine (see Figure 15.21).

Safer derivatives of BAL introduced have rather been most successful. DMSA and sodium 2,3-dimercaptopropane-1-sulfonate (DMPS) are water-soluble dithiols with safer drug profiles. The hydrophilic nature of DMSA facilitates good gastrointestinal

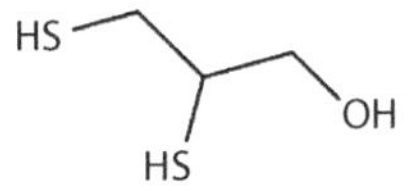

FIGURE 15.21 Chemical structures of BAL.

absorption thus allowing oral route of administration seen as a distinct advantage over BAL. DMPS is more effective than DMSA and BAL (28 times) in removing arsenic in animal models [184,185]. However, DMPS is slightly more toxic than DMSA; yet, both compounds are less toxic to BAL. Both these drugs show predominantly extracellular distribution with DMPS showing some intracellular distribution [186]. Hydrophilic property of DMSA responsible for its genesis ironically also is responsible for its most important limitation of chelating extracellular metal. Later results in limited application for acute and subchronic cases were the metal still resides in the extracellular compartment and has not deposited intracellularly. Thus, esters of DMSA synthesized with the aim to enhance tissue uptake for intracellular chelation were recently introduced. These mono- and diesters show higher therapeutic efficacy with monoesters preferred further due to lower toxicity. Monoisoamyl DMSA (MiADMSA), a C5-branched chain alkyl monoester of DMSA with higher lipophilicity compared to DMSA was found highly effective in reducing arsenic burden from various organs in chronically exposed animals [35,187]. MiADMSA thus has been identified as a promising drug candidate and is currently in its developmental phase. Preclinical data available highlights its efficacy and safety in arsenic acute and chronic toxicity in experimental animals. It is also found capable of mobilizing intracellularly bound cadmium [188]. The sulfhydryl groups have also been proposed to provide antioxidant functions. Safety of MiADMSA has been well established in preclinical in vitro and in vivo models with copper depletion as the only prominent reversible side effect [182].

15.3.2.2 Limitations and Newer Strategies

A brief description given earlier of the conventionally used chelating agents highlights their major limitations like adverse drug reactions, metal redistribution, and essential metal loss [189]. Further, most of the chelating agents considered safe (DMSA, etc.) show partial efficacy in the case of chronic metal exposure cases by virtue of their inability to cross physiological barriers. Thus, slow accumulation of metal inside the cells characteristic of chronic poisoning gets difficult to address. One such classical example is the failure of DMSA in the clinical trial held in Bangladesh in chronically arsenic-exposed patients [190]. Target metal specificity of any chelating agent accounts for another important criteria to avoid essential metal loss. Thus, newer therapeutic strategies for the management of arsenic poisoning need to be defined.

Our group has addressed the issue of identifying some therapeutic solutions like combination therapy, relevant nutritional supplementations like essential metals, natural and synthetic antioxidants with chelation benefits with the conventional chelation therapy. Combination therapy may be defined as prescribing more than one, structurally different chelating agent for more efficient and safer removal of metal from body. The two drugs would act through different mechanisms, chelating metal from different compartments, namely, hard and soft tissue or intra- and extracellular matrix (DMSA and MiADMSA); thus, additive or synergistic effects may be observed [191]. Moreover, such combination therapy, for example, with a lipophilic and lipophobic (MiADMSA and DMSA) would limit drawbacks like metal redistribution and form a safer regime due to lower doses prescribed [22,187,192,193]. Experimental evidence shows that such strategies result in not only better reduction in metal burden but also more effective recoveries in biomarkers, neurological defects, and molecular markers [21,71]. Supplementing conventional chelation therapy with antioxidants or essential metals has also been extensively investigated by our group. Antioxidants such as lipoic acid [194], N-acetyl cysteine, melatonin [187], gossypin [195], etc., show promising recoveries in heavy metal toxicity when coadministered with chelating agents. This may be further be supported by the fact that oxidative stress has been recognized as the major toxic mechanism for most heavy metals. Similarly, since toxic metals replace essential metals in the body as discussed in the previous sections, essential metal deficiency may lead to more potent toxic effects of heavy metals. Further, chelating agents by virtue of nonspecificity may also deplete some essential metals worsening the situation. Thus, coadministration of essential metals like iron, calcium, and zinc with the chelating agents has been investigated to reveal promising outcomes [196–199]. Therefore, it may be stated that newer therapeutic strategies have shown superior results compared to conventional chelation monotherapy (shown in Figures 15.22 and 15.23).

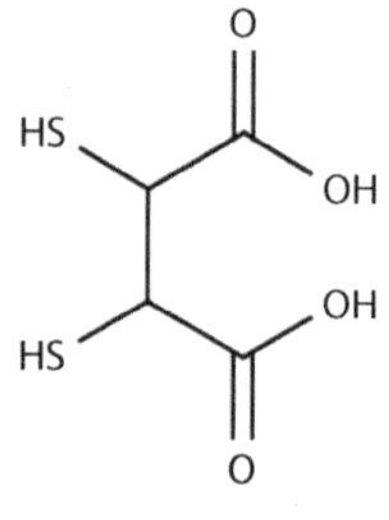

FIGURE 15.22   Chemical structures of DMSA.

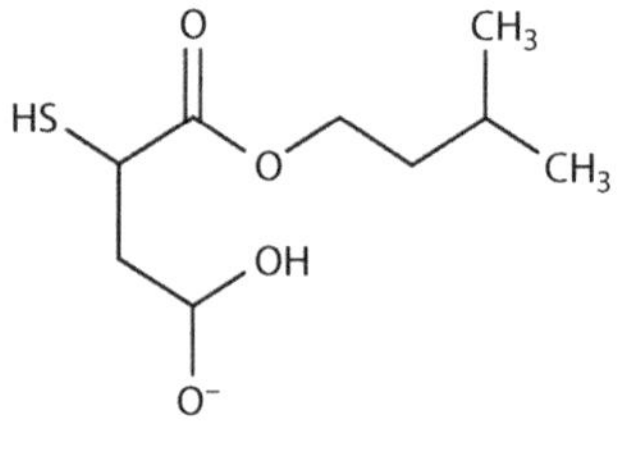

FIGURE 15.23   Chemical structure of MiADMSA.

15.4 CONCLUSIONS AND RECOMMENDATIONS

Arsenic causes serious toxic manifestations and now has gained an important attention worldwide owing to high concentration in drinking water and chemical exposure to humans. Much attention and focus on counteracting arsenic-induced toxic manifestations is the need of current scenario. Although a range of preventive and therapeutic measures are now available against arsenic intoxication such as employing antioxidants, flavonoids, herbal extracts, and functional food as nutraceuticals and chelators as therapy in removal of arsenic both from extracellular and intracellular spaces in the body. Newer therapeutic strategies in the management of chronic metal poisonings with more than one chelating agents (combination therapy) and coadministration of nutraceuticals need to be clinically refined and implemented.

REFERENCES

1. Kosnett M. *Arsenic Toxicity*. Atlanta, GA: Public Health Service, Agency for Toxic Substances and Disease Registry, 1990.
2. Nemeti B., Gregus Z. Reduction of arsenate to arsenite in hepatic cytosol. *Toxicol. Sci.* 70 (2002): 4–12.
3. Radabaugh T.R., Aposhian H.V. Enzymatic reduction of arsenic compounds in mammalian systems: Reduction of arsenate to arsenite by human liver arsenate reductase. *Chem. Res. Toxicol.* 13 (2000): 26–30.
4. Wester R.C., Hui X., Barbadillo S., Maibach H.I., Lowney Y.W., Schoof R.A., Holm S.E., Ruby M.V. In vivo percutaneous absorption of arsenic from water and CCA-treated wood residue. *Toxicol. Sci.* 79 (2004): 287–295.
5. Wildfang E., Radabaugh T.R., Vasken Aposhian H. Enzymatic methylation of arsenic compounds. IX. Liver arsenite methyltransferase and arsenate reductase activities in primates. *Toxicology* 168 (2001): 213–221.
6. Vahter M. Mechanism of arsenic biotransformation. *Toxicology* 181–182 (2002): 211–217.
7. Ferrario D., Croera C., Brustio R., Collotta A., Bowe G., Vahter M. et al. Toxicity of inorganic arsenic and its metabolites on haematopoietic progenitors "in vitro": Comparison between species and sexes. *Toxicology* 249 (2008): 102–108.
8. Valko M., Morris H., Cronin M.T.D. Metals, toxicity and oxidative stress. *Curr. Med. Chem.* 12 (2005): 1161–1208.
9. Roy D.R., Giri S., Chattaraj P.K. Arsenic toxicity: An atom counting and electrophilicity-based protocol. *Mol. Divers.* 13 (2009): 551–556.
10. Miller W.H., Schipper H.M., Lee J.S., Singer J., Waxman S. Mechanisms of action of arsenic trioxide. *Cancer Res.* 62 (2002): 3893–3903.
11. Navas-Acien A., Sharrett A.R., Silbergeld E.K., Schwartz B.S., Nachman K.E., Burke T.A. et al. Arsenic exposure and cardiovascular disease: A systematic review of the epidemiologic evidence. *Am. J. Epidemiol.* 162 (2005): 1037–1049.
12. Díaz-Villaseñor A., Burns A.L., Hiriart M., Cebrián M.E., Ostrosky-Wegman P. Arsenic-induced alteration in the expression of genes related to type 2 diabetes mellitus. *Toxicol. Appl. Pharmacol.* 225 (2007): 123–133.
13. Vahidnia A., Van der Voet G.B., de Wolff F.A. Arsenic neurotoxicity—A review. *Hum. Exp. Toxicol.* 26 (2007): 823–832.
14. Cohen S.M., Arnold L.L., Eldan M., Lewis A.S., Beck B.D. Methylated arsenicals: The implications of metabolism and carcinogenicity studies in rodents to human risk assessment. *Crit. Rev. Toxicol.* 36 (2006): 99–133.

15. McCarty K.M., Chen Y.C., Quamnuzzaman Q., Rahman M., Mahiuddin G., Hsueh Y.M. et al. Arsenic methylation, GSTT1, GSTM1, GSTP1 polymorphisms and skin lesions. *Environ. Health Perspect.* 115 (2007): 341–345.

16. Rossman T.G. Mechanism of arsenic carcinogenesis: An integrated approach. *Mutat. Res.* 533 (2003): 37–65.

17. Pi J., Diwan B.A., Sun Y., Liu J., Qu W., He Y. et al. Arsenic-induced malignant transformation of human keratinocytes: Involvement of Nrf2. *Free Radic. Biol. Med.* 45 (2008): 651–658.

18. States J.C., Srivastava S., Chen Y., Barchowsky A. Arsenic and cardiovascular disease. *Toxicol. Sci.* 107 (2009): 312–323.

19. Yang H.T., Chou H.J., Han B.C., Huang S.Y. Lifelong inorganic arsenic compounds consumption affected blood pressure in rats. *Food Chem. Toxicol.* 45 (2007): 2479–2487.

20. Flora S.J.S. Arsenic-induced oxidative stress and its reversibility. *Free Radic. Biol. Med.* 51 (2011): 257–281.

21. Bhadauria S., Flora S.J.S. Response of arsenic induced oxidative stress, DNA damage and metal imbalance to combined administration of DMSA and monoisoamyl DMSA during chronic arsenic poisoning in rats. *Cell Biol. Toxicol.* 23 (2007): 91–104.

22. Flora S.J.S., Bhadauria S., Kannan G.M., Singh N. Arsenic induced oxidative stress and the role of antioxidant supplementation during chelation: A review. *J. Environ. Biol.* 28 (2007): 333–347.

23. Chouhan S., Flora S.J.S. Arsenic and fluoride: Two major ground water pollutants. *Ind. J. Exp. Biol.* 48 (2010): 666–678.

24. Mishra D., Flora S.J.S. Quercetin administration during chelation therapy protects arsenic induced oxidative stress in mouse. *Biol. Trace Elem. Res.* 122 (2008): 137–147.

25. Halliwell B., Gutteridge J.M.C. *Free Radicals in Biology and Medicine*, 4th edn. Oxford, U.K.: Clarendon Press, 2007.

26. Gregus Z., Gyurasics A., Koszorus L. Interactions between selenium and group Va-metalloids (arsenic, antimony and bismuth) in the biliary excretion. *Environ. Toxicol. Pharmacol.* 5 (1998): 89–99.

27. Peraza M.A., Ayala-Fierro F., Barber D.S., Casarez E., Rael L.T. Effects of micronutrients on metal toxicity. *Environ. Health Perspect.* 106 (1998): 203–216.

28. Flora S.J.S., Tripathi N. Treatment of arsenic poisoning: An update. *Ind. J. Pharmacol.* 30 (1998): 209–217.

29. Chattopadhyay S., Ghosh S., Debnath J., Ghosh D. Protection of sodium arsenite-induced ovarian toxicity by coadministration of L-ascorbate (vitamin C) in mature Wistar strain rat. *Arch. Environ. Contam. Toxicol.* 41 (2001): 83–89.

30. Mahaffey K.R. Nutrition and lead: Strategies for public health. *Environ. Health Perspect.* 103 (1995): 191–196.

31. National Research Council (NRC). *Recommended Dietary Allowances*, 10th edn. Washington, DC: National Academy Press, 1989.

32. Chouchane S., Snow E.T. In vitro effects of arsenicals compounds on glutathione related enzymes. *Chem. Res. Toxicol.* 14 (2001): 517–522.

33. Lin S., Razo L.M.D., Styblo M., Wang C., Cullen W.R., Thomas D.J. Arsenicals inhibit thioredoxin reductase in cultured rat hepatocytes. *Chem. Res. Toxicol.* 14 (2001): 305–311.

34. Hopenhayn-Rich C., Biggs M.L., Kalman D.A., Moore L.E., Smith A.H. Arsenic methylation patterns before and after changing from high to lower concentrations of arsenic in drinking water. *Environ. Health Perspect.* 104 (1996): 1200–1207.

35. Flora S.J.S., Dubey R., Kannan G.M., Chauhan R.S., Pant B.P., Jaiswal D.K. Meso-2,3-dimercaptosuccinic acid (DMSA) and monoisoamyl DMSA effect on gallium arsenide induced pathological liver injury in rats. *Toxicol. Lett.* 132 (2002): 9–17.

36. Wuyi W., Linsheng Y., Shaofan H., Jianan T., Hairong L. Prevention of endemic arsen-ism with selenium. *Curr. Sci.* 81 (2001): 1215–1218.

37. Messarah M., Klibet F., Boumendjel A., Abdennour C., Bouzerna N., Boulakoud M.S. et al. Hepatoprotective role and antioxidant capacity of selenium on arsenic-induced liver injury in rats. *Exp. Toxicol. Pathol.* 64 (2012): 167–174.

38. Rayman M. The argument for increasing selenium intake. *Proc. Nutr. Soc.* 61 (2002): 203–215.

39. Hsueh Y.M., Ko Y.F., Huang Y.K., Chen H.W., Chiau H.Y., Huang Y.L. et al. Determinants of inorganic arsenic methylation capability among residents of the Lanyang Basin, Taiwan: Arsenic and selenium exposure and alcohol consumption. *Toxicol. Lett.* 137 (2003): 49–63.

40. Spallholtz J., Boylan L., Rhaman M. Environmental hypothesis is poor dietary selenium intake and underlying factor for arsenicosis and cancer in Bangladesh and West Bengal India? *Sci. Total Environ.* 323 (2004): 21–32.

41. Berry J.P., Galle P. Selenium-arsenic interaction in renal cells: Role of lysosomes. Electron microprobe study. *J. Submicrosc. Cytol. Pathol.* 26 (1994): 203–210.

42. Flora S.J.S., Kannan G.M., Kumar P. Selenium effects on gallium arsenide induced biochemical and immunological changes in male rats. *Chem. Biol. Interact.* 122 (1999): 1–13.

43. Kreppel H., Liu J., Liu Y., Reichl F.X., Klaassen C.D. Zinc-induced arsenite tolerance in mice. *Fundam. Appl. Toxicol.* 23 (1994): 32–37.

44. Glattre E., Mravcova A., Lener J., Vobecky M., Egertova E., Mysliveckova M. Study of distribution and interaction of arsenic and selenium in rat thyroid. *Biol. Trace Elem. Res.* 49 (1995): 177–186.

45. George C.M., Gamble M., Slavkovich V., Levy D., Ahmed A., Ahsan H. et al. A cross-sectional study of the impact of blood selenium on blood and urinary arsenic concentra-tions in Bangladesh. *Environ. Health* 1 (2013): 12–52.

46. Gailer J. Arsenic–selenium and mercury–selenium bonds in biology. *Coord. Chem. Rev.* 251 (2007): 234–254.

47. Cavar S., Bosnjak Z., Klapec T., Barisic K., Cepelak I., Jurasovic J. et al. Blood selenium, glutathione peroxidase activity and antioxidant supplementation of subjects exposed to arsenic via drinking water. *Environ. Toxicol. Pharmacol.* 29 (2010): 138–143.

48. Meno S.R., Nelson R., Hintze K.J., Self W.T. Exposure to monomethylarsonous acid (MMA(III)) leads to altered selenoprotein synthesis in a primary human lung cell model. *Toxicol. Appl. Pharmacol.* 239 (2009): 130–136.

49. Gailer J., George G.N., Pickering I.J., Prince R.C., Ringwald S.C., Pemberton J.E. et al. A metabolic link between arsenite and selenite: The seleno-bis (S-glutathionyl) arsin-ium ion. *J. Am. Chem. Soc.* 122 (2000): 4637–4639.

50. Jomova K., Valko M. Advances in metal-induced oxidative stress and human disease. *Toxicology* 10(283) (2011): 65–87.

51. Prasad A.S. Zinc: Role in immunity, oxidative stress and chronic inflammation. *Curr. Opin. Clin. Nutr. Metab. Care* 12 (2009): 646–652.

52. Prasad A.S., Beck F.W.J., Bao B., Fitzgerald J.T., Snell D.C., Steinberg J.D. et al. Zinc supplementation decreases incidence of infections in the elderly: Effect of zinc on gen-eration of cytokines and oxidative stress. *Am. J. Clin. Nutr.* 85 (2007): 837–844.

53. Prasad A.S. Clinical, immunological, anti-inflammatory and antioxidant roles of zinc. *Exp. Gerontol.* 43 (2008): 370–377.

54. Denduluri S., Langdon M., Chandra R.K. Effect of zinc administration on immune responses in mice. *J. Trace Elem. Exp. Med.* 10 (1997): 155–162.

55. Modi M., Gupta R., Prasad G.B.K.S., Flora S.J.S. Protective value of concomitant administration of trace elements against arsenic toxicity in rats. *J. Tissue Res.* 4 (2004): 257–262.

56. Maret W. Metallothionein redox biology in the cytoprotective and cytotoxic functions of zinc. *Exp. Gerontol.* 43 (2008): 363–369.
57. Flora S.J.S. Nutritional components modify metal absorption, toxic response and chelation therapy. *J. Nutr. Environ. Med.* 12 (2002): 53–67.
58. Shim H., Harris Z.L. Genetic defects in copper metabolism. *J. Nutr.* 133 (2003): 1527S–1531S.
59. Schmolke G., Elsenhans B., Ehtechami C., Forth W. Arsenic-copper interaction in the kidney of the rat. *Hum. Exp. Toxicol.* 11 (1992): 315–321.
60. Cheng Z., Li Y. What is responsible for the initiating chemistry of iron-mediated lipid peroxidation: An update. *Chem. Rev.* 107 (2007): 748–766.
61. Haas J.D., Brownlie T. Iron deficiency and reduced work capacity: A critical review of the research to determine a causal relationship. *J. Nutr.* 131 (2001): 691–696.
62. Bhaskaram P. Immunobiology of mild micronutrient deficiencies. *Brit. J. Nutr.* 85 (2001): 75–80.
63. Andrews N.C. Disorders of iron metabolism. *N. Engl. J. Med.* 341 (1999): 986–995.
64. Miret S., Simpson R.J., McKie A.T. Physiology and molecular biology of dietary iron absorption. *Annu. Rev. Nutr.* 23 (2003): 283–301.
65. Park H.S., Kim S.R., Lee Y.C. Impact of oxidative stress on lung diseases. *Respirology* 14 (2009): 27–38.
66. Ganz T. Hepcidin, a key regulator of iron metabolism and mediator of anemia of inflammation. *Blood* 102 (2003): 783–788.
67. Mittal M., Flora S.J.S. Vitamin E supplementation protects oxidative stress during arsenic and fluoride antagonism in male mice. *Drug Chem. Toxicol.* 30 (2007): 263–281.
68. Flora S.J.S. Structural, chemical and biological aspects of antioxidants for strategies against metal and metalloid exposure. *Oxid. Med. Cell. Long.* 2 (2009): 191–206.
69. Hall J.L. Cellular mechanisms for heavy metal detoxification and tolerance. *J. Exp. Bot.* 53 (2002): 1–11.
70. Jozefczak M., Remans T., Vangronsveld J., Cuypers A. Glutathione is a key player in metal-induced oxidative stress defences. *Int. J. Mol. Sci.* 13 (2012): 3145–3175.
71. Mishra D., Mehta A., Flora S.J.S. Reversal of arsenic-induced hepatic apoptosis with combined administration of DMSA and its analogues in guinea pigs: Role of glutathione and linked enzymes. *Chem. Res. Toxicol.* 21 (2008): 400–407.
72. Townsend D.M., Tew K.D., Tapiero H. The importance of glutathione in human disease. *Biomed. Pharmacother.* 57 (2003): 145–155.
73. Flora S.J.S., Shrivastava R., Mittal M. Chemistry and pharmacological properties of some natural and synthetic antioxidants for heavy metal toxicity. *Curr. Med. Chem.* 20 (2013): 4540–4574.
74. Ramanathan K., Shila S., Kumaran S., Panneerselvam C. Protective role of ascorbic acid and alpha-tocopherol on arsenic-induced microsomal dysfunctions. *Hum. Exp. Toxicol.* 22 (2003): 129–136.
75. You W.C., Zhang L., Gail M.H., Chang Y.S., Liu W.D., Ma J.L. et al. Gastric dysplasia and gastric cancer: *Helicobacter pylori*, serum vitamin C, and other risk factors. *Natl. Canc. Inst.* 92 (2000): 1607–1612.
76. Suh J., Zhu B.Z., Frei, B. Vitamin C: Basic metabolism and its function as an index of oxidative stress. *Free Radic. Biol. Med.* 34 (2003): 1306–1314.
77. Flora S.J.S. Pachauri V. Arsenic toxicity: Biochemical effects, mechanism of action and strategies for the prevention and treatment by chelating agents and herbal extracts. In: *Recent Advances in Herbal Drug Research and Therapy*, K. Gulati and A. Ray (Eds.). New Delhi, India: I.K., 2010, pp. 401–448.
78. Weber P., Bendich A., Machlin L.J. Vitamin E and human health: Rationale for determining recommended intake levels. *Nutrition* 13 (1997): 450–460.

79. Ibrahim I.A., Al-Joudi F.S. The angiotensin-converting enzyme inhibitor, captopril, alters some biochemical laboratory measurements in vitro. *Malaysian J. Biochem. Mol. Biol.* 17 (2009): 20–22.
80. Pechanova O. Contribution of captopril thiol group to the prevention of spontaneous hypertension. *Physiol. Res.* 56 (2007): S41–S48.
81. Andreoli S.P. Captopril scavenges hydrogen peroxide and reduces, but does not eliminate, oxidant-induced cell injury. *Am. J. Physiol.* 264 (1993): F120–F127.
82. Tzakos A.G., Naqvi N., Comporozos K., Pierattelli R., Theodorou V., Husainb A. et al. The molecular basis for the selection of captopril cis and trans conformations by angiotensin I converting enzyme. *Bioorg. Med. Chem. Lett.* 16 (2006): 5084–5087.
83. Bhatt K., Flora S.J.S. Oral co-administration of α-lipoic acid, quercetin and captopril prevents gallium arsenide toxicity in rats. *Environ. Toxicol. Pharmacol.* 28 (2009): 140–146.
84. Kalia K., Narula G., Kannan G.M., Flora S.J.S. Effects of combined administration of captopril and DMSA on arsenite induced oxidative stress and blood and tissue arsenic concentration in rats. *Comp. Biochem. Physiol. Part C Pharmacol. Toxicol.* 144 (2007): 372–379.
85. Nakagawa K., Uneno A., Nishikawa Y. Interaction between carnosine and captopril on free radical scavenging activity and angiotensine converting enzyme activity in vitro. *Yakugaku Zasshi* 126 (2006): 37–42.
86. Wojakowski W., Gminski J., Siemianowicz K., Goss M., Machalski M. The influence of angiotensin converting enzyme inhibitors on lipid peroxidation in sera and aorta of rabbits in diet-induced hypercholesterolemia. *Int. J. Mol. Med.* 5 (2000): 591–594.
87. Yonei Y., Hattori A., Tsutsui K., Okawa M., Ishizuka B. Effects of melatonin: Basics studies and clinical applications. *Anti-Aging Med.* 7 (2010): 85–91.
88. Hardeland R. Antioxidative protection by melatonin: Multiplicity of mechanisms from radical detoxification to radical avoidance. *Endocrine* 27 (2005) 119–130.
89. Tomas-Zapico C., Coto-Montes, A. Melatonin and its metabolites: New findings regarding their production and their radical scavenging actions. *J. Pineal. Res.* 39 (2005): 99–104.
90. Witt-Enderby P.A., Bennett J., Jarzynaka M.M., Firestine S., Melan M.A. Melatonin receptors and their regulation: Biochemical and structural mechanisms. *Life Sci.* 72 (2003): 2183–2198.
91. Reiter R.J. Actions of melatonin in the reduction of oxidative stress: A review. *Biomed. Sci.* 7 (2000): 444–458.
92. Agarwal S., Rao A.V. Carotenoids and chronic diseases. *Drug Metab. Drug Interact.* 17 (2000): 189–210.
93. Mortensen A., Skibsted L.H., Truscott T.G. The interaction of dietary carotenoids with radical species. *Arch. Biochem. Biophys.* 385 (2001): 13–19.
94. Rao A.V., Rao L.G. Carotenoids and human health. *Pharmacol. Res.* 55 (2007): 207–216.
95. Young A.J., Lowe G.M. Antioxidant and prooxidant properties of carotenoids. *Arch. Biochem. Biophys.* 385 (2001): 20–27.
96. Sindhu E.R., Preethi K.C., Kuttan R. Antioxidant activity of carotenoid lutein in vitro and in vivo. *Ind. J. Exp. Biol.* 48 (2010): 843–848.
97. Flora S.J.S., Chouhan S., Kannan G.M., Mittal M., Swarnakar H. Combined administration of taurine and monoisoamyl DMSA protects arsenic induced oxidative injury in rats. *Oxid. Med. Cell. Long.* 1 (2008): 39–45.
98. Ghosh J., Das J., Manna P., Sil P.C. Taurine prevents arsenic induced cardiac oxidative stress and apoptotic damage: Role of NF-kappa B, p38 and JNK MAPK pathway. *Toxicol. Appl. Pharmacol.* 240 (2009): 73–87.

99. Shinno E., Shimoji M., Imaizumi N., Kinoshita S., Sunakawa H., Aniya Y. Activation of rat liver microsomal glutathione S-transferase by gallic acid. *Life Sci.* 78 (2005): 99–106.

100. Lu Z., Nie G., Belton P.S., Tang H., Zhao B. Structure-activity relationship analysis of antioxidant ability and neuroprotective effect of gallic acid derivatives. *Neurochem. Int.* 48 (2006): 263–274.

101. Moini H., Packer L., Saris N.L. Antioxidant and prooxidant activities of α-lipoic acid and dihydrolipoic acid. *Toxicol. Appl. Pharmacol.* 182 (2002): 84–90.

102. Packer L., Witr E.H., Tritschler H.J. Alpha lipoic acid as a biological antioxidant. *Free Radic. Biol. Med.* 19 (1995): 227–250.

103. Halwagy M.E., Hossanin L. Alpha lipoic acid ameliorate changes occur in neurotransmitter and antioxidant enzymes that influenced by profenofos insecticide. *J. Egypt. Soc. Toxicol.* 34 (2006): 55–62.

104. Moon Y.J., Wang X., Morris M.E. Dietary flavonoids: Effects on xenobiotic and carcinogen metabolism. *Toxicol. In Vitro* 20 (2006): 187–210.

105. Kaur H., Mishra D., Bhatnagar P., Kaushik P., Flora S.J.S. Co-administration of α-lipoic acid and vitamin C protects liver and brain oxidative stress in mice exposed to arsenic contaminated water. *Water Qual. Exp. Health* 1 (2009): 135–144.

106. Modi M., Kaul R.K., Kannan G.M., Flora S.J.S. Co-administration of zinc and N-acetylcysteine prevents arsenic induced tissue oxidative stress in male rats. *J. Trace Elem. Med. Biol.* 20 (2006): 197–204.

107. Flora S.J.S. Arsenic induced oxidative stress and its turnover following combined administration of N-acetylcysteine and meso-2,3-dimercaptosuccinic acid in rats. *Clin. Exp. Pharmacol. Physiol.* 26 (1999): 865–869.

108. Madiha M.A., Abeer A.A., Hegazy A.M. Roles of N-acetylcysteine, methionine, vitamin C and vitamin E as antioxidants against lead toxicity in rats. *Aust. J. Basic Appl. Sci.* 5 (2011): 1178–1183.

109. Yedjou C.G., Tchounwou C.K., Haile S., Edwards F., Tchounwou P.B. N-acetyl-cysteine protects against DNA damage associated with lead toxicity in HepG2 cells. *Ethn. Dis.* 20 (2010): 101–103.

110. El-kadi A.O., Bleau A.M., Dumond I., Maurice H., Du Souich P. Does N-acetylcysteine increase the excretion of trace metals (calcium, magnesium, iron, zinc and copper) when given orally? *Drug Metab. Dispos.* 28 (2000): 1112–1120.

111. Fries G.R., Kapczinski F. N-acetylcysteine as a mitochondrial enhancer: A new class of psychoactive drugs? *Rev. Bras. Psiquiatr.* 33 (2011): 321–322.

112. Tomasi A. New acquisitions and therapeutic role of N-acetylcysteine in airways diseases caused by cigarette smoke and pollution. *Eur. Rev. Med. Pharmacol. Sci.* 8 (2004): 4–28.

113. Malasev D., Kuntic V. Investigation of metal–flavonoid chelates and the determination of flavonoids via metal-flavonoid complexing reactions. *J. Serb. Chem. Soc.* 72 (2007): 921–939.

114. Chen Y., Jiang Z., Qin W. Chemistry, composition and characteristics of sea buckthorn fruit and its oil. *Chem. Indust. Forest. Prod.* (*Chin.*) 10 (1990): 163–175.

115. Marchand L. Cancer preventive effects of flavonoids—A review. *Biomed. Pharmacother.* 56 (2002): 296–301.

116. Park Y.H., Chiou G.C. Structure-activity relationship (SAR) between some natural flavonoids and ocular blood flow in the rabbit. *J. Ocul. Pharmacol. Ther.* 20 (2004): 35–42.

117. Chow J.M., Shen S.C., Huan S.K., Lin H.Y., Chen Y.C. Quercetin, but not rutin and quercitrin prevention of H_2O_2-induced apoptosis via antioxidant activity and heme oxygenase 1 gene expression in macrophages. *Biochem. Pharmacol.* 69 (2005): 1839–1851.

118. Dwivedi N., Flora S.J.S. Dose dependent efficacy of quercetin in preventing arsenic induced oxidative stress in rat blood and liver. *J. Cell Tissue Res.* 11 (2011): 2605–2611.

119. Chouhan S., Yadav A., Kushwah P., Kaul R., Flora S.J.S. Silymarin and quercetin abrogates fluoride induced oxidative stress and toxic effects in rats. *Mol. Cell. Toxicol.* 7 (2011): 25–32.

120. Manna P., Sinha M., Pal P., Sil P.C. Arjunolic acid, a triterpenoid saponin, ameliorates arsenic-induced cytotoxicity in hepatocytes. *Chemico-Biol. Interact.* 170 (2007): 187–200.

121. Skottova N., Kazdova L., Oliyarnyk O., Vecera R., Sobolova L., Ulrichova J. Phenolics-rich extracts from *Silybum marianum* and *Prunella vulgaris* reduce a high-sucrose diet induced oxidative stress in hereditary hypertriglyceridemic rats. *Pharmacol. Res.* 50 (2004): 123–130.

122. Bongiovanni G.A., Soria E.A., Eynard A.R. Effects of the plant flavonoids silymarin and quercetin on arsenite-induced oxidative stress in CHO-K1 cells. *Food Chem. Toxicol.* 45 (2007): 971–976.

123. Soria E.A., Eynard A.R., Quiroga P.L., Bongiovanni G.A. Differential effects of quercetin and silymarin on arsenite-induced cytotoxicity in two human breast adenocarcinoma cell lines. *Life Sci.* 81 (2007): 1397–1402.

124. El-Kamary S.S., Shardell M.D., Abdel-Hamid M., Ismail S., El-Ateek M., Metwally M. et al. A randomized controlled trial to assess the safety and efficacy of silymarin on symptoms, signs and biomarkers of acute hepatitis. *Phytomed. Int. J. Phytother. Phytopharmacol.* 16 (2009): 391–400.

125. Nencini C., Giorgi G., Micheli L. Protective effect of silymarin on oxidative stress in rat brain. *Phytomedicine* 14 (2007): 129–135.

126. Jain A., Yadav A., Bozhkov A.I., Padalko V.I., Flora S.J.S. Therapeutic efficacy of silymarin and naringenin in reducing arsenic-induced hepatic damage in young rats. *Ecotoxicol. Environ. Safety* 74 (2011): 607–614.

127. Chtourou Y., Fetoui H., Sefi M., Trabelsi K., Barkallah M., Boudawara T., Kallel H., Zeghal N. Silymarin, a natural antioxidant, protects cerebral cortex against manganese-induced neurotoxicity in adult rats. *Biometals* 23 (2010): 985–996.

128. Jain A., Dwivedi N., Bhargava R., Flora S.J.S. Silymarin and naringenin protects nicotine induced oxidative stress in young rats. *Oxid. Antioxid. Med. Sci.* 1 (2012): 41–49.

129. Kanno S., Tomizawa A., Hiura T., Osanai Y., Shouji A., Ujibe M. et al. Inhibitory effects of naringenin on tumor growth in human cancer cell lines and sarcoma S-180-implanted mice. *Biol. Pharmaceut. Bull.* 28 (2005): 527–530.

130. Jung U.J., Kim H.J., Lee J.S., Lee M.K., Kim H.O., Park E.J. et al. Naringin supplementation lowers plasma lipids and enhances erythrocyte antioxidant enzyme activities in hypercholesterolemic subjects. *Clin. Nutr.* 22 (2003): 561–568.

131. Kimura M., Umegaki K., Kasuya Y., Sugisawa A., Higuchi M. The relation between single/double or repeated tea catechin ingestions and plasma antioxidant activity in humans. *Eur. J. Clin. Nutr.* 56 (2002): 1186–1193.

132. Janeiro P., Brett A.M.O. Catechin electrochemical oxidation mechanisms. *Anal. Chim. Acta* 518 (2004): 109–115.

133. Lindberg A.L., Ekstrom E.C., Nermell B., Rahman M., Lonnerdal B., Persson L.A. et al. Structure-activity relationship analysis of antioxidant ability and neuroprotective effect of gallic acid derivatives. *Neurochem. Int.* 48 (2006): 263–274.

134. Vahter M.E. Interactions between arsenic-induced toxicity and nutrition in early life. *J. Nutr.* 137 (2007): 2798–2804.

135. Sahu A.P., Paul B.N. The role of dietary whole sugar jaggery in prevention of respiratory toxicity of air toxics and in lung cancer. *Toxicol. Lett.* 95 (1998): 154.

136. Singh N., Kumar N., Raisuddin S., Sahu A.P. Genotoxic effects of arsenic: Prevention by functional food-jaggery. *Cancer Lett.* 268 (2008): 325–330.
137. Singh N., Kumar D., Lal K., Raisuddin S., Sahu A.P. Adverse health effects due to arsenic exposure: Modification by dietary supplementation of jaggery in mice. *Toxicol. Appl. Pharmacol.* 242 (2010): 247–255.
138. Gupta R., Flora S.J.S. Protective value of *Aloe vera* against some toxic effects of arsenic in rats. *Phytother. Res.* 19 (2005): 23–28.
139. Romieu I., Trenga C. Diet and obstructive lung diseases. *Epidemiol. Rev.* 23 (2001): 268–287.
140. Nayaka M.A.H., Sathisha U.V., Manohar M.P., Chandrashekar K.B., Dharmesh S.M. Cytoprotective and antioxidant activity studies of jaggery sugar. *Food Chem.* 115 (2009): 113–118.
141. Bogdanov S., Jurendic T., Sieber R., Gallmann P. Honey for nutrition and health: A review. *Am. J. Coll. Nutr.* 27 (2008): 677–689.
142. Khalil M.I., Sulaiman S.I., Boukraa L. Antioxidant properties of honey and its role in preventing health disorder. *Open Nutraceut. J.* 3 (2010): 6–16.
143. José M., Jambon P., Muñoz O., Manquián N., Bahamonde P., Neira M. Honey as a bio-indicator of arsenic contamination due to volcanic and mining activities in Chile. *Chil. J. Agric. Res.* 73 (2013).
144. Koleva I.I., Van Black T.A., Linssen J.P., De-Groot A., Evastatieva L.N. Screening of plant extracts for antioxidant activity: A comparative study on three testing methods. *Phytochem. Anal.* 13 (2002): 8–17.
145. Saxena G., Flora S.J.S. Changes in brain biogenic amines and heme-biosynthesis and their response to combined administration of succimers and *Centella asiatica* in lead poisoned rats. *J. Pharm. Pharmacol.* 58 (2006): 547–559.
146. Sharma P., Kumari P., Srivastava M.M., Srivastava S. Removal of cadmium from aqueous system by shelled *Moringa oleifera* Lam. seed powder. *Biores. Technol.* 97 (2006): 299–305.
147. Flora S.J.S., Mittal M., Mehta A. Heavy metal induced oxidative stress & its possible reversal by chelation therapy. *Ind. J. Med. Res.* 128 (2008): 501–523.
148. Bombardelli E. Approaches to the quality characteristics of medicinal plant derivatives. *Eur. Phytojournal* 1 (2001): 30–33.
149. Kumar M.H., Gupta Y.K. Effect of different extracts of *Centella asiatica* on cognition and markers of oxidative stress in rats. *J. Ethanopharmacol.* 79 (2002): 253–260.
150. Gupta R., Flora S.J.S. Effect of *Centella asiatica* on arsenic induced oxidative stress and metal distribution in rats. *J. Appl. Toxicol.* 26 (2006): 213–222.
151. Somchit M.N., Sulaiman M.R., Zuraini A., Samsuddin L., Somchit N., Israf D.A. et al. Antinociceptive and anti-inflammatory effects of *Centella asiatica*. *Ind. J. Pharmacol.* 36 (2004): 377–380.
152. Lee E.S., Park H., Jew S., Ryu J.H., Kim Y.C., Lee M.K. et al. *Yurngdong Book of Abstracts 219th ACS National Meeting*, San Francisco, CA, MEDI-137. Washington, DC: American Chemical Society, 2000.
153. Zainol M.K., Abd-Hamid A., Yosuf S., Muse R. Antioxidative activity and total phenolic compounds of leaf, root and petiole of four accessions of *Centella asiatica* (L.) Urban. *Food Chem.* 81 (2003): 575–581.
154. Duke J.A. *Handbook of Medicinal Herbs.* New York: CRC Press, 2001.
155. Park B.C., Bosire K.O., Lee E.S., Lee Y.S., Kim J.A. Asiatic acid induces apoptosis in SK-MEL-2 human melanoma cells. *Cancer Lett.* 218 (2005): 81–90.
156. Geetha K.R., Vasudevan D.M. Inhibition of lipid peroxidation by botanical extracts of *Ocimum sanctum*: In vivo and in vitro studies. *Life Sci.* 76 (2004): 21–28.

157. Sethi J., Sood S., Seth S., Talwar A. Protective effect of Tulasi (*Ocimum sanctum*) on lipid peroxidation in stress induced by anemic hypoxia in rabbits. *Ind. J. Physiol. Pharmacol.* 47 (2003): 115–119.
158. Shetty S., Udupa S., Udupa L. Evaluation of antioxidant and wound healing effects of alcoholic and aqueous extract of *Ocimum sanctum* Linn in rats. *Evid. Based Complem. Alternat. Med.* 5 (2007): 1–7.
159. Geetha S., Sai Ram M., Mongia S.S., Singh V., Ilavazhagan G., Sawhney R.C. Evaluation of antioxidant activity of leaf extract of seabuckthorn (*Hippophae rhamnoides* L.) on chromium(VI) induced oxidative stress in albino rats. *J. Ethanopharmacol.* 87 (2003): 247–251.
160. Geetha S., Sai Ram M., Singh V., Ilavazhagan G., Sawhney R.C. Antioxidant and immunomodulatory properties of seabuckthorn (*Hippophae rhamnoides*) an *in-vitro* study. *J. Ethanopharmacol.* 79 (2002): 373–378.
161. Rosch D. Structure-antioxidant efficiency relationships of phenolic compounds and their contribution to the antioxidant activity of sea buckthorn juice. *J. Agric. Food Chem.* 51 (2004): 4233–4239.
162. Banerjee S.K., Mukherjee P.K., Maulik S.K. Garlic as an antioxidant: The good, the bad and the ugly. *Phytother. Res.* 17 (2003): 97–106.
163. Roy Choudhury A., Das T., Sharma A., Talukder G. Dietary garlic extract in modifying clastogenic effects of inorganic arsenic in mice: Two-generation studies. *Mutat. Res.* 359 (1996): 165–170.
164. Roy Choudhury A., Das T., Sharma A. Mustard oil and garlic extract as inhibitors of sodium arsenite-induced chromosomal breaks in vivo. *Cancer Lett.* 121 (1997): 45–52.
165. Rajak S., Banerjee S.K., Sood S., Dinda A.K., Gupta Y.K., Gupta S.K. et al. *Emblica officinalis* causes myocardial adaptation and protects against oxidative stress in ischemic-reperfusion injury in rats. *Phytother. Res.* 8 (2004): 54–60.
166. Bhattacharya A., Kumar M., Ghosal S., Bhattacharya S.K. Effect of bioactive tannoid principles of *Emblica officinalis* on iron-induced hepatic toxicity in rats. *Phytomedicine* 7 (2000): 173–175.
167. Govindarajan R., Vijayakumar M., Pushpangadan P. Antioxidant approach to disease management and the role of 'Rasayana' herbs of Ayurveda. *J. Ethanopharmacol.* 2 (2005): 165–178.
168. Biswas S., Talukder G., Sharma A. Protection against cytotoxic effects of arsenic by dietary supplementation with crude extract of *Emblica officinalis* fruit. *Phytother. Res.* 13 (1999): 513–516.
169. Sairam M., Neetu D., Deepti P., Vandana M., Ilavazhagan G., Kumar D. et al. Cytoprotective activity of Amla (*Emblica officinalis*) against chromium(VI) induced oxidative injury in murine macrophages. *Phytother. Res.* 17 (2003): 430–433.
170. Sai Ram M., Neetu D., Yogesh B., Anju B., Dipti P., Pauline T. et al. Cyto-protective and immunomodulating properties of Amla (*Emblica officinalis*) on lymphocytes: An in-vitro study. *J. Ethnopharmacol.* 81 (2002): 5–10.
171. Liang G., Shao L., Wang Y., Zhao C., Chu Y., Xiao J. et al. Exploration and synthesis of curcumin analogues with improved structural stability both in vitro and in vivo as cytotoxic agents. *Bioorg. Med. Chem.* 17 (2009): 2623–2631.
172. Williams B., Granholm A.C., Sambamurti K. Age-dependent loss of NGF signalling in the rat basal forebrain is due to disrupted MAPK activation. *Neurosci. Lett.* 413 (2007): 110–114.
173. Sidhu G.S., Mani H., Gaddipati J.P., Singh A.K., Seth P., Banaudha K.K. et al. Curcumin enhances wound healing in streptozotocin induced diabetic rats and genetically diabetic mice. *Wound Repair Regener.* 7 (1999): 362–374.

174. Anand P., Kunnumakkara A.B., Newman R.A., Aggarwal B.B. Bioavailability of curcumin: Problems and promises. *Mol. Pharmacol.* 4 (2007): 807–818.
175. Yadav A., Lomash V., Samim M., Flora S.J.S. Curcumin encapsulated in chitosan nanoparticles: A novel strategy for the treatment of arsenic toxicity. *Chem. Biol. Interact.* 30 (2012): 49–61.
176. Srivastava P., Yadav R.S., Chandravanshi L.P., Shukla R., Dhuriya Y.K., Chauhan L.K.S. et al. Unraveling the mechanism of neuroprotection of curcumin in arsenic induced cholinergic dysfunctions in rats. *Toxicol. Appl. Pharmacol.* 279 (2014): 428–440.
177. Sinha M., Manna P., Sil P.C. Protective effect of arjunolic acid against arsenic-induced oxidative stress in mouse brain. *J. Biochem. Mol. Toxicol.* 22 (2008): 15–26.
178. Sinha M., Manna P., Sil P.C. Arjunolic acid attenuates arsenic-induced nephrotoxicity. *Pathophysiology* 15 (2008): 147–156.
179. Ndabigengesere A., Narasiah K.S. Quality of water treated by coagulation using *Moringa oleifera* seeds. *Water Res.* 32 (1998): 781–791.
180. Gupta R., Dubey D.K., Kannan G.M., Flora S.J.S. Biochemical study on the protective effects of seed powder of *Moringa oleifera* in arsenic toxicity. *Cell Biol. Interact.* 31 (2006): 44–56.
181. Morgan T., Gilbert T., Harry D.K. CLXII—Researches on residual affinity and co-ordination. Part II. Acetyl acetones of selenium and tellurium. *J. Chem. Soc.* 117 (1920): 1456–1465.
182. Flora S.J.S., Pachauri V. Chelation in metal intoxication. *Int. J. Environ. Res. Pub. Health* 7 (2010): 2745–2788.
183. Mandal B.K., Suzuki K.T. Arsenic round the world: A review. *Talanta* 58 (2002): 201–235.
184. Pingree S.D., Simmonds P.L., Woods J.S. Effects of 2,3-dimercapto-1-propanesulfonic acid (DMPS) on tissue and urine mercury levels following prolonged methylmercury exposure in rats. *Toxicol. Sci.* 61 (2001): 224–233.
185. Flora S.J.S., Bhattacharya R., Vijayaraghavan R. Combined therapeutic potential of meso-2,3-dimercaptosuccinic acid and calcium disodium edetate in the mobilization and distribution of lead in experimental lead intoxication in rats. *Fundam. Appl. Toxicol.* 25 (1995): 233–240.
186. Aposhian H.V. Enzymatic methylation of arsenic species and other new approaches to arsenic toxicity. *Annu. Rev. Pharmacol. Toxicol.* 37 (1997): 397–419.
187. Flora S.J.S., Pande M., Kannan G.M., Mehta A. Lead induced oxidative stress and its recovery following co-administration of melatonin or n-acetylcysteine during chelation with succimer in male rats. *Cell. Mol. Biol.* 50 (2004): 543–551.
188. Jones M.M., Singh P.K., Gale G.R, Smith A.B., Atkins L.M. Cadmium mobilization in vivo by intraperitoneal or oral administration of mono alkyl esters of meso-2,3-dimercaptosuccinic acid. *Pharmacol. Toxicol.* 70 (1992): 336–343.
189. Angle C.R. Chelation therapies for metal intoxication. In: Chang, L.W., ed. *Toxicology of Metals.* Boca Raton, FL: CRC Press, 1996, pp. 487–504.
190. Guha Mazumder D.N., Ghoshal U.C., Saha J., Santra A., De B.K., Chatterjee A., Dutta S., Angle C.R., Centeno J.A. Randomized placebo-controlled trial of 2,3-dimercapto succinic acid in therapy of chronic arsenicosis due to drinking arsenic contaminated subsoil water. *Clin. Toxicol.* 36 (1998): 683–690.
191. Flora S.J.S. Chelation therapy. In: *Comprehensive Inorganic Chemistry II*, J. Reedijk and K. Poeppelmeier (Eds.). Oxford, U.K.: Elsevier, 2013, pp. 987–1013.
192. Kostial K., Blanusa M., Plasek L.J., Samarzila M., Jones M.M., Singh P.K. Monoisoamyl and mono-n-hexyl-meso-2,3-dimercaptosuccinate in mobilizing Hg retention in relation to age of rats and route of administration. *J. Appl. Toxicol.* 15 (2001): 201–206.
193. Cory-Slechta D.A. Mobilisation of lead over the course of DMSA chelation therapy and long term efficacy. *J. Pharmacol. Exp. Therap.* 246 (1988): 84–91.

194. Pande M., Flora S.J.S. Lead induced oxidative damage and its response to combined administration of Lipoic acid and succimers in rats. *Toxicology* 177 (2002): 187–196.

195. Gautam P., Flora S.J.S. Oral supplementation of gossypin during lead exposure protects alteration in heme synthesis pathway and brain oxidative stress in rats. *Nutrition* 26 (2010): 563–570.

196. Flora S.J.S., Tandon S.K. Beneficial effects of zinc supplementation during chelation treatment of lead intoxication in rats. *Toxicology* 64 (1990): 129–139.

197. Flora S.J.S., Bhattacharya R., Sachan S.R.S. Dose dependent effects of zinc supplementation during chelation treatment of lead intoxication in rats. *Pharmacol. Toxicol.* 74 (1994): 330–333.

198. Modi M., Pathak U., Kalia K., Flora S.J.S. Arsenic antagonism studies with monoisoamyl DMSA and zinc in male mice. *Environ. Toxicol. Pharmacol.* 19 (2005): 131–138.

199. Joshi D., Mittal D.K., Shukla S., Srivastava A.K. Therapeuric potential of N-acetyl cysteine with antioxidants (Zn and Se) supplementation against dimethylmercury toxicity in male albino rats. *Exp. Toxicol. Pathol.* 64 (2010): 103–108.

16 Influence of Functional Foods in Combating Arsenic Toxicity

Debasis Bagchi, Anand Swaroop,
and Manashi Bagchi

CONTENTS

16.1 INTRODUCTION

Inorganic arsenic oxides including arsenite [trivalent arsenic, As(III)], the reduced form, and arsenate [pentavalent arsenic, As(V)] salts, the oxidized form, are the most toxic and harmful forms of arsenic, which are readily absorbed into the human body and accumulated in the tissues and fluids [1–6]. It is important to mention that As(III) is more toxic than As(V). Both of these arsenic oxides are integral constituents of geologic formation and are readily extracted into the ground water [1–6]. Approximately 137 million people in more than 70 countries have been reported to suffer from arsenic poisoning of drinking water [1–5]. Arsenic poisoning also occur from mining and ore smelting, however, the most serious problem occurs in water wells drilled into aquifers that have high concentrations of these arsenic oxides [2–6]. It is important to mention that organic arsenic is 500 times less harmful than inorganic arsenic [2–6]. Also, seafood is a common source of less toxic organic arsenic in the form of arsenobetaine [2–6].

World Health Organization recommends a limit of 0.01 mg/lit (10 ppb) of arsenic in drinking water. However, more recent findings demonstrate that consumption of water with levels as low as 0.00017 mg/lit (0.17 ppb) over long periods of time can lead to arsenicosis [7].

The metabolism of arsenic involves enzymatic and nonenzymatic methylation, and the excreted metabolite is dimethylarsinic acid (DMA) or cacodylic acid [8]. Arsenic poisoning starts with headaches, confusion, severe diarrhea, and drowsiness. Following repeated exposure, convulsions and changes in fingernail pigmentation take place [7]. Later, with continued prolonged exposure to arsenic poisoning, severe diarrhea, vomiting, blood in the urine, cramping muscles, hair loss, stomach pain, and more convulsions occur [7]. Arsenic poisoning is related to heart disease, cancer, stroke, night blindness, chronic lower respiratory diseases, and diabetes. Chronic arsenic poisoning is intricately associated with vitamin A deficiency. Arsenic toxicity occurs in diverse organs including the lungs, skin, kidneys, and liver. The ultimate result of arsenic poisoning is coma and death [7].

It is important to mention that high fat aggravates arsenic-induced oxidative stress in rat heart and liver [9]. Supplemental potassium also reported to reduce the risk of experiencing a life-threatening heart rhythm problem from arsenic trioxide [3,7].

16.2   FUNCTIONAL FOODS FOR COMBATING ARSENIC TOXICITY

Following are several functional foods and nutraceuticals that demonstrated significant benefits in ameliorating arsenic poisoning and toxicity.

16.2.1   Selenomethionine

Over the years, selenium supplements have demonstrated potential uses as a micronutrient in human health. Rodríguez-Sosa et al. examined the efficacy of dietary selenomethionine supplementation on arsenic toxicity in an in vivo mouse model [10]. The researchers conceptualized that arsenic induces pronounced genotoxicity and immunotoxicity in vivo through oxidative stress pathways. The protective ability of selenomethionine (0.2 and 2 ppm, respectively) was examined in C57BL/6N female mice exposed to sodium arsenite (3, 6, and 10 mg/kg) in tap water for 9 days [10]. It was reported that arsenic exposure increased selenomethionine excretion in the urine. Selenomethionine increased liver weight and decreased the concentration of total liver proteins in animals exposed to 10 mg/kg of arsenic. Furthermore, selenomethionine maintained a normal glutathione level in the hepatic tissue and increased glutathione peroxidase concentration. It is important to mention that the lipid peroxidation level was increased by selenomethionine even without arsenic exposure [10]. Selenomethionine also helped to maintain the CD4/CD8 ratio of lymphocytes in the spleen, although it increased the proportion of B cells. Selenomethionine supplementation prior to arsenic exposure increased the secretion of IL-4, IL-12, and interferon-γ and the stimulation index of the spleen cells in an in vitro assay. Selenomethionine improved the basal immunological parameters but did not reduce the damage caused by oxidative stress after low-dose arsenic exposure [10].

Chattopadhyay et al. assessed the protective ability of sodium selenite against sodium arsenite–induced ovarian and uterine disorders in mature albino rats. Sodium selenite exhibited some protective abilities [11].

16.2.2   DIETARY ZINC SUPPLEMENT

Researchers examined the efficacy of zinc in the removal of accumulated arsenic from different tissues in rats. Animals were given 400 μg arsenic/kg/day in drinking water over a period of 2 months followed by a cessation period of 1 month [12]. Animals were then given zinc supplement (2 mg/kg body weight/day) to both the untreated and arsenic-treated rats over the next month. Administration of zinc significantly removed the accumulated arsenic from liver, kidneys, spleen, and lung tissues significantly in vivo [12].

16.2.3   CURCUMIN

Researchers demonstrated that arsenic-induced elevation of serum alanine aminotransferase, aspartate aminotransferase, and hepatic malonaldehyde were significantly relieved by curcumin. Arsenic-induced reduction of blood and hepatic glutathione levels were all consistently ameliorated by curcumin. Dr. Gao et al. [13] also demonstrated that curcumin is instrumental in accelerating arsenic methylation and urinary elimination. Both hepatic Nrf2 protein, NADP(H) quinine oxidoreductase 1 (NQO1), and heme oxygenase-1 (HO-1) were consistently upregulated in curcumin-treated mice. Dr. Gao et al. concluded that curcumin may serve as a novel chemopreventive dietary agent in protecting against inorganic arsenic-induced hepatic toxicity in vivo [13].

16.2.4   LIPOIC ACID

Lobato et al. examined the efficacy of lipoic acid in modulating Cd and As toxicity in shrimp (*Litopenaeus vannamei*) [14]. Researchers exhibited that muscle from shrimp exposed to Cd alone or Cd+ As showed a remarkable decrease in glutathione levels, while it could be significantly attenuated following the pretreatment of shrimps with lipoic acid. Lipoic acid pretreatment also increased the glutathione-S-transferase (GST) activity in all groups. Overall, the authors demonstrated that lipoic acid may serve as a novel chemoprotectant against Cd- and As-induced oxidative stress in prawns [14].

16.2.5   PROBIOTICS

Monachese et al. demonstrated that probiotics including lactobacillus have some ability to ameliorate arsenic-induced toxicity in humans [15].

16.2.6 FOLIC ACID AND VITAMIN B12

Majumdar et al. [16] assessed the efficacy of folic acid and vitamin B_{12}, singly and in combination, against arsenic-induced hepatic oxidative stress and dysfunction. The authors demonstrated that folic acid and vitamin B_{12}, singly and in combination, can exhibit significant protection against arsenic-induced oxidative stress and apoptotic cell death [16]. Several biochemical, enzymatic, and oxidative stress markers as well as including iNOS protein expression, mitochondrial swelling, and cytochrome c oxidase and Ca^{2+}-ATPase activity, DNA fragmentation, Ca^{2+} content, and caspase-3 activity were assessed. Majumdar et al. clearly exhibited that folic acid and vitamin B_{12}, singly and in combination, may alleviate against arsenic-induced hepatic injury and mitochondria dysfunctions [16].

Gamble et al. hypothesized and demonstrated that inorganic arsenic (InAs) is methylated to monomethylarsonous acid (MMA) and DMA via folate-dependent one-carbon metabolism [17]. They also exhibited that impaired methylation is associated with adverse health outcomes. Therefore, folate level may influence arsenic methylation level and toxicity. A randomized, double-blind, placebo-controlled folic acid-supplementation trial was conducted in 200 adults (low plasma concentrations of folate ≤9 nmol/lit) in Bangladesh over a period of 12 weeks. The treatment group was given 400 µg folic acid/day. The total arsenic excreted as DMA in the folic acid group (72% before and 79% after supplementation) was significantly ($P < 0.0001$) greater than that of the placebo group. The reduction of total urinary arsenic excreted as MMA as 13% and 10%, respectively ($P < 0.0001$) and as inorganic arsenic as 15% and 11%, respectively ($P < 0.001$) [17]. The authors concluded that folic acid supplementation to the subjects with low plasma folate enhances arsenic methylation. Because people whose urine contains low proportions of dimethylarsinic acid and high proportions of MMA and ingested inorganic arsenic have been reported to be at greater risk of skin and bladder cancers and peripheral vascular disease. The results clearly demonstrate that folic acid may reduce the risk of arsenic-induced toxicity.

16.2.7 VITAMIN A

More than 50 years ago, Dr. Hall described the beneficial effect of oral supplementation of vitamin A (retinol) in the treatment of cutaneous arsenicosis [18]. Supplementation of oral vitamin A (150,000 USP units/day for 3 months) resulted in a partial regression of hyperpigmentation and heperkeratosis of palms in a 39-year-old male who had taken Fowler's solution (potassium arsenite) for the treatment of cholera [18]. More recently, Thianprasit presented a case series of nine patients with cutaneous arsenicosis who were treated for 2–7 months with oral synthetic retinoid. Clinical and histopathological improvement was noted in arsenical hyperkeratosis [19]. Other case reports of regression of arsenical keratosis with synthetic retinol have also been published.

16.2.8 VITAMIN E

Researchers examined the dose-dependent efficacy of vitamin E to long-term arsenic-exposed crossbred lactating goats. The protective efficacy of 100- and

150 IU/kg/day of body weight doses were assessed against a dose of 40 mg arsenic/kg body weight over a period of 12 months [20]. Vitamin E supplementation provided significant protection in several enzymatic and oxidative stress biomarkers including SOD, CAT, plasma total Ig, average lymphocyte stimulation index, and total antioxidant activity in arsenic-treated groups. Overall, vitamin E ameliorated arsenic-induced oxidative stress and toxicity in goats [20].

16.2.9   DIETARY PHOSPHATES

Researchers demonstrated that simultaneous administration of inorganic phosphate (6.56 M) and arsenic (6.07 mM) in the intestinal loops of rats caused significant reduction of arsenic toxicity. Short-term arsenic exposure (3 mg/kg body weight/day for 30 days) caused hepatic injury, inflammatory responses including iNOS expression, hepatic caspase-3 activity, and hepatic DNA fragmentation [21]. Inorganic phosphate exhibited significant protection against arsenic-induced hepatotoxicity [21].

16.2.10   METHIONINE

Researchers investigated the protective effects of methionine supplementation on arsenic-induced altered glucose homeostasis in rats. Pal and Chatterjee treated rats with sodium arsenite (5.55 mg/kg body weight/day, i.p.) over a period of 21 days, which caused a significant diminution in blood glucose level and fall in liver glycogen and pyruvic acid contents [22]. Dietary methionine supplementation significantly ameliorated arsenic poisoning and toxicity. Researchers concluded that hypoglycemia with reduced decreased glycolytic activity was induced by arsenic treatment, which can be partly attenuated by dietary methionine supplementation [22].

16.2.11   *MORINGA OLEIFERA* LEAVES

Sheikh et al. demonstrated that supplementation of *Moringa oleifera* leaves abrogated arsenic-induced elevation of triglyceride, glucose, total cholesterol, and urea. *M. oleifera* leaves ameliorated serum alkaline phosphatase, aspartate aminotransferase, alanine aminotransferase levels, and butyrylcholinesterase activity [23]. Furthermore, high density lipoprotein cholesterol level was significantly increased. Thus, *M. oleifera* leaves may serve as a novel therapeutic intervention against arsenic-induced toxicity [23].

16.2.12   JAGGERY

Singh et al. demonstrated that supplementation of jaggery and sugarcane juice can ameliorate arsenic-induced genotoxicity including chromosomal aberration in arsenic-treated mice [24,25]. Jaggery contains polyphenols, vitamin C, carotene, and other biologically active components. The serum levels of interleukin-1β, interleukin-6, and TNF-α were significantly increased in arsenic-exposed groups, while in the arsenic-exposed and jaggery-supplemented groups their levels were

TABLE 16.1

Functional Foods–Induced Prevention against Arsenic-Induced Toxicity and Poisoning

Functional Food	Experimental Model	Observation	References
Selenomethionine	Sodium arsenite–induced significant genotoxicity and immunotoxicity	Improved immunological parameters	Rodríguez-Sosa et al. [10]
Sodium selenite	Sodium arsenite–induced ovarian and uterine disorders in albino rats	Significant protection was observed by sodium selenite	Chattopadhyay et al. [11]
Dietary zinc supplement	Animals were given 400 µg arsenic/kg/day in drinking water	Zinc significantly removed accumulated arsenic from liver, kidneys, spleen, and lung tissues	Kamaluddin and Misbahuddin [12]
Curcumin	Arsenic-induced cellular and enzymatic disruption and oxidative injury was observed	Curcumin provided dramatic in vivo protection	Gao et al. [13]
Lipoic acid	Arsenic and cadmium-induced toxicity in shrimp (*L. vannamei*)	Lipoic acid increased the GST activity in shrimp	Lobato et al. [14]
Probiotics	Arsenic-induced toxicity in humans	Probiotics ameliorated arsenic-induced toxicity in humans	Monachese et al. [15]
Folic acid and vitamin B_{12}, singly and in combination	Arsenic-induced oxidative stress and apoptotic cell death	Protected against arsenic-induced hepatic injury and mitochondria dysfunctions	Majumdar et al. [16]
Folic acid (400 µg/day)	Arsenic-induced toxicity in humans	Reduces the risk of arsenic poisoning and toxicity in humans	Gamble et al. [17]
Vitamin A (retinol) (150,000 USP units/day for 3 months)	Arsenic-induced hyperpigmentation and heperkeratosis in human subject	Partially reduced hyperpigmentation and heperkeratosis	Hall [18]
Vitamin A (synthetic retinoid)	Nine patients suffering from cutaneous arsenicosis treated for 2–7 months	Significant improvement was observed in arsenical hyperkeratosis	Thianprasit [19]
Vitamin E (100–150 IU/kg/day for 12 months)	Arsenic-treated crossbred lactating goats (40 mg arsenic/kg body weight over a period of 12 months)	Provided significant protection in several enzymatic and oxidative stress biomarkers in goats	Das et al. [20]

(Continued)

TABLE 16.1 (*Continued*)
Functional Foods–Induced Prevention against Arsenic-Induced Toxicity and Poisoning

Functional Food	Experimental Model	Observation	References
Dietary phosphate	Simultaneous administration of inorganic phosphate (6.56 M) and arsenic (6.07 mM)	Inorganic phosphate provided significant protection against arsenic-induced hepatotoxicity	Majumdar et al. [21]
Methionine	Arsenic-induced altered glucose homeostasis in rats	Arsenic-induced hypoglycemia with reduced glycolytic activity was reversed by methionine	Pal and Chatterjee [22]
M. olifera leaves	Arsenic-induced elevation of triglyceride, glucose, total cholesterol, and urea in rats	*M. olifera*–induced significant protection	Sheikh et al. [23]
Jaggery	Arsenic-induced genotoxicity and chromosomal aberration in animals	Jaggery provided significant protection	Singh et al. [24,25]
Boron compounds	Arsenic toxicity in human blood sample in vitro	Boron compounds exerted significant protection	Turkez et al. [26]

normal [24,25]. The comet assay in bone marrow cells showed the genotoxic effects of arsenic, whereas combination with jaggery feeding reduced the DNA damage. Histopathologically, the lungs of arsenic-exposed mice showed necrosis and degenerative changes in bronchiolar epithelium with emphysema and thickening of alveolar septa, which was effectively antagonized by jaggery feeding [24,25]. These results demonstrate that jaggery effectively ameliorate the adverse effects of arsenic.

16.2.13 Boron Compounds

Several boron compounds including fructoborates exhibited significant potential in protecting against arsenic toxicity in an in vitro model in human blood samples (shown in Table 16.1).

16.3 CONCLUSION

These studies demonstrated that several functional foods and nutraceuticals have significant potential in protecting against arsenic-induced toxicity and poisoning.

REFERENCES

1. Stohs, S.J. and Bagchi, D. 1995. Oxidative mechanisms in the toxicity of metal ions. *Free Radic. Biol. Med.* 18:321–336.
2. USAToday.com. Arsenic in drinking water seen as threat, August 30, 2007.
3. (a) Mukherjee, A., Sengupta, M.K., Hossain, M.A., Ahamed, S., Das, B., Nayak, B., Lodh, D., Rahman, M.M., and Chakraborti, D. 2006. Arsenic contamination in groundwater: A global perspective with emphasis on the Asian scenario. *J. Health Popul. Nutr.* 24:142–163; (b) Shankar, S., Shanker, U., and Shikha. 2014. Arsenic contamination of groundwater: A review of sources, prevalence, health risks, and strategies for mitigation. *Scientific World Journal* 2014:304524. Published online 2014 Oct 14. doi: 10.1155/2014/304524.
4. Smedley, P.L. and Kinniburgh, D.G. 2002. A review of the source, behaviour and distribution of arsenic in natural waters. *Appl. Geochem.* 17:517–568.
5. Mukherjee, A., Sengupta, M.K., and Hossain, M.A. 2006. Arsenic contamination in groundwater: A global perspective with emphasis on the Asian scenario. *J. Health Populat. Nutr.* 24:142–163.
6. Sabbioni, E., Fischbach, M., Pozzi, G., Pietra, R., Gallorini, M., and Piette, J.L. 1991. Cellular retention, toxicity and carcinogenic potential of seafood arsenic. I. Lack of cytotoxicity and transforming activity of arsenobetaine in the BALB/3T3 cell line. *Carcinogenesis* 12:1287–1291.
7. (a) Watanabe, T. and Hirano, S. 2013. Metabolism of arsenic and its toxicological relevance. *Arch. Toxicol.* 87:969–979; (b) Montelescaut, E., Vermeersch, V., Commandeur, D., Huynh, S., Danguy des Deserts, M., Sapin, J., Ould-Ahmed, M. and Drouillard, I. 2014. Acute arsenic poisoning. *Ann. Biol. Clin. (Paris)* 72:735–738; (c) Burgdorf, W.H. and Hoenig, L.J. 2014. Arsenicosis: The greatest public health disaster in history. *JAMA Dermatol.* 2014 Nov;150(11):1151. doi: 10.1001/jamadermatol.2014.81.
8. Vigo, J.B. and Ellzey, J.T. 2006. Effects of arsenic toxicity at the cellular level: A review. *Texas J. Microsc.* 37:45–49.
9. Dutta, M., Ghosh, D., Ghosh, A.K., Bose, G., Chattopadhyay, A., Rudra, S., Dey, M., Bandopadhyay, A., Pattari, S.K., Mallick, S., and Bandyopadhyay, D. 2014. High fat diet aggravates arsenic induced oxidative stress in rat heart and liver. *Food Chem. Toxicol.* 66:262–277.
10. Rodriguez-Sosa, M., Garcia-Montalvo, E.A., Del Razo, L.M., and Vega, L. 2013. Effect of selenomethionine supplementation in food on the excretion and toxicity of arsenic exposure in female mice. *Biol. Trace Elem. Res.* 156:279–287.
11. Chattopadhyay, S., Pal Ghosh, S., Ghosh, D., and Debnath, J. 2003. Effect of dietary co-administration of sodium selenite on sodium arsenite-induced ovarian and uterine disorders in mature albino rats. *Toxicol. Sci.* 75:412–422.
12. Kamaluddin, M. and Misbahuddin, M. 2006. Zinc supplement on tissue arsenic concentration in rats. *Bangladesh Med. Res. Council Bull.* 32:87–91.
13. Gao, S., Duan, X., Wang, X., Dong, D., Liu, D., Li, X., Sun, G., and Li, B. 2013. Curcumin attenuates arsenic-induced hepatic injuries and oxidative stress in experimental mice through activation of Nrf_2 pathway, promotion of arsenic methylation and urinary excretion. *Food Chem. Toxicol.* 59:739–747.
14. Lobato, R.O., Nunes, S.M., Wasielesky, W., Fattorini, D., Regoli, F., Monserrat, J.M., and Ventura-Lima, J. 2013. The role of lipoic acid in the protection against of metallic pollutant effects in the shrimp *Litopenaeus vannamei* (Crustacea, Decapoda). *Comp. Biochem. Physiol. A: Mol. Integr. Physiol.* 165:491–497.
15. Monachese, M., Burton, J.P., and Reid, G. 2012. Bioremediation and tolerance of humans to heavy metals through microbial processes: A potential role for probiotics? *Appl. Environ. Microbiol.* 78:6397–6404.

16. Majumdar, S., Maiti, A., Karmakar, S., Das, A.S., Mukherjee, S., Das, D., and Mitra, C. 2012. Antiapoptotic efficacy of folic acid and vitamin B_{12} against arsenic-induced toxicity. *Environ. Toxicol.* 27:351–363.
17. Gamble, M.V., Liu, X., Ahsan, H., Pilsner, J.R., Ilievski, V., Slavkovich, V., Parvez, F., Chen, Y., Levy, D., Factor-Litvak, P., and Graziano, J.H. 2006. Folate and arsenic metabolism: A double-blind, placebo-controlled folic acid-supplementation trial in Bangladesh. *Am. J. Clin. Nutr.* 84:1093–1101.
18. Hall, A.F. 1946. Arsenical keratosis disappearing with vitamin A therapy. *Arch. Derm. Syphilol.* 53:154.
19. Thianprasit, M. 1984. Chronic cutaneous arsenism treated with aromatic retinoid. *J. Med. Assoc. Thailand* 67:93–100.
20. Das, T.K., Mani, V., Kaur, H., Kewalramani, N., Des, S., Hossain, A., Banerjee, D., and Datta, B.K. 2012. Effect of vitamin E supplementation on arsenic induced oxidative stress in goats. *Bull. Environ. Contam. Toxicol.* 89:61–66.
21. Majumdar, S., Karmakar, S., Maiti, A., Choudhury, M., Ghosh, A., Das, A.K., and Mitra, C. 2011. Arsenic-induced hepatic mitochondrial toxicity in rats and its amelioration by dietary phosphate. *Environ. Toxicol. Pharmacol.* 31:107–118.
22. Pal, S. and Chatterjee, A.K. 2004. Protective effect of methionine supplementation on arsenic-induced alteration of glucose homeostasis. *Food Chem. Toxicol.* 42:737–742.
23. Sheikh, A., Yeasmin, F., Agarwal, S., Rahman, M., Islam, K., Hossain, E., Hossain, S., Karim, M.R., Nikkon, F., Saud, Z.A., and Hossain, K. 2014. Protective effects of *Moringa oleifera* Lam leaves against arsenic-induced toxicity in mice. *Asian Pacific J. Trop. Biomed.* 4(Suppl. 1):S353–S358.
24. Singh, N., Kumar, D., Raisuddin, S., and Sahu, A.P. 2008. Genotoxic effects of arsenic: Prevention by functional food-jaggery. *Cancer Lett.* 268:325–330.
25. Singh, N., Kumar, D., Lal, K., Raisuddin, S., and Sahu, A.P. 2010. Adverse health effects due to arsenic exposure: Modification by dietary supplementation of jiggery in mice. *Toxicol. Appl. Pharmacol.* 242:247–255.
26. Turkez, H., Geyikoglu, F., Tatar, A., Keles, M.S., and Kaplan, I. 2012. The effects of some boron compounds against heavy metal toxicity in human blood. *Exp. Toxicol. Pathol.* 64:93–101.

A Running Commentary on the Remediation and Treatment of Arsenic Toxicity

Narayan Chakrabarty

This is a complete book about the toxicity of arsenic—its origin and source and its proliferation. Almost half of this world is now arsenic affected. Then its toxic nature comes into light. Why and how it is toxic? How it blocks heme synthesis pathway, how it affects the function of pyruvate dehydrogenase and the reduction of arsenic (V) to arsenic (III), etc., are all discussed [1].

In the next part, we discussed the filtration techniques for the removal of arsenic from groundwater and the shortcomings of these methods. Then we elaborately discussed the new techniques of bio- and phytoremediations [2]. The chemistry and process are also elaborately narrated.

Arsenic in foods, its sources, and removal to make these foods safe for consumption has been taken up in a chapter. Possible causes and sources have also been discussed [3].

In the treatment part, the biomarkers and their respective suitability are taken up, followed by classical treatments. The shortcomings of the processes are shown with the introduction of combination therapy. The success of combination therapy has been shown with pictures.

From this part, we enter into a comparatively newer remediation process. From the mechanism of arsenic toxicity in vivo, it is established beyond doubt that arsenic harms by oxidative process inside the body. So, antioxidants may be a good option to use as a weapon against arsenic toxicity.

Studies show that arsenic metabolism in cells produce reactive oxygen species (ROS) [4] and reactive nitrogen species (RNS). These are associated with the damage of lipids, DNA in cell and proteins. Arsenic also inhibits the formation of GSH and essential antioxidants [5]. Through the inhibition of the enzymatic activity of PDH by combining with sulfhydryl group, arsenic causes damage to gluconeogenesis and oxidative phosphorylation. Arsenic (V) not only reduces enzymatically to arsenic (III) causing the above damages but also substitute phosphorous atom in inorganic phosphate in glycolytic and cellular respiration pathway and most likely forms ADP–arsenate. Arsenolysis (due to easier bond hydrolysis of O–As than O–P) leads to premature uncomplexing oxidative phosphorylation. GSH, effective cellular antioxidant, acts as an electron donor for the reduction of arsenic (V) to arsenic (III) and arsenic (III) combines with GSH to render it physiologically ineffective.

It is a well-known fact that nutrition and food can influence arsenic metabolism and lack of various vitamins and antioxidants in economically weaker section of the people cause more havoc due to arsenic toxicity. Balanced nutrition causes a delaying effect on arsenic toxicity. Antioxidants like some vitamins; essential metals like zinc, iron, and selenium; flavonoids; unsaturated oils with phenolic OH group, etc.; and functional food like jiggery, honey, garlic, etc., have proven effect of combating arsenic toxicity. Selenium combines with some enzymes and proteins to acquire antioxidant properties (its excess intake have toxic effect). Selenium combines with glutathione (GSH), which in a secondary way, decomplex arsenic. This selenium-glutathione complex is also responsible for the excretion of arsenic from the body [6]. In short, the arsenic–protein combination inside the body is changed to selenium protein complex with the removal of arsenic. Zinc and iron both have the capacity to decomplex arsenic from arsenic–protein complex [7].

Nutritional antioxidants help in combating arsenic toxicity through different mechanisms and they are thoroughly discussed here. These antioxidants have one thing in common; they are all free-radical scavengers and thereby removes active oxygen and repair the oxidized membranes. The antioxidants discussed here are mainly polyphenols, amines, quinones, sulfide, and thiols. Of these, glutathione [8], vitamin C or ascorbic acid [9], vitamin E or alpha tocopherol [10], captopril [11], melatonin [12], carotenoids [13], taurine [14], gallic acid [15], alpha lipolic acid [16], N-acetylcysteine [17], and flavonoids (quercetine, silymarine, naringenin, and catechin) [18] are discussed in detail with the mechanism of action.

Among the functional foods jaggery and honey with their antioxidant characteristics and mechanistic part of their action are also discussed.

Among the herbal extracts aloe vera (with limited success), *Centella asiatica* [19], *Ocimum sanctum* [20], *Hippophae rhamnoide* [21], and *Allium sativum* (Garlic) [22] are discussed. Fruits of *Phyllanthus amblica* (rich in vitamin C), curcumin (an active ingredient in turmeric) [23], arjunolic acid (from arjun tree) [24], and *Moringa oleifera* [25], with their active components and the mechanism of action are thoroughly discussed.

The conventional treatments of chelation therapy using BAL, DMSA, and DMPS and their methylated derivatives like MiADMSA as chelators are most common. The limitations of the chelation therapy has been discussed and the best possible treatment for the remediation of arsenic toxicity has been suggested as combination therapy; that is, supplementing chelation therapy with essential metals and/or antioxidants. Some of these have already been tried with great results and these are also discussed in detail. Our suggestion is the replacement of conventional chelation monotherapy with combination therapy and it proves a phenomenal success and a bright ray of hope in the management of chronic arsenic poisoning. Further, doses of neutraceuticals and functional foods for administration are to be modified and precisely determined with clinical studies.

REFERENCES

1. Watanabe, T., Hirano, S. Metabolism of arsenic and its toxicological relevance. *Arch. Toxicol.* 87(6) (2013): 969–979
2. Zhao, F.J., Dunham, S.J., McGrath, S.P. Arsenic hyperaccumulator by different fern species. *New Phytol.* 156 (2002): 27–31.

3. Ackermann, J., Vetterlein, D., Kaiser, K., Mattusch, J., Jahn, R. The bioavailability of arsenic in floodplain soils: A simulation of water saturation. *Eur. J. Soil Sci.* 61 (2010): 84–96.

4. Valko, M., Morris, H., Cronin, M.T.D. Metals, toxicity and oxidative stress. *Curr. Med. Chem.* 12 (2005): 1161–1208.

5. Halliwell, B., Gutteridge, J.M.C. *Free Radicals in Biology and Medicine*, 4th edn. Clarendon Press, Oxford, U.K., 2007.

6. National Research Council (NRC). *Recommended Dietary Allowances*, 10th edn. National Academy Press, Washington, DC, 1989.

7. Prasad, A.S. Zinc: Role in immunity, oxidative stress and chronic inflammation. *Curr. Opin. Clin. Nutr. Metab. Care* 12 (2009): 646–652.

8. Flora, S.J.S., Shrivastava, R., Mittal, M. Chemistry and pharmacological properties of some natural and synthetic antioxidants for heavy metal toxicity. *Curr. Med. Chem.* 20 (2013): 4540–4574.

9. Ramanathan, K., Shila, S., Kumaran, S., Panneerselvam, C. Protective role of ascorbic acid and alpha-tocopherol on arsenic-induced microsomal dysfunctions. *Hum. Exp. Toxicol.* 22 (2003): 129–136.

10. Mittal, M., Flora, S.J.S. Vitamin E supplementation protects oxidative stress during arsenic and fluoride antagonism in male mice. *Drug Chem. Toxicol.* 30 (2007): 263–281.

11. Kalia, K., Narula, G., Kannan, G.M., Flora, S.J.S. Effects of combined administration of captopril and DMSA on arsenite induced oxidative stress and blood and tissue arsenic concentration in rats. *Comp. Biochem. Physiol. C: Pharmacol. Toxicol.* 144 (2007): 372–379.

12. Yonei, Y., Hattori, A., Tsutsui, K., Okawa, M., Ishizuka, B. Effects of melatonin: Basics studies and clinical applications. *Anti-Aging Med.* 7 (2010): 85–91.

13. Sindhu, E.R., Preethi, K.C., Kuttan, R. Antioxidant activity of carotenoid lutein in vitro and in vivo. *Ind. J. Exp. Biol.* 48 (2010): 843–848.

14. Flora, S.J.S., Chouhan, S., Kannan, G.M., Mittal, M., Swarnakar, H. Combined administration of taurine and monoisoamyl DMSA protects arsenic induced oxidative injury in rats. *Oxid. Med. Cell. Long.* 1 (2008): 39–45.

15. Lu, Z., Nie, G., Belton, P.S., Tang, H., Zhao, B. Structure-activity relationship analysis of antioxidant ability and neuroprotective effect of gallic acid derivatives. *Neurochem. Int.* 48 (2006): 263–274.

16. Moon, Y.J., Wang, X., Morris, M.E. Dietary flavonoids: Effects on xenobiotic and carcinogen metabolism. *Toxicol. In Vitro* 20 (2006): 187–210.

17. El-kadi, A.O., Bleau, A.M., Dumond, I., Maurice, H., Du Souich, P. Does N-acetylcysteine increase the excretion of trace metals (calcium, magnesium, iron, zinc and copper) when given orally? *Drug Metab. Dispos.* 28 (2000): 1112–1120.

18. Chen, Y., Jiang, Z., Qin, W. Chemistry, composition and characteristics of sea buckthorn fruit and its oil. *Chem. Indust. Forest Prod. (Chin.)* 10 (1990): 163–175.

19. Gupta, R., Flora, S.J.S. Effect of *Centella asiatica* on arsenic induced oxidative stress and metal distribution in rats. *J. Appl. Toxicol.* 26 (2006): 213–222.

20. Shetty, S., Udupa, S., Udupa, L. Evaluation of antioxidant and wound healing effects of alcoholic and aqueous extract of *Ocimum sanctum* Linn in rats. *Evid. Based Compl. Altern. Med.* 5 (2007): 1–7.

21. Geetha, S., Sai Ram, M., Singh, V., Ilavazhagan, G., Sawhney, R.C. Antioxidant and immunomodulatory properties of seabuckthorn (*Hippophae rhamnoides*) an in-vitro study. *J. Ethanopharmacol.* 79 (2002): 373–378.

22. Banerjee, S.K., Mukherjee, P.K., Maulik, S.K. Garlic as an antioxidant: The good, the bad and the ugly. *Phytother. Res.* 17 (2003): 97–106.

23. Anand, P., Kunnumakkara, A.B., Newman, R.A., Aggarwal, B.B. Bioavailability of curcumin: Problems and promises. *Mol. Pharmacol.* 4 (2007): 807–818.

24. Sinha, M., Manna, P., Sil, P.C. Protective effect of arjunolic acid against arsenic-induced oxidative stress in mouse brain. *J. Biochem. Mol. Toxicol.* 22 (2008a): 15–26.
25. Gupta, R., Dubey, D.K., Kannan, G.M., Flora, S.J.S. Biochemical study on the protective effects of seed powder of *Moringa oleifera* in arsenic toxicity. *Cell Biol. Interact.* 31 (2006): 44–56.

Glossary

Arsenic: A metalloid poison.

Arsenicosis: Pigmentary changes on the skin with or without skin lesions due to the consumption of arsenic.

ATP: Adenosine triphosphate, commonly known as currency of energy metabolism.

Bioremediation: Remediation involving suitable bacteria or live species.

Bowen's disease: A type of pigmentary changes due to arsenic poisoning.

Chelation: The phenomenon in which a metal forms a complex with an electron donor group known as a ligand.

Chelation therapy: A therapeutic treatment of arsenic poisoning using a suitable ligand.

Combination therapy: A therapeutic treatment of arsenic poisoning using a combination of chelation and the administration of nutraceuticals.

DMA/DMAA: Dimethylarsenic acid.

Functional food: A food that has other functional values.

Geochemistry: A study of the chemical parameters inside the soil.

GSH: One of the most powerful antioxidants.

Keratosis: Also under arsenicosis but accompanied by skin lesions.

MMA/MMAA: Monomethylarsenic acid.

Nanotechnology: A technology using nanosized particles.

Nutraceuticals: Naturally occurring substances or prepared species from them that have medicinal value and effect.

Phytoremediation: Remediation by plants or algae.

Phytovolatilization: Removal of arsenic by a plant body through aerial route by volatilizing arsenic as arsine.

ppb: Parts per billion, microgram per liter.

Rhizofiltration: Absorption of arsenic from the soil and in aquatic environment by a plant with its roots.

RNS: Reactive nitrogen species.

ROS: Reactive oxygen species.

Index

T - #0799 - 101024 - C536 - 234/156/24 - PB - 9781032098401 - Gloss Lamination